Genomic Architecture Handbook

Genomic Architecture Handbook

Edited by Paloma Young

Larsen & Keller
www.larsen-keller.com

Genomic Architecture Handbook
Edited by Paloma Young
ISBN: 979-8-88836-062-0 (Hardback)

Larsen & Keller
Published by Larsen and Keller Education,
5 Penn Plaza,
19th Floor,
New York, NY 10001, USA

Cataloging-in-Publication Data

Genomic architecture handbook / edited by Paloma Young.
p. cm.
Includes bibliographical references and index.
ISBN 979-8-88836-062-0
1. Genomics. 2. Genomics. 3. Molecular genetics. I. Young, Paloma.
QH447 .G48 2023
572.86--dc26

For more information regarding Larsen and Keller Education and its products, please visit the publisher's website www.larsen-keller.com

Table of Contents

Preface

A genome includes all of the genetic information of an organism. It is made up of the nucleotide sequences of DNA. The complete set of non-random arrangements of functional elements within the genome is known as the genomic architecture. The architecture of the genome is crucial in gene regulation. Utilizing high-throughput techniques for 3D interaction mapping and chromatin profiling produces extensive experimental data sets that describe the structure and dynamics of the genome. The study of genomic architecture is currently a very prominent area of research that has significant applications in the study of processes such as cell development, differentiation and carcinogenesis. This book explores all the diverse aspects of genomic architecture. It presents researches and studies performed by experts across the globe. The extensive content of this book provides the readers with a thorough understanding of the subject.

The information shared in this book is based on empirical researches made by veterans in this field of study. The elaborative information provided in this book will help the readers further their scope of knowledge leading to advancements in this field.

Finally, I would like to thank my fellow researchers who gave constructive feedback and my family members who supported me at every step of my research.

Editor

Familial Infertility (Azoospermia and Cryptozoospermia) in Two Brothers—Carriers of t(1;7) Complex Chromosomal Rearrangement (CCR): Molecular Cytogenetic Analysis

Marta Olszewska [1], Tomasz Stokowy [2], Nijole Pollock [3], Nataliya Huleyuk [4], Andrew Georgiadis [3], Svetlana Yatsenko [3], Danuta Zastavna [4,5], Alexander N. Yatsenko [3] and Maciej Kurpisz [1,*]

1 Institute of Human Genetics, Polish Academy of Sciences, Strzeszynska 32, 60-479 Poznan, Poland; marta.olszewska@igcz.poznan.pl

2 Department of Clinical Science, University of Bergen, Postboks 7804, 5020 Bergen, Norway; tomasz.stokowy@k2.uib.no

3 Department of OBGYN and Reproductive Science, Magee-Womens Research Institute, University of Pittsburgh, Pittsburgh, PA 15213, USA; pollocknc@mwri.magee.edu (N.P.); georgiadisa@mwri.magee.edu (A.G.); yatsenkosa@mail.magee.edu (S.Y.); yatsenkoan@mwri.magee.edu (A.N.Y.)

4 Institute of Hereditary Pathology, Ukrainian Academy of Medical Sciences, Lysenko Str. 31a, 79000 Lviv, Ukraine; huleyuk@yahoo.com (N.H.); zastavna.d@ihp.lviv.ua (D.Z.)

5 Department of Biotechnology and Bioinformatics, Faculty of Chemistry, Rzeszow University of Technology, Al. Powst. Warszawy 6, 35-959 Rzeszow, Poland

* Correspondence: maciej.kurpisz@igcz.poznan.pl

Abstract: Structural aberrations involving more than two breakpoints on two or more chromosomes are known as complex chromosomal rearrangements (CCRs). They can reduce fertility through gametogenesis arrest developed due to disrupted chromosomal pairing in the pachytene stage. We present a familial case of two infertile brothers (with azoospermia and cryptozoospermia) and their mother, carriers of an exceptional type of CCR involving chromosomes 1 and 7 and three breakpoints. The aim was to identify whether meiotic disruption was caused by CCR and/or genomic mutations. Additionally, we performed a literature survey for male CCR carriers with reproductive failures. The characterization of the CCR chromosomes and potential genomic aberrations was performed using: G-banding using trypsin and Giemsa staining (GTG banding), fluorescent in situ hybridization (FISH) (including multicolor FISH (mFISH) and bacterial artificial chromosome (BAC)-FISH), and genome-wide array comparative genomic hybridization (aCGH). The CCR description was established as: der(1)(1qter->1q42.3::1p21->1q42.3::7p14.3->7pter), der(7)(1pter->1p21::7p14.3->7qter). aCGH revealed three rare genes variants: *ASMT*, *GARNL3*, and *SESTD1*, which were ruled out due to unlikely biological functions. The aCGH analysis of three breakpoint CCR regions did not reveal copy number variations (CNVs) with biologically plausible genes. Synaptonemal complex evaluation (brother-1; spermatocytes II/oligobiopsy; the silver staining technique) showed incomplete conjugation of the chromosomes. Associations between CCR and the sex chromosomes (by FISH) were not found. A meiotic segregation pattern (brother-2; ejaculated spermatozoa; FISH) revealed 29.21% genetically normal/balanced spermatozoa. The aCGH analysis could not detect smaller intergenic CNVs of few kb or smaller (indels of single exons or few nucleotides). Since chromosomal aberrations frequently do not affect the phenotype of the carrier, in contrast to the negative influence on spermatogenesis, there is an obvious need for genomic sequencing to investigate the point mutations that may be responsible for the differences between the azoospermic and cryptozoospermic phenotypes observed in a family. Progeny from the same parents provide a unique opportunity to discover a novel genomic background of male infertility.

Keywords: azoospermia; cryptozoospermia; complex chromosomal rearrangement; chromosome translocation; sperm chromosomes; male infertility; male meiosis

---

1. Introduction

Infertility affects approximately 10–18% of couples of reproductive age [1–3]. It is estimated that approximately 7% of men and 12% of women worldwide are affected and approximately 40–60% of all infertility cases originate from the male side [4,5]. There are multiple reasons for male infertility, including genetic factors, which determine approximately 10–15% of revealed diminished fertility cases; and approximately 25% of infertile males are diagnosed as idiopathic (unexplained etiology) [1,5,6]. Apart from gene mutations that are clearly linked to male infertility [6–8], chromosomal aberrations may also underlie male reproductive problems. The frequency of all chromosomal aberrations detected in the lymphocytes of fertile males is about 0.7%, while in infertile men this increases five-fold [9].

The most common detectable aberrations are aneuploidy (the presence of an additional chromosome) and structural aberrations of the chromosomes (reciprocal chromosome translocations—RCTs; and Robertsonian chromosome translocations—RobTs (about 1/500–1/700 live births)) [3,10]. In a population of infertile males, the probability of chromosome aberrations is several times higher when compared to the whole population (i.e., for RCT: 1% of infertile males are carriers, while in oligozoospermia, RCTs are approximately 16% of all identified aberrations; in azoospermia cases, this value is about 4%) [3]. Unknown genetic and epigenetic issues, hormonal abnormalities, varicocele, cryptorchidism, reactive oxygen species (ROS), and environmental factors (i.e., pollution, lifestyle, drugs, smoking, etc.) are listed as causes of male reproductive failure [11]. A lack of spermatozoa in semen (azoospermia) may arise either from the obstruction of the vas deferens (obstructive azoospermia, OA, with preserved spermatogenesis in the testis; 15–40% of azoospermic men) or as a consequence of a spermatogenetic arrest (nonobstructive azoospermia, NOA), which appears in almost 1% of the total male population and increases to 15–20% of infertile males [2,12]. The frequency of chromosomal aberrations in males with azoospermia rises to 20% [2].

Structurally balanced or unbalanced aberrations involving more than two breakpoints on two or more chromosomes, known as complex chromosomal rearrangements (CCRs), have been described in over 250 papers [13–15]. Approximately 75% of CCRs appear *de novo* or are inherited maternally (70% of all familial cases) [16,17]. There are four types of CCR according to the combination of the number of chromosomes and breakpoints involved. Types I–III are genetically balanced, while type IV is related to the presence of indels or the disruption of a gene [17–20]. Type I is the simplest and is characterized by an equal number of chromosomes and breakpoints involved in the CCR. This is called a "three-way rearrangement" (three chromosomes with one breakpoint per chromosome), and is the most often observed among all CCR cases and is inherited maternally. Type II, the so called "exceptional complex chromosomal rearrangement," mainly appears *de novo* and is characterized by at least two breakpoints on the one of the chromosomes involved (i.e., reciprocal chromosomal translocation RCT, where one of the chromosomes consists of an inversion or insertion). The maximum number of chromosomes involved in type II is seven with a total number of 15 breakpoints [21]. Type III, also called the "double/triple two-way translocation," is the coincidental presence of two or three separate reciprocal or Robertsonian chromosome translocations [17]. Type IV, which is genetically imbalanced, and also called an "insertional translocation," is formed as a result of at least three breakpoints that lead to a variable number of copies of a DNA fragment [19].

CCR cases typically involve three (30%) or four (29%) chromosomal breakpoints. A higher number of breakpoints determines a higher risk of phenotypic abnormalities in a carrier (30–50%) or their offspring (20–90%) [13,14,20–24]. The high risk of abnormalities is estimated as 3.5% per breakpoint in CCR [14]. Drastic fertility reduction has been observed in CCR carriers, mostly from gametogenesis arrest, as a result of the disrupted pairings of chromosomes during the meiotic pachytene stage [13,24,25]. This is linked to the more complex character of the rearrangement, which involves more than two chromosomes, or a number of breakpoints that is higher than the number of chromosomes involved in the CCR. Chromosome pairing is linked to meiotic multivalent formation and possible impairment of the chromosome ends. Such impaired ends allow the conjugation of other chromosomes, mainly sex chromosomes, leading to diminished spermatogenesis manifested in decreased semen parameters; i.e., azoospermia (spermatogenetic block) or oligozoospermia (the elimination of genetically unbalanced gametes by molecular mechanisms) [26,27].

Additionally, some chromosomal fragments may stay asynapsed, leading to developmental block of spermatocytes. Alike simple RCTs, in CCRs, the behavior of multivalent formation also depends on the type of chromosomes involved (autosomes or gonosomes, acrocentrics). The chromosomal features describing the multivalent geometry (the sizes of interstitial and translocated fragments, the distances of the breakpoints to the centromeres, and the presence and number of interstitial chiasmata) determine the pathways of chromosomal disjunction [27,28]. Currently, only 161 balanced CCR male carriers have been described, and an evaluation of the reproductive status was performed in only 64 cases (summarized in Supplementary Table S1).

Researchers found that sperm cells of approximately 50% of genetically balanced male CCR carriers were able to fertilize (Supplementary Table S1) [14]. In several cases, the meiotic segregation of chromosomes involved in the rearrangement was examined, with the frequency of genetically unbalanced spermatozoa ranging from 69% to 88%. Such high values arise from the complexity of the rearrangement: i.e., simple type I CCR with three chromosomes involved, each with one breakpoint, generates 64 various unbalanced genotypes in spermatozoa, while in simple RCTs, a quadrivalent formation generates 16 different types of meiotic segregants. This fact clearly indicates an increased risk of reproductive failures in CCR carriers, mostly leading to miscarriages (Supplementary Table S1) [15,29].

Here, we present a familial case of two infertile brothers (azoospermic and cryptozoospermic) and their mother, all carriers of CCR involving chromosomes 1 and 7 and three breakpoints, resulting in the so-called exceptional type of CCR (type II). The purpose of the study was to identify whether meiotic disruption was caused by chromosomal rearrangement and/or by genomic mutations. Additionally, we performed a literature survey for male CCR carriers described according to their reproductive failures.

2. Results

The characteristics of chromosomes involved in CCR are shown in Figure 1, including: partial karyotypes from G-banding using trypsin and Giemsa staining (GTG banding), results from fluorescent in situ hybridization (FISH) and bacterial artificial chromosome-FISH (BAC-FISH), and ideograms of the chromosomes with marked breakpoint localizations. Multicolor FISH (mFISH) analysis excluded other chromosomal rearrangements. Detailed cytogenetic analysis showed that chromosome 1 had two breakpoints: (i) at the short arm 1p, the suggested breakpoint region was 1p21 (the GTG resolution showed band 1p21 as a dark band, while BAC-FISH revealed that the breakpoint was down to 1p21.3 locus 94,630,298), and (ii) at the long arm 1q in a band 1q42.3 (locus between 233,745,426—the normal position; and 234,256,411—translocated on 1p). In the case of chromosome 7, only one breakpoint was found in long arm 7p in a band 7p14.3 (between 31,667,326—translocated on 1q; and 34,388,646—the normal position). A fragment of the arm of 7p was translocated into 1q42.3, while the terminal fragment of 1q was translocated into 1p21. Thus, the chromosome description was established as: der(1)(1qter->1q42.3::1p21->1q42.3::7p14.3->7pter), der(7)(1pter->1p2 1::7p14.3->7qter), for both brothers and for their mother.

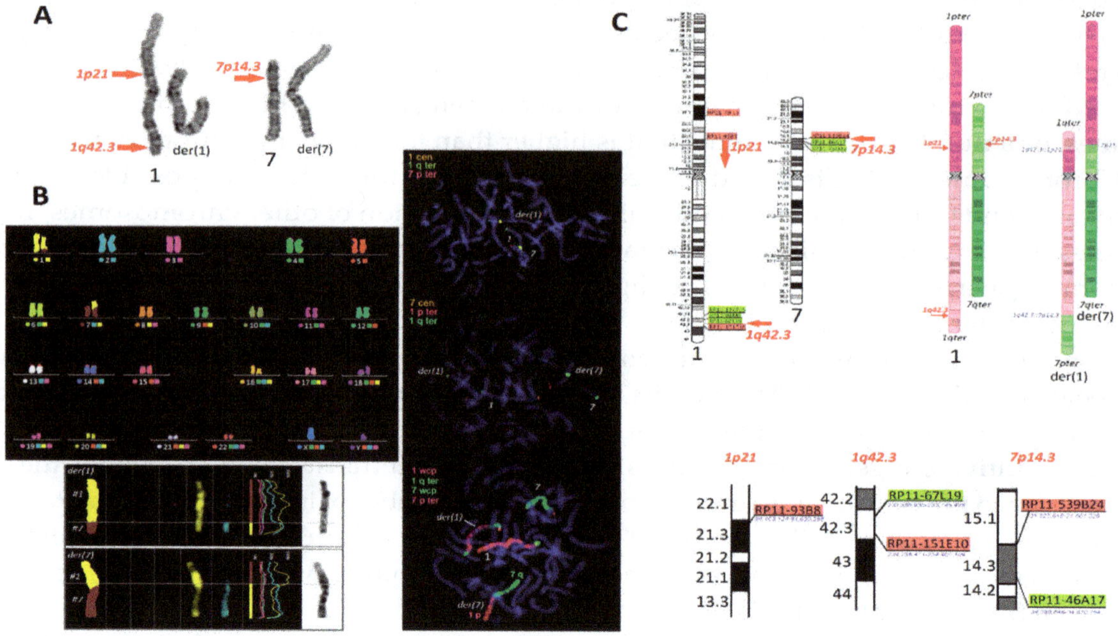

Figure 1. The characteristics of the chromosomes involved in complex chromosomal rearrangements (CCRs) t(1;7). (**A**) G-banding using trypsin and Giemsa staining (GTG banding) with marked breakpoints (UV–Vis microscope, Olympus BX41, magnification 1000×, software: Ikaros MetaSystems); (**B**) fluorescent in situ hybridization (FISH) staining, including various probe combinations (specific for: centromeres (cen), subtelomeres (ter), and whole chromosomes (wcp)), and multicolor FISH (mFISH)) (fluorescent microscope Zeiss AxioImager D1, magnification 1000×; software: ISIS MetaSystems); (**C**) a schematic representation of the bacterial artificial chromosome (BAC)-FISH) samples used (green—normal location; red—translocated), with marked breakpoint positions (arrows).

To identify potential unbalanced regions of the translocation and contributing genomic aberrations elsewhere, array comparative genomic hybridization (aCGH) experiments in the azoospermia patient (brother-1) and his mother were performed. Using a standard genome-wide 400 K array, high quality comparative genomic hybridization (CGH) data were obtained. Following standard genomic copy number variation (CNV) analysis of the aCGH data, 36 unique genomic CNVs were identified (Figure 2, Table 1). The steps included removing polymorphic variants via filtering the CNVs identified in the database of genomic variants (DGV db) (n = 5), annotation analysis for genes found in the remaining CNVs (n = 3) (Supplementary Table S2), and, finally, removal of those that did not show testis-specific expression and/or a plausible biological role in reproduction.

The final three nonpolymorphic CNVs in known genes: *ASMT* (acetylserotonin methyltransferase; OMIM: 300015; RefSeq: NM_004043.2), *GARNL3* (GTPase activating Rap/RanGAP domain like 3; RefSeq: NM_032293.5), and *SESTD1* (SEC14 and spectrin domain containing 1; RefSeq: NM_178123.5), were ruled out due to unlikely biological functions and a nonspecific testis expression profile. Thus, all identified candidate CNVs were ruled out (Supplementary Table S3). In addition, three breakpoint regions of the translocation were analyzed (Supplementary Figure S1) using a higher resolution analysis with individual CGH probes. For all breakpoint regions, 2804 probes were identified (Supplementary Tables S4 and S5). Respectively, each region's probe resolution was calculated, ranging between 6 and 8 kb.

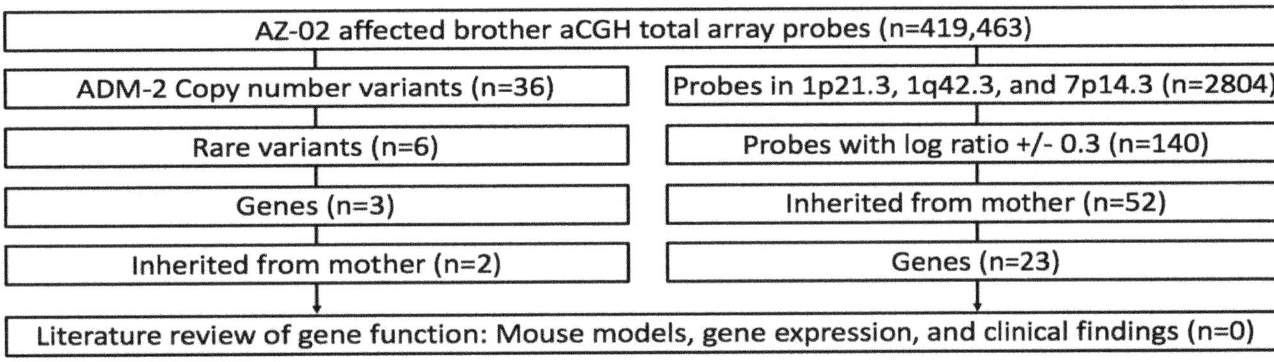

Figure 2. Workflow for genomic copy number variation (CNV) analysis of array comparative genomic hybridization (aCGH) data in the evaluated family. The major steps are shown in the left panel: (1) Low quality and noise probes removed, keeping real CNVs as described in the ADM-2 protocol in Materials and Methods ($n = 36$). (2) The confirmed variants were evaluated for rarity by a review of the database of genomic variants (DGV) for the reported CNV findings in healthy individuals ($n = 6$). CNVs that were reported more than five times in the DGV (fully overlapped and were the same type, amplification or deletion) were considered "polymorphic" variants and were excluded. (3) CNVs inherited from the mother were given higher priority ($n = 2$). (4) Rare CNVs with known genes were considered in a biological role ($n = 1$). (5) A gene review for gene function and testis-specific tissue expression was performed using the PubMed, OMIM, MGI, BioGPS, Unigene, and Aceview databases ($n = 0$). In the right panel, the major steps of the analysis of probes in the translocation regions are shown: (1) The probes were reviewed using the ADM-2 protocol for one probe ($n = 2804$). (2) All single probes with a log ratio of > 0.3 and < −0.3 were analyzed ($n = 140$). (3) CNVs that were also found in the mother were prioritized ($n = 52$). (4) CNVs for the genes analyzed ($n = 23$) and (5) rare genes were reviewed by a detailed literature and database search, as described above.

Next, we performed a revision of the single probes with amplification/deletion signal levels that had the potential to indicate variants too small to be detected by the aberration detection method 2 (ADM-2) algorithm. We identified 23 nonpolymorphic potential small CNVs with known genes. However, CNV filtering following the same biological annotation algorithm for confirmed CNVs did not reveal CNVs with biologically plausible genes (Supplementary Table S3). In addition, independent analysis of the CGH probe resolution (<10 kb) and RefSeq genes located in the translocation breakpoint regions ruled out the possibility of missing significant CNVs at the gene level resolution.

The synaptonemal complex analysis in brother-1 presented a spatial chromosomal formation curled into a ball (Figure 3A–C), which allowed the conjugation of proper fragments during meiosis. In only 1 of 53 (2%) analyzed cells, were chromosomes forming quadrivalents visible (Figure 3A), with a conjugation between the ends of the chromosomes involved (Figure 3B,C). Both the picture and a model also showed incomplete conjugation of chromosomal fragments. The quadrivalent figure was drawn following the proportions of the proper chromosome fragments (Figure 3B). The geometry was also confirmed by the 3D model of the synaptonemal complex (Figure 3C). No association between the CCR chromosomes and sex bivalence was observed (an example of FISH staining presented in Figure 3E).

Table 1. The CNV analysis report summary of the aCGH results for the AZ patient. Variants were identified and annotated using Cytogenomics software with the ADM-2 protocol criteria. All variants were then manually reviewed and categorized using the process described in Figure 3. The frequency was either polymorphic, labeled "pol," or rare, based on the DGV database (The Centre for Applied Genomics). All variants were compared to the mother's CGH results and variants were identified as maternally inherited ("Yes") or not ("No").

Event	Chr	Cytoband	#Probes	Log Ratio	Annotations	Gene Effect	Inherited from Mother
1	chr1:152556449–152581944	q21.3	6	0.617541	*LCE3C*	pol	Yes
2	chr1:196742735–196796220	q31.3	5	−0.978775	*CFHR3, CFHR1*	pol	No
3	chr2:34697718–34730142	p22.3	7	−0.636087		pol	Yes
4	chr2:78709955–78721280	p12	4	−0.852013		pol	No
5	chr2:87392136–87801877	p11.2	11	0.508958	*NCRNA00152*	pol	Yes
6	chr2:132205346–132217492	q21.1	3	−0.809717		pol	No
7	chr2:180069605–180070733	q31.2	3	1.002822	***SESTD1*, 1MB dup**	rare	Yes
8	chr3:162556223–162619141	q26.1	7	0.471521		pol	No
9	chr4:69387056–69483277	q13.2	12	−0.742031	*UGT2B17, UGT2B15*	pol	No
10	chr5:17345455–17353452	p15.1	3	−0.926486		pol	Yes
11	chr5:140223256–140236399	q31.3	4	−0.878128	*PCDHA1, PCDHA2, PCDHA3*	pol	Yes
12	chr6:259881–287425	p25.3	6	0.523226		pol	Yes
13	chr6:32450699–32493043	p21.32	6	−1.853271	*HLA-DRB5*	pol	No
14	chr6:165725547–165737665	q27	3	0.948835		rare	No
15	chr7:141750430–141792094	q34	9	−0.511123	*MGAM,*	pol	Yes
16	chr8:39234992–39386158	p11.22	28	0.821278	*ADAM5P, ADAM3A*, 151 kb dup	pol	Yes
17	chr9:130041553–130145721	q33.3	23	−0.981719	***GARNL3*, 104 kb del, het**	rare	No
18	chr11:18949929–18960666	p15.1	3	0.804263	*MRGPRX1*, 10 kb dup het	pol	No
19	chr11:25635357–25764082	p14.3	9	0.689271		rare	Yes
20	chr11:55368154–55450788	q11	16	0.436620	*OR4C11, OR4P4, OR4S2*	pol	Yes
21	chr12:9637323–9718846	p13.31	11	1.538928		pol	Yes
22	chr12:10583558–10593748	p13.2	3	−0.864549	*KLRC2*,10 kb del het, gene cluster	pol	No
23	chr12:11218244–11225675	p13.2	3	−1.376538	*PRR4, PRH1*	pol	No
24	chr12:11230835–11249210	p13.2	3	−4.773687	*PRR4, PRH1, TAS2R43*	pol	Yes
26	chr14:74001651–74012568	q24.3	3	−5.04754	*HEATR4, ACOT1*, 1.1MB del	pol	Yes
27	chr16:70174866–70193942	q22.1	4	−0.996207	*PDPR*	pol	No
28	chr16:74394080–74407341	q23.1	3	−0.943902	*LOC283922*	pol	No
29	chr16:78372097–78381281	q23.1	3	−5.242076	*WWOX*	pol	No
30	chr17:34437475–34475514	q12	7	0.847688		pol	Yes
31	chr19:53522243–53550020	q13.41	6	0.783781		pol	No
32	chr20:1563715–1577359	p13	4	−3.864894	*SIRPB1*, 13 kb homo del	pol	Yes
33	chr22:18889039–19010562	q11.21	20	0.522787	*DGCR6, PRODH, DGCR5 …*	pol	Yes
34	chr22:24347959–24395353	q11.23	10	0.607834	*LOC391322, GSTT1, GSTTP2*	pol	No
35	chr22:25664618–25919542	q11.23–q12.1	54	0.578645	*IGLL3, LRP5L*	pol	No
36	chrX:1731610–1752284	p22.33	10	0.513510	***ASMT***	rare	No

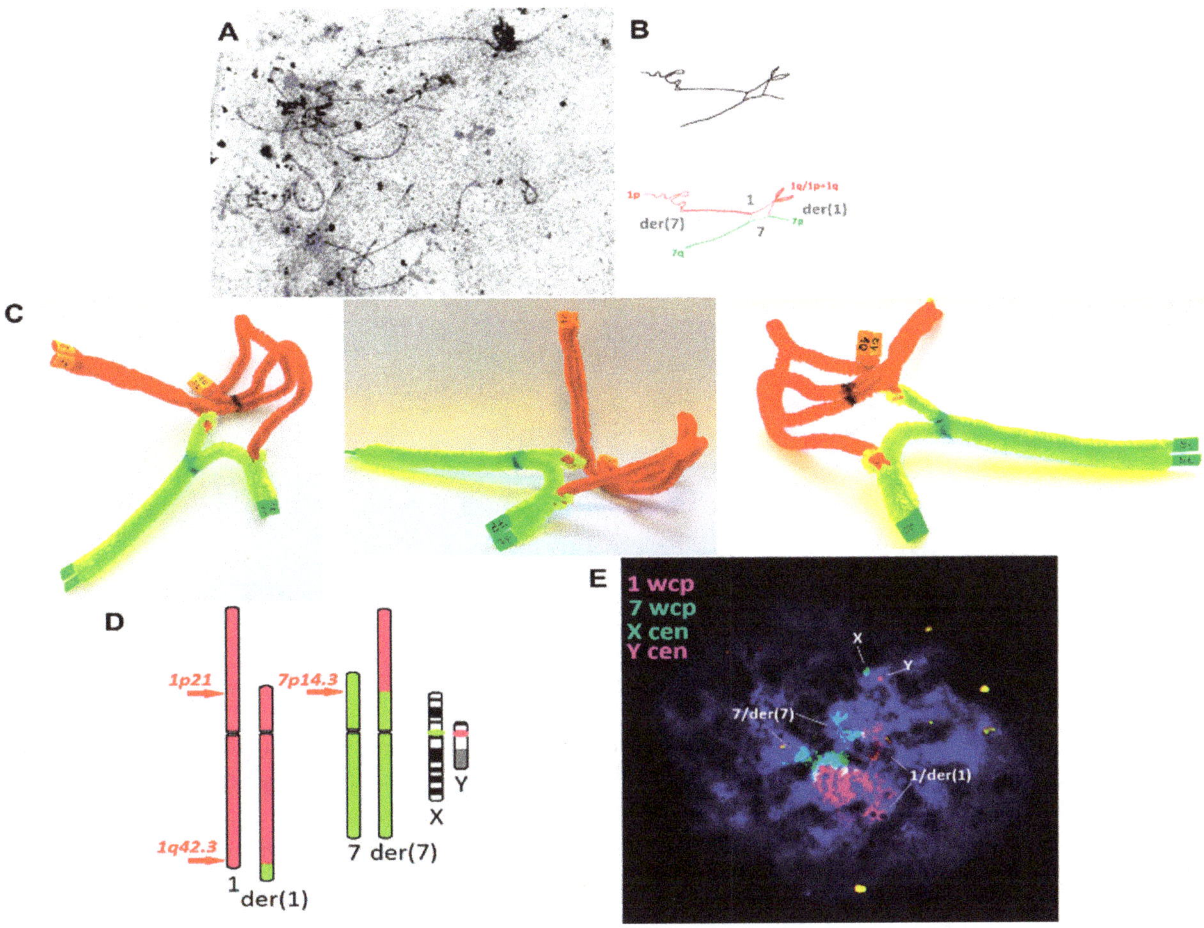

Figure 3. Analysis of the chromosomes at the spermatocyte II stage (brother-1 with azoospermia). (**A**) Synaptonemal complex staining (silver staining; UV–Vis microscope Olympus BX41, magnification 1000×, software: CellB Olympus); (**B**) a schematic suggestion of the involvement of chromosomes 1 and 7 in a meiotic quadrivalent form, following the length of the observed fragments; (**C**) a 3D model of CCR chromosome pairing, following the length of the normal and derivative chromosomes; red—chromosome 1; green—chromosome 7; navy blue—centromeres; brown—near the 1q–1p21 breakpoint; (**D**) the FISH staining scheme for the analysis of the association between XY-bivalent and tetravalent 1/7/der(1)/der(7); (**E**) the FISH result on spermatocyte II showing no association between XY-bivalent and tetravalent 1/7/der(1)/der(7) (fluorescent microscope Zeiss AxioImager D1, magnification 1000×; software: ISIS MetaSystems).

In brother-2, the evaluation of the meiotic segregation pattern was performed on ejaculated spermatozoa. A schematic FISH representation of the labelling and the frequencies of particular segregation types is shown in Figure 4. The frequency of genetically normal/balanced spermatozoa was 29.21%. Among genetically unbalanced gametes, the highest percentage was observed after segregation, 3:1 type (38.21%), followed by adjacent I (20.22%), adjacent II (5.25%), and 4:0 (0.37%). Sperm cells with untypical FISH signals had a frequency of 6.74%.

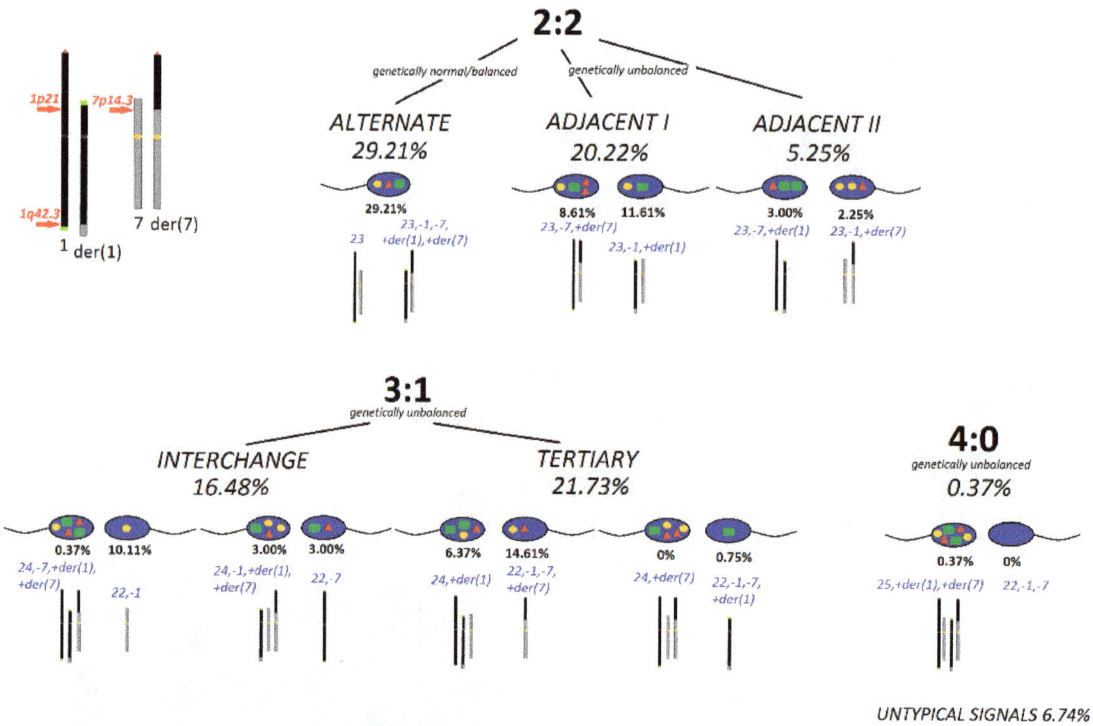

Figure 4. A schematic FISH representation of the labelling and frequency of particular segregation types for meiotic segregation pattern analysis performed for the spermatozoa of brother-2. FISH probes used: subtelomere specific for 1pter (red triangle), for 1qter (green square), and a centromere specific for chromosome 7 (yellow ellipse = red + green) (as described in the Materials and Methods section). The number of evaluated spermatozoa: $n = 267$.

3. Discussion

In this study, we investigated the molecular characteristics of a genetically balanced exceptional CCR involving chromosome 1 (with two breakpoints) and 7 (one breakpoint) found in the family of two infertile brothers and their mother. We narrowed down the breakpoint positions using FISH experiments with commercially available FISH probes, and experimental BAC probes for selected regions of interest. Additionally, we performed aCGH evaluation to determine whether, in the regions of the breakpoints, any CNVs with known genes crucial for spermatogenesis were present, but no plausible candidates were found.

Within more than 250 of papers related to the CCR cases published so far, 161 cases concerned male CCR carriers. To our knowledge, only 64 cases were evaluated for reproductive problems (including one Klinefelter case) (Supplementary Table S1). In 31 cases, the spermatozoa were able to fertilize, leading to reproductive failures and reproductive successes, while the other 31 patients remained infertile. In two CCR cases, the fertility history was unknown. Two infertile CCR cases performed the intracytoplasmic sperm injection (ICSI) procedure, which resulted in genetically balanced offspring with CCRs. Within the group of patients with reproductive failures, in 25 cases, repeated abortions (RAs), miscarriages, and/or multiple congenital abnormalities (MCAs) were observed in the offspring. Only four CCR carriers had no reproductive failures and fathered children with normal (2), abnormal (1), or unknown (1) karyotypes (Supplementary Table S1). When considering the semen parameters among those 64 male CCR carriers, in 15 cases (23.4%) azoospermia was observed, followed by cryptozoospermia and oligozoospermia (C, O; 3/64—4.8%, each), oligoasthenozoospermia (OA; 4/64—6.3%), oligoasthenoteratozoospermia (OAT; 7/64—10.9%), isolated asthenozoospermia (A; 1/64—1.6%), or others. Normal semen parameters were noted for nine males only (N; 14.1%), while in 10 cases, the seminal analysis was not performed (Supplementary Table S1).

As mentioned previously, there are four types of CCR depending on the number of chromosomes and breakpoints involved in the rearrangement. Within the 63 reviewed cases (except the Klinefelter case), type I—the simplest and most common, was found in 39.7% of CCR males with reproductive problems (Supplementary Table S1). Type II was observed in 30.2% of cases (Supplementary Table S1). Our CCR case described in the present study belongs to the type II group, with two breakpoints on chromosome 1 and one breakpoint on chromosome 7. Type III—the coincidental presence of two or three chromosomal translocations, was noted in 17.7% of CCR males with reproductive problems (Supplementary Table S1). The last type of CCR, type IV—the genetically imbalanced type, was observed in 9.5% of CCR males with reproductive failures.

Another aspect of CCRs is the involvement frequency of particular chromosomes in CCR. When considering the data concerning 63 CCR males with reproductive failures (Supplementary Table S1), we found that the total number of chromosomal breaks was 208. Among these, 19 breakpoints were found in chromosome 1, followed by chromosome 9 (17), chromosomes 13 and 3 (17), and chromosomes and 14 and 4 (16). No breakpoints were found for chromosomes: X, 17, and 20. Such observations may lead to two conclusions. First, the larger the chromosome that is involved in the CCR, the higher the probability for rearrangement. Additionally, there is a correlation between the group of chromosomes involved and the proper type of meiotic segregation [14]. Chromosomes from groups A–C (the large ones) were segregated mostly in adjacent I mode (76%), followed by the 4:2 mode with chromosomes from the D–G groups (the small ones).

In CCR carriers with recurrent abortions, the involvement of chromosomes from groups A–C was often observed (84%). Second, chromosome abnormalities have been widely observed in cases with reproductive problems, which are prone to breakpoint appearances. The best example is chromosome 9 and the acrocentric ones for which the involvement in CCR is clearly linked to the 4:2 segregation mode [14]. Our CCR case described in the present study concerned chromosomes 1 and 7. As described above, chromosome 1 is the most prone to breakpoint occurrence, while chromosome 7 is in the group of medium risk with 11 breakpoints noted (Supplementary Table S1). Thus, the involvement of both chromosomes in CCR is not surprising. The mean number of breaks per rearrangement did not differ between fertile and infertile male CCR carriers, even considering that the risk of reproductive failures and infertility increased with the complexity of the CCR (Supplementary Table S1, [14,30]).

The origin of the CCRs in the majority of the reviewed cases (36/64) was unknown, followed by 16 *de novo* cases, seven maternally inherited (also our cases described in this study), two paternally inherited, two mixed, and one inherited but not defined. Certain familial CCRs can change from one generation to another. For example, carriers of double two-way translocations may lead to a more complex rearrangement in their offspring [31–34], or carriers of, e.g., three-way translocations may father children with two simple reciprocal translocations [35]. Such rebuilding and rearrangement, leading to simpler or more complicated aberrations, arises in approximately 45% of CCR cases [17]. An example is a case of a fertile CCR carrier of der(Y;15),rob(13;14) who inherited his CCR from both parents: der(Y) form his father and rob(13;14) from his mother, and fathered by himself a healthy daughter (46,XX) [36]. In the rest of the CCR cases, the rearrangement remained the same (also as described in our study).

Another interesting point is the varied fertility status in family members with the same chromosomal aberration (i.e., fathers and sons, brothers, or cousins). Johanisson et al. described a familial case of family members with the same CCR t(9;12;13), including the father, three daughters, and two sons [37]. Among the sons, one was infertile, while the second one was subfertile. Both brothers were evaluated meiotically according to the possible associations between hexavalent and XY bivalent. The results for both brothers were similar—in 80–90% of spermatocytes, no association was found. The reasons for the different phenotypes remained unknown [37]. Another example is a case of two brothers and a cousin, where brother 1 was a CCR carrier of double two-way translocation t(8;9), t(1;16) and revealed infertility, while his brother and cousin carried the same simple RCT t(8;9) and had progeny [34]. Likely, the accumulation of two rearrangements simultaneously, followed by the increased frequency of genetically unbalanced spermatozoa, resulted in a lack of conception. Similarly, in RCT carriers, there are also known cases of varied fertility statuses in male family members [38–46]. The reasons for such situation remain unclear. The answer may be hidden in the genomes of particular patients, and next-generation sequencing (NGS) methods will allow researchers to identify them.

Next, we discuss the inheriting of CCR between sexes, i.e., from a mother to son, and the differences in the (in)fertility between parents and their offspring. An example is a case of an infertile oligoasthenozoospermic male with CCR involving chromosomes 1, 3, and 13, with a lack of conception [47]. He inherited CCR from his mother who had no fertility problems. Similarly, in a case of t(1;3;6) described by Hornak et al. [48], the normozoospermic CCR carrier was infertile, while his mother and her sister (also CCR carriers) were fertile [48]. The detailed genomic explanation in this case remained unknown; however, it was able to be linked to the familial transmission from mother to son.

Oogenesis lacks chromosome-mediated checkpoint control at the metaphase/anaphase stage, due to the absence of a sex vesicle (a unique element for spermatogenesis, representing a specific chromatin state of the sex chromosomes) [49]. This leads to the high rate of meiotic nondisjunction in females, and is responsible for the increased chromosomal error rate in oogenesis [49–51]. The carriership of CCR mostly leads to chromosomal asynapsed regions, which in males are prone to the association of XY bivalent and then to the disruption of spermatogenesis [26,52]. Heterozygous females are predominantly fertile and have phenotypically normal offspring, while male carriers are subfertile or even sterile. Our case presented in this study seemed to confirm this observation. Namely, the mother of the brothers was fertile, even with two reproductive failures in her reproductive history, but the both evaluated brothers remained infertile due to the lack or extremely low number of spermatozoa.

In the case of the brother B1 with azoospermia, as described in the present study, synaptonemal complex analysis revealed a spatial curled configuration of the quadrivalent with the asynapsed interstitial fragments. However, no association with XY bivalent was observed. Of course, such associations cannot be excluded due to the small number of cells analyzed ($n = 24$). The curled formation of the quadrivalent was so sophisticated that it hindered the procession of the further steps of meiosis; thus, we can assume that the spermatogenetic block appeared at the pachytene stage of meiosis.

What about the cryptozoospermia phenotype of the brother B2 with the same CCR? In this case, the spermatogenesis went further in a very restricted manner—few sperm cells were found in the ejaculate, and they were genetically imbalanced (71%). Factors such as the age or environmental items were excluded as potentially influencing the state of spermatogenesis in both brothers. Both of them worked in normal conditions and were similar in age. Additionally, both brothers performed several sperm analyses within the last 7 years, but the results were the same: azoospermia (B1) and cryptozoospermia (B2). We can suppose two reasons for that.

First, various levels of association between the asynapsed regions and XY body between the two brothers. However, (i) in brother B1 no associations were observed, but the number of cells analyzed was too small to conclude without hesitation. Additionally, we had no possibility to analyze material from the testis from the brother B2. Therefore, we can neither accept nor exclude this hypothesis due to the lack of material. The limited volume of testicular oligobiopsy derived from brother-1 and the lack of such sample from brother-2 constituted the primary limiting factor for performing any additional examination of meiosis. Testicular oligobiopsy is difficult to accept in many individuals. Additionally, in circumstances in which there are some spermatozoa available in the ejaculate, potentially allowing for *in vitro* fertilization (IVF) participation (as in brother-2 with cryptozoospermia), there is no recommendation for gonadal open biopsy sampling.

The second reason is that the answer is likely hidden in genetic mutations, independent from the cytogenetic characteristics of the described cases. To come closer to the answer, we performed aCGH screening, which revealed variations in 26 genes, including three rare variants in genes: *ASMT*, *GARNL3*, and *SESTD1*. However, further filtering according to established criteria, excluded the potential gene candidates for being responsible for various phenotypes between both brothers. Thus, the problem concerning the differences in phenotypes remained unsolved. aCGH analysis could not detect smaller intergenic CNV of few kb or smaller, such as indels of single exons or a few nucleotides. This suggests the need for whole exome or genome sequencing (WES/WGS) to investigate the presented familial case further, especially as genomic sequencing studies suggested that 85% of known genetic variants causing diseases are represented in the human coding sequence by point mutations causing the respective disorders [53–55].

In summary, chromosomal aberrations mostly do not affect the phenotype of the carrier, in contrast to the disruptive influence on spermatogenesis, which affects the gamete quantity (oligo-, azoospermia) and/or quality (generation of chromosomally abnormal sperm) of the carrier. Therefore, the detailed characteristics of each CCR case constitute a highly valuable source of data for genetic counselling in cases with reproductive failures. This also supports the necessity of careful genomic analysis in cases with the possibility of novel point mutations and supports the addition of the new information to existing databases of male infertility.

4. Materials and Methods

4.1. Patients

Pedigree information for the family CCR case is presented in Figure 5. The samples analyzed consisted of material from three members of one family: the mother (M, II-5, age 59), brother-1 (III-3, age 31) with azoospermia, and brother-2 with cryptozoospermia (III-5, age 27). All of them were carriers of a non-mosaic complex chromosome rearrangement involving two chromosomes and three breakpoints (chromosome 1—two breakpoints, chromosome 7—one breakpoint). The chromosomes involved in CCR and the breakpoints were identified by classic cytogenetic methods, including GTG banding, and then confirmed with FISH. Both brothers presented infertility (lack of conception; their wives have a normal karyotype 46,XX), and their multiple semen analyses demonstrated nonobstructive azoospermia (brother-1; no spermatozoa, semen volume: 1.5–2.0 mL) and cryptozoospermia (brother-2; sperm concentration: 0.1×10^6/mL, semen volume: 1.8–2.0 mL, single round cells after centrifugation) (according to guidelines of the World Health Organization WHO, 2010, [56]).

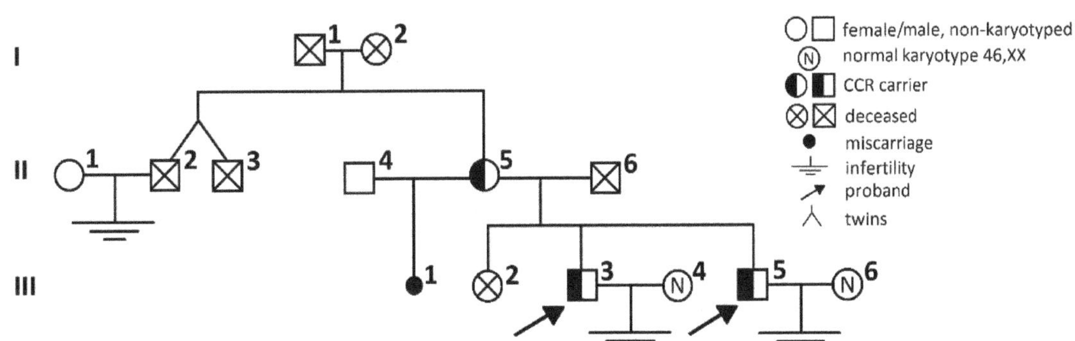

Figure 5. Pedigree information for the CCR t(1;7) family. Two probands are indicated for the brothers (marked with arrows): with azoospermia (brother-1, III-3) and 10 years of lasting infertility, and with cryptozoospermia (brother-2, III-5) and 3 years of lasting infertility. Their mother (II-5), a carrier of the same CCR, revealed one miscarriage at the 12th week of gestation (III-1), and one child death at nine months (III-2), likely due to myelodysplasia (but this was not documented). An infertility case was also noted in another family member (II-2, 20 years of a lack of conception).

No Y chromosome microdeletions in the sex-determining region Y (*SRY*, RefSeq: NM_003140) and azoospermia factor region (AZF) (according to European Academy of Andrology (EAA) guidelines), and no mutations in the *CFTR* gene (RefSeq: NM_000492) were identified. Hormone evaluation was performed only in brother-1 and revealed increased levels of follicle-stimulating hormone (FSH; 15.0; reference value 1–14.0 mo/L) and luteinizing hormone (LH; 12.0; 0.7–7.4 mo/L), while the level of testosterone was normal (4.8; 0.3–12.0 ng/mL). The biological samples included DNA extracted from the blood (the mother and both brothers), lymphocytes fixed according to classical 3:1 ice-cold methanol:acetic acid fixative (the mother and both brothers), testicular oligobiopsy, fixed with 4% paraformaldehyde solution (PFA/1× PBS) for synaptonemal complex analysis (brother-1), and ejaculate, fixed with 3:1 ice-cold methanol:acetic acid fixative (brother-2). All analyzed patients were notified of the purpose of the research and written informed consent was obtained, according to guidelines of the Local Bioethical Committee, Poznan University of Medical Sciences (approval no. 772/15; approved on 01 October 2015).

4.2. Fluorescence In Situ Hybridization (FISH)

The FISH experiments were prepared with combinations of directly labeled probes and/or with BAC probes prepared for the proper chromosomal regions. FISH probe combinations were used for:

(i) The characterization of chromosomes involved in CCR. Centromere-specific: 1cen (*locus* D1Z1, catalogue number LPE01R/G) and 7 cen (D7Z1, LPE07R/G). Subtelomere: 1pter (clone CEB108, LPT01pR), 1qter (clone 160H23, LPT01qG), and 7pter (clone 109a6, LPT07pR, Cytocell, UK). Whole chromosome painting: 1wcp (catalogue number XCP1R), 7wcp (XCP7G), and mFISH (multicolour FISH) (MetaSystems, Altlussheim, Germany). For each combination, at least 50 metaphase plates were analyzed, and 1000 interphase cells (FISH with cen/subtel probes) were counted to exclude the mosaicism.

(ii) The translocation breakpoint analysis—BAC probes for the proper chromosomal regions are listed in Supplementary Table S6 and in Figure 1. BAC clones were chosen from the RPCI-11 library collection (BACPAC Resource Center, the Children's Hospital Oakland Research Institute) based on hg19 (human genome reference GRCh37 h37). At least 10 metaphase plates were analyzed for each round of FISH.

(iii) The analysis of the association between the chromosomes involved in CCR and sex chromosome bivalent (spermatocytes II from the testicular oligobiopsy of brother-1). wcp and cen-specific probes: 1wcp (XCP1R), 7wcp (XCP7G), Xcen (DXZ1, LPE0XG), and Ycen (DYZ3, LPE0YcR) (MetaSystems, Germany; Cytocell, UK); $n = 24$.

(iv) The meiotic segregation pattern in the sperm cells from the ejaculate of brother-2. A 3-colour combination of cen-specific and subtel-specific probes: 1pter (LPT01pR), 1qter (LPT01qG), and 7cen (D7Z1, LPE07R/G) (Cytocell, UK); $n = 267$.

4.3. BAC Preparation

The BAC DNA was isolated using Qiaprep Spin Miniprep kit and protocol (Qiagen). BAC labelling was prepared according to the manufacturer's instructions with a labelling kit (Platinum Bright: Nucleic Acid Labeling Kit, Kreatech; catalogue number: GLK-001 for green, GLK-002 for red/orange). Briefly, 1 μg of BAC-DNA was mixed with 2 μL of ULS (universal linkage system labelling DNA by binding to the N7 position of the guanine), and 2 μL of 10× labelling solution, to a final volume of 20 μL. Next, the prepared mix was incubated at 85 °C for 30 min, and then put on ice and spun down (6000 rpm, 1 min). Labelled BAC was purified from the excess dye using a spin column. Then, repetitive α-satellite sequences were blocked, and non-specific binding was reduced by adding 25× excess of C_0T DNA (Kreatech) and ssDNA (Invitrogen; catalogue number 15632-011). The BAC sample was precipitated with 1/4 vol. of 10 M NH_4Ac and 2.5 vol. of 100% EtOH, −20 °C, overnight. Next, the sample was centrifuged (15000 rpm, 30 min., 0 °C), and the resulting pellet was air-dried. The BAC-DNA was dissolved in 50% hybridization solution and frozen at −20 °C until the FISH procedure.

4.4. Hybridization

The fixed semen sample from brother-2 was spread onto slides, washed in PBS, and incubated in a decondensation solution (10 mM DTT, 100 mM TRIS-HCl; pH 8.5, 43 °C) for 7 min. Next, the slides were rinsed in 2× SSC (pH 7.0), air-dried, and then stored in a freezer at −20 °C until the FISH procedure.

Fixed lymphocyte cultures were spread onto slides directly before FISH. FISH was performed following the manufacturer's protocol (Cytocell, Cambridge, UK) with modifications described previously [57]. The hybridization mixtures contained various volumes of probes depending on their specificity: centromere—2.0 μL; subtelomere—3.0 μL; wcp/mFISH—10.0 μL; and BAC—5.0 μL. If needed, the mixes were filled with hybridization solution to a final volume of 10 or 20 μL. The FISH efficiency was approximately 98%. For analysis, a Zeiss AxioImager D1 microscope equipped with the necessary filters (DAPI/FITC/SpO/TR/Cy5/DEAC/Triple) and objectives (20×, 100× immersion) was used. Images were acquired with a CCD camera (Jenoptik, Germany) and processed using ISIS software (MetaSystems, Altlussheim, Germany).

4.5. Genomic Microarray CGH

To test for potential genomic aberrations, genome-wide aCGH was performed using SurePrint G3 Human CGH 2 × 400 k Oligo Microarrays (Agilent Technologies, Santa Clara, CA, USA). CGH analysis was performed according to the manufacturer's protocol (Agilent Technologies). The array design featured 420,288 total distinct biological features, or probes. Briefly, genomic DNA for brother-1 and the mother was extracted from the patients' peripheral blood leukocytes (Puregene; Qiagen). Reference male DNA was purchased from Promega (Madison, WI). Experimental DNA was enzymatically fragmented and labeled with fluorescent dye Cyanine-5; the reference DNA was labeled with Cyanine-3 dye. The labeled DNA was hybridized to the CGH probes for 40 h at 66 °C. After washing, the array slides were scanned (SureScan Microarray C scanner, Agilent) and analyzed using CytoGenomic Workbench software (Agilent). The CGH array quality was assessed as good, since 419,463 biological features (99.8%) were identified successfully.

Copy number variant (CNV) calling was performed using the ADM-2 protocol with a minimal region of three consecutive probes, a mean log ratio >0.25 or <−0.25, and a direct signal intensity range of 100–1000 units. Variant calls that met the protocol criteria were reviewed initially to exclude variants of uncertain quality, such as calls with large areas of missing probe coverage (>100 kb). To verify, if the resultant CNVs were polymorphic, the CNVs were checked for frequency in the DGV database (Database of Genomic Variants, The Centre for Applied Genomics). Variants that were fully overlapped with >5 reported CNVs of the same type (amplification or deletion) in the DGV were considered "polymorphic." Gene function annotation was performed with the OMIM (Online Mendelian Inheritance in Man), MGI (Mouse Genome Informatics, Jackson Laboratories), and RefSeq databases. The expression annotation was done with BioGPS (Scripps Research Institute) and Aceview (NCBI).

4.6. Synaptonemal Complex Analysis

The evaluation of synaptonemal complexes was performed on secondary spermatocytes from a testicular oligobiopsy of B1. After the collection of the testis sample via oligobiopsy, germline cells were placed in a Petri dish with 10 drops of 1× PBS (pH 7.4; Lonza, Walkersville, MD, USA). Next, three drops of 0.4% KCl were added and the sample was macerated with forceps to obtain a suspension of single cells (all under a light microscope inspection). The resuspended cells were placed onto slides, air-dried, and fixed with 4% PFA/1× PBS for 10 min. at room temperature. Next, the slides were rinsed with 0.5% Triton X-100 (Sigma), air-dried, and incubated in 70% EtOH for 10 min. The fixed biopsy samples were stored at 4 °C until further use.

The synaptonemal complexes were visualized using the silver staining technique described previously [58,59]. Briefly, the slide was heated at 80 °C for 5 min. Then, freshly made buffers were applied: four drops of 50% $AgNO_3$/water and two drops of a solution of 1% formic acid in 2% gelatin/water. The covered sample was incubated at 80 °C for 2–3 min., until the silver precipitated (dark brown color). Next, the slide was rinsed in distilled water and incubated in 4% Giemsy staining solution (Merck) for 10–12 min. After rinsing in water and air-drying, our microscopic analysis was performed (Olympus BX41, magnification 1000×, oil-immersion, software: CellSense Dimensions, Olympus). The number of analyzed cells was $n = 53$.

Supplementary Materials: Supplementary Materials can be found at http://www.mdpi.com/1422-0067/21/12/4559/s1. Supplementary Figure S1. CGH probe position on chromosomes 1 and 7. Supplementary Table S1 Male CCR carriers with evaluated reproductive status, according to the literature data published so far. Supplementary Table S2 Non-polymorphic variants found in aCGH screening of brother-1 and mother samples. Supplementary Table S3. Final result of aCGH screening—list of genes found, incl. their role. Supplementary Table S4 Coordinates of 3 breakpoint regions of the translocation analyzed using a higher resolution analysis with individual aCGH probes. Supplementary Table S5 aCGH results obtained for 3 breakpoint regions of the translocation. Supplementary Table S6 BAC clones used in breakpoint mapping of the chromosomes involved in CCR t(1;7).

Author Contributions: M.O.: designed the study, prepared and analyzed FISH and BAC-FISH breakpoint mapping, prepared the synaptonemal complex and sperm FISH analyses, interpreted results, and drafted the manuscript; T.S.: performed bioinformatic analysis; N.P.: performed aCGH analysis and BAC preparation, and drafted the aCGH part of the manuscript; N.H.: collected material, family data, and identified rearrangement (GTG); A.G. and S.Y.: prepared BAC samples and performed part of BAC-FISH breakpoint mapping; D.Z.: recruited the patients; A.N.Y.: interpreted results of aCGH and drafted the manuscript; M.K.: collected funds, interpreted results, and finalized the manuscript. All authors have read and agreed to the published version of the manuscript.

Acknowledgments: Authors would like to thank Ewa Wiland (Institue of Human Genetics PAS, Poznan) for consultations concerning the possible breakpoint positions of the rearrangement and their behavior during synaptonemal complex formation.

References

1. Sharma, A. Male Infertility; Evidences, Risk Factors, Causes, Diagnosis and Management in Human. *Ann. Clin. Lab. Res.* **2017**, *5*, 188. [CrossRef]
2. Matzuk, M.M.; Lamb, D.J. The biology of infertility: Research advances and clinical challenges. *Nat. Med.* **2008**, *14*, 1197–1213. [CrossRef] [PubMed]
3. Mau-Holzmann, U.A. Somatic chromosomal abnormalities in infertile men and women. *Cytogenet. Genome Res.* **2005**, *111*, 317–336. [CrossRef] [PubMed]
4. Tüttelmann, F.; Ruckert, C.; Röpke, A. Disorders of spermatogenesis: Perspectives for novel genetic diagnostics after 20 years of unchanged routine. *Med. Genet.* **2018**, *30*, 12–20. [CrossRef] [PubMed]
5. Nieschlag, E.; Behre, H.M.; Nieschlag, S. *Andrology. Male Reproductive Health and Dysfunction*, 3rd ed.; Nieschlag, E., Behre, H.M., Nieschlag, S., Eds.; Springer-Verlag: Berlin/Heidelberg, Germany, 2010.
6. Krausz, C.; Riera-Escamilla, A. Monogenic Forms of Male Infertility. In *Genetics of Endocrine Diseases and Syndromes. Experientia Supplementum*; Igaz, P., Patócs, A., Eds.; Springer Nature: Cham, Switzerland, 2020; Volume 111, pp. 341–366.
7. Cannarella, R.; Condorelli, R.A.; Mongioì, L.M.; La Vignera, S.; Calogero, A.E. Molecular Biology of Spermatogenesis: Novel Targets of Apparently Idiopathic Male Infertility. *Int. J. Mol. Sci.* **2020**, *21*, 1728. [CrossRef]
8. Hwang, K.; Yatsenko, A.N.; Jorgez, C.J.; Mukherjee, S.; Nalam, R.L.; Matzuk, M.M.; Lamb, D.J. Mendelian genetics of male infertility. *Ann. N. Y. Acad. Sci.* **2010**, *1214*, E1–E17. [CrossRef]
9. Templado, C.; Bosch, M.; Benet, J. Frequency and distribution of chromosome abnormalities in human spermatozoa. *Cytogenet. Genome Res.* **2005**, *111*, 199–205. [CrossRef]
10. Harton, G.L.; Tempest, H.G. Chromosomal disorders and male infertility. *Asian J. Androl.* **2012**, *14*, 32–39. [CrossRef]
11. Kefer, J.C.; French, D.B. Azoospermia: Diagnosis and Management. In *Male Infertility. Current Clinical Urology*; Sabanegh, E.S., Ed.; Humana Press: Totowa, NJ, USA, 2011; pp. 23–30.
12. Lee, J.Y.; Dada, R.; Sabanegh, E.; Carpi, A.; Agarwal, A. Role of genetics in azoospermia. *Urology* **2011**, *77*, 598–601. [CrossRef]
13. Ergul, E.; Liehr, T.; Mrasek, K.; Sazci, A. A *de novo* complex chromosome rearrangement involving three chromosomes (2, 13, and 18) in an oligozoospermic male. *Fertil. Steril.* **2009**, *92*, 391.e9–391.e12. [CrossRef]

14. Madan, K. Balanced complex chromosome rearrangements: Reproductive aspects. A review. *Am. J. Med. Genetics* **2012**, *158A*, 947–963. [CrossRef] [PubMed]
15. Trpchevska, N.; Dimova, I.; Arabadji, T.; Milachich, T.; Angelova, S.; Dimitrova, M.; Hristova-Savova, M.; Andreeva, P.; Timeva, T.; Shterev, A. A family study of complex chromosome rearrangement involving chromosomes 1, 8, and 11 and its reproductive consequences. *J. Assist. Reprod. Genet.* **2017**, *34*, 659–669. [CrossRef] [PubMed]
16. Gruchy, N.; Barreau, M.; Kessler, K.; Gourdier, D.; Leporrier, N. A paternally transmitted complex chromosomal rearrangement (CCR) involving chromosomes 2, 6, and 18 includes eight breakpoints and five insertional translocations (ITs) through three generations. *Am. J. Med. Genet.* **2010**, *152A*, 185–190. [CrossRef]
17. Pellestor, F.; Anahory, T.; Lefort, G.; Puechberty, J.; Liehr, T.; Hedon, B.; Sarda, P. Complex chromosomal rearrangements: Origin and meiotic behavior. *Hum. Reprod. Update* **2011**, *17*, 479–494. [CrossRef] [PubMed]
18. Pellestor, F.; Puechberty, J.; Weise, A.; Lefort, G.; Anahory, T.; Liehr, T.; Sarda, P. Meiotic segregation of complex reciprocal translocations: Direct analysis of the spermatozoa of a t(5;13;14) carrier. *Fertil. Steril.* **2011**, *95*, e17–e22. [CrossRef]
19. Kang, S.-H.L.; Shaw, C.; Ou, Z.; Eng, P.A.; Cooper, M.L.; Pursley, A.N.; Sahoo, T.; Bacino, C.A.; Chinaut, C.; Stankiewicz, P.; et al. Insertional translocation detected using FISH confirmation of array-comparative genomic hybridization (aCGH) results. *Am. J. Med. Genet.* **2010**, *152A*, 1111–1126. [CrossRef]
20. Poot, M.; Haaf, T. Mechanisms of origin, phenotypic effects and diagnostic implications of complex chromosome rearrangements. *Mol. Syndromol.* **2015**, *6*, 110–134. [CrossRef]
21. Houge, G.; Liehr, T.; Schoumans, J.; Ness, G.O.; Solland, K.; Starke, H.; Claussen, U.; Strømme, P.; Akre, B.; Vermeulen, S. Ten years follow up of a boy with a complex chromosomal rearrangement: Going from a > 5 to 15-breakpoint CCR. *Am. J. Med. Genet. A* **2003**, *118A*, 235–240. [CrossRef]
22. Gorski, J.L.; Kistenmacher, M.N.; Punnett, H.H.; Zackai, E.H.; Emanuel, B.S. Reproductive risks for carriers of complex chromosome rearrangements: Analysis of 25 families. *Am. J. Med. Genet.* **1988**, *29A*, 247–261. [CrossRef]
23. Madan, K.; Nieuwint, A.W.M.; van Bever, Y. Recombination in a balanced complex translocation of a mother leading to a balanced reciprocal translocation in the child. Review of 60 cases of balanced complex translocations. *Hum. Genet.* **1997**, *99*, 806–815. [CrossRef]
24. Lee, I.W.; Su, M.T.; Hsu, C.C.; Lin, Y.H.; Chen, P.Y.; Kuo, P.L. Constitutional Complex Chromosomal Rearrangements In Azoospermic Men—Case Report And Literature Review. *Urology* **2006**, *68*, 1343.E5–1343.E8. [CrossRef] [PubMed]
25. Bartels, I.; Starke, H.; Argyriou, L.; Sauter, S.M.; Zoll, B.; Liehr, T. An exceptional complex chromosomal rearrangement (CCR) with eight breakpoints involving four chromosomes (1;3;9;14) in an azoospermic male with normal phenotype. *Eur. J. Med. Genet.* **2007**, *50*, 133–138. [CrossRef] [PubMed]
26. Hunt, P.A.; Hassold, T.J. Sex matters in meiosis. *Science* **2002**, *296*, 2181–2183. [CrossRef] [PubMed]
27. Oliver-Bonet, M.; Benet, J.; Sun, F.; Navarro, J.; Abad, C.; Liehr, T.; Starke, H.; Greene, C.; Ko, E.; Martin, R.H. Meiotic studies in two human reciprocal translocations and their association with spermatogenic failure. *Hum. Reprod.* **2005**, *20*, 683–688. [CrossRef]
28. Faraut, T.; Mermet, M.A.; Demongeot, J.; Cohen, O. Cooperation of selection and meiotic mechanisms in the production of imbalances in reciprocal translocations. *Cytogenet. Cell Genet.* **2000**, *88*, 15–21. [CrossRef]
29. Batista, D.A.; Pai, G.S.; Stetten, G. Molecular analysis of a complex chromosomal rearrangement and a review of familial cases. *Am. J. Med. Genet.* **1994**, *53*, 255–263. [CrossRef]
30. Kim, J.W.; Chang, E.M.; Song, S.H.; Park, S.H.; Yoon, T.K.; Shim, A.H. Complex chromosomal rearrangements in infertile males: Complexity of rearrangement affects spermatogenesis. *Fertil. Steril.* **2011**, *95*, 349–352. [CrossRef]
31. Walker, S.; Howard, P.J.; Hunter, D. Familial complex autosomal translocations involving chromosomes 7, 8, and 9 exhibiting male and female transmission with segregation and recombination. *J. Med. Genet.* **1985**, *22*, 484–491. [CrossRef]
32. Zahed, L.; Der Kaloustian, V.; Batanian, J.R. Familial Complex Chromosome Rearrangement Giving Rise to Balanced and Unbalanced Recombination Products. *Am. J. Med. Genet.* **1998**, *79*, 30–34. [CrossRef]
33. Soler, A.; Sanchez, A.; Carrio, A.; Badenas, C.; Mila, M.; Margarit, E.; Borrell, A. 2005 Recombination in a male carrier of two reciprocal translocations involving chromosomes 14, 140, 15, and 21 leading to balanced and unbalanced rearrangements in offspring. *Am. J. Med. Genet.* **2005**, *134A*, 309–314. [CrossRef]

34. Ferfouri, F.; Boitrelle, F.; Clement, P.; Gomes, D.M.; Selva, J.; Vialard, F. Can one translocation impact the meiotic segregation of another translocation? A sperm-FISH analysis of a 46,XY,t(1;16)(q21;p11.2), t(8;9)(q24.3;p24) patient and his 46,XY,t(8;9)(q24.3;p24) brother and cousin. *Mol. Hum. Reprod.* **2013**, *19*, 109–117. [CrossRef]
35. Bass, H.N.; Sparkes, R.S.; Lessner, M.M.; Fox, M.; Phoenix, B.; Bernar, J. A family with three independent autosomal translocations associated with 7q32->7qter syndrome. *J. Med. Genet.* **1985**, *22*, 59–63. [CrossRef] [PubMed]
36. Ferfouri, F.; Boitrelle, F.; Clement, P.; Molina, G.D.; Selva, J.; Vialard, F. 2014 Sperm FISH analysis of a 44,X,der(Y),t(Y;15)(q12;q10)pat, rob (13;14)(q10;q10)mat complex chromosome rearrangement. *Andrologia* **2014**, *46*, 576–582. [CrossRef] [PubMed]
37. Johannisson, R.; Lohrs, U.; Passarge, E. Pachytene analysis in males heterozygous for a familial translocation (9;12;13)(q22;q22;q32) ascer- tained through a child with partial trisomy 9. *Cytogenet. Cell. Genet.* **1988**, *47*, 160–166. [CrossRef] [PubMed]
38. Johannisson, R.; Lohrs, U.; Wolff, H.H.; Schwinger, E. Two different XY- quadrivalent associations and impairment of fertility in men. *Cytogenet. Cell. Genet.* **1987**, *45*, 222–230. [CrossRef] [PubMed]
39. Estop, A.M.; Levinson, F.; Cieply, K.; Vankirk, V. The segregation of a translocation t(1;4) in two male carriers heterozygous for the translocation. *Hum. Genet.* **1992**, *89*, 425–429. [CrossRef]
40. Rousseaux, S.; Chevret, E.; Monteli, M.; Cozzi, J.; Pelletier, R.; Devillard, F.; Lespinasse, J.; Sèle, B. Meiotic segregation in males heterozygote for reciprocal translocations: Analysis of sperm nuclei by two and three colour fluorescence in situ hybridization. *Cytogenet. Cell. Genet.* **1995**, *71*, 240–246. [CrossRef]
41. Cora, T.; Acar, H.; Kaynak, M. Molecular cytogenetic detection of meiotic segregation patterns in sperm nuclei of carriers of 46,XY,t(15;17)(q21; q25). *J. Androl.* **2002**, *23*, 793–798.
42. Anton, E.; Vidal, F.; Egozcue, J.; Blanco, J. Preferential alternate segregation in the common t(11;22)(q23;q11) reciprocal translocation: Sperm FISH analysis in two brothers. *Reprod. BioMed. Online.* **2004**, *9*, 637–644. [CrossRef]
43. Morel, F.; Douet-Guilbert, N.; Roux, C.; Tripogney, C.; Le Bris, M.J.; De Braekeleer, M.; Bresson, J.-L. Meiotic segregation of a t(7;8)(q11.21;cen) translocation in two carrier brothers. *Fertil. Steril.* **2004**, *81*, 682–685. [CrossRef]
44. Wiland, E.; Midro, A.T.; Panasiuk, B.; Kurpisz, M. The analysis of meiotic segregation patterns and aneuploidy in the spermatozoa of father and son with translocation t(4;5)(p15.1;p12) and the prediction of the individual probability rate for unbalanced progeny at birth. *J. Androl.* **2007**, *28*, 262–272. [CrossRef] [PubMed]
45. Vozdova, M.; Oracova, E.; Horinova, V.; Rubes, J. Sperm fluorescence in situhybridization study of meiotic segregation and an interchromosomal effect in carriers of t(11;18). *Hum. Reprod.* **2008**, *23*, 581–588. [CrossRef] [PubMed]
46. Olszewska, M.; Wiland, E.; Huleyuk, N.; Fraczek, M.; Midro, A.T.; Zastavna, D.; Kurpisz, M. Chromosome (re)positioning in spermatozoa of fathers and sons—Carriers of reciprocal chromosome translocation (RCT). *BMC Med. Genom.* **2019**, *12*, 30. [CrossRef]
47. Li, L.; Heng, X.; Yun, W.; Zheng, S.; Zhang, J.; Fan, W. Familial complex chromosome rearrangement (CCR) involving 5 breakpoints on chromosomes 1, 3 and 13 in a severe oligozoospermic patient. *J. Assist. Reprod. Genet.* **2013**, *30*, 423–429. [CrossRef]
48. Hornak, M.; Vozdova, M.; Musilova, P.; Prinosilova, P.; Oracova, E.; Linkova, V.; Vesela, K.; Rubes, J. Comprehensive meiotic segregation analysis of a 4-breakpoint t(1;3;6) complex chromosome rearrangement using single sperm array comparative genomic hybridization and FISH. *Reprod. BioMed. Online* **2014**, *29*, 499–508. [CrossRef] [PubMed]
49. LeMaire-Adkins, R.; Radke, K.; Hunt, P.A. Lack of checkpoint control at the metaphase/anaphase transition: A mechanism of meiotic nondisjunction in mammalian females. *J. Cell Biol.* **1997**, *139*, 1611–1619. [CrossRef] [PubMed]
50. Kurahashi, H.; Bolor, H.; Kato, T.; Kogo, H.; Tsutsumi, M.; Inagaki, H.; Ohye, T. Recent advance in our understanding of the molecular nature of chromosomal abnormalities. *J. Hum. Genet.* **2009**, *54*, 253–260. [CrossRef] [PubMed]
51. Steuerwald, N. Meiotic spindle checkpoints for assessment of aneuploid oocytes. *Cytogenet. Genome. Res.* **2005**, *111*, 256–259. [CrossRef]
52. Roeder, G.S.; Bailis, J.M. The pachytene checkpoint. *Trends. Genet.* **2000**, *16*, 395–403. [CrossRef]

53. Ku, C.S.; Cooper, D.N.; Polychronakos, C.; Naidoo, N.; Wu, M.; Soong, R. Exome sequencing: Dual role as a discovery and diagnostic tool. *Ann. Neurol.* **2012**, *71*, 5–14. [CrossRef]
54. Botstein, D.; Risch, N. Discovering genotypes underlying human phenotypes: Past successes for mendelian disease, future approaches for complex disease. *Nat. Genet.* **2003**, *33*, 228–237. [CrossRef] [PubMed]
55. Eisfeldt, J.; Pettersson, M.; VezziI, F.; Wincent, J.; Kaller, M.; Gruselius, J.; Nilsson, D.; Syk Lundberg, E.; Carvalho, C.M.B.; Lindstrand, A. Comprehensive structural variation genome map of individuals carrying complex chromosomal rearrangements. *PLoS Genet.* **2019**, *15*, e1007858. [CrossRef] [PubMed]
56. World Health Organization. Reference values and semen nomen- clature. In *WHO Laboratory Manual for the Examination and Processing of Human Semen*, 5th ed.; WHO Press: Geneva, Switzerland, 2010; pp. 223–225.
57. Olszewska, M.; Fraczek, M.; Huleyuk, N.; Czernikiewicz, A.; Wiland, E.; Boksa, M.; Zastavna, D.; Panasiuk, B.; Midro, A.T.; Kurpisz, M. Chromatin structure analysis of spermatozoa from reciprocal chromosome translocation carriers (RCT) with known meiotic segregation patterns. *Reprod. Biol.* **2013**, *13*, 209–220. [CrossRef] [PubMed]
58. Howell, W.M.; Black, D.A. Controlled silver-staining of nucleolus organizer regions with a protective colloidal developer: A 1-step method. *Experientia* **1980**, *36*, 1014–1015. [CrossRef]
59. Latos-Bielenska, A. Badanie chromosomów mitotycznych i mejotycznych techniką immunocytochemicznej detekcji bromodezoksyurydyny wbudowanej do chromosomowego DNA. Praca Habilitacyjna (D.Sc. Thesis), Wydawnictwo Akademii Medycznej im., K. Marcinkowskiego w Poznaniu, 1991.

2

CRISPR Gene Editing of Murine Blood Stem and Progenitor Cells Induces MLL-AF9 Chromosomal Translocation and MLL-AF9 Leukaemogenesis

Evgenia Sarrou [1], Laura Richmond [1], Ruaidhrí J. Carmody [2], Brenda Gibson [3] and Karen Keeshan [1,*]

1 Paul O'Gorman Leukaemia Research Centre, Institute of Cancer Sciences, University of Glasgow, Glasgow G12 0YN, UK; t.sarrou@gmail.com (E.S.); Laura.Richmond@glasgow.ac.uk (L.R.)
2 Centre for Immunobiology, Institute of Infection, Immunity & Inflammation, College of Medicine, Veterinary and Life Sciences, University of Glasgow, Glasgow G12 8TA, UK; ruaidhri.carmody@glasgow.ac.uk
3 Royal Hospital for Children, Glasgow G51 4TF, UK; Brenda.Gibson@ggc.scot.nhs.uk
* Correspondence: Karen.keeshan@glasgow.ac.uk

Abstract: Chromosomal rearrangements of the mixed lineage leukaemia (*MLL*, also known as *KMT2A*) gene on chromosome 11q23 are amongst the most common genetic abnormalities observed in human acute leukaemias. *MLL* rearrangements (*MLLr*) are the most common cytogenetic abnormalities in infant and childhood acute myeloid leukaemia (AML) and acute lymphocytic leukaemia (ALL) and do not normally acquire secondary mutations compared to other leukaemias. To model these leukaemias, we have used clustered regularly interspaced short palindromic repeats (CRISPR)/Cas9 gene editing to induce MLL-AF9 (MA9) chromosomal rearrangements in murine hematopoietic stem and progenitor cell lines and primary cells. By utilizing a dual-single guide RNA (sgRNA) approach targeting the breakpoint cluster region of murine Mll and Af9 equivalent to that in human MA9 rearrangements, we show efficient de novo generation of MA9 fusion product at the DNA and RNA levels in the bulk population. The leukaemic features of MA9-induced disease were observed including increased clonogenicity, enrichment of c-Kit-positive leukaemic stem cells and increased MA9 target gene expression. This approach provided a rapid and reliable means of de novo generation of Mll-Af9 genetic rearrangements in murine haematopoietic stem and progenitor cells (HSPCs), using CRISPR/Cas9 technology to produce a cellular model of MA9 leukaemias which faithfully reproduces many features of the human disease in vitro.

Keywords: CRISPR/Cas9; chromosomal translocation; mixed lineage leukaemia; tumourigenesis; stem cells

1. Introduction

Rearrangements of the mixed lineage leukaemia gene (*MLLr*) are the most common cytogenetic abnormalities found in infant and childhood acute lymphocytic leukaemia (ALL) and acute myeloid leukaemia (AML) [1] and are found in 80% and 35–50% of all paediatric ALL and AML cases respectively [1]. These leukaemias do not typically acquire many secondary mutations [2] and are thought to result from inappropriate repair of DNA double strand breaks (DSBs) via nonhomologous end joining (NHEJ) in haematopoietic cells [3]. As the 8.3-kb breakpoint cluster region of mixed lineage leukaemia (*MLL*) is located between exons 8 and 13 for almost all *MLLr* cases [4], these fusion genes contain the first 8–13 exons of *MLL* and a variable number of exons from the fusion partner genes (FPGs). Up to 135 *MLL* FPGs have been identified and are well described elsewhere [5–7].

These include cytoplasmic proteins (AF6, GAS7 and EEN), septins (SEPT2, SEPT5, SEPT6, SEPT9 and SEPT11) and histone acetyltransferases (CBP and P300) as well as the largest subset, the nuclear proteins. This category contains the most common FPGs such as *ENL*, *ELL*, *AF4*, *AF9* and *AF10*, which account for approximately 80% of all MLL fusions [5]. Of these, MLL-AF9 (MA9) t(9;11)(p22;q23) translocations are found in 30% of *MLLr* AML cases [5] irrespective of age and are commonly found in a subset of adult patients with therapy-related AML [8].

Although the precise functional properties of AF9 are unknown, it has been demonstrated to form multimer complexes with members of the polycomb repressive complex 1 (PRC1) complex and disruptor of telomere silencing (*DOT1L*) via a conserved ANC1 homology domain (AHD), indicating a role in transcriptional regulation [9,10]. These properties appear to play a critical role in leukaemic transformation [11,12]. Oncogenic properties of MLL fusion proteins include interference with transcriptional elongation, resulting in dysregulation of target genes [13,14]. Aberrant expression of MLL target genes is strongly dependent on DOT1L-mediated H3K79 methylation [15], which is a critical requirement for MA9 leukaemogensis [16]. Members of the Homeobox (*HOX*) family are commonly overexpressed in *MLLr* leukaemias as a direct target of MLL fusion proteins [4,17]. The HOX family are critical factors in the self-renewing properties of haematopoietic stem cells (HSCs) and their overexpression results in a differentiation block and an increase in self-renewal of immature myeloid progenitor cells. The *HOX* cofactors *MEIS1* [18,19] and *PBX3* [20] are also direct targets of MLL fusion proteins [21] frequently found upregulated in coordination with HOX targets. Importantly, there is evidence to suggest that MA9-mediated leukaemia can progress even in the absence of *HOX* genes such as *HOXA9* [22], indicating an important role of non-*HOX* family genes. This includes *CDK6* [23], *EVI-1* [24,25] and *EYA1* [26,27], which are also found upregulated in *MLLr* leukaemias as a result of direct binding of MLL-FPG proteins to gene promoters [26,28,29], and play a vital role in leukaemic transformation including cellular immortalization [26], hyperproliferation [23,24], chemoresistance [24] and dysregulated self-renewal [25,27]. CDK6 specifically has been highlighted as a critical effector of *MLLr* leukemogenesis as its depletion in mice with MA9-driven AML was shown to overcome the myeloid differentiation block and to prolong survival in vivo [29].

Numerous murine models of *MLLr* have been generated and have each offered valuable insights. Limitations with murine models are in large part due to the technology used to mimic *MLL* chromosomal rearrangements. An early knock-in model [30] utilized homologous recombination (HR) to generate an *Mll-Af9* fusion gene and displayed a myeloproliferative disorder (MPD), leading primarily to AML characterized by expansion of immature myeloid populations with a small percentage developing B-cell ALL (B-ALL) [31]; thus, this model resembles the phenotypic heterogeneity of *MLLr* leukaemias, which can manifest as AML; ALL; or in a minority of cases, mixed phenotype acute leukaemia (MPAL). However, this system lacked tissue-specific control of MA9 expression throughout development and mice were prone to developmental defects as a result of heterozygosity for wild type (WT) *Mll* [32]. To gain more insight into cell-type specific effects, a subsequent approach involved enrichment of haematopoietic populations from these HR-generated MA9 knock-in mice, followed by secondary transplantation into WT recipients [33]. This also demonstrated efficient generation of AML, especially when Lin−Sca-1+c-Kit+ (LSK) cells, which include HSCs and haematopoietic stem and progenitor (HSPCs), were transplanted. An advantage of these methods is the expression of the MA9 transgene from the endogenous *Mll* promoter, which results in physiological transgene expression levels. This model cannot however address the leukaemia stem cell origin with respect to de novo AML initiation.

Translocator models utilizing the Cre-loxP system have generated MA9 translocations via loxP sites inserted into those intronic regions of endogenous *Mll* and *Af9* genes where breakpoints are most frequently found in human *MLLr* patients [34]. Using Lmo2-Cre to express Cre recombinase, Mll-Af9 was generated in pluripotent stem cells and led exclusively to myeloid leukaemia whereas T-cell-restricted expression using lck-Cre did not lead to leukomogeneis [35]. The advantage of this Cre-loxP system is that MA9 expression can be driven by lineage-restricted promoters from the endogenous loci [35], reflecting physiological expression. Further models targeting Mll-Af9 expression to specific cells within the haematopoietic system have utilized retrovirus-driven expression followed by transduction and transplantation approaches. These methods benefit from its speed and ease of use [36,37] and has led to the identification of both the HSC and the granulocyte and macrophage progenitor (GMP) as potential leukaemic stem cells-of-origin with divergent clinical features [38]. Retrovirus-mediated expression models however drive transgene expressions that are not physiological. To gain greater control over the level of cell-restricted transgene expression, Dox-inducible transgenic models have since been developed [39,40] enabling close-to-physiological levels and reversible transgene expression in a Dox dose-dependent manner. Using this model, it was demonstrated that long-term HSC (LT-HSC) populations result in an invasive and chemoresistant AML with a primitive progenitor phenotype and a distinct stemness-related gene expression pattern [39]. Dox-inducible transgene expression is however not controlled by an endogenous promoter. "Leaky" Cre expression from tissue-specific promoters together with the expensive and time-consuming nature of tissue-restricted strain generation are major limitations with the Cre-loxP and Dox-inducible transgenic model systems. More recent attempts to circumvent some of the issues highlighted above have utilized transcription activator like effector nucleases (TALEN) technology to generate endogenous MA9 [41,42]. These cells exhibited a significantly higher clonogenic potential with colony morphologies consistent with an immature cell type and development of AML, ALL and MPAL upon xenotransplantation [41].

Clustered regularly interspaced short palindromic repeats (CRISPR)/Cas9 technology facilitates creation of DSBs at almost any genomic sequence of interest technology and has revolutionised the way in which we are able to manipulate the genome, affording a more flexible and readily applicable method for generating novel disease models, including chromosomal translocations, with an ease and simplicity far outweighing prior technologies such as TALEN and Cre-LoxP. CRISPR technology has facilitated generation of chromosomal deletions [43] and rearrangements, including inversions [44] and translocations [45–48]. Here, we demonstrate efficient CRISPR-mediated generation of an endogenous *Mll-Af9* reciprocal translocation in a murine cell line and primary HSPCs. By utilizing a dual-single guide RNA (sgRNA) approach to target the most common breakpoints in human MA9 rearrangements, we show rapid, reliable generation of simultaneous breakpoints and a fusion product at the DNA and RNA levels. The resultant transformed cells reliably recapitulated features of human *MLLr* leukaemias. These include increased expression of MA9 target genes, reduced IL3 dependency, higher clonogenic potential and self-renewal in primary murine HSPCs. This multiplexed CRISPR technology offers a robust and malleable approach for generating murine models of the multitude of human translocation-associated leukaemias.

2. Results

2.1. A CRISPR/Cas9 System for Mll-Af9 Genome Editing (302)

We utilized a CRISPR/Cas9 system for the generation of murine t(4;9) MA9 chromosomal translocations in murine HSPCs, utilizing simultaneous targeting of the *Mll* and *Af9* genes by CRISPR sgRNAs directed to chromosomes. Based on prior investigation of common *MLL* and *AF9* breakpoint regions in AML patient samples [49] and murine models of MA9 [35], we designed guides targeting frequently translocated intronic regions between *Mll* exons 8 and 9 (Mll sg1 and 2) and exons 10 and 11 (Mll sg3 and 4) and between *Af9* exons 8 and 9 (*Af9* sg1, 2 and 3) (Figure 1A) to generate two potential MLL-AF9 fusion products (Figure 1B). Guide sequences were generated using the CRISPR.MIT web tool (http://crispr.mit.edu) [50,51], which scores each sequence based on a combination of on- and off-target potentials. We selected those guides with the highest scores, indicating an appropriate efficiency of on-target cleavage with minimal off-target potential. Single guide sequences were cloned into a lentiviral CRISPR-green fluorescent protein (GFP) vector (Supplementary Figure S1), and the resultant high-titre lentivirus was used to transduce the murine HSPC myeloid progenitor 32D cell line, which is IL3-dependent. At day 5 post-transduction, cells were fluorescent activated cell sorting (FACS) sorted for GFP expression and indel formation in GFP$^+$ populations were assessed by a surveyor assay (Figure 1C).

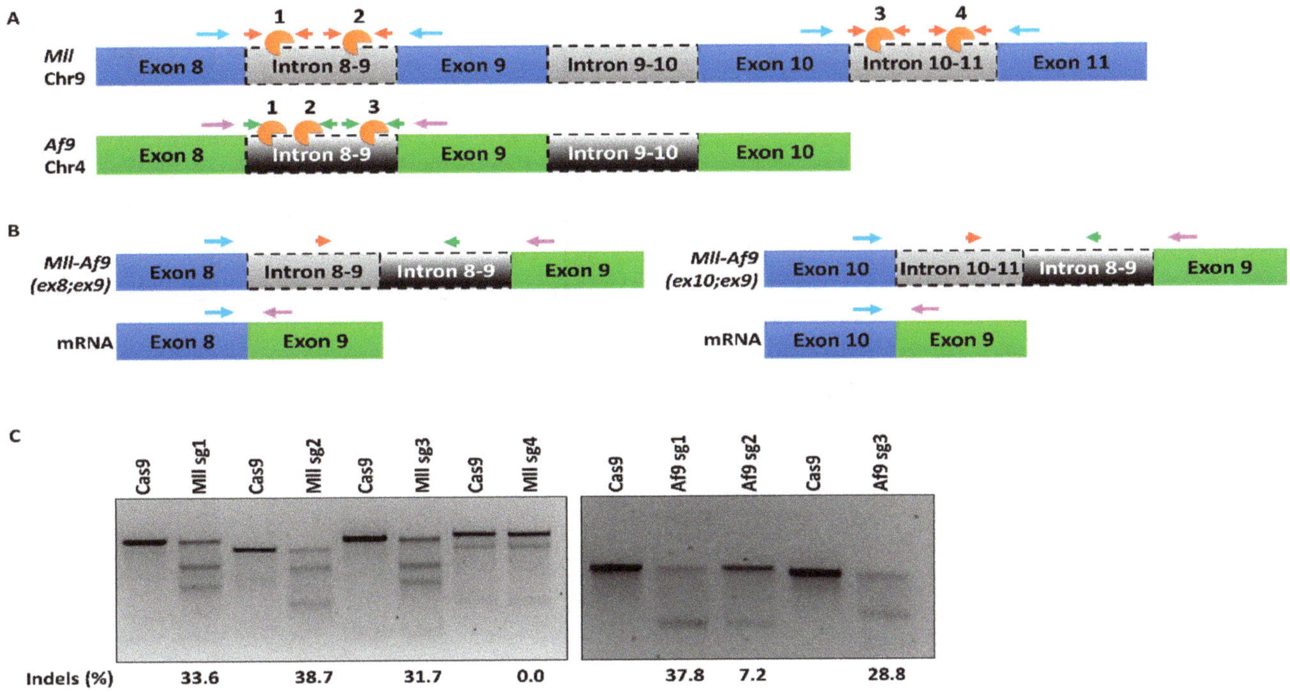

Figure 1. Design and validation of clustered regularly interspaced short palindromic repeats (CRISPR)/Cas9 gene editing tools targeting murine mixed lineage leukaemia (*Mll*) and *Af9* genes: (**A**) Schematic representation of partial *Mll* and *Af9* genomic regions with sgRNA location. Four guides were designed targeting either introns 8 and 9 (*Mll* sg1 and -2) or introns 10 and 11 (*Mll* sg3 and -4) of the *Mll* gene, and three were designed to target introns 8 and 9 (AF9 sg1–3) of Af9. sgRNA locations/Cas9 cleavage sites are shown in orange. Red (*Mll*) and green (*Af9*) arrows represent PCR primers flanking cleavage sites. Blue and purple arrows represent RT-PCR primers on MLL and AF9 transcripts, respectively. (**B**) Representation of two predicted *Mll-Af9* fusion gene products following editing with either set of *Mll* and *Af9*-targeting sgRNAs. (**C**) Surveyor assay of 32D cells transduced with a CRISPR vector expressing the indicated sgRNAs. PCR was performed on genomic DNA from transfected GFP$^+$ cell populations using primers as indicated in Figure 1A. Products were assayed by the surveyor and analysed by 2% gel electrophoresis. Percentage indel formation was calculated by quantification of each DNA band. Cells transduced with an empty CRISPR vector (Cas9 only) were used as the control sample for each.

Of those guides targeting *Mll* introns, Mll sg1-3 showed approximately equal indel efficiencies of 31–38%, with no cleavage seen using sg4 (Figure 1C and Figure S2). Similarly, *Af9*-targeting sgRNA1 and -3 showed approximately equal efficiencies of 28–37%, while sg2 gave a significantly poorer indel frequency of 7% (Figure 1C). These data demonstrate efficient cleavage of *Mll* and *Af9* loci using a CRISPR system detectable in the bulk population. Based on these results, Mll sg2 was selected for targeting of *Mll* introns 8 and 9, sg3 was selected for introns 10 and 11, and Af9 sg1 and –3 were selected for *Af9* introns 8 and 9.

2.2. Endogenous Generation of MA9 Translocation in Murine HSPCs (368)

To generate MLL-AF9 translocations, simultaneous DNA DSBs in the *Mll* and *Af9* genes of murine 32D HSPCs were created (Figure 2A). Cells were transduced with a CRISPR lentivirus targeting either the *Mll* or *Af9* loci or both in combination and were FACS sorted for GFP at day 5 post-transduction. In 32D murine HSPCs, GFP expression > 20% was noted with each combination of dual guides (Figure 2B) and GFP$^+$ populations were sorted for verification of the MA9 translocation and downstream functional analysis.

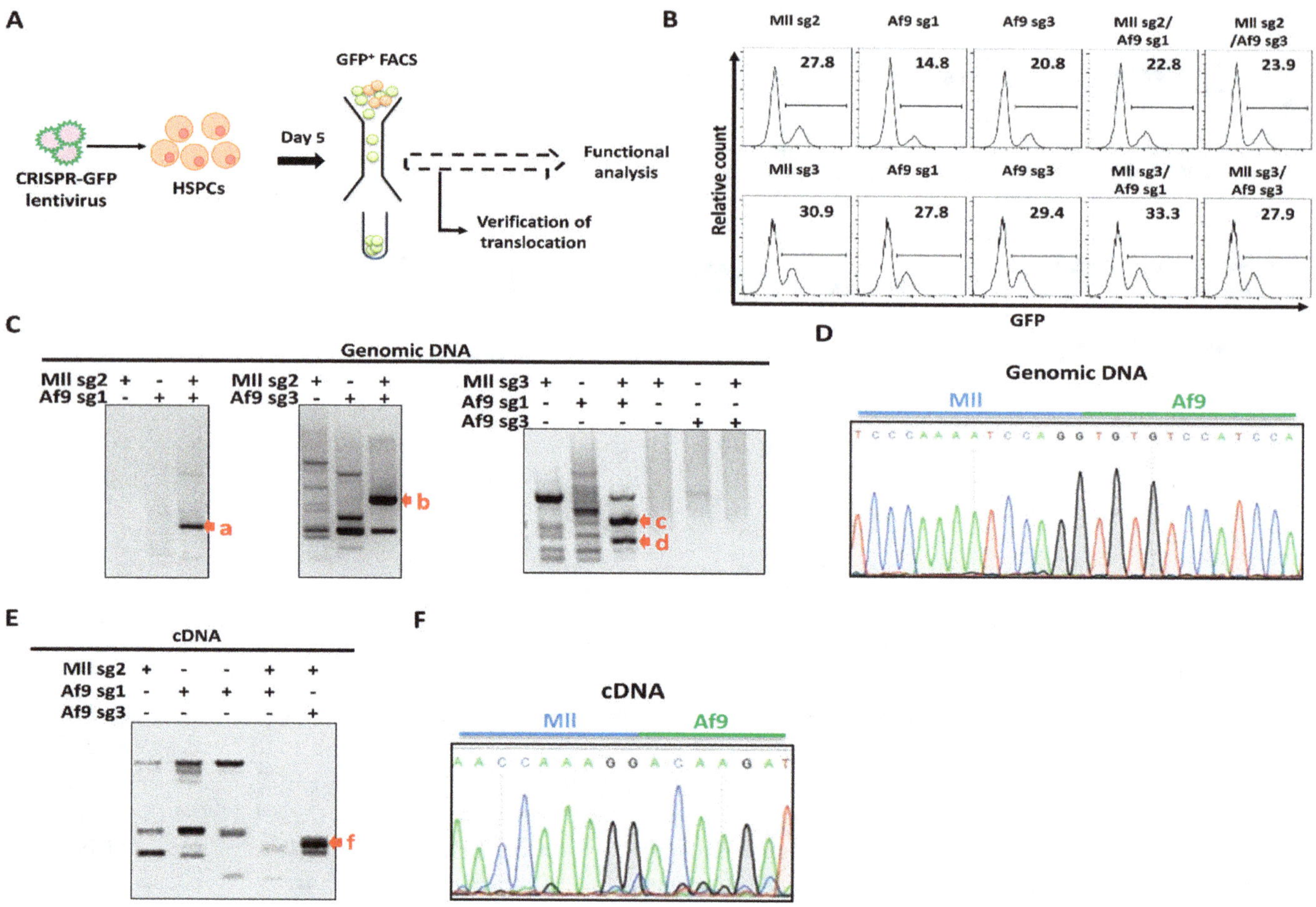

Figure 2. Endogenous generation of MLL-AF9 (MA9) translocation in murine haematopoietic cells: (**A**) Workflow schematic for generation of t(4;9) translocated 32D HSPCs. Cells were transduced with single or dual CRISPR/Cas9 lentivirus and sorted for GFP at day 5 for downstream analysis. (**B**) Flow cytometry histograms showing percentage of GFP$^+$ cells at day 5 post-transduction. (**C**) Validation of genomic MA9 translocation: Genomic DNA was extracted from sorted GFP$^+$ populations and PCR amplified using primers flanking the MA9 breakpoint. PCR products were then resolved by agarose gel electrophoresis. (**D**) Sanger sequencing of the PCR product from the sg2/sg3 sample (red "**b**" arrow, part **C**): Alignment of the sequence to reference genome verified the alignment with murine *Mll* and *Af9*. (**E**) RT-PCR analysis of the MA9 breakpoint region verifying MA9 translocation at RNA resolution: cDNA was reverse transcribed from RNA and PCR amplified using primers flanking the MA9 breakpoint. (**F**) Sanger sequencing of the sg2/sg3 RT-PCR product: Alignment of the sequence to reference RNA showed alignment to murine *Mll* and *Af9* transcripts.

Verification of the MA9 translocation was carried out in GFP$^+$ 32D cells at both the DNA and transcript levels by utilizing PCR primers flanking the breakpoint region (Figure 2A) followed by Sanger sequencing of PCR amplicons. Unique banding patterns were seen following electrophoretic analysis of PCR products from 3/4 *Mll* and *Af9* guide combinations (Figure 2C and Figure S3), with sg3/sg3 as the only guide combination which did not generate a product. Amplicons as indicated in Figure 2C were further analysed by Sanger sequencing and alignment to reference *Mll* and *Af9* genomic sequences, which identified the sg2/sg3-transduced sample (band b, Figure 2C) as having correct alignment to both murine *Mll* (NCBI gene ID: 214162) and *Af9* (NCBI gene ID: 70122), validating generation of the MA9 translocation at the genomic level (Figure 2D). Sequence alignment of amplicons from samples transduced with alternative guide combinations (Figure 2C, red arrows excluding b) identified matches in exonic and intronic regions of the mouse genome, including a noncoding region of chromosome 14 and the *Pigz* gene on chromosome 16. This may be indicative of a potential off-target effect of these sgRNA combinations, in line with current knowledge regarding the off-target potential of Cas9 cleavage. RT-PCR and subsequent electrophoresis of RNA from the dual sg2/sg3-transduced sample further validated the translocation at the transcript level; utilizing primers flanking the mRNA breakpoint region (Figure 2B), we generated an amplicon unique to this sample (Figure 2E) which correctly aligned to murine *Mll* (NM_001357549.1) and *Af9* (NM_027326.3) reference transcripts, thus supporting formation of the MA9 mRNA fusion (Figure 2F). Taken together, these data indicate that CRISPR/Cas9-mediated genome editing can be used to induce formation of a murine t(4;9) MA9 translocation in 32D HSPCs and that this fusion gene product is transcribed appropriately.

2.3. CRISPR-Mediated MA9 Translocations Recapitulate Features of MA9 Leukaemia In Vitro (622)

Multifactorial functional analysis was carried out firstly on MA9-validated 32D murine HSPCs, in which we examined proliferative capability, IL3-dependency and expression of MA9-target genes. Over 30 days in liquid culture, we saw no significant difference in proliferation between 32D cells transduced with the sg2/sg3 guide combination compared to those transduced with either single guide (Figure 3A), indicating that the MA9 translocation does not confer a demonstrable proliferative advantage in this already immortalised cell line. MA9 leukemic cells have previously shown only moderate IL3 dependency in contrast to 32D cells which are strictly IL3 dependent [52]. Thus, we sought to examine the effects of IL3 starvation on MA9 32D cells in culture. Over 24 h of IL3 starvation, we noted a slight but nonsignificant growth advantage and resistance to starvation-induced stress in the MA9 cells when compared with single guide controls (Figure 3B), suggesting a decrease in IL-3 dependency conferred by CRISPR-generated MA9 translocation in this cell type, consistent with previous data [52]. Analysis of the expression of common MA9-associated genes in sg2/sg3 double-transduced cells revealed significant upregulation of *Dot1l*, *Cdk6* and *Sirt1* mRNA (Figure 3C). Both DOT1l and CDK6 have critical and well-established roles in MLLr function and leukemogenesis [16,29], and *Sirt1* has been shown to be upregulated in response to MA9 expression in murine models [53]. Taken together, these data strongly support that CRISPR-mediated generation of MA9 using the endogenous loci can recapitulate the functional properties of MA9 leukaemic features.

Following verification of the MA9 translocation in 32D murine HSPCs, we next sought to use our approach in primary murine HSPCs. C-Kit$^+$ bone marrow (BM) cells were isolated from WT mice and transduced using the validated sg2/sg3 sgRNA singly or in combination. At day 5 post-transduction, cells were sorted for GFP$^+$ and expanded in liquid culture for an additional 5 days prior to beginning functional analysis of the translocation in the primary bulk cell population. Equal numbers of cells from liquid cultures were assessed in colony forming cell (CFC) assays to determine self-renewal, proliferation and differentiation capabilities. Following 20 days in a CFC assay, cells transduced with the sg2/sg3 combination of guides displayed significantly increased cell numbers and formed more colonies when compared to single transduced samples (Figure 3D,E). We next carried out immunophenotyping of day 20 CFC colonies in order to assess the myeloid differentiation status of the cells. We saw differences neither in the live/viable population (Figure 3G) nor in the CD11bGr1- or CD11b+F4/80+ myeloid fractions in any of the samples (Figure 3F,I,J), suggesting that the MA9 translocation in this model does not have a negative impact on cell viability or myeloid differentiation. However, the sg2/sg3 MA9 cells showed a significantly larger population of cells expressing the c-Kit progenitor marker (Figure 3F,H), which has been associated with the leukaemia stem cell (LSC) in AMLs including MA9 [36,37]. This highlights not only a proliferative advantage in the double-transduced MA9 population but also an enrichment of a self-renewing LSC-type population. These data indicate that the CRISPR-mediated generation of the MA9 translocation in murine HSPCs mediates leukaemic transformation, indicated by a proliferative advantage and an increase in self-renewal properties.

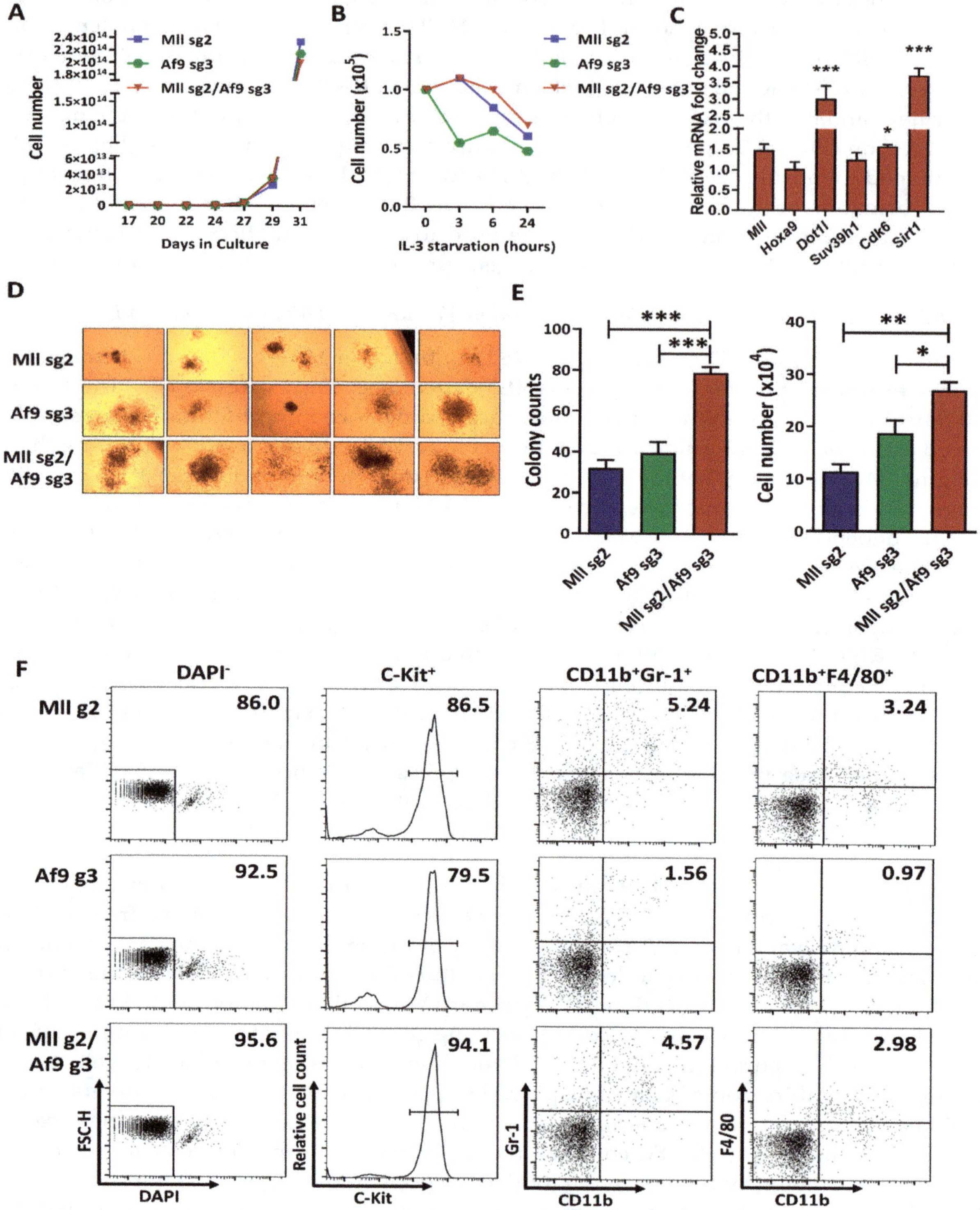

Figure 3. *Cont.*

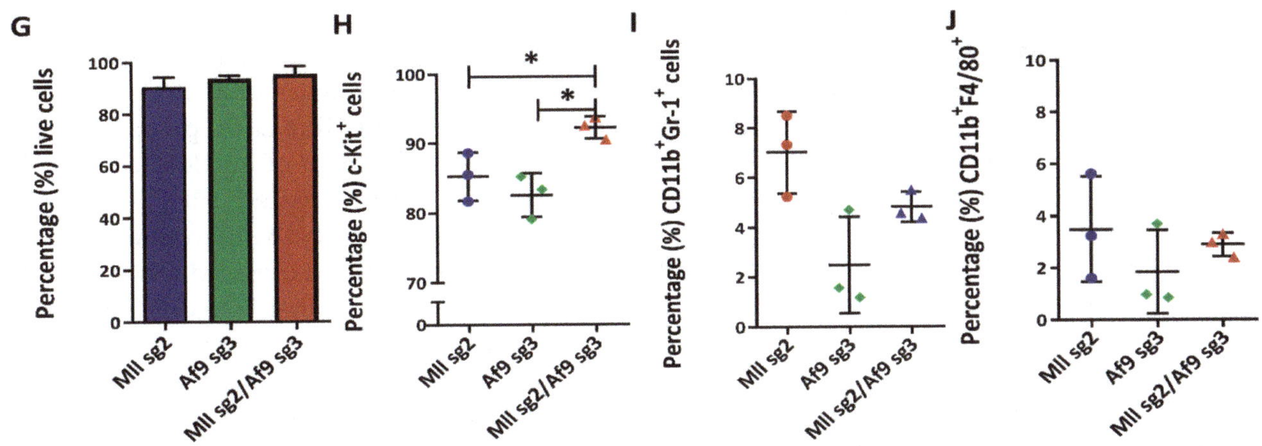

Figure 3. CRISPR-mediated MA9 translocation recapitulates features of MA9 leukaemia: (**A**) Cumulative cell number and growth curve of 32D HSPCs transduced with either *Mll* sg2 or *Af9* sg3 or combination. Cells did not show any significant difference in proliferation following MA9 translocation. (**B**) Cell number and growth curve of 32D HSPCs transduced with *Mll* and/or *Af9* guides as indicated with IL-3 starvation: MA9-translocated cells displayed reduced levels of IL-3 dependency compared to single-guide controls. (**C**) Expression of MA9 target genes: ΔΔCT values of the sg2/sg3 sample are normalised to *Mll* sg2 single guide control. The graph shows mean ± SD of representative experiments. All samples were analysed in technical duplicates. Statistical significance was calculated by one-way ANOVA followed by Bonferroni posttest. $p < 0.001$ is represented as ***, and $p < 0.05$ is represented as *. (**D**) Representative colony pictures (400×) of CFC colonies after ten days in culture (left panel): Samples were plated in technical triplicates. (**E**) Mean ± SD of colony counts (left) and cell numbers (right). (**F**) Representative flow cytometric plots from samples in Figure 3D with the percentage of relevant populations indicated numerically. (**G–J**) Percentage of each population in samples shown as (**G**) live (DAPI$^-$), (**H**) c-Kit$^+$, (**I**) CD11b$^+$Gr-1$^+$ and (**J**) CD11b$^+$F4/80$^+$. Each sample is shown as mean ± SD from three technical triplicates. Statistical significance was calculated by one-way ANOVA followed by Bonferroni posttest. $p < 0.001$ is represented as ***, $p < 0.01$ is represented as ** and $p < 0.05$ is represented as *.

3. Discussion

Herein, we have demonstrated generation of a de novo MA9 translocation in murine HSPCs using state-of-the-art CRISPR/Cas9 technology to create a novel murine leukaemia model which faithfully recapitulates multiple features of human MA9 leukaemia. By rational design of sgRNA flanking breakpoint regions with common occurrence in *MLLr* AML patients and prior MA9 transgenic models, we identified optimal combinations of sgRNA sequences with which to induce simultaneous DSBs at the specific loci. Each guide was validated for individual cleavage efficiency, and subsequently, optimal combinations were identified for multiplexed CRISPR genome editing to induce an MA9 chromosomal translocation in HSPCs. Following CRISPR editing, the presence of a successful MA9 translocation was verified at both the genomic and transcriptional levels by PCR analysis and sequencing. Analysis of proliferation, clonogenicity, differentiation and gene expression robustly demonstrated the functional ways in which this model recapitulates features of the disease in vitro. This novel approach circumvents common caveats of previous methods, with no requirement for transgenic or immunocompromised mice or primary human material. We have demonstrated the ease of use of this approach in cell line and primary murine HSPCs which can be applied to other chromosomal translocations.

Increased clonogenicity and enrichment of progenitor cells with increased self-renewal capacity is a fundamental characteristic of leukaemias with *MLL* rearrangements. Although we saw no proliferative advantage in MA9 32D HSPCs, we did observe a modest reduction in growth factor dependence as previously demonstrated. Normal primary HSPC cells have limited growth potential in vitro, which showed increased colony growth and number when expressing MA9. Increased clonogenicity is associated with increased c-Kit expression, a cell surface marker found in myeloblasts in up to 80% of AML cases [54]. C-Kit expression has also been strongly associated with poor prognoses in other instances of chromosomal translocations such as t(8;21) [55]. We saw a significant enrichment of c-Kit-expressing cells, indicating an enrichment of colony-forming LSC-like early progenitor cells in MA9-expressing cells consistent with leukaemic transformation. Assessment of myeloid differentiation did not reveal a block in expression of differentiation markers in MA9-mediated cells, indicating that, although cells have acquired self-renewal properties, they maintain the ability to differentiate toward the myeloid lineage. A block in terminal myeloid differentiation cannot be ruled out, and whether cells can complete myeloid differentiation remains to be investigated. Cells that are unable to differentiate to the GMP stage of differentiation in murine models of *MLLr* do not undergo leukaemic transformation [56], and so, it is understood that some level of differentiation down the myeloid pathway is required for *MLLr* AML transformation. Data produced by Stavropoulou et al. utilizing a Dox-inducible MA9 model show that surface marker analysis of colonies derived from iMLL-AF9 LT-HSC and GMP subpopulations showed cells expressing the Mac-1 and Gr-1 differentiation markers in addition to c-Kit in the presence of Dox [39].

MA9-initiated oncogenesis has been shown to depend on MA9 modification of transcriptional elongation, resulting in dysregulated expression of target genes. We demonstrated an increase in expression of *Dot1* and *Cdk6*, both of which are essential for the oncogenic function of MA9-rearranged cells [29,57]. Interestingly, we also noted a significant increase in expression of *Sirt1*, which is found significantly overexpressed at the protein level in LSC populations from AML patients [53], where it acts to maintain their survival, growth and drug resistance. The lack of increased *HoxA9* expression observed in our cells may be due to low frequency of translocation within the bulk population analysed, as has been previously noted in TALEN-focused studies [41,42] and CRISPR-generated models [58].

Recently, work by Jeong et al. [58] similarly demonstrated efficient generation of MA9 translocations by CRISPR/Cas9 technology, utilizing methodologies in line with those described herein. The authors used the human system to generate MA9 translocation in human CD34+ cord blood (CB) cells in contrast to the murine system we have presented here. Consistent with our results, human MA9-expressing CB cells in that study demonstrated a growth advantage in liquid culture and clonal expansion in CFC analysis and induction of AML, ALL and MPAL following transplantation in immunocompromised mice. The translocation efficiency achieved by Jeong et al. [58] is considerably higher than in previous TALEN-generated models of MA9 at around 0.5–1%. However, it is important to note that this efficiency remains low overall and so should be taken into consideration when analysing the bulk population. In contrast to the single-cell clonal analysis performed by Jeong et al. [58], we are able to demonstrate the leukaemic transformation effects of CRISPR-generated MA9 in the bulk population that may be masking some effects such as increased Hox gene expression. A considerable advantage of our methodology which allows effects to be seen in the bulk population is the reduction in time in culture of primary HSPCs that is required with single-clone selection. Further single-cell analysis from bulk populations in both the 32D cell line and primary HSPCS would give further insight into individual aspects of MA9-mediated and context-specific gene regulation. A limitation with using primary human HSPCs is the requirement for immunocompromised mice for xenografts. Our work is the first to show the use of this technology to generate the murine MA9 translocation from the endogenous loci faithfully recapitulating MA9-mediated transformation in murine HSPCs. A significant advantage of the murine system is the potential for investigation into immune-, gene- and cell-specific contexts using the myriad of murine genetic models and xenotransplantation approaches.

Our methodology has several advantages for use in genetic models. The tractability of this system is such that we can readily study MA9 translocations in combination with other genetic or epigenetic abnormalities. Generation of new sgRNAs targeting sequences of interest are quickly and easily cloned into the CRISPR vector validated in this work and can be multiplexed as required. Our sgRNA target sequences were also selected not only for their commonality within *MLLr* patients but also for their prior use in alterative murine models of MA9, and as such, our approach offers not only a robust model of human disease but also a means of comparison to alterative methodologies [35,41,42]. The work presented here and by others [48,58] exemplifies the ongoing shift towards precise genome editing with CRISPR technologies to quickly and accurately recapitulate human disease in model systems. We have demonstrated the tractability and potential of this methodology in modelling human MA9 leukaemias, which represents an important advancement in the study of MA9-driven leukaemia.

4. Materials and Methods

4.1. Cell Culture

All cell cultures were grown at 37 °C, 5% CO_2. Human embryonic kidney (HEK)-239T and murine embryonic fibroblast NIH-3T3 cells were maintained in Dulbecco Modified Eagle Medium (DMEM; Gibco, Paisley, UK) supplemented with 10% heat-inactivated foetal bovine serum (FBS; Gibco, Paisley, UK), 2 mM L-glutamine (Gibco, Paisley, UK) and 1% Penicillin/ Streptomycin (Pen/Strep; Gibco, Paisley, UK). 32D murine myeloid progenitor cell line HSPCs were maintained in Roswell Park Memorial Institute-1640 (RPMI-1640; Gibco, Paisley, UK) supplemented with 10% FBS, 10% WEHI-3B conditioned media, 2 mM L-glutamine and 1% Pen/Strep. Primary murine HSPCs were cultured in DMEM pre-stimulation media supplemented with 15 % FBS, 2 mM L-glutamine, 1% Pen/Strep, 10 ng/mL recombinant murine interleukin-3 (rmIL3; PeproTech, London, UK), 10 ng/mL recombinant murine interleukin-6 (rmIL6; PeproTech, London, UK) and 100 ng/mL recombinant murine stem cell factor (rmSCF; PeproTech, London, UK) immediately following isolation and were transferred to DMEM activation media supplemented with 15% FBS, 2 mM L-glutamine, 1% Pen/Strep, 10 ng/mL rmIL3, 10 ng/mL rmIL6, 100 ng/mL rmSCF and 4 μg/mL polybrene (Sigma-Aldrich, Dorset, UK) for transduction.

4.2. Murine Bone Marrow (BM) Harvest and c-Kit Enrichment

C57BL/6 mice were purchased from Charles River UK and housed in the Beatson Biological Service and Research Units. All mouse experiments were approved by the local animal welfare ethical review body (AWERB) committee and United Kingdom (UK) home office and performed according to UK Home Office project license 60/4512 approved 25 April 2013(Animal Scientific Procedures Act 1986) guidelines. Femur, tibia and pelvic girdle of wild-type (WT) C57BL/6 mice were crushed in phosphate buffered saline (PBS)/2% FBS and passed through a 40-μM cell strainer (Fisher Scientific, Loughborough, UK) to harvest BM. C-kit enrichment was carried out by magnetic activated cell sorting (MACS) separation of BM samples following incubation with anti-CD117 microbeads (Miltenyi Biotech, Bergisch Gladbach, Germany).

4.3. sgRNA Design, Synthesis and CRISPR Transduction

Reference mRNA transcripts of genes of interest (http://www.ncbi.nlm.nih.gov/nucleotide/) were used for identification of suitable protospacer sequences using the CRISPR MIT tool (previously available at http://crispr.mit.edu; accessed August 2015) [50,51]. Top and bottom strand protospacers were purchased as single stranded DNA (ssDNA) oligos (IDT, Surrey, UK) (Supplementary Table S1) and cloned into pL.CRISPR.EFS.GFP (Supplementary Methods and Figure S1). pL-CRISPR.EFS.GFP was a gift from Benjamin Ebert (Addgene plasmid # 57818) [59]. The all-in-one CRISPR lentivirus was used to constitutively express sgRNAs and Cas9 under the control of human U6 (hU6) and elongation factor 1α (EF-1α) promotors, respectively. HEK-293T and NIH-3T3 cells were used for lentiviral production and titration, respectively, and 32D or primary murine HSPCs were transduced with high-titre CRISPR lentivirus as previously described [60].

4.4. Flow Cytometry and Cell Sorting

Samples were stained and analysed in PBS or PBS/2% FBS or sorted for GFP in the same buffer. Single cells were gated by forward scatter area (FSC-A) vs height (FSC-H), and then viable cells were gated by FSC-A vs side scatter area (SSC-A). For viability staining, analysis was performed only in single-cell gates and dead cells were excluded by 4′,6-diamidino-2-phenylindole (DAPI; Sigma-Aldrich, Dorset, UK) staining. Compensation was performed using unstained cells, single stained cells and single stained UltraComp eBeads (Thermo Scientific, Paisley, UK). A complete list of antibodies and dyes can be found in Supplementary Table S2.

4.5. Surveyor Assay

Genomic DNA was extracted using QuickExtract™ DNA Extraction Solution (Epicentre Biotechnologies, Madison, WI, USA), and analysis with Surveyor Nuclease S was used to calculate indel frequency. Full methods and primer sequences can be found in the Supplementary Methods and Table S3.

4.6. Genomic Target Amplification (GTA)

Genomic DNA was extracted from GFP-sorted cells (day 5 post-sorting) using Epicentre QuickExtract™ DNA Extraction Solution. PCR amplification of target loci was performed using Q5 High-Fidelity DNA Polymerase (New England Biolabs; NEB, Herts, UK), and the products were analysed on 2% agarose (Sigma-Aldrich, Dorset, UK) gel. Bands of interest were excised using the GeneJET Gel Extraction and DNA Cleanup Micro Kit (Thermo Scientific, Paisley, UK) before analysis by Sanger sequencing. Primer sequences can be found in Supplementary Table S3.

4.7. Reverse Transcription PCR

Total RNA was isolated from GFP-sorted cells (day 5 post-sorting) by either Arcturus PicoPure RNA Isolation Kit (Thermo Scientific, Paisley, UK) or RNeasy Mini Kit (QIAGEN, Manchester, UK)), and target-specific cDNA synthesis was carried out by Superscript III (Thermo Scientific, Paisley, UK) using primers specific for locus of interest. cDNA was amplified using Taq DNA polymerase (NEB, Herts, UK) and target-specific primers. Purified products were analysed on 2% agarose gel, and bands of interest were excised and analysed by Sanger sequencing. Primer sequences can be found in Supplementary Table S4.

4.8. Gene Expression Analysis (qRT-PCR)

Total RNA was isolated from GFP-sorted cells, and gene expression analysis was carried out as previously described [60]. Results were calculated by the ΔΔCt method with normalization to the *Gusb* housekeeping gene. Primer sequences can be found in Supplementary Table S5.

4.9. Colony-Forming Cell (CFC) Assay

Methocult GF M3434 (Stem Cell Technologies, Vancouver, BC, Canada) methylcellulose media containing recombinant murine stem cell factor (rmSCF), recombinant murine interleukin-3 (rmIL-3), recombinant human interleukin-6 (rhIL-6) and recombinant human erythropoietin (rhEPO) was used. Equal numbers of BM c-kit-enriched, GFP-sorted cells were prepared in triplicate and resuspended in culture media. Cell suspension was added to methylcellulose and transferred to 3 × 35 mm dishes (Greiner-Bio-One, Gloucestershire, UK) for 10–20 days. Colonies were scored, and cells were analysed by flow cytometry. If required, cells were replated to a second round of CFC following the same protocol.

4.10. Statistics

GraphPad Prism (version 5.00.288; LA Jolla, CA, USA) was used for statistical analysis and graphing. Statistical significance was determined by one-way ANOVA with Bonferroni posttest.

Author Contributions: E.S. designed and performed the experiments, analysed and interpreted the data, and made all figures. L.R. and R.J.C. provided data analysis and cowrote the manuscript. B.G. supervised the study. K.K. cowrote the manuscript, designed the research and supervised the study. All authors have read and agreed to the published version of the manuscript.

Acknowledgments: We acknowledge the technical assistance of Jennifer Cassells and Karen Dunn for flow cytometry and animal experiments respectively. We thank all staff at the Paul O'Gorman Leukaemia Research Centre for their support and advice.

References

1. Slany, R.K. The molecular mechanics of mixed lineage leukemia. *Oncogene* **2016**, *35*, 5215–5223. [CrossRef] [PubMed]
2. Andersson, A.K.; Ma, J.; Wang, J.; Chen, X.; Gedman, A.L.; Dang, J.; Nakitandwe, J.; Holmfeldt, L.; Parker, M.; Easton, J.; et al. The landscape of somatic mutations in infant MLL-rearranged acute lymphoblastic leukemias. *Nat. Genet.* **2015**, *47*, 330–337. [CrossRef] [PubMed]
3. Richardson, C.; Jasin, M. Frequent chromosomal translocations induced by DNA double-strand breaks. *Nature* **2000**, *405*, 697–700. [CrossRef] [PubMed]
4. Slany, R.K. When epigenetics kills: MLL fusion proteins in leukemia. *Hematol. Oncol.* **2005**, *23*, 1–9. [CrossRef]
5. Meyer, C.; Burmeister, T.; Gröger, D.; Tsaur, G.; Fechina, L.; Renneville, A.; Sutton, R.; Venn, N.C.; Emerenciano, M.; Pombo-De-Oliveira, M.S.; et al. The MLL recombinome of acute leukemias in 2017. *Leukemia* **2018**, *32*, 273–284. [CrossRef]
6. Meyer, C.; Kowarz, E.; Hofmann, J.; Renneville, A.; Zuna, J.; Trka, J.; Ben Abdelali, R.; Macintyre, E.; De Braekeleer, E.; De Braekeleer, M.; et al. New insights to the MLL recombinome of acute leukemias. *Leukemia* **2009**, *23*, 1490–1499. [CrossRef]
7. Huret, J.L.; Ahmad, M.; Arsaban, M.; Bernheim, A.; Cigna, J.; Desangles, F.; Guignard, J.C.; Jacquemot-Perbal, M.C.; Labarussias, M.; Leberre, V.; et al. Atlas of genetics and cytogenetics in oncology and haematology in 2013. *Nucleic Acids Res.* **2013**, *41*, D920–D924. [CrossRef]
8. Super, H.J.G.; McCabe, N.R.; Thirman, M.J.; Larson, R.A.; Le Beau, M.M.; Pedersen- Bjergaard, J.; Philip, P.; Diaz, M.O.; Rowley, J.D. Rearrangements of the MLL gene in therapy-related acute myeloid leukemia in patients previously treated with agents targeting DNA-topoisomerase II. *Blood* **1993**, *82*, 3705–3711. [CrossRef]
9. Malik, B.; Hemenway, C.S. CBX8, a component of the Polycomb PRC1 complex, modulates DOT1L-mediated gene expression through AF9/MLLT3. *FEBS Lett.* **2013**, *587*, 3038–3044. [CrossRef]
10. Zhang, W.; Xia, X.; Reisenauer, M.R.; Hemenway, C.S.; Kone, B.C. Dot1a-AF9 complex mediates histone H3 Lys-79 hypermethylation and repression of ENaCα in an aldosterone-sensitive manner. *J. Biol. Chem.* **2006**, *281*, 18059–18068. [CrossRef]
11. Erfurth, F.; Hemenway, C.S.; de Erkenez, A.C.; Domer, P.H. MLL fusion partners AF4 and AF9 interact at subnuclear foci. *Leukemia* **2004**, *18*, 92–102. [CrossRef] [PubMed]
12. Leach, B.I.; Kuntimaddi, A.; Schmidt, C.R.; Cierpicki, T.; Johnson, S.A.; Bushweller, J.H. Leukemia fusion target AF9 is an intrinsically disordered transcriptional regulator that recruits multiple partners via coupled folding and binding. *Structure* **2013**, *21*, 176–183. [CrossRef] [PubMed]
13. Mueller, D.; Bach, C.; Zeisig, D.; Garcia-Cuellar, M.P.; Monroe, S.; Sreekumar, A.; Zhou, R.; Nesvizhskii, A.; Chinnaiyan, A.; Hess, J.L.; et al. A role for the MLL fusion partner ENL in transcriptional elongation and chromatin modification. *Blood* **2007**, *110*, 4445–4454. [CrossRef] [PubMed]
14. Mueller, D.; García-Cuéllar, M.P.; Bach, C.; Buhl, S.; Maethner, E.; Slany, R.K. Misguided transcriptional elongation causes mixed lineage leukemia. *PLoS Biol.* **2009**, *7*, e1000249. [CrossRef]
15. Kuntimaddi, A.; Achille, N.J.; Thorpe, J.; Lokken, A.A.; Singh, R.; Hemenway, C.S.; Adli, M.; Zeleznik-Le, N.J.; Bushweller, J.H. Degree of Recruitment of DOT1L to MLL-AF9 Defines Level of H3K79 Di- and Tri-methylation on Target Genes and Transformation Potential. *Cell Rep.* **2015**, *11*, 808–820. [CrossRef]

16. Nguyen, A.T.; Taranova, O.; He, J.; Zhang, Y. DOT1L, the H3K79 methyltransferase, is required for MLL-AF9—Mediated leukemogenesis. *Blood* **2011**, *117*, 6912–6922. [CrossRef]
17. Yokoyama, A.; Somervaille, T.C.P.; Smith, K.S.; Rozenblatt-Rosen, O.; Meyerson, M.; Cleary, M.L. The menin tumor suppressor protein is an essential oncogenic cofactor for MLL-associated leukemogenesis. *Cell* **2005**, *123*, 207–218. [CrossRef]
18. Armstrong, S.A.; Staunton, J.E.; Silverman, L.B.; Pieters, R.; Den Boer, M.L.; Minden, M.D.; Sallan, S.E.; Lander, E.S.; Golub, T.R.; Korsmeyer, S.J. MLL translocations specify a distinct gene expression profile that distinguishes a unique leukemia. *Nat. Genet.* **2002**, *30*, 41–47. [CrossRef]
19. Zeisig, B.B.; Milne, T.; Garcia-Cuellar, M.-P.; Schreiner, S.; Martin, M.-E.; Fuchs, U.; Borkhardt, A.; Chanda, S.K.; Walker, J.; Soden, R.; et al. Hoxa9 and Meis1 Are Key Targets for MLL-ENL-Mediated Cellular Immortalization. *Mol. Cell. Biol.* **2004**, *24*, 617–628. [CrossRef]
20. Guo, H.; Chu, Y.; Wang, L.; Chen, X.; Chen, Y.; Cheng, H.; Zhang, L.; Zhou, Y.; Yang, F.C.; Cheng, T.; et al. PBX3 is essential for leukemia stem cell maintenance in MLL-rearranged leukemia. *Int. J. Cancer* **2017**, *141*, 324–335. [CrossRef]
21. Prange, K.H.M.; Mandoli, A.; Kuznetsova, T.; Wang, S.Y.; Sotoca, A.M.; Marneth, A.E.; Van Der Reijden, B.A.; Stunnenberg, H.G.; Martens, J.H.A. MLL-AF9 and MLL-AF4 oncofusion proteins bind a distinct enhancer repertoire and target the RUNX1 program in 11q23 acute myeloid leukemia. *Oncogene* **2017**, *36*, 3346–3356. [CrossRef] [PubMed]
22. Kumar, A.R.; Hudson, W.A.; Chen, W.; Nishiuchi, R.; Yao, Q.; Kersey, J.H. Hoxa9 influences the phenotype but not the incidence of Mll-AF9 fusion gene leukemia. *Blood* **2004**, *103*, 1823–1828. [CrossRef] [PubMed]
23. Van Der Linden, M.H.; Willekes, M.; Roon, E.; Seslija, L.; Schneider, P.; Pieters, R.; Stam, R.W. MLL fusion-driven activation of CDK6 potentiates proliferation in MLL-rearranged infant ALL. *Cell Cycle* **2014**, *13*, 834–844. [CrossRef]
24. Bindels, E.M.J.; Havermans, M.; Lugthart, S.; Erpelinck, C.; Wocjtowicz, E.; Krivtsov, A.V.; Rombouts, E.; Armstrong, S.A.; Taskesen, E.; Haanstra, J.R.; et al. EVI1 is critical for the pathogenesis of a subset of MLL-AF9-rearranged AMLs. *Blood* **2012**, *119*, 5838–5849. [CrossRef] [PubMed]
25. Laricchia-Robbio, L.; Nucifora, G. Significant increase of self-renewal in hematopoietic cells after forced expression of EVI1. *Blood Cells Mol. Dis.* **2008**, *40*, 141–147. [CrossRef] [PubMed]
26. Wang, Q.F.; Wu, G.; Mi, S.; He, F.; Wu, J.; Dong, J.; Luo, R.T.; Mattison, R.; Kaberlein, J.J.; Prabhakar, S.; et al. MLL fusion proteins preferentially regulate a subset of wild-type MLL target genes in the leukemic genome. *Blood* **2011**, *117*, 6895–6905. [CrossRef]
27. Takacova, S.; Luzna, P.; Stranecky, V.; Divoky, V. The Potential Role of the Six1/Eya1 Pathway in the Establishment of Leukemia Stem Cells in MLL-ENL—Induced Leukemia. *Blood* **2011**, *118*, 1362. [CrossRef]
28. Arai, S.; Yoshimi, A.; Shimabe, M.; Ichikawa, M.; Nakagawa, M.; Imai, Y.; Goyama, S.; Kurokawa, M. Evi-1 is a transcriptional target of mixed-lineage leukemia oncoproteins in hematopoietic stem cells. *Blood* **2011**, *117*, 6304–6314. [CrossRef]
29. Placke, T.; Faber, K.; Nonami, A.; Putwain, S.L.; Salih, H.R.; Heidel, F.H.; Krämer, A.; Root, D.E.; Barbie, D.A.; Krivtsov, A.V.; et al. Requirement for CDK6 in MLL-rearranged acute myeloid leukemia. *Blood* **2014**, *124*, 13–23. [CrossRef]
30. Corral, J.; Lavenir, I.; Impey, H.; Warren, A.J.; Forster, A.; Larson, T.A.; Bell, S.; McKenzie, A.N.J.; King, G.; Rabbitts, T.H. An Mll-AF9 fusion gene made by homologous recombination causes acute leukemia in chimeric mice: A method to create fusion oncogenes. *Cell* **1996**, *85*, 853–861. [CrossRef]
31. Dobson, C.L.; Warren, A.J.; Pannell, R.; Forster, A.; Lavenir, I.; Corral, J.; Smith, A.J.H.; Rabbitts, T.H. The Mll-AF9 gene fusion in mice controls myeloproliferation and specifies acute myeloid leukaemogenesis. *EMBO J.* **1999**, *18*, 3564–3574. [CrossRef] [PubMed]
32. Chen, W.; Li, Q.; Hudson, W.A.; Kumar, A.; Kirchhof, N.; Kersey, J.H. Amurine Mll-AF4 knock-in model results in lymphoid and myeloid deregulation and hematologic malignancy. *Blood* **2006**, *108*, 669–677. [CrossRef] [PubMed]
33. Chen, W.; Kumar, A.R.; Hudson, W.A.; Li, Q.; Wu, B.; Staggs, R.A.; Lund, E.A.; Sam, T.N.; Kersey, J.H. Malignant Transformation Initiated by Mll-AF9: Gene Dosage and Critical Target Cells. *Cancer Cell* **2008**, *13*, 432–440. [CrossRef] [PubMed]
34. Collins, E.C.; Pannell, R.; Simpson, E.M.; Forster, A.; Rabbitts, T.H. Inter-chromosomal recombination of Mll and Af9 genes mediated by cre-lox in mouse development. *EMBO Rep.* **2000**, *1*, 127–132. [CrossRef]

35. Drynan, L.F.; Pannell, R.; Forster, A.; Chan, N.M.M.; Cano, F.; Daser, A.; Rabbitts, T.H. Mll fusions generated by Cre-loxP-mediated de novo translocations can induce lineage reassignment in tumorigenesis. *EMBO J.* **2005**, *24*, 3136–3146. [CrossRef]
36. Krivtsov, A.V.; Twomey, D.; Feng, Z.; Stubbs, M.C.; Wang, Y.; Faber, J.; Levine, J.E.; Wang, J.; Hahn, W.C.; Gilliland, D.G.; et al. Transformation from committed progenitor to leukaemia stem cell initiated by MLL-AF9. *Nature* **2006**, *442*, 818–822. [CrossRef] [PubMed]
37. Somervaille, T.C.P.; Cleary, M.L. Identification and characterization of leukemia stem cells in murine MLL-AF9 acute myeloid leukemia. *Cancer Cell* **2006**, *10*, 257–268. [CrossRef] [PubMed]
38. Krivtsov, A.V.; Figueroa, M.E.; Sinha, A.U.; Stubbs, M.C.; Feng, Z.; Valk, P.J.M.; Delwel, R.; Döhner, K.; Bullinger, L.; Kung, A.L.; et al. Cell of origin determines clinically relevant subtypes of MLL-rearranged AML. *Leukemia* **2013**, *27*, 852–860. [CrossRef]
39. Stavropoulou, V.; Kaspar, S.; Brault, L.; Sanders, M.A.; Juge, S.; Morettini, S.; Tzankov, A.; Iacovino, M.; Lau, I.J.; Milne, T.A.; et al. MLL-AF9 Expression in Hematopoietic Stem Cells Drives a Highly Invasive AML Expressing EMT-Related Genes Linked to Poor Outcome. *Cancer Cell* **2016**, *30*, 43–58. [CrossRef]
40. Chen, X.; Burkhardt, D.B.; Hartman, A.A.; Hu, X.; Eastman, A.E.; Sun, C.; Wang, X.; Zhong, M.; Krishnaswamy, S.; Guo, S. MLL-AF9 initiates transformation from fast-proliferating myeloid progenitors. *Nat. Commun.* **2019**, *10*, 5767. [CrossRef] [PubMed]
41. Buechele, C.; Breese, E.H.; Schneidawind, D.; Lin, C.H.; Jeong, J.; Duque-Afonso, J.; Wong, S.H.K.; Smith, K.S.; Negrin, R.S.; Porteus, M.; et al. MLL leukemia induction by genome editing of human CD34+ hematopoietic cells. *Blood* **2015**, *126*, 1683–1694. [CrossRef] [PubMed]
42. Breese, E.H.; Buechele, C.; Dawson, C.; Cleary, M.L.; Porteus, M.H. Use of genome engineering to create patient specific MLL translocations in primary human hematopoietic stem and progenitor cells. *PLoS ONE* **2015**, *10*, e0136644. [CrossRef] [PubMed]
43. Adikusuma, F.; Williams, N.; Grutzner, F.; Hughes, J.; Thomas, P. Targeted Deletion of an Entire Chromosome Using CRISPR/Cas9. *Mol. Ther.* **2017**, *25*, 1736–1738. [CrossRef] [PubMed]
44. Maddalo, D.; Manchado, E.; Concepcion, C.P.; Bonetti, C.; Vidigal, J.A.; Han, Y.C.; Ogrodowski, P.; Crippa, A.; Rekhtman, N.; De Stanchina, E.; et al. In vivo engineering of oncogenic chromosomal rearrangements with the CRISPR/Cas9 system. *Nature* **2014**, *516*, 423–427. [CrossRef]
45. Torres, R.; Martin, M.C.; Garcia, A.; Cigudosa, J.C.; Ramirez, J.C.; Rodriguez-Perales, S. Engineering human tumour-associated chromosomal translocations with the RNA-guided CRISPR-Cas9 system. *Nat. Commun.* **2014**, *5*, 3964. [CrossRef]
46. Jiang, J.; Zhang, L.; Zhou, X.; Chen, X.; Huang, G.; Li, F.; Wang, R.; Wu, N.; Yan, Y.; Tong, C.; et al. Induction of site-specific chromosomal translocations in embryonic stem cells by CRISPR/Cas9. *Sci. Rep.* **2016**, *6*, 21918. [CrossRef]
47. Reimer, J.; Knöß, S.; Labuhn, M.; Charpentier, E.M.; Göhring, G.; Schlegelberger, B.; Klusmann, J.H.; Heckl, D. CRISPR-Cas9-induced t(11;19)/MLL-ENL translocations initiate leukemia in human hematopoietic progenitor cells in vivo. *Haematologica* **2017**, *102*, 1558–1566. [CrossRef]
48. Schneidawind, C.; Jeong, J.; Schneidawind, D.; Kim, I.S.; Duque-Afonso, J.; Wong, S.H.K.; Iwasaki, M.; Breese, E.H.; Zehnder, J.L.; Porteus, M.; et al. MLL leukemia induction by t(9;11) chromosomal translocation in human hematopoietic stem cells using genome editing. *Blood Adv.* **2018**, *2*, 832–845. [CrossRef]
49. Jansen, M.W.J.C.; van der Velden, V.H.J.; van Dongen, J.J.M. Efficient and easy detection of MLL-AF4, MLL-AF9 and MLL-ENL fusion gene transcripts by multiplex real-time quantitative RT-PCR in TaqMan and LightCycler [11,12]. *Leukemia* **2005**, *19*, 2016–2018. [CrossRef] [PubMed]
50. Cong, L.; Ran, F.A.; Cox, D.; Lin, S.; Barretto, R.; Habib, N.; Hsu, P.D.; Wu, X.; Jiang, W.; Marraffini, L.A.; et al. Multiplex genome engineering using CRISPR/Cas systems. *Science* **2013**, *339*, 819–823. [CrossRef]
51. Hsu, P.D.; Scott, D.A.; Weinstein, J.A.; Ran, F.A.; Konermann, S.; Agarwala, V.; Li, Y.; Fine, E.J.; Wu, X.; Shalem, O.; et al. DNA targeting specificity of RNA-guided Cas9 nucleases. *Nat. Biotechnol.* **2013**, *31*, 827–832. [CrossRef] [PubMed]
52. Metcalf, D.; Glaser, S.P.; Xu, Z.; Di Rago, L.; Mifsud, S. Reversible growth factor dependency and autonomy during murine myelomonocytic leukemia induced by oncogenes. *Proc. Natl. Acad. Sci. USA* **2013**, *110*, 17029–17034. [CrossRef] [PubMed]

53. Sasca, D.; Hähnel, P.S.; Szybinski, J.; Khawaja, K.; Kriege, O.; Pante, S.V.; Bullinger, L.; Strand, S.; Strand, D.; Theobald, M.; et al. SIRT1 prevents genotoxic stress-induced p53 activation in acute myeloid leukemia. *Blood* **2014**, *124*, 121–133. [CrossRef] [PubMed]
54. Ikeda, H.; Kanakura, Y.; Tamaki, T.; Kuriu, A.; Kitayama, H.; Ishikawa, J.; Kanayama, Y.; Yonezawa, T.; Tarui, S.; Griffin, J.D. Expression and functional role of the proto-oncogene c-kit in acute myeloblastic leukemia cells. *Blood* **1991**, *78*, 2962–2968. [CrossRef]
55. Paschka, P.; Marcucci, G.; Ruppert, A.S.; Mrózek, K.; Chen, H.; Kittles, R.A.; Vukosavljevic, T.; Perrotti, D.; Vardiman, J.W.; Carroll, A.J.; et al. Adverse prognostic significance of KIT mutations in adult acute myeloid leukemia with inv(16) and t(8;21): A Cancer and Leukemia Group B study. *J. Clin. Oncol.* **2006**, *24*, 3904–3911. [CrossRef]
56. Ohlsson, E.; Hasemann, M.S.; Willer, A.; Lauridsen, F.K.B.; Rapin, N.; Jendholm, J.; Porse, B.T. Initiation of MLL-rearranged AML is dependent on C/EBPα. *J. Exp. Med.* **2014**, *211*, 5–13. [CrossRef]
57. Chang, M.J.; Wu, H.; Achille, N.J.; Reisenauer, M.R.; Chou, C.W.; Zeleznik-Le, N.J.; Hemenway, C.S.; Zhang, W. Histone H3 Lysine 79 Methyltransferase Dot1 Is Required for Immortalization by MLL Oncogenes. *Cancer Res.* **2010**, *70*, 10234–10242. [CrossRef]
58. Jeong, J.; Jager, A.; Domizi, P.; Pavel-Dinu, M.; Gojenola, L.; Iwasaki, M.; Wei, M.C.; Pan, F.; Zehnder, J.L.; Porteus, M.H.; et al. High-efficiency CRISPR induction of t(9;11) chromosomal translocations and acute leukemias in human blood stem cells. *Blood Adv.* **2019**, *3*, 2825–2835. [CrossRef]
59. Heckl, D.; Kowalczyk, M.S.; Yudovich, D.; Belizaire, R.; Puram, R.V.; McConkey, M.E.; Thielke, A.; Aster, J.C.; Regev, A.; Ebert, B.L. Generation of mouse models of myeloid malignancy with combinatorial genetic lesions using CRISPR-Cas9 genome editing. *Nat. Biotechnol.* **2014**, *32*, 941–946. [CrossRef]
60. O'Connor, C.; Yalla, K.; Salomé, M.; Ananyambica Moka, H.; Gómez Castañeda, E.; Eyers, P.A.; Keeshan, K. Trib2 expression in granulocyte-monocyte progenitors drives a highly drug resistant acute myeloid leukaemia linked to elevated Bcl2. *Oncotarget* **2018**, *9*, 14977–14992.

3

Turtle Insights into the Evolution of the Reptilian Karyotype and the Genomic Architecture of Sex Determination

Basanta Bista * and Nicole Valenzuela *

Department of Ecology, Evolution, and Organismal Biology, Iowa State University, Ames, IA 50011, USA
* Correspondence: bbista@iastate.edu (B.B.); nvalenzu@iastate.edu (N.V.)

Abstract: Sex chromosome evolution remains an evolutionary puzzle despite its importance in understanding sexual development and genome evolution. The seemingly random distribution of sex-determining systems in reptiles offers a unique opportunity to study sex chromosome evolution not afforded by mammals or birds. These reptilian systems derive from multiple transitions in sex determination, some independent, some convergent, that lead to the birth and death of sex chromosomes in various lineages. Here we focus on turtles, an emerging model group with growing genomic resources. We review karyotypic changes that accompanied the evolution of chromosomal systems of genotypic sex determination (GSD) in chelonians from systems under the control of environmental temperature (TSD). These transitions gave rise to 31 GSD species identified thus far (out of 101 turtles with known sex determination), 27 with a characterized sex chromosome system (13 of those karyotypically). These sex chromosomes are varied in terms of the ancestral autosome they co-opted and thus in their homology, as well as in their size (some are macro-, some are micro-chromosomes), heterogamety (some are XX/XY, some ZZ/ZW), dimorphism (some are virtually homomorphic, some heteromorphic with larger-X, larger W, or smaller-Y), age (the oldest system could be ~195 My old and the youngest < 25 My old). Combined, all data indicate that turtles follow some tenets of classic theoretical models of sex chromosome evolution while countering others. Finally, although the study of dosage compensation and molecular divergence of turtle sex chromosomes has lagged behind research on other aspects of their evolution, this gap is rapidly decreasing with the acceleration of ongoing research and growing genomic resources in this group.

Keywords: sex chromosome evolution; karyotypic and molecular evolution; genomic architecture of sexual development; adaptation and natural selection; genome organization and function; nucleolar organizing region; dosage compensation; faster-X and faster-Z; climate change and global warming; reptilian vertebrates

---

1. Introduction

A paramount event in the history of life is the early evolution of sexual reproduction. Sex, which is nearly universal in eukaryotes, joins half of two parental genomes from gametes produced by meiosis into unique combinations, contributing to the phenotypic diversity that is naturally selected during adaptive evolution [1]. However, the evolution of gametes into two types (small and mobile versus large and immobile) that led to the evolution of male and female functions meant that a developmental decision is necessary for individuals to become sperm-producing males, egg-producing females, or hermaphrodites that produce both. In multicellular organisms, where separated sexes evolved often, and particularly in animals, this decision is controlled most frequently by a pair of specialized chromosomes (the sex chromosomes) [2]. In other animals, this decision is plastic and occurs in response to environmental cues (ESD), of which the most common in vertebrates is temperature (TSD) [3]. However, a clear explanation for why some vertebrate lineages rely exclusively on sex chromosomes, some exclusively on TSD, and some combine both [4–6], remains elusive, as are the mechanistic trajectories by which species transition between TSD and sex chromosomes. Reptiles are ideal to answer these questions because these evolutionary transitions have occurred repeatedly in this group, leading to varied sex-determining mechanisms. Namely, reptilian sex determination encompasses TSD in crocodilians and tuatara, genotypic sex determination (GSD) in snakes, and either TSD or GSD in turtles and lizards (some lizards even combine GSD and TSD) [3]. Furthermore, TSD and GSD is not the same in all reptiles [3,7]. Rather, TSD reptiles respond to temperature in one of three patterns: (a) by producing males at colder and females at warmer conditions, (b) the opposite, or (c) by producing males at intermediate values and females above and below (reviewed in [7]). GSD reptiles also vary, as some possess female- and others male-heterogametic sex chromosomes, including homomorphic or heteromorphic ZZ/ZW, XX/XY, X_1X_2Y, Z_1Z_2W, and ZW_1W_2 systems [3,8]. Here, we concentrate on turtles, an emerging model group with growing genomic resources, and explore the karyotypic changes that accompanied the evolution of the genomic architecture of sexual development during transitions between TSD and GSD in this group. Hereafter, we will refer to XX/XY ZZ/ZW systems as XY and ZW, respectively.

2. Sex Chromosome Evolution

Theoretical models suggest that vertebrate sex chromosomes evolved from ESD or from polygenic sex determination [2] by co-opting an autosomal pair of chromosomes harboring a sex-determining locus. This initial step is expected to trigger a cascade of events that leads to the eventual degeneration of the heterogametic Y or W due to their unusual mode of transmission through a single sex [2]. The model proposes that first, recombination is reduced adaptively via chromosomal inversions or via selection on a modifier locus favoring recombination suppression [9], to preserve the linkage disequilibrium between the sex-determining locus and sexually antagonistic genes [10–12] (e.g., Y genes that favor males but are harmful if they were expressed in females [13]). Consequently, mildly deleterious mutations accumulate in the non-recombining region via Muller's ratchet or genetic drift, causing Y or W genes to lose their function and ultimately disappear altogether [14]. Moreover, strong selection acting on the sex-determining region can induce background selection, genetic hitchhiking and selective sweeps that reduce genetic diversity in adjacent regions [2]. This degenerative process is prevented in the pseudo-autosomal regions of the sex chromosome where recombination remains intact [2]. Such extensive degeneration of the W and Y may constitute an evolutionary trap from which TSD evolution is difficult [15], because this transition requires traversing a valley in the fitness landscape (sensu [16]) where individuals are produced that carry suboptimal or lethal WW or YY genotypes. This problem may be averted when sex chromosomes are virtually homomorphic [2,17]. But do turtles follow this theoretical evolutionary trajectory? Valuable existing information sheds light on the evolutionary history of turtle sex chromosomes, despite how relatively little is known about their content compared to mammals or birds, partly because only a single GSD turtle genome assembly has been published (*Pelodiscus sinensis* [18]), and its sex chromosome scaffolds remain unmapped.

3. Sex Chromosomes were Gained and Lost Multiple Times in Turtles

Most turtles possess TSD, a system that appears ancestral to turtles, reptiles, and likely to all amniotes based on the most complete phylogenetic comparative analyses possible to date given the existing information [15,19–21]. These species-level phylogenetic analyses revealed that the history of turtle sex determination is marked by the retention of an ancestral TSD mechanism in most chelonian lineages, punctuated by few transitions to sex chromosomes (five so far identified), and two potential reversals from GSD back to TSD where sex chromosomes may have been lost [19,22] (Figure 1). However, these evolutionary transitions are not all created equal, and instead, they have followed unique trajectories accompanied by profound genomic modifications, as will be described.

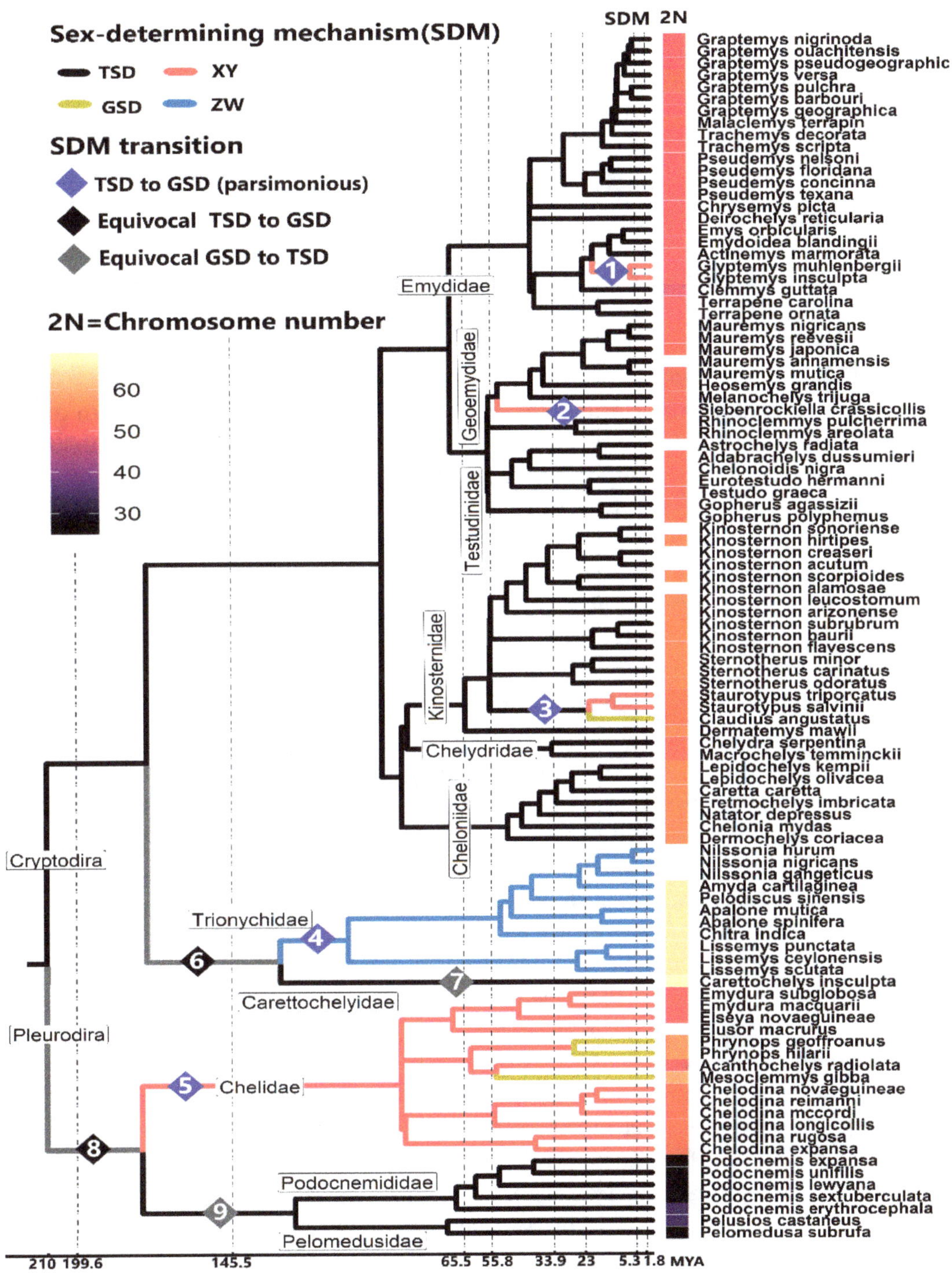

Figure 1. Phylogenetic relationships of turtles with a known sex-determining mechanism (SDM) and diploid number (2N). Timing of hypothesized gains and losses of sex chromosomes correspond to the timing of split of the colored branches. Diamonds indicate hypothesized transitions in sex determination. Transitions 1–5 are the most parsimonious. Transitions 6–7 and 8–9 represent alternative hypotheses to transitions 4 and 5, respectively, proposed based on ancestral reconstruction using maximum likelihood [19]. Data from [3,19,23–26].

Indeed, of the >355 recognized species of turtles to date [27], sex determination is known in only 101 of them. These include 31 GSD species, of which 27 have a characterized sex chromosome system that vary in age, heterogamety, homology, shape, and size (turtles possess macro and micro chromosomes [28] (Figures 1 and 2). Four other GSD turtles were identified as such because they produce

1:1 sex ratios across incubation temperatures, thus ruling out TSD [4], but their heterogamety remains unknown. This gap exists because few studied turtles have large heteromorphic sex chromosomes easily visualized using classical cytogenetic techniques [29–31]. Thus, the detection of virtually homomorphic sex chromosomes in other turtles requires higher-resolution molecular cytogenetic approaches, such as comparative genome hybridization (CGH), which only became available recently [25,26,32–36]. Male or female heterogamety has been identified recently in some GSD turtles by PCR or qPCR amplification of molecular markers first developed in closely related taxa [23,24,37].

From the distribution of these sex chromosome systems in the turtle tree of life, male heterogamety was inferred to have evolved at least three times in the suborder Cryptodira (the turtles who hide their necks inside their shell) and at least once in the suborder Pleurodira (the turtles who bend their necks to the side outside their shell) (Figure 1). These events gave rise to the 16 turtle XY systems identified thus far: 11 in the pleurodirans within a single family Chelidae—*Acanthochelys radiolata, Elseya novaeguineae*, six in the genus *Chelodina*, and two in *Emydura*, [25,31–33], plus five from three cryptodiran families—*Siebenrockiella crassicollis*, two in *Glyptemys*, and two in *Staurotypus* [23,29,30,36,38].

Curiously, unlike the repeated independent evolution of XY in chelonians, a single origin of female heterogamety is known, in the cryptodiran family of softshell turtles (Tryonichidae). Indeed, ZW has been documented in 10 soft-shell turtles, two cases (*P. sinensis* and *Apalone spinifera*) by molecular cytogenetic data [35,39], and eight others by PCR/qPCR— three species in the genus *Lissemys*, three in *Nilssonia*, plus *Chitra indica* and *Amyda cartilaginea* [24]. The report of a non-softshell turtle (*Pangshura smithii*) having an independently evolved heteromorphic ZW was debunked recently, as it was due to a karyotyping error [40,41]. Therefore, it is unclear if *P. smithii* has GSD with homomorphic sex chromosomes or TSD. Similarly, the co-existence of sex chromosomes and TSD was empirically refuted in *P. sinensis* and *Chrysemys picta* [42,43], whereas thermal sex reversals occur in several lizards [44–48]. However, why XY systems are more likely to evolve in turtles than ZW systems is unclear, although this pattern matches theoretical models (reviewed in [2]). Some of these models predict that when sex chromosomes arise during the evolution of separate sexes in a population of hermaphrodites, selection favors first the spread of a recessive male-sterility mutation that produces females when homozygous, and then selection favors the spread of a dominant female sterility mutation in the hermaphrodites that produces males, leading to the formation of a XY system (reviewed in [2]). Alternatively, sexual selection, which is stronger in males in general, may favor the fixation of a major sex-determining factor beneficial to the heterogametic sex, thus favoring the evolution of an XY system (reviewed in [2]).

From a karyotypic perspective, the first difference that is noticeable between the 13 turtle sex chromosome systems for which cytogenetic data are available (Figure 2), is that five are micro-sex-chromosomes (μSC) (in the softshells *A. spinifera, P. sinensis* and in the Australian chelids of the genus *Chelodina*), whereas the other eight are macro-sex-chromosomes (MSC). The second notable karyotypic characteristic, is that four independent evolutionary events resulted in heterogametic sex chromosomes (Y or W) in turtles that are larger than the X or Z (in the genera *Emydura, Glyptemys, Siebenrockiella*, and in the family Trionychidae), whereas only two instances are known that resulted in a smaller Y than X (in the genera *Acanthochelys* and *Staurotypus*), and one case is known of the evolution of virtually homomorphic XY (in the genus *Chelodina*) (Figure 2). Cytogenetic data revealed that the larger size of the Y or W compared to the X or Z is due to the accumulation of repeat sequences, but the particular details vary among lineages. For instance, the repeats involved in this enlargement of the Y or Z encompass an expanded Z-linked NOR in trionychids [35], expanded heterochromatin (C- and G-bands) in the Y of *Glyptemys* [36] and *Siebenrockiella* [30], and accumulation of microsatellites in the Y of *Emydura* and *Elseya* [25,26,49]. Notably, the male-specific region of the *Emydura* Y likely evolved by an Y-autosome fusion that resulted in the translocation of an ancestral micro-Y chromosome present in *Chelodina* [25], which contributes to the larger Y in *Emydura*. This is the only case of Y-autosome fusion known in chelonians, whereas Y-autosome fusions are well documented in squamate reptiles and fish, and tend to be more common than X-autosome, Z-autosome, or W-autosome fusions [8]. Also remarkably, the larger size of *Staurotypus* X is not due to the degeneration of the Y but to the translocation of the NOR to the X, such that the Y in this turtle represents the ancestral condition and the X is the derived sex chromosome [50]. Taken together, the current data indicate that turtle sex chromosome differentiation does not always involve the shrinkage of the Y or W proposed in classical models of sex chromosome evolution, and add support to the alternative that sex chromosomes of varying degrees and type of differentiation are evolutionary stable states [1].

Sub-Order	Family	Species	SC Type	SC Size	NOR linkage	Refs
Pleurodira	Chelidae	*Acanthochelys radiolata*	X Y	Macro	Auto	[31]
	Chelidae	*Chelodina longicollis, Chelodina expansa, Chelodina novaeguineae*	X Y	Micro	Auto	[32, 26]
	Chelidae	*Emydura macquarii, Emydura subglobosa*	X Y	Micro	Auto	[33, 25]
	Chelidae	*Elseya novaeguineae*	X Y	Micro	Auto	[26]
Cryptodira	Trionychidae	*Apalone spinifera*	Z W	Micro	W	[35]
	Trionychidae	*Pelodiscus sinensis*	Z W	Micro	W	[39]
	Emydidae	*Glyptemys insculpta*	X Y	Macro	Auto	[36]
	Kinosternidae	*Staurotypus triporcatus, Staurotypus salvinii*	X Y	Macro	X	[29]
	Geoemydidae	*Siebenrockiella crassicollis*	X Y	Micro	Auto	[30]

Figure 2. Karyotypic characteristics of turtle sex chromosomes (SC) from species studied cytogenetically. Macro = macro-chromosomes; Micro = micro-chromosomes; Streaked regions = centromeres; blue = nucleolar organizing region (NOR); red = male-specific regions in the Y or female-specific region in the W, detected via CGH; as described in the text and reviewed in [28].

Despite the multiple gains of sex chromosomes identified in turtles, some evidence suggests that reversals back to TSD from GSD may have occurred also where sex chromosomes would have been lost, once in each chelonian suborder. In particular, two putative evolutionary reversals from GSD to TSD were identified by a maximum likelihood reconstruction of ancestral sex-determining mechanisms (SDMs): (1) one in the monotypic family Carettochelyidae (suborder Cryptodira) which split from the ZW softshell family Trionychidae in the Mesozoic ~148 Mya (http://www.timetree.org/), and (2) another in the superfamily Pelomedusoidea (suborder Pleurodira) which split from the XY family Chelidae at around a similar time ~144 Mya (http://www.timetree.org/) [19] (Figure 1). If true, this would mean that the TSD system in *Carettochelys insculpta* and Pelomedusoidea are secondary and independently derived traits, and might be expected to differ mechanistically from the ancestral TSD retained in most other turtles. However, these reversals are questionable given that their inference is not parsimonious [20], and that reanalysis of the same dataset by maximum likelihood reconstruction of ancestral SDMs using an updated software package did not recover the same result [21]. Yet, the significant intensification of the rate of molecular evolution of sexual development genes observed in both Carettochelyidae and Podocnemididae [22], above and beyond the already accelerated rate seen in their GSD sister lineages, supports the notion that GSD-to-TSD reversals might have taken place. Moreover, the branching of these lineages suggests that these reversals might have occurred soon after GSD evolved in the common ancestor of Carettochelyidae and Trionychidae, and in the common ancestor of Podocnemididae and Chelidae, likely before extensive differentiation of their sex chromosomes accrued that would have been harder to overcome [1,15]. Nonetheless, the support for these putative reversals remains relatively scant and stronger inferences require further research.

4. Independent and Convergent Evolution of Turtle Sex Chromosomes

The existing data indicate that not all turtle sex chromosomes are homologous to each other [28]. Indeed, partial data on the gene content of the sex chromosomes of some cryptodiran turtles for which information is available (*Staurotypus triporcatus, Glyptemys insculpta, Siebenrockiella crassicollis*, and softshells such as *A. spinifera* and *P. sinensis*) revealed that the evolution of their XY and ZW chromosomes occurred by the co-option of different ancestral reptilian autosomes [28]. Yet, these turtle sex chromosomes may share a deeper homology with blocks of a more ancient proto sex chromosome [28]. In particular, gene content indicates that *S. triporcatus* XY share homology with chicken's Z (GGA-Z) and the ZW of *Gekko hokouensis* lizards, and with a block of the frog *Xenopus tropicalis* chromosome 1 (XTR1) [28,34,51]. This XTR1 block contains *Dmrt1*, chicken's sex-determining gene, whose action is dosage-dependent [28,51]. On the other hand, the ZW of softshells (e.g., *A. spinifera* and *P. sinensis*), which derive from a single gain in the common ancestor of the softshell family *Trionychidae* [24,35], are homologous to each other, to chicken GGA15, and to the X of *Anolis carolinensis* lizards [28,38], and share partial homology with a second block of XTR1 [28].

In yet another evolutionary twist, the only two turtle lineages known to have recruited the same pair of ancestral autosomes, *G. insculpta* and *S. crassicollis*, did so independently [28]. Indeed, the XY in these two species are homologous to each other, to GGA5, and to XTR5, which contains the male development gene *Wt1* [28,36]. The notion that these two XY systems represent convergent evolution and followed independent trajectories is also supported by a secondary homology shared between the short arm of *G. insculpta* (but not *S. crassicollis'*) XY and GGA26, which surprisingly, is homologous to a third block of XTR1 [28,36]. To make matters even more intriguing, another (fourth) block of XTR1 shares homology with the sex chromosomes of other vertebrates, namely, the X of therian mammals including humans, and the Z of the lizard *Takydromus sexlineatus* [28,52]. Moreover, XTR1 is also homologous to the sex chromosomes of three other anuran amphibians [53].

Thus, the origin of these independently derived sex chromosomes appears to be non-random. Instead, all the data combined suggest that certain ancestral chromosomes constitute proto-sex chromosomes that are more likely to take on a role as sex chromosomes in various derived taxa (e.g., XTR1 or a putative ancestral XTR1 + XTR5 from which many turtle chromosomes derive) [28]. It is noticeable that a common ancestral chromosome can follow evolutionary trajectories leading to both male or female heterogamety (e.g., softshell turtles ZW versus *Anolis* lizard XY; *Staurotypus* turtles XY versus chicken's and *Gekko* lizards' ZW, etc.). Perhaps chromosomes which are rich in genes involved in sexual development are better suited to take on a role as sex chromosomes, and perhaps both the evolution of these sex chromosomes and their fixation in populations may be facilitated by genomic rearrangements [12]. Additionally, chromosomes with genes resilient to changes in dosage may also be favored for co-option as sex chromosomes as they may be buffered naturally from the degeneration and gene content depletion that may be inevitable in the evolving Y or W [28,54,55].

Taken together, all evidence permits drawing parallels and contrasts between the sex chromosomes of chelonians and other reptilian lineages. Namely, like other reptiles, turtle sex chromosomes vary in the degree of heteromorphism (Figure 2), and some of them carry the genes of the nucleolar organizing region (NOR) [28]. Just like turtles, both lizards and snakes co-opted various ancestral autosomes as sex chromosomes independently [56–58]. Unlike turtles, however, the repeated evolution of sex chromosomes in squamates (lizards and snakes) occurred not only as transitions from TSD but also as transitions between male and female heterogamety, a process that has not been detected in turtles. For instance, the recent discovery of convergent XY sex chromosomes in pythons and boas refuted the long-held notion that they share the ZW system previously thought to be ubiquitous and homologous across snakes [57]. Also, no case of multiple sex chromosomes is known in turtles, whereas examples exist in squamates [3,58].

5. Ecological and Karyotypic Correlates of the Birth and Death of Turtle Sex Chromosomes

What are the causes and consequences of the gain or loss of turtle sex chromosomes? The answer to this question is not fully resolved, yet significant strides have been made towards solving this mystery. Some ecological and karyotypic correlates of sex chromosome evolution exist in turtles, including diploid number, climate change, population sex ratios, and longevity. In particular, turtles exhibit high variance in the diploid number of chromosomes ranging from $2N = 28$ to 68 (Figure 1). The genomic rearrangements responsible for generating this diversity occurred at a rate 20× higher in turtle lineages that underwent an evolutionary transition in their sex-determining mechanism than in lineages whose sex determination remained static [19]. However, which is the cause and which the consequence remains unclear. For instance, perhaps the chromosomal rearrangements responsible for changes in chromosome number are fertile grounds for changes in sex determination, as they could affect gene expression through direct disruption of genes, changes in regulatory elements or altered positional effects [59]. Further, while chromosomal fusion or fission are large-scale changes that alter chromosome number, smaller-scale rearrangements can also have significant ramifications. For example, inversions induce the repression of recombination between homologs of a chromosomal pair, which is a hallmark of sex chromosome evolution [60]. Such inversions are known in turtles, and include multiple parts of the male-specific region of Y chromosome in *G. insculpta* that contains the male development gene *Wt1*, which might have facilitated the transition from ESD to GSD in this lineage [36]. But some other identified inversions and translocations involving sexual development genes (*Dax1*, *Fhl2*, *Fgf9*, *Sf1* and *Rspo1*) across TSD and GSD turtles do not correlate with transitions in sex determination [61]. Alternatively, perhaps transitions in sex determination trigger molecular changes that render the genome unstable and more susceptible to chromosomal fusion or fission which, in turn, will alter the chromosome number [19]. These pending questions remain the foci of active research.

Climate change is an ecological factor that may trigger such transitions in sex determination in turtles, particularly, the evolution of sex chromosomes from TSD, as an adaptive response to alleviate extreme biased sex ratios caused by steady warming or cooling events [19,62]. Specifically, a theoretical model proposes that ESD could be adaptive in species inhabiting patchy or heterogenous environments where some environments increase the fitness of one sex and other environments increase the fitness of the other sex (provided that offspring cannot predict the environment they enter – nor can their parents, and that individuals from all environments mate at random) [63]. Under these circumstances, the plasticity provided by ESD permits individuals to develop into the sex that is better suited in each environment. However, if environments become homogenous, or in the case of TSD, if climate change becomes directionally warmer or cooler over time, population sex ratios may be skewed. Then, natural selection could favor the evolution of sex chromosomes via the evolution of a masculinizing or a feminizing locus, resulting in a XY or ZW system, respectively [2], which will restore sex ratios closer to parity. (This model assumes that 1:1 sex ratio is optimal, which is not always the case [62,64,65]. This scenario is supported in turtles by the observation that some lineages known to have experienced a transition in sex determination split from their sister lineages at geological times when global temperatures were near a peak [19]. However, relatively few transitions in sex determination are documented in turtles across repeated bouts of past climate change compared to squamates, the other reptilian clade with labile sex determination. This difference may be attributed to their contrasting life history, particularly their differences in lifespan [21]. Indeed, the greater longevity of turtles would enable them to withstand longer stretches of biased sex ratios caused by directional climate change compared to lizards and snakes, thus explaining the frequent retention of TSD in turtles and the evolution of GSD more readily in squamates [21]. However, the speed of contemporary climate change and the predicted warmer average temperatures that fluctuate more widely pose a challenge to extant TSD turtles in unprecedented ways [66,67].

6. The Architecture of Sex Determination with and without Sex Chromosomes

The regulatory gene network that underpins the development of males and females is composed of many common elements across vertebrates, but TSD and GSD lineages differ in the level of plasticity of this regulation to environmental inputs [68,69]. Indeed, the signal which initiates the cascade of events leading to sex differentiation is hard coded in the genome within the sex chromosomes in the form of a master sex-determining gene(s), whereas in TSD this developmental decision is triggered by an environmental factor (i.e., temperature) [4]. Thus, because sex chromosomes contain the only consistent genomic differences between the sexes, and reptilian sex chromosomes are diverse, their study illuminates the evolution of master sex-determining genes, and more generally, our understanding of the genetic architecture underlying sexual development in vertebrates. For instance, in therian mammals, the Y-linked *Sry* initiates male differentiation [70], whereas sex determination in birds relies on the dosage of the Z-linked gene *Dmrt1* in ZZ-males versus ZW-females [71]. However, no master sex-determining gene(s) has been discovered in reptiles. Some candidates exist in reptiles, including in chelonians [72–77], and the list is growing thanks to the advent of third generation sequencing and advanced molecular cytogenetics (e.g.,) [28,61,74–78]. Namely, the XY of *S. triporcatus* turtles contains *Dmrt1* [28,78], a gene whose molecular evolution is linked to transitions in sex determination in reptiles [79] and which displays sexually dimorphic expression in TSD turtles ([80] and references therein). Notably, while *Sry* is exclusive to therian mammals, *Dmrt1* or its orthologs or paralogs are master triggers of sex determination in various vertebrates (reviewed in [80]). However, whether *Dmrt1* is the master sex-determining gene in *Staurotypus* is untested. Nonetheless, *Dmrt1* retention in both the X and Y in *S. triporcatus* and in the Z and W of *G. hokouensis* lizards [28] indicates that *Dmrt1*'s mode of action likely differs between birds and these reptiles. *Dmrt1* was demonstrated to be important in testicular differentiation of the GSD softshell *P. sinensis* [81], but its autosomal nature [61] invalidates its potential role as a sex-linked master sex-determining gene. Another example of a putative master sex-determining gene in GSD turtles is *Sf1*, a male-development gene that translocated to the ZW of *Apalone* softshells lineage from its ancestral autosomal location [61]. *Sf1* was proposed as a potential candidate for this role based on its monomorphic expression by temperature in *Apalone* softshell turtles, in contrast to its differential expression at male- versus female-producing temperature in the TSD turtle *C. picta* [72,73,76]. A third candidate comes from the XY of *G. insculpta* and *S. crassicollis* turtles, which contain the male development gene *Wt1* [28,36], a transcription factor whose relic thermosensitive expression in early embryos of *Apalone* is countered by the immediate downstream action of *Sf1* [82,83]. Noticeably, the relic thermosensitive expression in *Apalone* of another important sexual development gene, *Dax1*, is also countered by the immediate downstream action of *Sf1* [83].

The identification of top regulators of sexual development remains even more elusive in TSD turtles, because no genomic differences exist between the sexes, and finding candidates requires the analysis of the response of gene regulatory pathways and networks to incubation temperature [82]. Candidate gene approaches, along with transcriptome and methylome analyses, plus initial functional assays have provided a number of candidates in TSD turtles for an upstream function in sexual development (e.g., [74–77], ruling out elements whose sexually dimorphic expression occurs too late in development to act as top master regulators (e.g., [73,76,82–89]). To complicate matters further, this body of evidence has uncovered significant divergence in the transcriptional patterns of genes in this regulatory network [73,80]. As this growing body of work cannot be summarized properly in the limited space available here, we direct the reader to some excellent reviews on this area (e.g., [68,90,91]), and highlight here only some salient results. Comparative data from *C. picta* (TSD) and *Apalone mutica* or *A. spinifera* (GSD) have provided important insights. In particular, early dimorphic transcription in *C. picta* of genes responsible for the formation of the bipotential gonad and later testicular development in vertebrates (*Sf1* and *Wt1*), rendered these elements as potential activators of *C. picta*'s thermosensitive period (when temperature exerts its strongest effect on population sex ratios) [72,73,82]. Their early expression in *C. picta* coincides with the time when the expression of genes involved in chromatin organization and chromatin modification is enriched [92,93] but also with the dimorphic expression of female-development regulators, such as *Ctnnb1* (*β-catenin*) and its downstream target *Fst* [93]. On the contrary, the later differential expression of *Dax1* [83], *Dmrt1* [80], *Sox9* and *Aromatase* [73] ruled them out for a role as upstream TSD thermal sensors/activators in *C. picta*. An interesting candidate TSD gene is *CIRBP*, which exhibits allelic-specific expression at male- and female-producing temperature in *Chelydra serpentina* turtles (TSD) [75], but which shows thermo-insensitive expression in *C. picta* and *A. spinifera* [93]. Epigenetic regulation of TSD also appears important. For instance, DNA methylation regulates *Aromatase* in *Trachemys scripta* turtles, a key ovarian development gene that is also affected by histone modifications [94,95]. Methylome analysis indicated that some of the thermosensitive responses of the regulatory network of sexual development in TSD turtles is mediated by DNA methylation of additional components other than *Aromatase* [92,96]. And other epigenetic modifications also influence sexual development. For instance, the histone demethylase KDM6B induces the transcription of *Dmrt1* in *T. scripta* [89], a gene important in testicular formation during the thermosensitive period in this turtle [87]. Furthermore, transcriptomic analysis of epigenetic machinery genes suggests that TSD, at least in *C. picta*, is potentially mediated by hormonally controlled epigenetic processes, or by epi-genetically controlled hormonal pathways (via acetylation, methylation, and ncRNAs) [77]. Importantly, the response of the epigenetic machinery genes to temperature indicate that differences in key epigenetic events before the onset of the thermosensitive period may define the divide between TSD and GSD, as represented by *C. picta* and *A. spinifera* [77]. It should be noted that while these growing efforts are helping to resolve the position of these factors in the sexual development cascade, further research is still needed.

7. Consequences of Sex Chromosome Evolution—Dosage Compensation and Faster Molecular Evolution

The degeneration of the heterogametic sex chromosomes Y and W over evolutionary time [97,98] described earlier can have ill fitness consequences unless these effects are balanced by the evolution of counter mechanisms. For instance, in human, chicken, and multiple animal species with sex chromosomes, only a small fraction of genes remain active in Y or W compared to X or Z [99,100]. This loss of function or physical loss of genes in the Y or W generates a gene dosage imbalance between autosomes and sex chromosomes, and between males and females (i.e., the homogametic sex will have twice as much dosage of X- or Z-genes compared to the heterogametic sex) [2]. However, the balance of dosage is very important for genes to maintain their proper function as they are part of genetic networks. An adaptive solution to this problem evolved in the form of dosage compensation mechanisms that modify the transcription of genes with differential dosage so that their expression is balanced fully or partially, either along the entire sex chromosome (global dosage compensation) or only for some important genes (local dosage compensation) [2,14,101,102].

Earlier animal data suggested that global and complete dosage compensation is more predominant in XY taxa (such as therian mammals), while local and partial dosage compensation is more common in ZW taxa (such as birds), but numerous exceptions discovered over time called this conclusion into question [101–103]. Very little is known about dosage compensation in reptiles, with only three studies published on squamates and none on turtles thus far, despite the fact that the diversity of independently evolved sex chromosomes in closely related reptiles renders them an ideal group to address this issue. Existing reptilian data indicate that *A. carolinensis* lizards utilize a mixed pattern of dosage compensation [104]. Namely, some regions of the X chromosome in *A. carolinensis* show complete dosage compensation and some regions lack complete compensation, indicating that the evolution of complete dosage compensation might be an ongoing process [104]. On the other hand, incomplete dosage compensation is seen in two ZW squamates, i.e., snakes and the Komodo dragon [56,105].

Sex chromosomes and autosomes also differ in ways that influence the rate of divergence of sex-linked genes compared to their autosomal counterparts. Specifically, the Y or W are hemizygous, experience reduced recombination and are inherited through one sex only, whereas the X or Z are inherited through both sexes but spend twice as much time in one sex [1,2]. These differences lead to striking differences in their molecular evolution.

Indeed, faster-Z and faster-X divergence is expected because reduced recombination of sex chromosomes facilitates the accumulation of beneficial mutations and the removal of deleterious mutations at higher rates than in autosomes, and because their smaller population size renders them more susceptible to genetic drift [106–108].

While faster rates of divergence of coding sequences are documented in X and Z of many species, including birds (e.g., *Gallus gallus*, *Taeniopygia guttata*), mammals (*Mus castaneus* and *Homo sapiens*) and fruit flies (*Drosophila melanogaster*) [107–112], research in reptiles is in its infancy. Namely, initial studies in turtles examining a very small set of sex-linked genes detected faster-Z in *P. sinensis* but not in *A. spinifera* (both softshell turtles), and slower-X in *S. triporcatus* turtles compared to orthologs in sister taxa where these genes are autosomal [76].

Additionally, a comparison of turtles, crocodilians, squamates, birds and mammals, detected three genes (*Dmrt1*, *Ctnnb1*, *Ar*) with faster amino acid substitution rates when they are Z-linked (*Dmrt1* and *Ctnnb1*, Z-linked in birds and snakes) but not when they are X-linked (*Ar*, X-linked in mammals), compared to when they are autosomal [22]. Similar to turtles, faster-Z is observed in chicken and other birds [107] whose sex chromosomes are homologous to *Staurotypus* X, and faster-X is supported in *A. carolinensis* [104] whose sex chromosomes are homologous to *Apalone*'s Z. On the contrary, the report of faster-Z in snakes [56] needs revisiting, since it is based on data from *Boa* whose purported ZW chromosomes are now known to be autosomes [57]. The major roadblock to study molecular evolution of sex chromosomes in turtles and other reptiles should be alleviated with the publication of additional chelonian genomes with mapped sex chromosomes.

8. Conclusions

Reptiles have diverse systems of sex determination as well as sex chromosomes which are unmatched in mammals and birds. Although sex chromosomes and sex determination in many species of reptiles have been studied, the genomic basis of sexual development has yet to be fully characterized in reptiles. Turtles represent a clade where multiple sex chromosomes evolved independently (XY and ZW) from ancestral TSD systems (or were perhaps lost during GSD to TSD reversals), giving rise to sex chromosomes of with varying size, age and homology. Unlike lizards, there is no evidence reported for the influence of environmental factors overriding sex chromosomes in GSD turtles. We argue that the characterization of sex chromosomes and sexual development in GSD turtles might be the key to identifying major players of sexual development including the master sex determining gene(s) in reptiles. Convergent evolution of sex chromosome in *G. insculpta* and *S. crassicollis* where XY systems evolved independently from the same ancestral autosome would offer a great opportunity to identify genes that are co-opted as sex determining genes. Taken together, existing data indicate that turtles support some tenets of classic theoretical models of sex chromosome evolution, while other tenets are countered. For instance, the evolution of some turtle sex chromosomes involves chromosomal rearrangements such as the translocation of sexual development genes, or inversions that may contribute to their divergence, yet that divergence does not always result in a morphologically degenerate Y or W. Finally, advanced technologies like whole genome sequencing, transcriptomics, methylomics, and other epigenetic approaches should improve our understanding of chromosomal structure and content, global and local gene expression and epigenetic signatures, all of which will be vital to decipher the enigmatic evolutionary trajectories of sex chromosomes and to develop a model of sexual development in turtles, reptiles, and vertebrates.

Author Contributions: Conceptualization, investigation, writing, review and editing, B.B. and N.V.; visualization, B.B.; supervision and funding acquisition, N.V. All authors have read and agreed to the published version of the manuscript.

References

1. Bachtrog, D.; Mank, J.E.; Peichel, C.L.; Kirkpatrick, M.; Otto, S.P.; Ashman, T.L.; Hahn, M.W.; Kitano, J.; Mayrose, I.; Ming, R.; et al. Sex determination: Why so many ways of doing it? *PLoS Biol.* **2014**, *12*, e1001899. [CrossRef] [PubMed]
2. Bachtrog, D.; Kirkpatrick, M.; Mank, J.E.; McDaniel, S.F.; Pires, J.C.; Rice, W.; Valenzuela, N. Are all sex chromosomes created equal? *Trends Genet.* **2011**, *27*, 350–357. [CrossRef] [PubMed]
3. Tree of Sex Consortium. Tree of Sex: A database of sexual systems. *Sci. Data* **2014**, *1*, 140015. [CrossRef] [PubMed]
4. Valenzuela, N.; Adams, D.C.; Janzen, F.J. Pattern does not equal process: Exactly when is sex environmentally determined? *Am. Nat.* **2003**, *161*, 676–683. [CrossRef]
5. Sarre, S.D.; Georges, A.; Quinn, A. The ends of a continuum: Genetic and temperature-dependent sex determination in reptiles. *Bioessays* **2004**, *26*, 639–645. [CrossRef]
6. Valenzuela, N. Co-evolution of genomic structure and selective forces underlying sexual development and reproduction. *Cytogenet. Genome Res.* **2009**, *127*, 232–241. [CrossRef]
7. Valenzuela, N.; Lance, V. *Temperature-Dependent Sex Determination in Vertebrates*; Smithsonian Books: Washington, DC, USA, 2004.
8. Pennell, M.W.; Kirkpatrick, M.; Otto, S.P.; Vamosi, J.C.; Peichel, C.L.; Valenzuela, N.; Kitano, J. Y fuse? Sex chromosome fusions in fishes and reptiles. *PLoS Genet.* **2015**, *11*, e1005237. [CrossRef]
9. Fridolfsson, A.K.; Cheng, H.; Copeland, N.G.; Jenkins, N.A.; Liu, H.C.; Raudsepp, T.; Woodage, T.; Chowdhary, B.; Halverson, J.; Ellegren, H. Evolution of the avian sex chromosomes from an ancestral pair of autosomes. *Proc. Natl. Acad. Sci. USA* **1998**, *95*, 8147–8152. [CrossRef]
10. Rice, W.R. The accumulation of sexually antagonistic genes as a selective agent promoting the evolution of reduced recombination between primitive sex chromosomes. *Evolution* **1987**, *41*, 911–914. [CrossRef]
11. Rice, W.R. Evolution of the Y sex chromosome in animals. *Bioscience* **1996**, *46*, 331–343. [CrossRef]
12. Kirkpatrick, M. The evolution of genome structure by natural and sexual selection. *J. Hered.* **2017**, *108*, 3–11. [CrossRef] [PubMed]
13. Gibson, J.R.; Chippindale, A.K.; Rice, W.R. The X chromosome is a hot spot for sexually antagonistic fitness variation. *Proc. Biol. Sci.* **2002**, *269*, 499–505. [CrossRef] [PubMed]
14. Mank, J.E. The W, X, Y and Z of sex-chromosome dosage compensation. *Trends Genet.* **2009**, *25*, 226–233. [CrossRef] [PubMed]
15. Pokorna, M.; Kratochvíl, L. Phylogeny of sex-determining mechanisms in squamate reptiles: Are sex chromosomes an evolutionary trap? *Zool. J. Linn. Soc.* **2009**, *156*, 168–183. [CrossRef]
16. Wright, S. The Roles of Mutation, Inbreeding, Crossbreeding, and Selection in Evolution. Genetics. In Proceedings of the Sixth International Congress on, New York, NY, USA, 24 August 1932; Volume 1, pp. 356–366.
17. Perrin, N. Sex reversal: A fountain of youth for sex chromosomes? *Evolution* **2009**, *63*, 3043–3049. [CrossRef] [PubMed]

18. Wang, Z.; Pascual-Anaya, J.; Zadissa, A.; Li, W.; Niimura, Y.; Huang, Z.; Li, C.; White, S.; Xiong, Z.; Fang, D.; et al. The draft genomes of soft-shell turtle and green sea turtle yield insights into the development and evolution of the turtle-specific body plan. *Nat. Genet.* **2013**, *45*, 701–706. [CrossRef]
19. Valenzuela, N.; Adams, D.C. Chromosome number and sex determination coevolve in turtles. *Evolution* **2011**, *65*, 1808–1813. [CrossRef]
20. Pokorna, M.J.; Kratochvil, L. What was the ancestral sex-determining mechanism in amniote vertebrates? *Biol. Rev.* **2016**, *91*, 1–12. [CrossRef]
21. Sabath, N.; Itescu, Y.; Feldman, A.; Meiri, S.; Mayrose, I.; Valenzuela, N. Sex determination, longevity, and the birth and death of reptilian species. *Ecol. Evol.* **2016**, *6*, 5207–5220. [CrossRef]
22. Literman, R.; Burrett, A.; Bista, B.; Valenzuela, N. Putative Independent Evolutionary Reversals from Genotypic to Temperature-Dependent Sex Determination are Associated with Accelerated Evolution of Sex-Determining Genes in Turtles. *J. Mol. Evol.* **2018**, *86*, 11–26. [CrossRef]
23. Literman, R.; Radhakrishnan, S.; Tamplin, J.; Burke, R.; Dresser, C.; Valenzuela, N. Development of sexing primers in *Glyptemys insculpta* and *Apalone spinifera* turtles uncovers an XX/XY sex-determining system in the critically-endangered bog turtle *Glyptemys muhlenbergii*. *Conserv. Genet. Resour.* **2017**, *9*, 651–658. [CrossRef]
24. Rovatsos, M.; Praschag, P.; Fritz, U.; Kratochvsl, L. Stable Cretaceous sex chromosomes enable molecular sexing in softshell turtles (Testudines: Trionychidae). *Sci. Rep.* **2017**, *7*, 42150. [CrossRef] [PubMed]
25. Lee, L.; Montiel, E.E.; Valenzuela, N. Discovery of Putative XX/XY Male Heterogamety in *Emydura subglobosa* Turtles Exposes a Novel Trajectory of Sex Chromosome Evolution in *Emydura*. *Cytogenet. Genome Res.* **2019**, *158*, 160–169. [CrossRef] [PubMed]
26. Mazzoleni, S.; Augstenova, B.; Clemente, L.; Auer, M.; Fritz, U.; Praschag, P.; Protiva, T.; Velensky, P.; Kratochvil, L.; Rovatsos, M. Sex is determined by XX/XY sex chromosomes in Australasian side-necked turtles (Testudines: Chelidae). *Sci. Rep.* **2020**, *10*, 4276. [CrossRef] [PubMed]
27. Stanford, C.B.; Rhodin, A.G.J.; van Dijk, P.P.; Horne, B.D.; Blanck, T.; Goode, E.V.; Hudson, R.; Mittermeier, R.A.; Currylow, A.; Eisemberg, C.; et al. (Eds.) Turtle Conservation Coalition. Turtles in trouble: Turtles in Trouble: The World's 25+ Most Endangered Tortoises and Freshwater Turtles-2018. In *Chelonian Research Foundation, Conservation International*; Wildlife Conservation Society, and San Diego Zoo Global: Ojai, CA, USA, 2018; Volume 80, pp. 1–84.
28. Montiel, E.E.; Badenhorst, D.; Lee, L.S.; Literman, R.; Trifonov, V.; Valenzuela, N. Cytogenetic Insights into the Evolution of Chromosomes and Sex Determination Reveal Striking Homology of Turtle Sex Chromosomes to Amphibian Autosomes. *Cytogenet. Genome Res.* **2016**, *148*, 292–304. [CrossRef] [PubMed]
29. Bull, J.J.; Moon, R.G.; Legler, J.M. Male heterogamety in kinosternid turtles (genus *Staurotypus*). *Cytogenet. Cell Genet.* **1974**, *13*, 419–425. [CrossRef] [PubMed]
30. Carr, J.L.; Bickham, J.W. Sex chromosomes of the Asian black pond turtle, *Siebenrockiella crassicollis* (Testudines: Emydidae). *Cytogenet. Cell Genet.* **1981**, *31*, 178–183. [CrossRef]
31. McBee, K.; Bickham, J.W.; Rhodin, A.G.J.; Mittermeier, R.A. Karyotypic variation in the genus *Platemys* (Testudines, Pleurodira). *Copeia* **1985**, *1985*, 445–449. [CrossRef]
32. Ezaz, T.; Valenzuela, N.; Grutzner, F.; Miura, I.; Georges, A.; Burke, R.L.; Graves, J.A. An XX/XY sex microchromosome system in a freshwater turtle, *Chelodina longicollis* (Testudines: Chelidae) with genetic sex determination. *Chromosome Res.* **2006**, *14*, 139–150. [CrossRef]
33. Martinez, P.A.; Ezaz, T.; Valenzuela, N.; Georges, A.; Marshall Graves, J.A. An XX/XY heteromorphic sex chromosome system in the Australian chelid turtle *Emydura macquarii*: A new piece in the puzzle of sex chromosome evolution in turtles. *Chromosome Res.* **2008**, *16*, 815–825. [CrossRef]
34. Kawai, A.; Ishijima, J.; Nishida, C.; Kosaka, A.; Ota, H.; Kohno, S.; Matsuda, Y. The ZW sex chromosomes of Gekko hokouensis (Gekkonidae, Squamata) represent highly conserved homology with those of avian species. *Chromosoma* **2009**, *118*, 43–51. [CrossRef] [PubMed]
35. Badenhorst, D.; Stanyon, R.; Engstrom, T.; Valenzuela, N. A ZZ/ZW microchromosome system in the spiny softshell turtle, *Apalone spinifera*, reveals an intriguing sex chromosome conservation in Trionychidae. *Chromosome Res.* **2013**, *21*, 137–147. [CrossRef] [PubMed]
36. Montiel, E.E.; Badenhorst, D.; Tamplin, J.; Burke, R.L.; Valenzuela, N. Discovery of the youngest sex chromosomes reveals first case of convergent co-option of ancestral autosomes in turtles. *Chromosoma* **2017**, *126*, 105–113. [CrossRef] [PubMed]

37. Literman, R.; Badenhorst, D.; Valenzuela, N. qPCR-based molecular sexing by copy number variation in rRNA genes and its utility for sex identification in soft-shell turtles. *Methods Ecol. Evol.* **2014**, *5*, 872–880. [CrossRef]
38. Kawagoshi, T.; Nishida, C.; Matsuda, Y. The origin and differentiation process of X and Y chromosomes of the black marsh turtle (*Siebenrockiella crassicollis*, Geoemydidae, Testudines). *Chromosome Res.* **2012**, *20*, 95–110. [CrossRef]
39. Kawai, A.; Nishida-Umehara, C.; Ishijima, J.; Tsuda, Y.; Ota, H.; Matsuda, Y. Different origins of bird and reptile sex chromosomes inferred from comparative mapping of chicken Z-linked genes. *Cytogenet. Genome Res.* **2007**, *117*, 92–102. [CrossRef]
40. Sharma, G.; Kaur, P.; Nakhasi, U. Female heterogamety in the Indian cryptodiran chelonian, Kachuga smithi Gray. In *Dr BS Chuahah Commemoration Volume*; Zoological Society of India: Orissa, India, 1975; pp. 359–368.
41. Mazzoleni, S.; Augstenova, B.; Clemente, L.; Auer, M.; Fritz, U.; Praschag, P.; Protiva, T.; Velensky, P.; Kratochvil, L.; Rovatsos, M. Turtles of the genera *Geoemyda* and *Pangshura* (Testudines: Geoemydidae) lack differentiated sex chromosomes: The end of a 40-year error cascade for Pangshura. *PeerJ* **2019**, *7*, e6241. [CrossRef]
42. Valenzuela, N.; Badenhorst, D.; Montiel, E.E.; Literman, R. Molecular cytogenetic search for cryptic sex chromosomes in painted turtles Chrysemys picta. *Cytogenet. Genome Res.* **2014**, *144*, 39–46. [CrossRef]
43. Mu, Y.; Zhao, B.; Tang, W.Q.; Sun, B.J.; Zeng, Z.G.; Valenzuela, N.; Du, W.G. Temperature-dependent sex determination ruled out in the Chinese soft-shelled turtle (*Pelodiscus sinensis*) via molecular cytogenetics and incubation experiments across populations. *Sex. Dev.* **2015**, *9*, 111–117. [CrossRef]
44. Shine, R.; Elphick, M.J.; Donnellan, S. Co-occurrence of multiple, supposedly incompatible modes of sex determination in a lizard population. *Ecol. Lett.* **2002**, *5*, 486–489. [CrossRef]
45. Quinn, A.E.; Georges, A.; Sarre, S.D.; Guarino, F.; Ezaz, T.; Graves, J.A. Temperature sex reversal implies sex gene dosage in a reptile. *Science* **2007**, *316*, 411. [CrossRef] [PubMed]
46. Radder, R.S.; Quinn, A.E.; Georges, A.; Sarre, S.D.; Shine, R. Genetic evidence for co-occurrence of chromosomal and thermal sex-determining systems in a lizard. *Biol. Lett.* **2008**, *4*, 176–178. [CrossRef] [PubMed]
47. Quinn, A.E.; Radder, R.S.; Sarre, S.D.; Georges, A.; Ezaz, T.; Shine, R. Isolation and development of a molecular sex marker for *Bassiana duperreyi*, a lizard with XX/XY sex chromosomes and temperature-induced sex reversal. *Mol. Genet. Genom.* **2009**, *281*, 665–672. [CrossRef] [PubMed]
48. Holleley, C.E.; O'Meally, D.; Sarre, S.D.; Marshall Graves, J.A.; Ezaz, T.; Matsubara, K.; Azad, B.; Zhang, X.; Georges, A. Sex reversal triggers the rapid transition from genetic to temperature-dependent sex. *Nature* **2015**, *523*, 79–82. [CrossRef] [PubMed]
49. Matsubara, K.; O'Meally, D.; Azad, B.; Georges, A.; Sarre, S.D.; Graves, J.A.; Matsuda, Y.; Ezaz, T. Amplification of microsatellite repeat motifs is associated with the evolutionary differentiation and heterochromatinization of sex chromosomes in Sauropsida. *Chromosoma* **2016**, *125*, 111–123. [CrossRef]
50. Sites, J.W., Jr.; Bickham, J.W.; Haiduk, M.W. Derived X chromosome in the turtle genus *Staurotypus*. *Science* **1979**, *206*, 1410–1412. [CrossRef]
51. Hirst, C.E.; Major, A.T.; Ayers, K.L.; Brown, R.J.; Mariette, M.; Sackton, T.B.; Smith, C.A. Sex Reversal and Comparative Data Undermine the W Chromosome and Support Z-linked DMRT1 as the Regulator of Gonadal Sex Differentiation in Birds. *Endocrinology* **2017**, *158*, 2970–2987. [CrossRef]
52. Rovatsos, M.; Vukic, J.; Kratochvil, L. Mammalian X homolog acts as sex chromosome in lacertid lizards. *Heredity* **2016**, *117*, 8–13. [CrossRef]
53. Brelsford, A.; Stock, M.; Betto-Colliard, C.; Dubey, S.; Dufresnes, C.; Jourdan-Pineau, H.; Rodrigues, N.; Savary, R.; Sermier, R.; Perrin, N. Homologous sex chromosomes in three deeply divergent anuran species. *Evolution* **2013**, *67*, 2434–2440. [CrossRef]
54. O'Meally, D.; Ezaz, T.; Georges, A.; Sarre, S.D.; Graves, J.A. Are some chromosomes particularly good at sex? Insights from amniotes. *Chromosome Res.* **2012**, *20*, 7–19. [CrossRef]
55. Ezaz, T.; Srikulnath, K.; Graves, J.A. Origin of Amniote Sex Chromosomes: An Ancestral Super-Sex Chromosome, or Common Requirements? *J. Hered.* **2017**, *108*, 94–105. [CrossRef] [PubMed]
56. Vicoso, B.; Emerson, J.J.; Zektser, Y.; Mahajan, S.; Bachtrog, D. Comparative sex chromosome genomics in snakes: Differentiation, evolutionary strata, and lack of global dosage compensation. *PLoS Biol.* **2013**, *11*, e1001643. [CrossRef] [PubMed]

57. Gamble, T.; Castoe, T.A.; Nielsen, S.V.; Banks, J.L.; Card, D.C.; Schield, D.R.; Schuett, G.W.; Booth, W. The Discovery of XY Sex Chromosomes in a Boa and Python. *Curr. Biol.* **2017**, *27*, 2148–2153 e2144. [CrossRef] [PubMed]
58. Alam, S.M.I.; Sarre, S.D.; Gleeson, D.; Georges, A.; Ezaz, T. Did Lizards Follow Unique Pathways in Sex Chromosome Evolution? *Genes* **2018**, *9*, 239. [CrossRef] [PubMed]
59. Harewood, L.; Fraser, P. The impact of chromosomal rearrangements on regulation of gene expression. *Hum. Mol. Genet.* **2014**, *23*, R76–R82. [CrossRef] [PubMed]
60. Charlesworth, D. Evolution of recombination rates between sex chromosomes. *Philos. Trans. R. Soc. Lond. B Biol. Sci.* **2017**, *372*, 20160456. [CrossRef]
61. Lee, L.; Montiel, E.E.; Navarro-Dominguez, B.M.; Valenzuela, N. Chromosomal Rearrangements during Turtle Evolution Altered the Synteny of Genes Involved in Vertebrate Sex Determination. *Cytogenet. Genome Res.* **2019**, *157*, 77–88. [CrossRef]
62. Valenzuela, N. Evolution and maintenance of temperature-dependent sex determination. In *Temperature-Dependent Sex Determination in Vertebrates*; Valenzuela, N., Lance, V.A., Eds.; Smithsonian Books: Washington, DC, USA, 2004; Volume 131, pp. 131–147.
63. Charnov, E.L.; Bull, J. When is sex environmentally determined? *Nature* **1977**, *266*, 828–830. [CrossRef]
64. Ohno, S. *Sex Chromosomes and Sex-Linked Genes*; Monographs on Endocrinology; Springer: Berlin, Germany, 1967; Volume 1.
65. Bull, J.J. *Evolution of Sex Determining Mechanisms*; The Benjamin/Cummings Publishing Company, Inc.: Menlo Park, CA, USA, 1983.
66. Neuwald, J.L.; Valenzuela, N. The lesser known challenge of climate change: Thermal variance and sex-reversal in vertebrates with temperature-dependent sex determination. *PLoS ONE* **2011**, *6*, e18117. [CrossRef]
67. Valenzuela, N.; Literman, R.; Neuwald, J.L.; Mizoguchi, B.; Iverson, J.B.; Riley, J.L.; Litzgus, J.D. Extreme thermal fluctuations from climate change unexpectedly accelerate demographic collapse of vertebrates with temperature-dependent sex determination. *Sci. Rep.* **2019**, *9*, 4254. [CrossRef]
68. Capel, B. Vertebrate sex determination: Evolutionary plasticity of a fundamental switch. *Nat. Rev. Genet.* **2017**, *18*, 675–689. [CrossRef] [PubMed]
69. Valenzuela, N. Causes and Consequences of Evolutionary Transitions in the Level of Phenotypic Plasticity of Reptilian Sex Determination. In *Transitions Between Sexual Systems*; Springer Nature Switzerland AG: Cham, Switzerland, 2018; Volume 60, pp. 345–363.
70. Schafer, A.J.; Goodfellow, P.N. Sex determination in humans. *Bioessays* **1996**, *18*, 955–963. [CrossRef] [PubMed]
71. Smith, C.A.; Roeszler, K.N.; Ohnesorg, T.; Cummins, D.M.; Farlie, P.G.; Doran, T.J.; Sinclair, A.H. The avian Z-linked gene DMRT1 is required for male sex determination in the chicken. *Nature* **2009**, *461*, 267–271. [CrossRef] [PubMed]
72. Valenzuela, N.; LeClere, A.; Shikano, T. Comparative gene expression of steroidogenic factor 1 in *Chrysemys picta* and *Apalone mutica* turtles with temperature-dependent and genotypic sex determination. *Evol. Dev.* **2006**, *8*, 424–432. [CrossRef] [PubMed]
73. Valenzuela, N.; Neuwald, J.L.; Literman, R. Transcriptional evolution underlying vertebrate sexual development. *Dev. Dyn.* **2013**, *242*, 307–319. [CrossRef] [PubMed]
74. Czerwinski, M.; Natarajan, A.; Barske, L.; Looger, L.L.; Capel, B. A timecourse analysis of systemic and gonadal effects of temperature on sexual development of the red-eared slider turtle *Trachemys scripta elegans*. *Dev. Biol.* **2016**, *420*, 166–177. [CrossRef]
75. Schroeder, A.L.; Metzger, K.J.; Miller, A.; Rhen, T. A Novel Candidate Gene for Temperature-Dependent Sex Determination in the Common Snapping Turtle. *Genetics* **2016**, *203*, 557–571. [CrossRef]
76. Radhakrishnan, S.; Valenzuela, N. Chromosomal Context Affects the Molecular Evolution of Sex-linked Genes and Their Autosomal Counterparts in Turtles and Other Vertebrates. *J. Hered.* **2017**, *108*, 720–730. [CrossRef]
77. Radhakrishnan, S.; Literman, R.; Neuwald, J.L.; Valenzuela, N. Thermal Response of Epigenetic Genes Informs Turtle Sex Determination with and without Sex Chromosomes. *Sex. Dev.* **2018**, *12*, 308–319. [CrossRef]

78. Kawagoshi, T.; Uno, Y.; Nishida, C.; Matsuda, Y. The *Staurotypus* turtles and aves share the same origin of sex chromosomes but evolved different types of heterogametic sex determination. *PLoS ONE* **2014**, *9*, e105315. [CrossRef]
79. Janes, D.E.; Organ, C.L.; Stiglec, R.; O'Meally, D.; Sarre, S.D.; Georges, A.; Graves, J.A.; Valenzuela, N.; Literman, R.A.; Rutherford, K.; et al. Molecular evolution of Dmrt1 accompanies change of sex-determining mechanisms in reptilia. *Biol. Lett.* **2014**, *10*, 20140809. [CrossRef] [PubMed]
80. Mizoguchi, B.; Valenzuela, N. Alternative splicing and thermosensitive expression of Dmrt1 during urogenital development in the painted turtle, *Chrysemys picta*. *PeerJ* **2020**, *8*, e8639. [CrossRef] [PubMed]
81. Sun, W.; Cai, H.; Zhang, G.; Zhang, H.; Bao, H.; Wang, L.; Ye, J.; Qian, G.; Ge, C. Dmrt1 is required for primary male sexual differentiation in Chinese soft-shelled turtle *Pelodiscus sinensis*. *Sci. Rep.* **2017**, *7*, 4433. [CrossRef] [PubMed]
82. Valenzuela, N. Relic thermosensitive gene expression in a turtle with genotypic sex determination. *Evolution* **2008**, *62*, 234–240. [CrossRef] [PubMed]
83. Valenzuela, N. Evolution of the gene network underlying gonadogenesis in turtles with temperature-dependent and genotypic sex determination. *Integr. Comp. Biol.* **2008**, *48*, 476–485. [CrossRef] [PubMed]
84. Rhen, T.; Metzger, K.; Schroeder, A.; Woodward, R. Expression of putative sex-determining genes during the thermosensitive period of gonad development in the snapping turtle, *Chelydra serpentina*. *Sex. Dev.* **2007**, *1*, 255–270. [CrossRef]
85. Valenzuela, N.; Shikano, T. Embryological ontogeny of aromatase gene expression in *Chrysemys picta* and *Apalone mutica* turtles: Comparative patterns within and across temperature-dependent and genotypic sex-determining mechanisms. *Dev. Genes Evol.* **2007**, *217*, 55–62. [CrossRef]
86. Valenzuela, N. Multivariate expression analysis of the gene network underlying sexual development in turtle embryos with temperature-dependent and genotypic sex determination. *Sex. Dev.* **2010**, *4*, 39–49. [CrossRef]
87. Ge, C.; Ye, J.; Zhang, H.; Zhang, Y.; Sun, W.; Sang, Y.; Capel, B.; Qian, G. Dmrt1 induces the male pathway in a turtle species with temperature-dependent sex determination. *Development* **2017**, *144*, 2222–2233. [CrossRef]
88. Tang, W.Q.; Mu, Y.; Valenzuela, N.; Du, W.G. Effects of Incubation Temperature on the Expression of Sex-Related Genes in the Chinese Pond Turtle, *Mauremys reevesii*. *Sex. Dev.* **2017**, *11*, 307–319. [CrossRef]
89. Ge, C.; Ye, J.; Weber, C.; Sun, W.; Zhang, H.; Zhou, Y.; Cai, C.; Qian, G.; Capel, B. The histone demethylase KDM6B regulates temperature-dependent sex determination in a turtle species. *Science* **2018**, *360*, 645–648. [CrossRef] [PubMed]
90. Rhen, T.; Schroeder, A. Molecular mechanisms of sex determination in reptiles. *Sex. Dev.* **2010**, *4*, 16–28. [CrossRef] [PubMed]
91. Merchant-Larios, H.; Diaz-Hernandez, V. Environmental sex determination mechanisms in reptiles. *Sex. Dev.* **2013**, *7*, 95–103. [CrossRef] [PubMed]
92. Radhakrishnan, S.; Literman, R.; Mizoguchi, B.; Valenzuela, N. MeDIP-seq and nCpG analyses illuminate sexually dimorphic methylation of gonadal development genes with high historic methylation in turtle hatchlings with temperature-dependent sex determination. *Epigenet. Chromatin* **2017**, *10*, 28. [CrossRef] [PubMed]
93. Radhakrishnan, S.; Literman, R.; Neuwald, J.; Severin, A.; Valenzuela, N. Transcriptomic responses to environmental temperature by turtles with temperature-dependent and genotypic sex determination assessed by RNAseq inform the genetic architecture of embryonic gonadal development. *PLoS ONE* **2017**, *12*, e0172044. [CrossRef] [PubMed]
94. Matsumoto, Y.; Buemio, A.; Chu, R.; Vafaee, M.; Crews, D. Epigenetic control of gonadal aromatase (cyp19a1) in temperature-dependent sex determination of red-eared slider turtles. *PLoS ONE* **2013**, *8*, e63599. [CrossRef]
95. Matsumoto, Y.; Hannigan, B.; Crews, D. Temperature Shift Alters DNA Methylation and Histone Modification Patterns in Gonadal Aromatase (cyp19a1) Gene in Species with Temperature-Dependent Sex Determination. *PLoS ONE* **2016**, *11*, e0167362. [CrossRef]
96. Venegas, D.; Marmolejo-Valencia, A.; Valdes-Quezada, C.; Govenzensky, T.; Recillas-Targa, F.; Merchant-Larios, H. Dimorphic DNA methylation during temperature-dependent sex determination in the sea turtle *Lepidochelys olivacea*. *Gen. Comp. Endocrinol.* **2016**, *236*, 35–41. [CrossRef]
97. Charlesworth, B. The evolution of chromosomal sex determination and dosage compensation. *Curr. Biol.* **1996**, *6*, 149–162. [CrossRef]

98. Mank, J.E. Small but mighty: The evolutionary dynamics of W and Y sex chromosomes. *Chromosome Res.* **2012**, *20*, 21–33. [CrossRef]
99. Graves, J.A.M. The rise and fall of SRY. *Trends Genet.* **2002**, *18*, 259–264. [CrossRef]
100. Handley, L.J.; Ceplitis, H.; Ellegren, H. Evolutionary strata on the chicken Z chromosome: Implications for sex chromosome evolution. *Genetics* **2004**, *167*, 367–376. [CrossRef] [PubMed]
101. Chandler, C.H. When and why does sex chromosome dosage compensation evolve? *Ann. N. Y. Acad. Sci.* **2017**, *1389*, 37–51. [CrossRef] [PubMed]
102. Gu, L.; Walters, J.R. Evolution of Sex Chromosome Dosage Compensation in Animals: A Beautiful Theory, Undermined by Facts and Bedeviled by Details. *Genome Biol. Evol.* **2017**, *9*, 2461–2476. [CrossRef]
103. Mank, J.E. Sex chromosome dosage compensation: Definitely not for everyone. *Trends Genet.* **2013**, *29*, 677–683. [CrossRef]
104. Rupp, S.M.; Webster, T.H.; Olney, K.C.; Hutchins, E.D.; Kusumi, K.; Wilson Sayres, M.A. Evolution of Dosage Compensation in Anolis carolinensis, a Reptile with XX/XY Chromosomal Sex Determination. *Genome Biol. Evol.* **2017**, *9*, 231–240. [CrossRef]
105. Rovatsos, M.; Rehak, I.; Velensky, P.; Kratochvil, L. Shared Ancient Sex Chromosomes in Varanids, Beaded Lizards, and Alligator Lizards. *Mol. Biol. Evol.* **2019**, *36*, 1113–1120. [CrossRef]
106. Charlesworth, B.; Coyne, J.A.; Barton, N.H. The relative rates of evolution of sex chromosomes and autosomes. *Am. Nat.* **1987**, *130*, 113–146. [CrossRef]
107. Mank, J.E.; Axelsson, E.; Ellegren, H. Fast-X on the Z: Rapid evolution of sex-linked genes in birds. *Genome Res.* **2007**, *17*, 618–624. [CrossRef]
108. Mank, J.E.; Nam, K.; Ellegren, H. Faster-Z evolution is predominantly due to genetic drift. *Mol. Biol. Evol.* **2010**, *27*, 661–670. [CrossRef]
109. Grath, S.; Parsch, J. Rate of Amino Acid Substitution Is Influenced by the Degree and Conservation of Male-Biased Transcription Over 50 Myr of *Drosophila* Evolution. *Genome Biol. Evol.* **2012**, *4*, 346–359. [CrossRef] [PubMed]
110. Kousathanas, A.; Halligan, D.L.; Keightley, P.D. Faster-X adaptive protein evolution in house mice. *Genetics* **2014**, *196*, 1131–1143. [CrossRef] [PubMed]
111. Wright, A.E.; Harrison, P.W.; Zimmer, F.; Montgomery, S.H.; Pointer, M.A.; Mank, J.E. Variation in promiscuity and sexual selection drives avian rate of Faster-Z evolution. *Mol. Ecol.* **2015**, *24*, 1218–1235. [CrossRef] [PubMed]
112. Lu, J.; Wu, C.I. Weak selection revealed by the whole-genome comparison of the X chromosome and autosomes of human and chimpanzee. *Proc. Natl. Acad. Sci. USA* **2005**, *102*, 4063–4067. [CrossRef]

4

Meiotic Chromosome Contacts as a Plausible Prelude for Robertsonian Translocations

Sergey Matveevsky [1,*], Oxana Kolomiets [1], Aleksey Bogdanov [2], Elena Alpeeva [2] and Irina Bakloushinskaya [2]

[1] Vavilov Institute of General Genetics, Russian Academy of Sciences, 119991 Moscow, Russia; olkolomiets@mail.ru
[2] Koltzov Institute of Developmental Biology, Russian Academy of Sciences, 119334 Moscow, Russia; bogdalst@yahoo.com (A.B.); alpeeva_l@mail.ru (E.A.); irina.bakl@gmail.com (I.B.)
* Correspondence: sergey8585@mail.ru

Abstract: Robertsonian translocations are common chromosomal alterations. Chromosome variability affects human health and natural evolution. Despite the significance of such mutations, no mechanisms explaining the emergence of such translocations have yet been demonstrated. Several models have explored possible changes in interphase nuclei. Evidence for non-homologous chromosomes end joining in meiosis is scarce, and is often limited to uncovering mechanisms in damaged cells only. This study presents a primarily qualitative analysis of contacts of non-homologous chromosomes by short arms, during meiotic prophase I in the mole vole, *Ellobius alaicus,* a species with a variable karyotype, due to Robertsonian translocations. Immunocytochemical staining of spermatocytes demonstrated the presence of four contact types for non-homologous chromosomes in meiotic prophase I: (1) proximity, (2) touching, (3) anchoring/tethering, and (4) fusion. Our results suggest distinct mechanisms for chromosomal interactions in meiosis. Thus, we propose to change the translocation mechanism model from 'contact first' to 'contact first in meiosis'.

Keywords: *Ellobius alaicus*; translocation; non-homologous chromosome connections; meiosis; synaptonemal complex

1. Introduction

Chromosomal stability, number and positioning are essential factors for correct genome functionality and inheritance. At the end of 19th century, Rabl hypothesized the non-random, three-dimensional location of chromosomes in the nucleus, whereas in the last two decades, these observations have been advanced with data from advanced technological methods, including fluorescence in situ hybridization (FISH), immunocytochemistry and others [1–3]. Information on the tissue-specific positioning of chromosomes has revealed functional nuclear regulation [4,5], or altered states in cancer cells [6,7]. Chromosomal changes at the individual development are usually highlighted as catastrophic genomic events, while such chromosomal alterations often result in carcinogenesis and infertility [8,9]. Genomes of carcinogenetic cells are highly dynamic, and in some cases, chromosomal changes, either induced or spontaneous, lead to therapeutic resistance [10]. Chromoanagenesis, encompasses chromothripsis, chromoanasynthesis and chromoplexy, and was recently described as a possible mechanism contributing to chromosomal evolution [11–13]. Even though a rearranged genome may suffer maladaptive modifications, some variations are beneficial for organisms or species in terms of advantageous natural selection. Karyotypic diversity, structural variations of autosomes and sex chromosomes exemplify the importance of chromosomal changes throughout evolution [14–19].

Distinct drivers for chromosomal change have been identified, e.g., mobile Penelope elements implicated in *Drosophila* speciation [20], LINE-1 elements in mammals [21], and noncoding RNAs [22]

etc. These factors destabilize genomes, initiate DNA damage, and provoke the abnormal linking of chromosomes. At least two translocation models based on non-random chromosome distribution in the nucleus have been proposed [23].

An initial step in "breakage-first" models are double-strand breaks (DSBs). Then potential partners occasionally tie up and produce changed chromosomes, which can obtain distinct fragments of non-homologous chromosomes, up to the whole arm, e.g., Robertsonian translocations (Rbs) or Whole-Arm Reciprocal Translocations (WARTs) [23]. Translocation probability rates could be higher if chromosomes were located close to each other in the nuclear space [24]. For intermingling chromosomes, DSBs in contact zones can lead to non-homologous linking and translocations [25,26]. In this context, the evolutionary integrative breakage model [27] stressed determining the genomic distribution of evolutionary breakpoints due to particular DNA sequence composition and the nucleome, combined with alteration of gene expression due to genome reshuffling.

In the 'contact-first' models, chromosomes should be broken, but DSBs start in colocalized chromatids inside specific protein complexes [28]. Firstly, chromosomes come together, then undergo DSBs and join with other partners. Chromosome region mobility is different during the cell cycle; it is higher in the early G1 phase, but decreases in the S phase. In an interphase nucleus, chromatin status, positioning, and cytoskeleton mechanical forces, all influence chromosome movement [29]. The combined variant of both models was also proposed. If DSBs occur in G1 and are not repaired until G2, they may cluster and form translocations [30].

In all models, the fate of small acrocentric chromosomal arms is uncertain; but most probably, they are eliminated. It is important to highlight these models were developed for interphase nuclei when chromosomes were decondensed and occupied specific chromosomal territories. Moreover, the tissue-specific positioning of chromosomes in interphase nuclei [31], enables the formation of distinctive carcinogenic translocations [8], which better fit the first model.

When we investigate the altered three-dimensional organization of somatic cells with re-arranged chromosomes, we cannot immediately determine the evolutionary consequences. How will such translocations pass to the next generations? Therefore, we must look for genome rearrangements in the germline. De novo chromosome rearrangements can arise during germ cell proliferation, meiosis, and in haploid sperm or eggs [32]. Any rearrangement provoking genomic instability may be beneficial for diversification and genetic speciation.

A specific fusion between acrocentric chromosomes, ended by metacentric chromosomes was first described by Robertson [33], and later named Robertsonian translocations (Rbs). The frequency of such translocations in humans is high: approximately 1 in 1000 individuals [34]. The most common rob(13q14q) and rob(14q21q) translocations originate during oogenesis [35]. A breakpoint diversity exemplified distinct ways for Rb formation, the significant input of the pre-meiotic replication and proper meiotic recombination [36,37]. Probable mechanisms for the formation of Rbs involving telomere changes were suggested [38]. One of the potential mechanisms may be a loss of p-arm telomeres when chromosome breakage occurs within minor satellite sequences [39,40]; another way is a fusion without any losses, and inactivation of telomeres [41]; the one more way may operate via the deletion/inactivation of the telomerase RNA gene which induces telomere shortening [42]. Data on meiosis are scarce and are limited to non-homologous end joining (NHEJ) in mouse spermatocytes after gamma radiation [43], however telocentric chromosome associations in the pachytene are observed for *Mus domesticus* [44].

The high frequency of translocations in humans and the evolutionary input of Rbs requires exploration of non-model species to reveal origin and maintenance mechanisms. Several mammalian species have demonstrated natural variability's in chromosomal numbers, including Rbs. *Mus, Sorex, Ellobius* species, and some others exhibit changes in diploid numbers, along with stable fundamental numbers due to whole branch fusions [45–48]. Recently, using chromosome painting, we described karyotype structures in three cryptic *Ellobius* species, *E. talpinus*, *E. tancrei*, and *E. alaicus*; we demonstrated a homology of re-arranged chromosomes, and showed the existence of XX sex chromosomes in males and females [49–51]. We hypothesized that a neocentromere origin in one pair of chromosomes was an initial disturbance event for the *E. tancrei* (2n = 54–30) genome, in contrast to stable *E. talpinus* (2n = 54) [49,52]. *E. alaicus* is very close to *E. tancrei*, a translocation Rb(2.11) emerged in both species; other Rbs are species-specific ones. *E. alaicus* demonstrates rapid chromosomal changes in nature, such as the fixation of Rbs in the large population in the Pamir-Alay [50]. In this study, we analyzed the meiotic sustainability of species with rapidly evolving genome.

2. Materials and Methods

2.1. Material and Mitotic Chromosomes

We used samples of 6 specimens of *E. alaicus*, kept in the cytogenetic collection (a part of the Joint collection of wildlife tissues for fundamental, applied and environmental researches of the Koltzov Institute of Developmental Biology RAS, Core Centrum of the Koltzov Institute of Developmental Biology RAS, state registration number 6868145). Two males and two females (collection numbers; 27353, 27357, 27354, 27356) were karyotyped using bone marrow suspensions [53]. Tissue from another male (27532) and female (27351) were also used to derive somatic cell cultures. C-band staining of mitotic metaphase plates was performed according to Sumner [54].

We followed international, national, and institutional guidelines for animal care. Studies were approved by the Ethics Committee for Animal Research of the Koltzov Institute of Developmental Biology RAS and the Vavilov Institute of General Genetics RAS.

2.2. Cell Culture

Chondrocyte and fibroblast cell lines were obtained from the Cell culture collection of the Koltzov Institute of Developmental Biology RAS. Chondrocytes were cryopreserved after the first subcultivation and fibroblasts after the third subcultivation and were stored at −196 °C in liquid nitrogen using cultivation media with 10% DMSO for cryopreservation. Karyotype evaluation was performed after the cells were recovered after freezing.

2.3. Meiotic Chromosome Studies and Immunostaining

Samples from three *E. alaicus* (27352, 27353 and 27357) adult males were used for the meiotic study. Synaptonemal complex (SC) preparations were made and fixed according to Peters et al. [55], with some modifications [56].

Primary antibodies used for immunostaining: rabbit anti-synaptonemal complex protein 3 (SYCP3) antibody (diluted 1:250, Abcam, Cambridge, UK); human anti-centromere Calcinosis Raynaud's phenomenon, Esophageal dysmotility, Sclerodactyly, and Telangiectasia (CREST) antibody (CREST, 1:250, Fitzgerald Industries International, USA); mouse anti-phospho-histone H2AX (diluted 1:250–500, Abcam) (also known as γH2AFX); rabbit anti-H3K9me3 antibody (1:100, Abcam; kindly provided by Dr. Jesus Page).

As secondary antibodies we used goat anti-rabbit IgG, Alexa Fluor 488-conjugate (Invitrogen, Carlsbad, CA, USA); goat anti-human IgG, Alexa Fluor 546-conjugate (Invitrogen); goat anti-mouse IgG, Alexa Fluor 546-conjugate (Invitrogen, USA) (diluted 1:250–500). Slides were washed in phosphate-buffered saline (PBS) and placed into Vectashield, with 4′,6-diamidino-2-phenylindole (DAPI) (Vector Laboratories, USA). Slides were analyzed using a fluorescence light microscope, Axio Imager D1 (Carl Zeiss, Jena, Germany). Immunostaining was described previously [57,58]. We immunostained H3K9me3 histones using two approaches; 1) the first round of SYCP3 staining, then a second round—H3K9me3 or 2) the first round of H3K9me3, then a second round of SYCP3 staining.

SCs measurements in 41 spermatocytes were performed using the MicroMeasure program (Colorado State University, CO, USA).

3. Results

3.1. Mitotic Chromosomes and SC Karyotype of E. alaicus

All animals demonstrated normal for *E. alaicus* karyotypes; 2n = 52. the fundamental number of chromosome arms (NF) was 56 (Figure S1a–f), consisting of one pair of submetacentrics (№7), characteristic of *E. tancrei* and *E. alaicus*, and one pair of large Robertsonian metacentrics 2(Rb2.11), typical of *E. alaicus* [50]. Mitotic metaphases from bone marrow suspensions showed no visible alterations (Figure S1b). Mitotic metaphases from fibroblast cultures demonstrated a stable karyotype in one male (Figure S1a), and a small number of deviations in one female, i.e., associations and polyploid cells. In chondrocyte cultures of the same specimens, more associations and polyploid cells were identified in female cells (Figure S1d–f), alongside with a large number of micronuclei, disturbed anaphases and massive chromosomal changes (Figures S1g and S2).

A total of 302 spermatocytes at different prophase I stages in three males were analyzed. As expected, in the pachytene stage we revealed 25 fully synapsed autosomal bivalents (large Robertsonian metacentric Rb(2.11)), one mid-size submetacentric №7 with neocentromere [49,52]; 23 acrocentrics and an XX sex bivalent (Figure 1), formed by two large acrocentrics. Analyzing large cell numbers made it possible to distinguish a range of gradually decreasing in size acrocentrics.

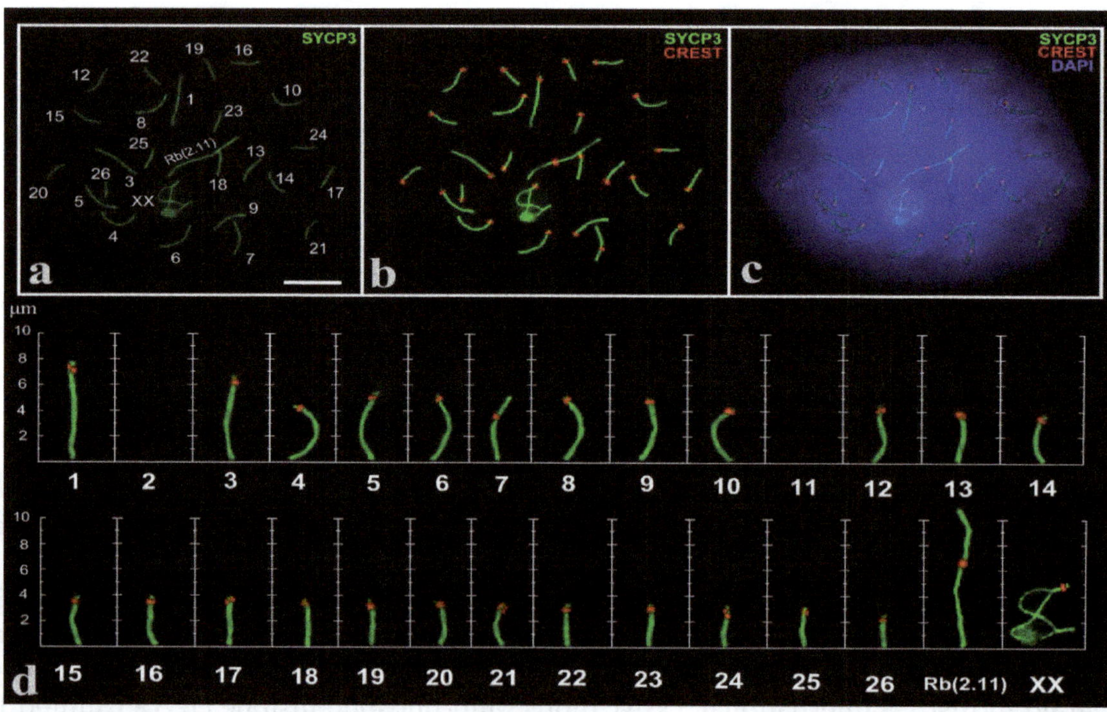

Figure 1. Pachytene spermatocyte (**a**–**c**) and synaptonemal complex (SC) karyotype (**d**) of the Alay mole vole, *E. alaicus*. SCs were immunostained with antibodies against synaptonemal complex protein 3 (SYCP3) (green) and centromeres—with antibodies to kinetochores (Calcinosis Raynaud's phenomenon, Esophageal dysmotility, Sclerodactyly, and Telangiectasia (CREST), red). The 4′,6-diamidino-2-phenylindole (DAPI)-stained the chromatin (blue). SC numbers corresponded to the metaphase karyotype. Rb(2.11)—Robertsonian submetacentric. Chromosome №7 is non-Robertsonian submetacentric. XX—male sex chromosomes. Bar (**a**–**c**) = 5 μm.

3.2. Prophase I Stages in E. alaicus

All prophase I stages were demonstrated for *E. alaicus*. Thin short SYCP3 fragments and SYCP3 conglomerates were formed in the leptotene stage (Figure 2a). In the early zygotene stage, long axial elements were visible, and SYCP3 blocks were kept (Figure 2b). As the zygotene progressed, the axial elements of homologous chromosomes began to contact each other (Figure S3), and synapsed more frequently from telomere areas to the central part (Figure 2c), less often in the opposite way (Figure 2d). In the pachytene stage, chromosomes were completely synapsed (Figure 2f). In the diplotene stage, chromosome desynapsis might be different within a single cell, from telomere sites to the center, or vice versa (Figure 2g). Then SYCP3 degraded (Figure 2h), and it was preserved as dots in the diakinesis stage (Figure 2i).

Sex (XX) chromosomes from early to mid zygotene were detected as separate axial elements (Figure S3). From the mid-late zygote, the XX chromosomes became well visible (Figure 2c). The sex bivalent is similar to those of the cryptic species, *E. talpinus* [59] and *E. tancrei* [60,61]. In the middle pachytene, the XX-bivalent was usually shifted to the periphery of the meiotic nucleus and had two telomeric synaptic regions, a wide asynaptic region, and chromatin bodies on axial elements (Figure 2f). It should be noted that sometimes chromatin (nucleolus-like) bodies in sex bivalents were SYCP3-positive (Figure 1, Figure 2e,f), which has not been previously observed for the other two cryptic species and *E. alaicus* (2n = 48) from the Pamir–Alay [50].

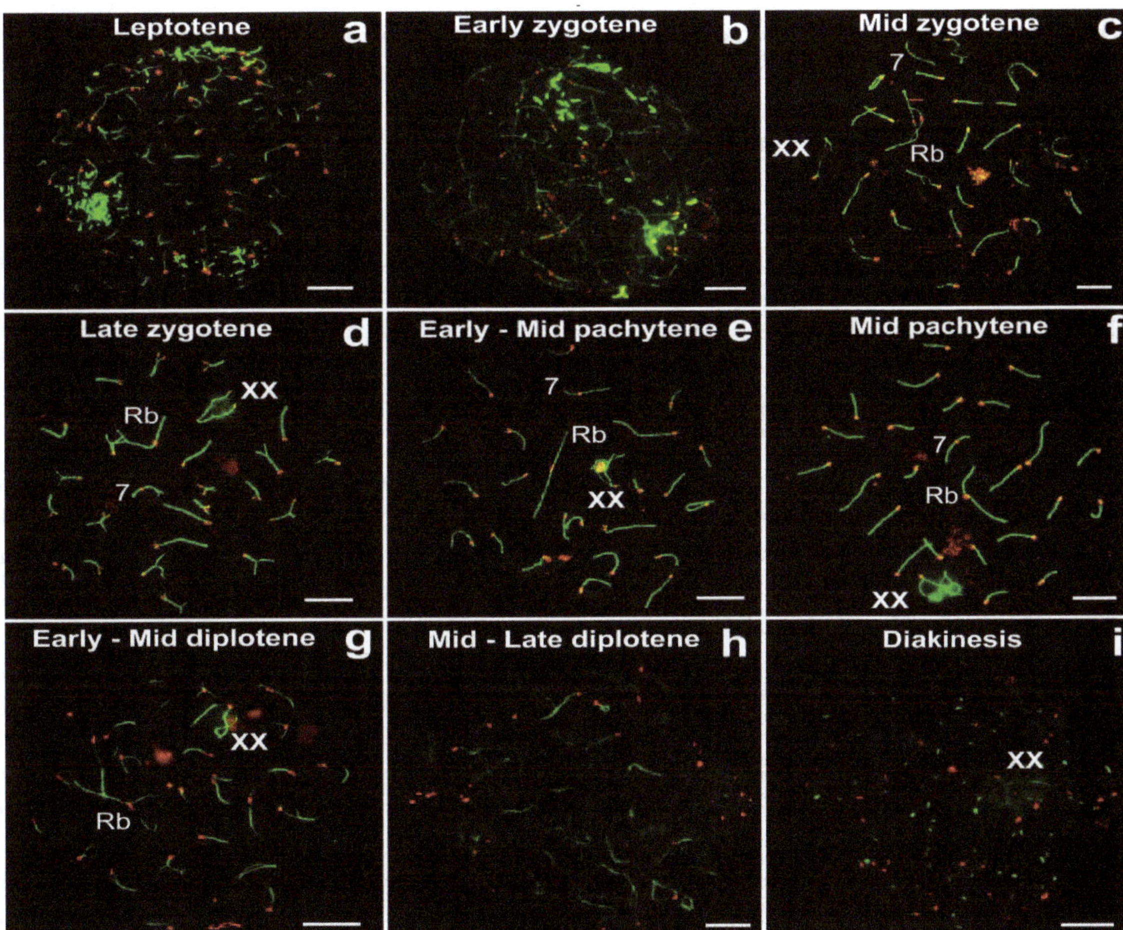

Figure 2. Prophase I stages in *E. alaicus* males. Axial/lateral elements were identified using anti-SYCP3 antibodies (green) and kinetochores (red) using CREST. During the leptotene stage, numerous thin SYCP3 fragments and SYCP3 conglomerates were formed (**a**), formed axial elements (**b**) were visible in the zygotene. Chromosome synapsis began with either distal (telomeric) (**c**) or central chromosome parts (**d**). At the pachytene stage, 25 autosomal SCs and sex (XX) bivalents were formed (**e**,**f**). The XX bivalent demonstrated synapsis at telomeric areas and asynapsis at the central region (**e**,**f**). In the mid-late diplotene, desynapsis progressed from both the central and telomeric areas (**g**). SYCP3 was degraded (**h**) and retained as dots (**i**). Rb—Robertsonian submetacentric. Chromosome 7 is non-Robertsonian submetacentric. Bar (**a**–**i**) = 5 µm.

3.3. *Types of Non-Homologous Acrocentric Connections*

Various non-homologous chromosome contacts were identified in meiotic prophase I, during zygotene-early pachytene stages. We revealed at least four sequential contact and link types of chromosomes (Figure 3, Figure 4, Figure 5, Figure 6, Figure 7).

3.3.1. Proximity

Acrocentrics occupied spatial positions, intending to get their centromeres closer (pink squares and pink numbers of chromosomes in Figure 3, Figure 7, Figures S4 and S5). The distance between the chromosomes was about 1–2 microns. Centromere regions of chromosomes were located around the H3K9me3-domain (Figure 6 and Figures S5–S7). This type precedes true meiotic contacts (see the following types). This 'proximity' type was somewhat subjective, because it can easily be confused with closely located chromosomes.

3.3.2. Touching

Acrocentrics moved closer to each other. In each contact acrocentric, one axial element of the short arm was extended, reaching out to one another as if touching each other with their ends (blue squares and blue numbers of chromosomes in Figure 3, Figure 5, Figure 7, Figures S4 and S5). Usually, the distance between the chromosomes is less than 1 micron.

3.3.3. Anchoring/Tethering

One of the axial elements in the short arms of non-homologous acrocentrics were linked to each other by SYCP3-filament (yellow squares and yellow numbers of chromosomes in Figure 3, Figure 4, Figure 5, Figure 7, Figures S4 and S5). Other axial elements of the short arms of the two non-homologous partners were not connected (see blue arrowheads in Figures 4 and 5c).

3.3.4. Fusion

The other two axial elements in the short arm of the non-homologous acrocentrics were tightly adjacent to each other, or possibly connected entirely. The two acrocentrics likely represented a single bivalent with two centromeres (Figure 4e,f, Figure 5c). Centromeric regions were closer to each other compared to other connection types. Such dicentric bivalents were observed in some low-chromosomal forms in *E. tancrei* (Figure 7 and Figure S8).

Such linkage was evidenced by immunostaining; two thin SYCP3-positive filaments (Figure 4e,f). The difference between 'anchoring/tethering' (see filaments between chromosomes 5 and 6 in Figure S4e) and 'fusion' (Figure 4e,f; see chromosomes 14 and 18 in Figure S4e) can be determined by the presence of a complete bridge, with a short distance between two contacting non-homologous chromosomes. In all cases, the contact was made by the short arms of acrocentric chromosomes by SYCP3-filaments (see enlarged fragments of cells; Figure 4).

51% of pachytene cells had meiotic contacts including 'touching', 'anchoring/tethering' and 'fusion' types and excluding 'proximity' (Figure 5a). The patterns of proportion of different connection types were similar in three males: the number of 'touching' was greater than the number of 'anchoring/tethering', the number of 'fusion' was less than two other types (Figure 5b). The 'fusion' type was not found in one male, №27353 (Figure 5b). The rarity of 'fusion' type may be due to the fact that more molecular events should precede the joining/fusion of the axial elements of the short arms of two non-homologous acrocentrics.

Chromosome combinations in pachytene spermatocytes of *E. alaicus* were numerous. We were unable to determine the trend in the frequencies of contact between certain chromosomes. However, chromosomes 1,4,5,20–22 came into contact more often. For example, chromosome №1, which was the most regularly seen in all three types of true interactions ('touching', 'anchoring/tethering', and 'fusion'). Rb(1.3) was described in *E. alaicus* with 2n = 50 [50].

3.4. Histone H3K9me3 in Prophase I and Meiotic Chromosomes Contacts

We investigated H3K9me3 (trimethylation of H3 lysine 9) immunolocalization and distribution in contacting chromosomes. H3K9me3 is an epigenetic marker for heterochromatin allocation in prophase I [62].

During the zygotene stage, large clouds of H3K9me3 were localized at pericentromeric regions of axial elements, or at fully formed SCs (Figure S6a–c). In the pachytene stage, clouds of H3K9me3 were reduced in size and were clearly localized to SC pericentromeric regions (Figure 6a–c, Figure S6d–i). Rb-metacentrics had minimal H3K9me3 levels in the centromeric region (Figure 6a–c, Figure S6d–i), or were absent. Non-Robertsonian submetacentric №7 did not demonstrate clear H3K9me3 signals in all studied cells (Figures S6e–h and S7a–c). H3K9me3 usually shrouded one of the axial elements, and a chromatin body inside the sex bivalent (XX) (Figure 6a–d); less often H3K9me3 covered it entirely (Figure S5g,h), whilst γH2AFX totally enclosed XX (Figures 6c and 3b,e). A more detailed description of XX sex chromosomes in *E. alaicus* will be discussed in future work.

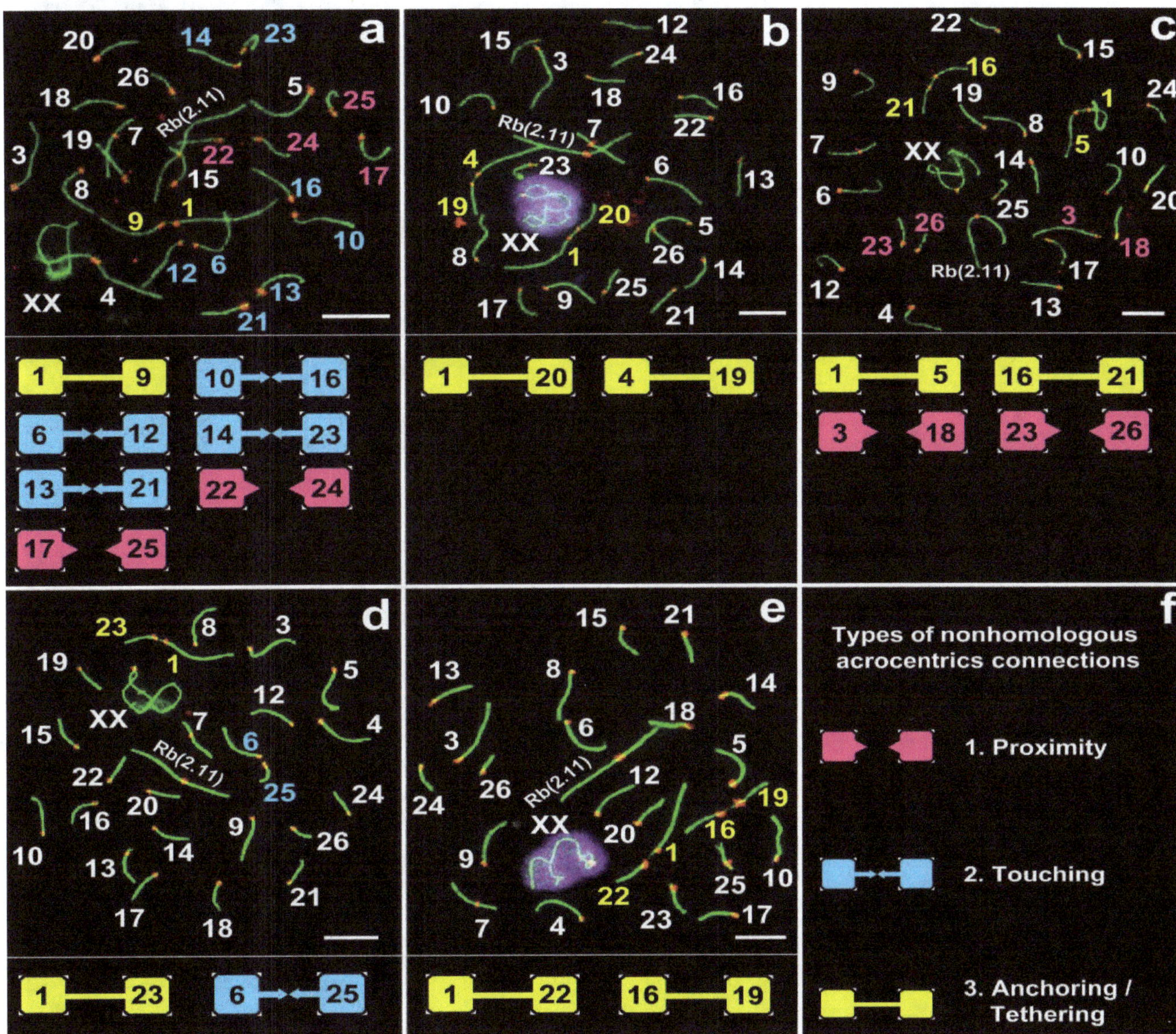

Figure 3. Different chromosome combinations in pachytene spermatocytes of *E. alaicus* (**a**–**f**). Axial elements were identified using anti-SYCP3 antibodies (green), kinetochores using CREST antibodies (red), and anti-γH2AFX (violet) was used as a marker of chromatin inactivation. Pink, blue, and yellow squares correspond to different types of non-homologous acrocentric connections (**f**). Bar (**a**–**e**) = 5 μm.

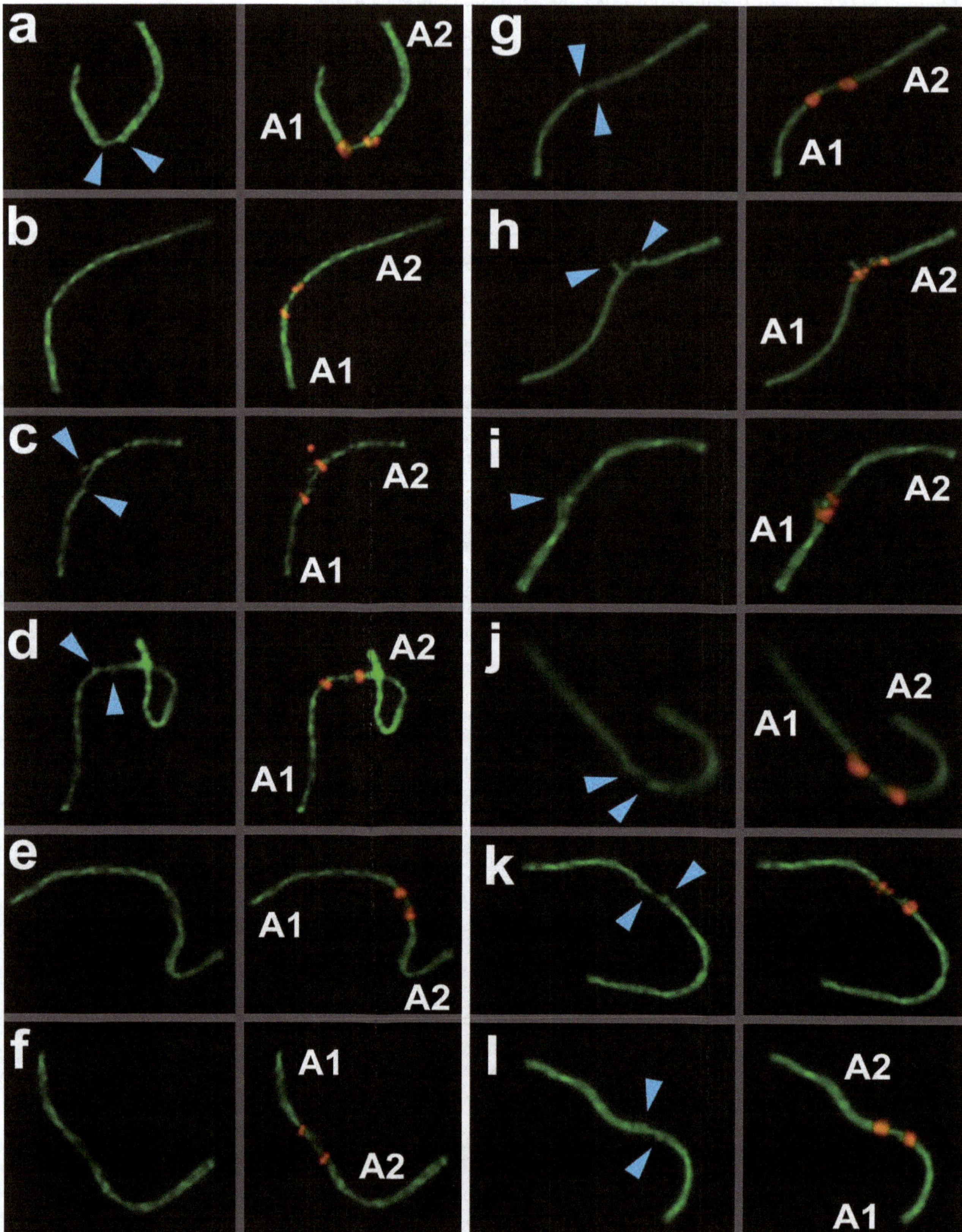

Figure 4. Anchoring/tethering and fusion of non-homologous acrocentric chromosomes in *E. alaicus* spermatocytes (**a**–**l**). Lateral elements were identified using anti-SYCP3 antibodies (green), and CREST antibodies for kinetochores (red). A1 and A2—contacting non-homologous acrocentrics. Blue arrowheads show the free ends of lateral elements of non-homologous acrocentrics.

In non-homologous acrocentric contacts, H3K9me3 was involved in varying volumes, usually in centromeric regions. We detected a large H3K9me3 cloud between chromosomes at the 'proximity' type (pinks points in Figure 6a,c,e–g,i). For closer chromosome contacts, i.e., 'touching' and 'anchoring/tethering', H3K9me3 distribution ranged from average to insignificant (small) (yellow points in Figure 6a,d,e,g,i). If contacts involved shorter and longer acrocentrics, H3K9me3 signals were located closer to centromeric regions of shorter acrocentrics (Figure 6g,i). For other cases, H3K9me3 signals were clearly identified between the two centromeres of contacting chromosomes (Figure 6d,h).

3.5. Mitotic Cells

In contrast to meiotic cells demonstrating 'touching', 'anchoring', and 'fusion' of different acrocentrics during meiotic prophase I (Figure 5), in mitotic metaphases, we detected a single association in most cells, which were similar to the 'fusion' type. In total, we checked 693 cells: 532 cells had normal karyotype, 32 cells demonstrated chromosomal associations (5.6%), and 126 were aberrant ones (micronuclei, polyploid cells, disturbed anaphases, chromothripsis, etc.). C-banding exposed small blocks of pericentromeric heterochromatin and several intercalary blocks (for two pairs of acrocentrics) (Figure S1c). Small blocks of C-heterochromatin were visible in associated chromosomes, but such blocks were not always merged.

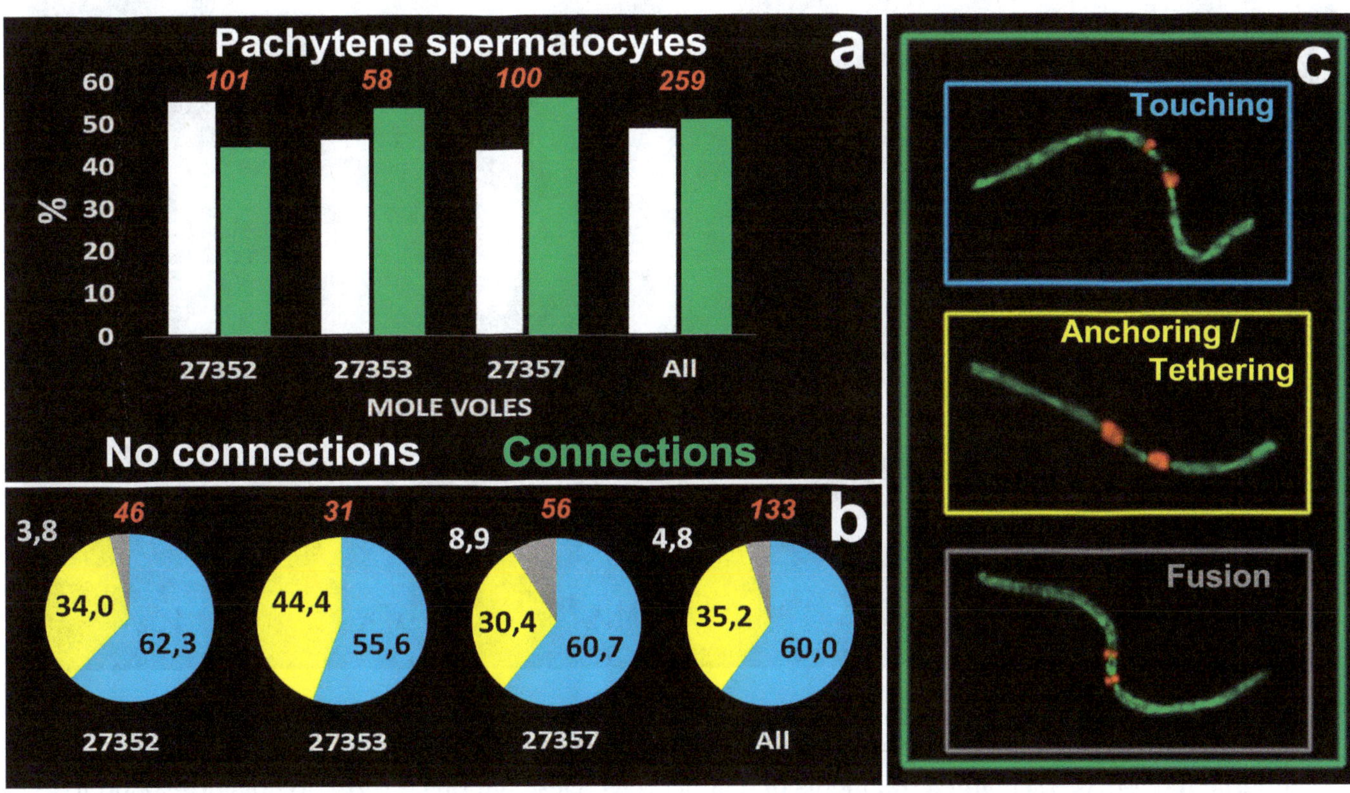

Figure 5. Quantitative characteristics of meiotic contacts in *E. alaicus*. (**a**). The diagram shows the percentage of pachytene cells with or without meiotic connections in three mole vole males (№27352, 27353, 27357). 'Touching', 'anchoring/tethering' and 'fusion' types were counted in cells with connections only. (**b**). Proportions of connection types. Blue color corresponds to 'touching', yellow—to 'anchoring/tethering' and gray—to "fusion". Red numbers are counted pachytene spermatocytes (**a**,**b**). (**c**). Types of contacts that are included in the diagrams. The color of the frames corresponds to the types of contacts (**b**).

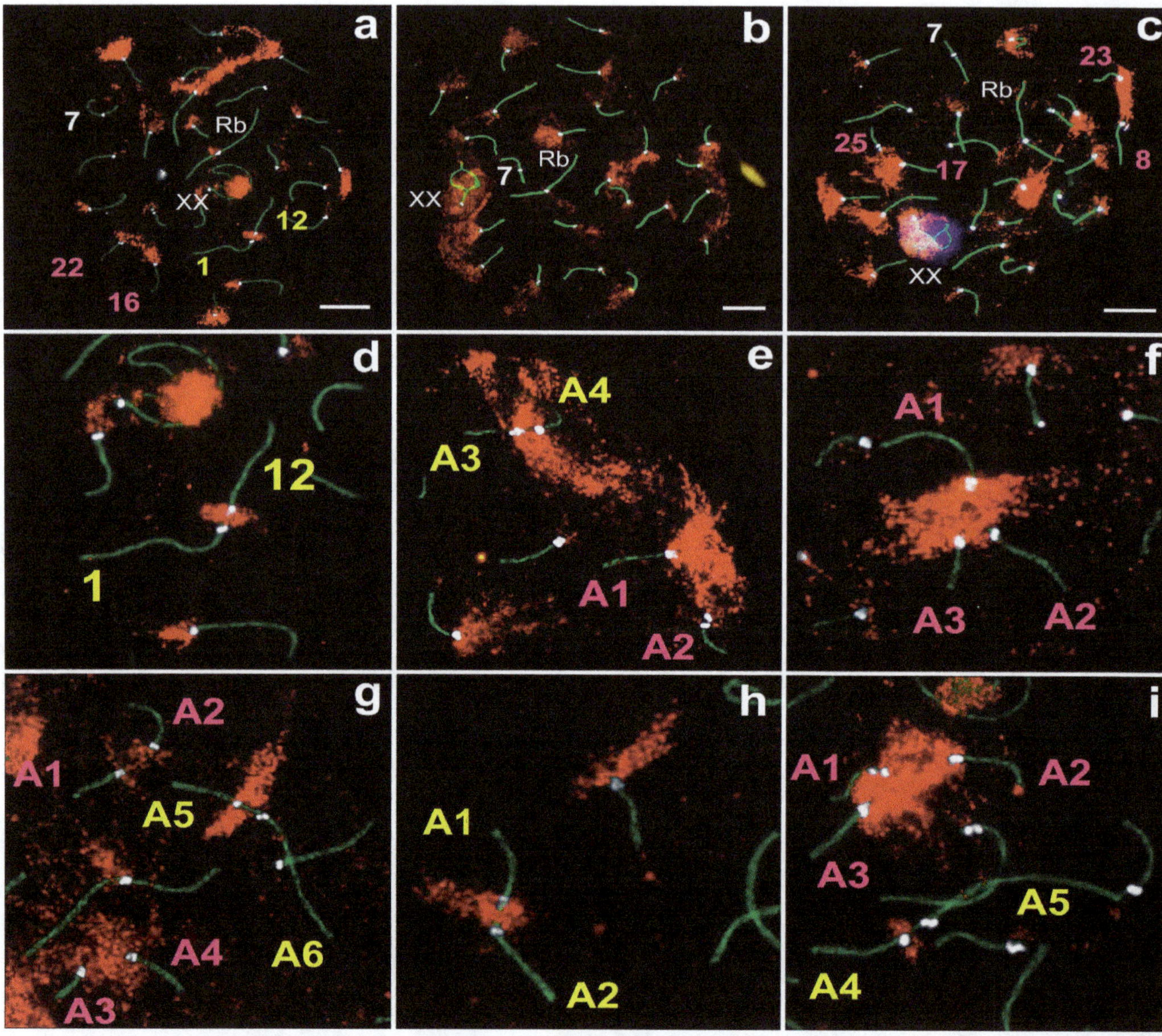

Figure 6. Location of SYCP3 (green), H3K9me3 (red), CREST (white) and gamma-H2AFX (violet) in the pachytene stage of prophase I of *E. alaicus* spermatocytes (**a**–**i**). Size bar = 5 μm (**a**–**c**). (**d**) an enlarged fragment of (**a**), (**e**–**i**) enlarged fragments of distinct cells. Designations A1–A6 do not correspond to numbers of chromosomes as in Figure 1. See the explanation in the text.

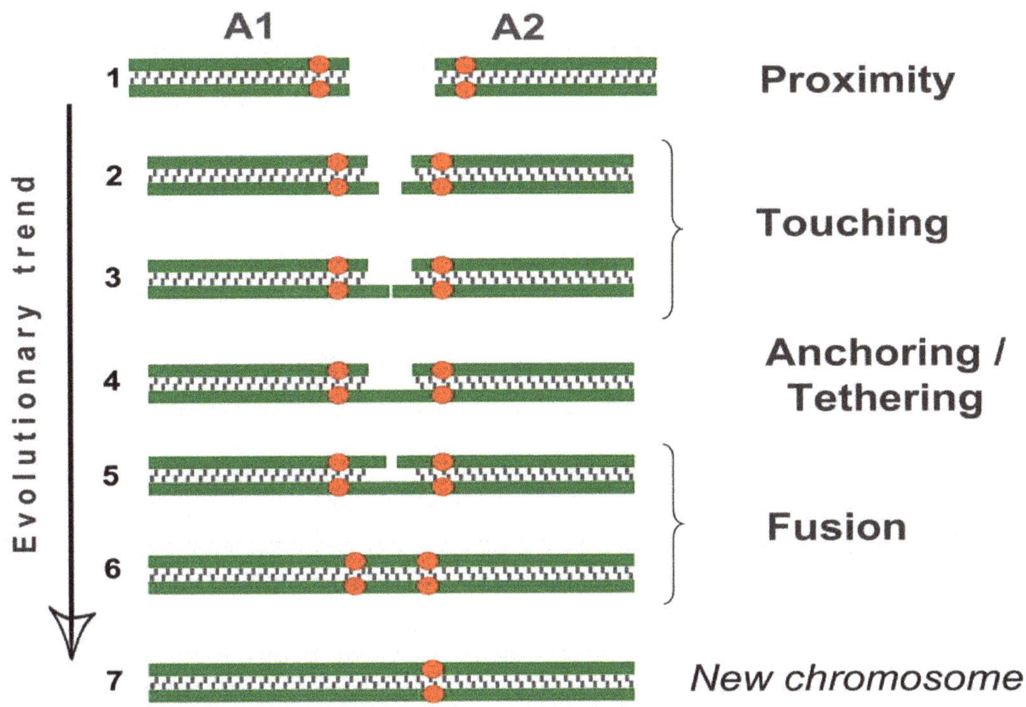

Figure 7. Different types of chromosome connections in meiotic prophase I. The scheme is based on observations of chromosome behaviors during *E. alaicus* meiosis. Green corresponds to SC axial/lateral elements, red dots refer to centromeres. Non-homologous acrocentrics (A1 and A2) occupy positions in the nucleus, intending to get closer (proximity, position 1). One of two lateral elements in the short arm A1 and A2 are extended (position 2), touching each other (position 3) (touching) and eventually bind together (anchoring/tethering, position 4). The other two axial elements are tightened to each other. (Position 5) and probably joined together (position 6) (fusion). After the probable binding of the two axial elements, the centromeres are located somewhat closer to each other (compare positions 5 and 6). Such dicentric bivalents were found in all oocytes of low-chromosomal forms of *E. tancrei*, 2n = 34 (Figure S8). In such dicentric chromosomes, one centromere may be inactivated, with a new chromosome emerging (position 7).

4. Discussion

In this work, we describe the variety of chromosomal interactions in male meiosis of *E. alaicus*. Immunocytochemical staining of spermatocytes demonstrated at least four contact types for non-homologous acrocentrics in meiotic prophase I. Starting from clustering inside the heterochromatic cloud, chromosomes demonstrate touching by the elongated axial element of the short arms, then tether by SYCP3 filaments and complete the process when tightly adjacent to each other. As a result, two acrocentrics likely represents a single bivalent with two centromeric regions.

Recently, the first report on connections between non-homologous chromosomes and SC structures was demonstrated by investigating meiosis in CD-1 male mice after irradiation [43]. The authors of this work stressed that the formation of chromosome bridges between non-homologous chromosomes differed from normal endogenous DSB interactions, which did not require connections between axial elements of homologous chromosomes. The wild ancestors of CD-1 mice were bred in a Swiss laboratory in the 1920s. Genomic studies revealed that CD-1 was mostly derived from *M. domesticus* [63], a species with enormous chromosome variability [64]. In this species, specific major and minor satDNA tandem repeats, which are oriented head-to-tail at centromeres [65], may facilitate fusion of mono-armed chromosomes, and build up bi-armed ones. Repeat polarity occurs in telocentric and Rb chromosomes, therefore it was assumed that tandem repetitive satDNA in *M. domesticus* may have been a universal pattern of chromosomal evolution. Numerous Rbs, characteristic for the species, and presumed random associations of bivalents in the wild type, *M. domesticus* (2n = 40) spermatocytes, exemplified a wide spectrum of fusions due to the universal structure of pericentromeric satDNA in this species.

We did not analyze pericentromeric heterochromatin in *Ellobius*, although the numerous associations (at 'proximity' type) were similar to *M. domesticus* [43]. Pericentromeric heterochromatin evidently participates in chromosome contacts and fusions. A prelude at the zygotene (Figure S3) may be assigned as an example. Formation of Rbs may depend on telomere changes, especially shortening and inactivation [38]. Recently, the study of wild house mice demonstrated that telomere shortening and the number of critically short telomeres are likely to result in Rb formation [66]. Earlier, we did not detect the telomeric sequences in centromeric regions of numerous Rbs, including Rb(2.11), in *E. tancrei* and *E. talpinus* hybrids [49]. The lack of signal may be explained by elimination as well as inactivation of telomere fragments. The study should be continued in *E. alaicus*.

Contrary to mouse experiments [43], where chromosome contacts were discovered after gamma radiation, *E. alaicus* came from intact natural habitats, close to terra typica of the species. Numerous and various chromosomal contacts, the enormous number of cells with such chromosome 'dance' in all studied males demonstrated an internal background for evolutionary changes. In natural *E. alaicus* populations, we detected [50] four different Rb translocations Rb(2.11), Rb(1.3), Rb(4.9), Rb(3.10) in distinct combinations, with 2n from 52 to 48. A small number of chromosomal associations in mitotic metaphases, which we demonstrated now for animals with a single pair of Rb(2.11), indirectly confirmed the leading role of changes in the meiotic prophase I. We revealed differences in mitotic cell divisions for fibroblast and chondrocyte cultures. Cell culture often demonstrates unstable karyotype structures and distinct abnormality types and frequencies [67,68]. This assumption was our rationale for comparing data on chromosome numbers and behaviors using bone marrow slides, and two different cell culture approaches. Fibroblasts mostly retained 2n = 52, and numerous polyploid cells. Chondrocyte cultures appeared to be more fragile and demonstrated distinct changes as micronuclei, disturbed anaphases, massive chromosome changes, and associations (Figures S1 and S2). The evolutionary significance of such mutations is incredible if such events appear in the germline [32,69]. The picture of chromosomal re-assemblage (Figure S1) may be an example of chromothripsis, which combines chromosome shattering and fusion [70].

In mitotic cells, we observed two centromeres and two blocks of heterochromatin in the association of two small acrocentrics (Figure S1c,e). Previously, in the pachytene, we observed a small metacentric with two centromeres in all studied animals from natural *E. tancrei* populations, 2n = 34 (as in Figure S8). *E. tancrei* demonstrates a wide spectrum of homologous and non-homologous Robertsonian metacentrics in natural populations [51]. We suggest that dicentric chromosome formation in natural populations of this species, confirms the 'contact first in meiosis' model.

Although we had no meiotic data to prove double-stranded DNA breaks and chromosome reassembling, we demonstrated that the 'contact-first' scheme may be applicable to meiotic translocation mechanisms. Another question—the fate of small arms of acrocentric chromosomes, which are apparently kept in fused chromosomes (Figure 4e,f and Figure 5c)—is closely related to the role and evolution of the centromere. As previously mentioned, we supposed the neocentromere origin was a crucial step for *E. tancrei* and *E. alaicus* genome evolution. In cases of 'fused' chromosomes of *E. alaicus*, we revealed two distinct centromere signals using CREST immunostaining. The question of the further fate of two centromeres and chromosome fragments between them is now open, and the study should be continued.

Considering that mitosis and meiosis data are consistent, we can argue for the essential role of heterochromatin. In mouse spermatocytes, heterochromatin preserves the association of homologous centromeres and promotes faithful chromosome segregation in meiosis I [71]. Heterochromatin, in maintaining genome stability [72], may be a focal point for changes. We demonstrated that a 'proximity' to fuse correlated with the heterochromatin affinity of non-homologous chromosomes. Chromosomes were located around the H3K9me3-heterochromatin cloud (Figure 6f,i), similar to telocentric associations in mice [44]. Clustering of heterochromatic regions inside the H3K9me3 clouds was demonstrated for different sets of acrocentrics in *Mus musculus domesticus* either for the wild type 2n = 40 or for different forms with Rbs [44,73,74]. This clustering could be the background of the translocations. However, no data on contacting chromosomes were published, except the gamma radiation treatment caused the formation of bridges between chromosomes [43]. Earlier, we described the participation of heterochromatin of short chromosome arms in the formation of SC chains in mole voles heterozygous for multiple Rbs [75].

High natural variability along with the rapid fixation of new translocations and specific contacts of non-homologous chromosomes in meiosis, which we demonstrated for *E. alaicus*, built a promising background to change the model for the mechanism of translocations from the 'contact first' to the 'contact first in meiosis'.

The evolutionary significance of meiosis as a tool to increase recombinational diversity is balanced by its function as a mechanism for purifying selection [76]. Meiotic checkpoints block cell cycle progression in response to defects and preclude abnormal chromosome segregation. Nevertheless, creative meiosis became apparent if we apply the 'contact-first' model for meiotic prophase I, as we demonstrated for a unique rodent species, which entered the stage of Robertsonian translocations emergence.

Supplementary Materials
Figure S1: Mitotic chromosomes of *Ellobius alaicus*, Figure S2: Aberrations in mitosis of *Ellobius alaicus*, Figure S3: Zygotene spermatocyte of *E.alaicus*, Figure S4: Different chromosome combinations in pachytene spermatocytes of *E.alaicus*, Figure S5: Different chromosome combinations in pachytene spermatocytes of *E.alaicus*, Figure S6: Location of SYCP3, H3K9me3, CREST and chromatin configuration (DAPI) in zygotene—pachytene stages of prophase I of *E. alaicus* spermatocytes, Figure S7: Location of SYCP3, H3K9me3, CREST, and chromatin configuration (DAPI) in pachytene—diakinesis stages of the prophase I of *E. alaicus* spermatocytes, Figure S8: A pachytene oocyte of *E. tancrei*, 2n = 34, NF = 56.

Author Contributions: Conceptualization, S.M., O.K., I.B.; methodology, O.K., I.B., S.M.; investigation, S.M., I.B., E.A. and A.B.; writing, I.B., S.M., E.A., A.B., and O.K.; visualization, S.M., and I.B.; funding acquisition, S.M., I.B.; supervision, O.K., and I.B. All the authors read and approved the final text of the submitted manuscript and the supporting information. All authors have read and agreed to the published version of the manuscript.

Acknowledgments: We thank the Genetic Polymorphisms Core Facility of the Vavilov Institute of General Genetics of the Russian Academy of Sciences, Moscow, for the possibility to use their microscopes. We are grateful to Core Centrum of the Koltzov Institute of Developmental Biology RAS for the possibility of using samples from the Cell culture collection and the Joint collection of wildlife tissues for fundamental, applied, and environmental researches.

References

1. Cremer, T.; Cremer, C. Chromosome territories, nuclear architecture and gene regulation in mammalian cells. *Nat Rev. Genet.* **2001**, *2*, 292–301. [CrossRef]
2. Cremer, T.; Cremer, C. Rise, fall and resurrection of chromosome territories: A historical perspective. Part I. The rise of chromosome territories. *Eur. J. Histochem.* **2006**, *50*, 161–176. [PubMed]
3. Cremer, T.; Cremer, M. Chromosome territories. *Cold Spring Harb. Perspect. Biol.* **2010**, *2*, a003889. [CrossRef]
4. Solovei, I.; Kreysing, M.; Lanctôt, C.; Kösem, S.; Peichl, L.; Cremer, T.; Guck, J.; Joffe, B. Nuclear architecture of rod photoreceptor cells adapts to vision in mammalian evolution. *Cell* **2009**, *137*, 356–368. [CrossRef] [PubMed]
5. Tan, L.; Xing, D.; Chang, C.H.; Li, H.; Xie, X.S. Three-dimensional genome structures of single diploid human cells. *Science* **2018**, *361*, 924–928. [CrossRef] [PubMed]
6. Boveri, T. Concerning the origin of malignant tumours by Theodor Boveri. Translated and annotated by Henry Harris. *J. Cell Sci.* **2008**, *121* (Suppl. 1), 1–84. [CrossRef] [PubMed]

7. Mai, S. The three-dimensional cancer nucleus. *Genes Chromosomes Cancer* **2019**, *58*, 462–473. [CrossRef] [PubMed]
8. Meaburn, K.J.; Misteli, T.; Soutoglou, E. Spatial genome organization in the formation of chromosomal translocations. *Semin. Cancer Biol.* **2007**, *17*, 80–90. [CrossRef]
9. Heyer, E.E.; Deveson, I.W.; Wooi, D.; Selinger, C.I.; Lyons, R.J.; Hayes, V.M.; O'Toole, S.A.; Ballinger, M.L.; Gill, D.; Thomas, D.M.; et al. Diagnosis of fusion genes using targeted RNA sequencing. *Nat. Commun.* **2019**, *10*, 1388. [CrossRef]
10. Salgueiro, L.; Buccitelli, C.; Rowald, K.; Somogyi, K.; Kandala, S.; Korbel, J.O.; Sotillo, R. Acquisition of chromosome instability is a mechanism to evade oncogene addiction. *EMBO Mol. Med.* **2020**, *12*, e10941. [CrossRef]
11. Maciejowski, J.; Li, Y.; Bosco, N.; Campbell, P.J.; de Lange, T. Chromothripsis and kataegis induced by telomere crisis. *Cell* **2015**, *163*, 1641–1654. [CrossRef] [PubMed]
12. Pellestor, F. Chromothripsis and the Macroevolution Theory. *Methods Mol. Biol.* **2018**, *1769*, 43–49. [PubMed]
13. Pellestor, F.; Gatinois, V. Chromoanagenesis: A piece of the macroevolution scenario. *Mol. Cytogenet.* **2020**, *13*, 3. [CrossRef] [PubMed]
14. Matthey, R. *Les Chromosomes des Vértebrés*; F. Rouge: Lausanne, Suisse, 1949.
15. White, M.J.D. *Modes of Speciation*; W.H. Freeman and Co.: New York, NY, USA, 1978.
16. King, M. *Species Evolution: The Role of Chromosome Change*; Cambridge University Press: Cambridge, UK, 1993.
17. Faria, R.; Navarro, A. Chromosomal speciation revisited: Rearranging theory with pieces of evidence. *Trends Ecol. Evol.* **2010**, *25*, 660–669. [CrossRef]
18. Ruiz-Herrera, A.; Farré, M.; Robinson, T.J. Molecular cytogenetic and genomic insights into chromosomal evolution. *Heredity* **2012**, *108*, 28–36. [CrossRef]
19. Graves, J.A. Did sex chromosome turnover promote divergence of the major mammal groups? De novo sex chromosomes and drastic rearrangements may have posed reproductive barriers between monotremes, marsupials and placental mammals. *Bioessays* **2016**, *38*, 734–743. [CrossRef]
20. Evgen'ev, M.B.; Zelentsova, H.; Poluectova, H.; Lyozin, G.T.; Veleikodvorskaja, V.; Pyatkov, K.I.; Zhivotovsky, L.A.; Kidwell, M.G. Mobile elements and chromosomal evolution in the virilis group of Drosophila. *Proc. Natl. Acad. Sci. USA* **2000**, *97*, 11337–11342. [CrossRef]
21. Rodriguez-Martin, B.; Alvarez, E.G.; Baez-Ortega, A.; Zamora, J.; Supek, F.; Demeulemeester, J.; Santamarina, M.; Ju, Y.S.; Temes, J.; Garcia-Souto, D.; et al. Pan-cancer analysis of whole genomes identifies driver rearrangements promoted by LINE-1 retrotransposition. *Nat. Genet.* **2020**, *52*, 306–319. [CrossRef]
22. Brown, J.D.; Mitchell, S.E.; O'Neill, R.J. Making a long story short: Noncoding RNAs and chromosome change. *Heredity* **2012**, *108*, 42–49. [CrossRef]
23. Aten, J.A.; Stap, J.; Krawczyk, P.M.; van Oven, C.H.; Hoebe, R.A.; Essers, J.; Kanaar, R. Dynamics of DNA double-strand breaks revealed by clustering of damaged chromosome domains. *Science* **2004**, *303*, 92–95. [CrossRef]
24. McCord, R.P.; Balajee, A. 3D genome organization influences the chromosome translocation pattern. In *Chromosome Translocation*; Springer: Singapore, 2018; pp. 113–133.
25. Branco, M.R.; Pombo, A. Intermingling of chromosome territories in interphase suggests role in translocations and transcription-dependent associations. *PLoS Biol.* **2006**, *4*, e138. [CrossRef] [PubMed]
26. Szczepińska, T.; Rusek, A.M.; Plewczynski, D. Intermingling of chromosome territories. *Genes Chromosomes Cancer* **2019**, *58*, 500–506. [CrossRef] [PubMed]
27. Farré, M.; Robinson, T.J.; Ruiz-Herrera, A. An Integrative Breakage Model of genome architecture, reshuffling and evolution: The Integrative Breakage Model of genome evolution, a novel multidisciplinary hypothesis for the study of genome plasticity. *BioEssays* **2015**, *37*, 479–488. [CrossRef] [PubMed]
28. Klein, I.A.; Resch, W.; Jankovic, M.; Oliveira, T.; Yamane, A.; Nakahashi, H.; Di Virgilio, M.; Bothmer, A.; Nussenzweig, A.; Robbiani, D.F.; et al. Translocation-capture sequencing reveals the extent and nature of chromosomal rearrangements in B lymphocytes. *Cell* **2011**, *147*, 95–106. [CrossRef]
29. Gothe, H.J.; Minneker, V.; Roukos, V. Dynamics of double-strand breaks: Implications for the formation of chromosome translocations. In *Chromosome Translocation*; Springer: Singapore, 2018; pp. 27–38.

30. Schrank, B.; Gautier, J. Assembling nuclear domains: Lessons from DNA repair. *J. Cell Biol.* **2019**, *218*, 2444–2455. [CrossRef]
31. Finn, E.H.; Misteli, T. Molecular basis and biological function of variability in spatial genome organization. *Science* **2019**, *365*, eaaw9498. [CrossRef]
32. Kim, S.; Peterson, S.E.; Jasin, M.; Keeney, S. Mechanisms of germ line genome instability. *Semin. Cell Dev. Biol.* **2016**, *54*, 177–187. [CrossRef]
33. Robertson, W.R.B. Chromosome studies. I. Taxonomic relationships shown in the chromosomes of Tettigidae and Acrididae: V-shaped chromosomes and their significance in Acrididae, Locustidae, and Gryllidae: Chromosomes and variation. *J. Morphol.* **1916**, *27*, 179–331. [CrossRef]
34. Hamerton, J.L.; Canning, N.; Ray, M.; Smith, S. A cytogenetic survey of 14,069 newborn infants: I. Incidence of chromosome abnormalities. *Clin. Genet.* **1975**, *8*, 223–243. [CrossRef]
35. Page, S.L.; Shatter, L.G. Nonhomologous Robertsonian translocations form predominantly during female meiosis. *Nat. Genet.* **1997**, *15*, 231–232. [CrossRef]
36. Page, S.L.; Shin, J.C.; Han, J.Y.; Andy Choo, K.H.; Shaffer, L.G. Breakpoint diversity illustrates distinct mechanisms for Robertsonian translocation formation. *Hum. Mol. Genet.* **1996**, *5*, 1279–1288. [CrossRef] [PubMed]
37. Richardson, C.; Moynahan, M.E.; Jasin, M. Double-strand break repair by interchromosomal recombination: Suppression of chromosomal translocations. *Genes Dev.* **1998**, *12*, 3831–3842. [CrossRef] [PubMed]
38. Slijepcevic, P. Telomeres and mechanisms of Robertsonian fusion. *Chromosoma* **1998**, *107*, 136–140. [CrossRef] [PubMed]
39. Garagna, S.; Broccoli, D.; Redi, C.A.; Searle, J.B.; Cooke, H.J.; Capanna, E. Robertsonian metacentrics of the mouse lose telomeric sequences but retain some minor satellite DNA in the pericentromeric area. *Chromosoma* **1995**, *103*, 685–692. [CrossRef] [PubMed]
40. Nanda, I.; Schneider-Rasp, S.; Winking, H.; Schmid, M. Loss of telomeric sites in the chromosomes of Mus musculus domesticus (Rodentia: Muridae) during Robertsonian rearrangements. *Chromosom. Res.* **1995**, *3*, 399–409. [CrossRef] [PubMed]
41. Meyne, J.; Baker, R.J.; Hobart, H.H.; Hsu, T.C.; Ryder, O.A.; Ward, O.G.; Wiley, J.E.; Wurster-Hill, D.H.; Yates, T.L.; Moyzis, R.K. Distribution of nontelomeric sites of (TTAGGG)n telomeric sequences in vertebrate chromosomes. *Chromosoma* **1990**, *99*, 3–10. [CrossRef]
42. Blasco, M.A.; Lee, H.; Hande, M.P.; Samper, E.; Lansdorp, P.M.; De Pinho, R.A.; Greider, C.W. Telomere shortening and tumor formation by mouse cells lacking telomerase RNA. *Cell* **1997**, *91*, 25–34. [CrossRef]
43. Enguita-Marruedo, A.; Martín-Ruiz, M.; García, E.; Gil-Fernández, A.; Parra, M.T.; Viera, A.; Rufas, J.S.; Page, J. Transition from a meiotic to a somatic-like DNA damage response during the pachytene stage in mouse meiosis. *PLoS Genet.* **2019**, *15*, e1007439. [CrossRef]
44. Berríos, S.; Manterola, M.; Prieto, Z.; López-Fenner, J.; Page, J.; Fernández-Donoso, R. Model of chromosome associations in *Mus domesticus* spermatocytes. *Biol. Res.* **2010**, *43*, 275–285. [CrossRef]
45. Garagna, S.; Page, J.; Fernandez-Donoso, R.; Zuccotti, M.; Searle, J.B. The Robertsonian phenomenon in the house mouse: Mutation, meiosis and speciation. *Chromosoma* **2014**, *123*, 529–544. [CrossRef]
46. Searle, J.B.; Polly, P.D.; Zima, J. (Eds.) *Shrews, Chromosomes and Speciation*; Cambridge University Press: Cambridge, UK, 2019; p. 488.
47. Lyapunova, E.A.; Vorontsov, N.N.; Korobitsina, K.V.; Ivanitskaya, E.Y.; Borisov, Y.M.; Yakimenko, L.V.; Dovgal, V.Y. A Robertsonian fan in *Ellobius talpinus*. *Genetica* **1980**, *52/53*, 239–247. [CrossRef]
48. Bakloushinskaya, I.; Matveevsky, S. Unusual ways to lost Y chromo-some and survive with changed autosomes: A story of mole voles Ellobius (Mammalia, Rodentia). *OBM Genet.* **2018**, *2*, 023. [CrossRef]
49. Bakloushinskaya, I.Y.; Matveevsky, S.N.; Romanenko, S.A.; Serdukova, N.A.; Kolomiets, O.L.; Spangenberg, V.E.; Lyapunova, E.A.; Graphodatsky, A.S. A comparative analysis of the mole vole sibling species *Ellobius tancrei* and *E. talpinus* (Cricetidae, Rodentia) through chromosome painting and examination of synaptonemal complex structures in hybrids. *Cytogenet. Genome Res.* **2012**, *136*, 199–207. [CrossRef] [PubMed]
50. Bakloushinskaya, I.; Lyapunova, E.A.; Saidov, A.S.; Romanenko, S.A.; O'Brien, P.C.; Serdyukova, N.A.; Ferguson-Smith, M.A.; Matveevsky, S.; Bogdanov, A.S. Rapid chromosomal evolution in enigmatic mammal with XX in both sexes, the Alay mole vole *Ellobius alaicus* Vorontsov et al., 1969 (Mammalia, Rodentia). *Comp. Cytogenet.* **2019**, *13*, 147. [CrossRef]

51. Romanenko, S.A.; Lyapunova, E.A.; Saidov, A.S.; O'Brien, P.; Serdyukova, N.A.; Ferguson-Smith, M.A.; Graphodatsky, A.S.; Bakloushinskaya, I. Chromosome Translocations as a Driver of Diversification in Mole Voles Ellobius (Rodentia, Mammalia). *IJMS* **2019**, *20*, 4466. [CrossRef]
52. Matveevsky, S.N. Signs of Sexual Dimorphism in Meiosis and Karyotype Variability of Mole Vole *Ellobius* (Rodentia, Mammalia). Ph.D. Thesis, NI Vavilov Institute of General Genetics of Russian Academy of Science, Moscow, Russia, 2011; pp. 1–172. (In Russian).
53. Ford, C.E.; Hamerton, J.L. A colchicine, hypotonic citrate, squash sequence for mammalian chromosomes. *Stain Technol.* **1956**, *31*, 247–251. [CrossRef]
54. Sumner, A.T. A simple technique for demonstrating centromeric heterochromatin. *Exp. Cell Res.* **1972**, *75*, 304–306. [CrossRef]
55. Peters, A.H.F.M.; Plug, A.W.; van Vugt, M.J.; de Boer, P. A drying-down technique for the spreading of mammalian meiocytes from the male and female germ line. *Chromosome Res.* **1997**, *5*, 66–71. [CrossRef]
56. Page, J.; Berríos, S.; Rufas, J.S.; Parra, M.T.; Suja, J.Á.; Heyting, C.; Fernández-Donoso, R. The pairing of X and Y chromosomes during meiotic prophase in the marsupial species *Thylamys elegans* is maintained by a dense plate developed from their axial elements. *J. Cell Sci.* **2003**, *116*, 551–560. [CrossRef]
57. Matveevsky, S.; Bakloushinskaya, I.; Tambovtseva, V.; Romanenko, S.; Kolomiets, O. Analysis of meiotic chromosome structure and behavior in Robertsonian heterozygotes of *Ellobius tancrei* (Rodentia, Cricetidae): A case of monobrachial homology. *Comp. Cytogenet.* **2015**, *9*, 691. [CrossRef]
58. Matveevsky, S.; Kolomiets, O.; Bogdanov, A.; Hakhverdyan, M.; Bakloushinskaya, I. Chromosomal evolution in mole voles *Ellobius* (Cricetidae, Rodentia): Bizarre sex chromosomes, variable autosomes and meiosis. *Genes* **2017**, *8*, 306. [CrossRef] [PubMed]
59. Kolomiets, O.L.; Matveevsky, S.N.; Bakloushinskaya, I.Y. Sexual dimorphism in prophase I of meiosis in the Northern mole vole (*Ellobius talpinus* Pallas, 1770) with isomorphic (XX) chromosomes in males and females. *Comp. Cytogenet.* **2010**, *4*, 55–66. [CrossRef]
60. Kolomiets, O.L.; Vorontsov, N.N.; Lyapunova, E.A.; Mazurova, T.F. Ultrastructure, meiotic behavior, and evolution of sex chromosomes of the genus *Ellobius*. *Genet.* **1991**, *84*, 179–189. [CrossRef]
61. Matveevsky, S.; Bakloushinskaya, I.; Kolomiets, O. Unique sex chromosome systems in *Ellobius*: How do male XX chromosomes recombine and undergo pachytene chromatin inactivation? *Sci. Rep.* **2016**, *6*, 29949. [CrossRef] [PubMed]
62. Nakayama, J.; Rice, J.C.; Strahl, B.D.; Allis, C.D.; Grewal, S.I. Role of histone H3 lysine 9 methylation in epigenetic control of heterochromatin assembly. *Science* **2001**, *292*, 110–113. [CrossRef] [PubMed]
63. Aldinger, K.A.; Sokoloff, G.; Rosenberg, D.M.; Palmer, A.A.; Millen, K.J. Genetic variation and population substructure in outbred CD-1 mice: Implications for genome-wide association studies. *PLoS ONE* **2009**, *4*, e4729. [CrossRef] [PubMed]
64. Castiglia, R. Sympatric sister species in rodents are more chromosomally differentiated than allopatric ones: Implications for the role of chromosomal rearrangements in speciation. *Mammal Rev.* **2014**, *44*, 1–4. [CrossRef]
65. Garagna, S.; Marziliano, N.; Zuccotti, M.; Searle, J.B.; Capanna, E.; Redi, C.A. Pericentromeric organization at the fusion point of mouse Robertsonian translocation chromosomes. *Proc. Natl. Acad. Sci. USA* **2001**, *98*, 171–175. [CrossRef]
66. Sánchez-Guillén, R.A.; Capilla, L.; Reig-Viader, R.; Martínez-Plana, M.; Pardo-Camacho, C.; Andrés-Nieto, M.; Ventura, J.; Ruiz-Herrera, A. On the origin of Robertsonian fusions in nature: Evidence of telomere shortening in wild house mice. *J. Evol. Biol.* **2015**, *28*, 241–249. [CrossRef]
67. Palumbo, E.; Russo, A. Common fragile site instability in normal cells: Lessons and perspectives. *Genes Chromosomes Cancer* **2019**, *58*, 260–269. [CrossRef]
68. Guadarrama-Ponce, R.; Aranda-Anzaldo, A. The epicenter of chromosomal fragility of Fra14A2, the mouse ortholog of human FRA3B common fragile site, is largely attached to the nuclear matrix in lymphocytes but not in other cell types that do not express such a fragility. *J. Cell. Biochem.* **2020**, *121*, 2209–2224. [CrossRef] [PubMed]
69. Fukami, M.; Shima, H.; Suzuki, E.; Ogata, T.; Matsubara, K.; Kamimaki, T. Catastrophic cellular events leading to complex chromosomal rearrangements in the germline. *Clin. Genet.* **2017**, *91*, 653–660. [CrossRef] [PubMed]

70. Stephens, P.J.; Greenman, C.D.; Fu, B.; Yang, F.; Bignell, G.R.; Mudie, L.J.; Pleasance, E.D.; Lau, K.W.; Beare, D.; Stebbings, L.A.; et al. Massive genomic rearrangement acquired in a single catastrophic event during cancer development. *Cell* **2011**, *144*, 27–40. [CrossRef] [PubMed]
71. Eyster, C.; Chuong, H.H.; Lee, C.Y.; Pezza, R.J.; Dawson, D. The pericentromeric heterochromatin of homologous chromosomes remains associated after centromere pairing dissolves in mouse spermatocyte meiosis. *Chromosoma* **2019**, *128*, 355–367. [CrossRef]
72. Janssen, A.; Colmenares, S.U.; Karpen, G.H. Heterochromatin: Guardian of the genome. *Annu. Rev. Cell Dev. Biol.* **2018**, *34*, 265–288. [CrossRef]
73. Manterola, M.; Page, J.; Vasco, C.; Berríos, S.; Parra, M.T.; Viera, A.; Rufas, J.S.; Zuccotti, M.; Garagna, S.; Fernández-Donoso, R. A high incidence of meiotic silencing of unsynapsed chromatin is not associated with substantial pachytene loss in heterozygous male mice carrying multiple simple robertsonian translocations. *PLoS Genet.* **2009**, *5*, e1000625. [CrossRef]
74. Capilla, L.; Medarde, N.; Alemany-Schmidt, A.; Oliver-Bonet, M.; Ventura, J.; Ruiz-Herrera, A. Genetic recombination variation in wild Robertsonian mice: On the role of chromosomal fusions and Prdm9 allelic background. *Proc. R. Soc. Lond. B Biol. Sci.* **2014**, *281*, 20140297. [CrossRef]
75. Kolomiets, O.L.; Lyapunova, E.A.; Mazurova, T.F.; Yanina, I.Y.; Bogdanov, Y.F. Participation of heterochromatin in formation of synaptonemal complex chains in animals heterozygous for multiple Robertsonian translocation. *Russ. J. Genet.* **1986**, *22*, 273–283. (In Russian)
76. Lenormand, T.; Engelstädter, J.; Johnston, S.E.; Wijnker, E.; Haag, C.R. Evolutionary mysteries in meiosis. *Phil. Trans. R. Soc. B* **2016**, *371*, 20160001. [CrossRef]

5

Chromosomics: Bridging the Gap between Genomes and Chromosomes

Janine E. Deakin [1,*], Sally Potter [2,3,†], Rachel O'Neill [4,†], Aurora Ruiz-Herrera [5,6,†], Marcelo B. Cioffi [7], Mark D.B. Eldridge [3], Kichi Fukui [8], Jennifer A. Marshall Graves [1,9], Darren Griffin [10], Frank Grutzner [11], Lukáš Kratochvíl [12], Ikuo Miura [13], Michail Rovatsos [11], Kornsorn Srikulnath [14], Erik Wapstra [15] and Tariq Ezaz [1,*]

1 Institute for Applied Ecology, University of Canberra, Canberra, ACT 2617, Australia
2 Research School of Biology, Australian National University, Acton, ACT 2601, Australia
3 Australian Museum Research Institute, Australian Museum, 1 William St Sydney, NSW 2010, Australia
4 Institute for Systems Genomics and Department of Molecular and Cell Biology, University of Connecticut, Storrs, CT 06269, USA
5 Departament de Biologia Cel·lular, Fisiologia i Immunologia, Universitat Autònoma de Barcelona, 08193 Cerdanyola del Vallès, Spain
6 Genome Integrity and Instability Group, Institut de Biotecnologia i Biomedicina, Universitat Autònoma de Barcelona, 08193 Cerdanyola del Vallès, Spain
7 Laboratório de Citogenética de Peixes, Departamento de Genética e Evolução, Universidade Federal de São Carlos, São Carlos, SP 13565-905, Brazil
8 Graduate School of Pharmaceutical Sciences, Osaka University, Suita 565-0871, Osaka, Japan
9 School of Life Sciences, LaTrobe University, Melbourne, VIC 3168, Australia
10 School of Biosciences, University of Kent, Canterbury CT2 7NJ, UK
11 School of Biological Sciences, The University of Adelaide, Adelaide, SA 5005, Australia
12 Department of Ecology, Faculty of Science, Charles University, Viničná 7, 128 44 Prague 2, Czech Republic
13 Amphibian Research Center, Hiroshima University, Higashi-Hiroshima 739-8526, Japan
14 Laboratory of Animal Cytogenetics & Comparative Genomics (ACCG), Department of Genetics, Faculty of Science, Kasetsart University, Bangkok 10900, Thailand
15 School of Natural Sciences, University of Tasmania, Hobart 7000, Australia
* Correspondence: Janine.Deakin@ecanberra.edu.au (J.E.D.); Tariq.Ezaz@canberra.edu.au (T.E.)
† These authors contributed equally to this work.

Abstract: The recent advances in DNA sequencing technology are enabling a rapid increase in the number of genomes being sequenced. However, many fundamental questions in genome biology remain unanswered, because sequence data alone is unable to provide insight into how the genome is organised into chromosomes, the position and interaction of those chromosomes in the cell, and how chromosomes and their interactions with each other change in response to environmental stimuli or over time. The intimate relationship between DNA sequence and chromosome structure and function highlights the need to integrate genomic and cytogenetic data to more comprehensively understand the role genome architecture plays in genome plasticity. We propose adoption of the term 'chromosomics' as an approach encompassing genome sequencing, cytogenetics and cell biology, and present examples of where chromosomics has already led to novel discoveries, such as the sex-determining gene in eutherian mammals. More importantly, we look to the future and the questions that could be answered as we enter into the chromosomics revolution, such as the role of chromosome rearrangements in speciation and the role more rapidly evolving regions of the genome, like centromeres, play in genome plasticity. However, for chromosomics to reach its full potential, we need to address several challenges, particularly the training of a new generation of cytogeneticists, and the commitment to a closer union among the research areas of genomics, cytogenetics, cell biology and bioinformatics. Overcoming these challenges will lead to ground-breaking discoveries in understanding genome evolution and function.

Keywords: cytogenetics; sex chromosomes; chromosome rearrangements; genome plasticity; centromere; genome biology; evolution

1. Introduction

Advances in technology have made sequencing the entire genome of an organism essentially routine. However, DNA sequence is only one relatively static component of the highly dynamic entities within the nucleus of a cell—chromosomes. Where a particular sequence is located on a chromosome and how it interacts with other parts of the genome are important aspects of genome biology often overlooked in genome sequencing projects. We propose a new framework for studying genome biology that integrates approaches in genome sequencing, cytogenetics and cell biology, as well as a renewed focus on training the next generation of genome biologists in the skills required for the integration of these data. We propose the adoption of the term 'chromosomics', which combines the original definition of cytogenetics (chromosomes and cytology) with genomics (gene content, structure and function for an entire organism), to ensure a closer integration of these fields. The term chromosomics was originally proposed by Uwe Claussen to introduce the branch of cytogenetics that deals with the three-dimensional structure of chromosomes and their associated gene regulation [1]. However, we propose that this term encompass the integration of the latest advances in cytogenetics, genome sequencing, epigenomics and cell biology. The adoption of a chromosomics approach to answering the big fundamental questions in biology will undoubtedly lead to major discoveries that were previously beyond reach.

In recent times, the field of genomics has largely distanced itself from cytogenetics, the field providing insight into chromosome structure, function and evolution. This separation of the fields has been to the detriment of a full understanding of how the genome works in the cell. These two fields were never intended to work in isolation. In 1920, Hans Winkler coined the term 'genome' to combine the study of genes and chromosomes [2], yet in modern interpretations of 'genome', chromosomes are often forgotten and the focus is solely on the DNA sequence. Similarly, Walter Sutton in 1902 (no published record) used the term 'cytogenetics' to combine cytology (the study of cell structure and function) with genetics (the study of genes, genetic variations and heredity). However, the cytological aspects of cytogenetics are largely ignored by most modern cytogenetic studies. As these respective fields have narrowed their focus, the result has been the development of technological and methodological advancements (examples in Table 1) that could allow us to more fully capture the dynamic nature and evolution of chromosomes from potentially any species to provide insight into fundamental biological questions.

Table 1. Recent Technological Advances in Cytogenetics and Genomics.

Technique	Uses	Example References
Chromosome microdissection and sequencing	Sequencing individual chromosomes/chromosome segments and haplotyping	[3–5]
Flow sorting of chromosomes and sequencing	Assigning sequences to chromosomes	[6,7]
Fibre-fluorescence in situ hybridisation (FISH)	Ordering of sequences large insert clones on a DNA fibre	[8]
Universal probe set and multiprobe slides	Rapid bacterial artificial chromosome (BAC) mapping across multiple species	[9]

Table 1. *Cont.*

Homologue-specific oligopaints	Visually distinguish single copy regions of homologous chromosomes	[10]
Super-resolution microscopy	Imaging of chromatin and nuclear organisation	[11]
BioNano	Genome mapping to improve assemblies and detect structural variations	[12]
Long-read sequencing (e.g., PacBio, Oxford Nanopore)	Improving genome assemblies, identifying structural variants	[13–15]
Linked-read sequencing (10X Chromium)	Phasing and improving scaffolding of genome assemblies	[16]
Hi-C sequencing and CHiA-PET	Chromatin interactions and improving genome assemblies (Hi-C)	[17–19]

Chromosomes play a vital role in the nucleus, as they are essential for DNA to replicate and segregate during cell division. They are not randomly positioned in the nucleus, but organised into specific areas called chromosomal territories [20] that change during the cell cycle [21,22] and development [23–25]. Maintenance of these territories is important for proper cell functioning, replication, and the accurate division and differentiation of cells. If we repack the DNA into a chromosome, we see that the DNA is wrapped around a nucleosome consisting of eight histone proteins to produce a chromatin fibre, which is attached to a backbone of non-histone proteins called the chromosome scaffold (Figure 1). The dynamic nature of the chromosome throughout the cell cycle and in response to environmental influences is enabled by the ability of the chromatin fibre to vary the level of DNA compaction and histone composition (epigenetics), and the ability of the scaffold proteins to follow the changes of the chromatin fibre [26]. Chromatin remodelling and changes in chromatin conformation affect interactions between sequences in different genomic regions and can influence gene regulation. The close connection between DNA, chromosome structure and the position of chromosomes in the nucleus highlights the need to integrate genomic data to better understand chromosome architecture and function. This information will provide a more comprehensive understanding of the evolutionary plasticity and organisational functions of genome architecture and mechanisms of faithful transmission of the genome to offspring.

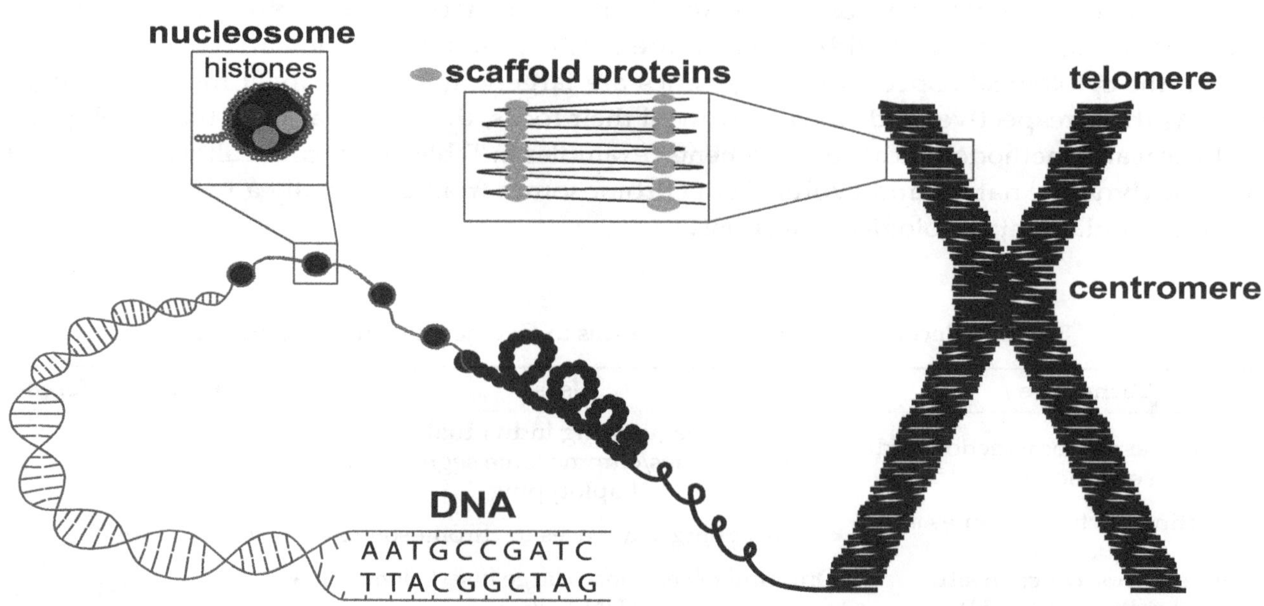

Figure 1. Repacking the DNA into a chromosome. The double-stranded DNA helix is wrapped around a nucleosome consisting of eight histone proteins to produce a chromatin fibre, which is attached to a backbone of non-histone proteins called scaffold proteins which form the chromosome scaffold.

In the past, incremental advances in understanding genome biology have been made through combining information from cytology, cytogenetics and genomics, often from data gathered by different groups focused on one particular question and over many years (e.g., the discovery the Philadelphia chromosome causing chronic myelogenous leukemia or the discovery of the sex-determining gene, *SRY*; Figure 2). Now is the time to reunite cytogenetic and sequencing approaches. Not only has the resolution of chromosomes under various forms of microscopy greatly accelerated (e.g., deconvolution system [27], structured illumination microscopy [26] and super-resolution microscopy [28]), but new sequencing technologies are now promising to make genome assemblies close to chromosome level a reality. In addition, new sequence-based techniques for chromosome conformation capture promise to fill in the details in our cytological picture of how active and inactive chromatin is assembled and arranged into functional units in the interphase nucleus. Collectively, these advances afford the capability to answer key fundamental questions in genome biology.

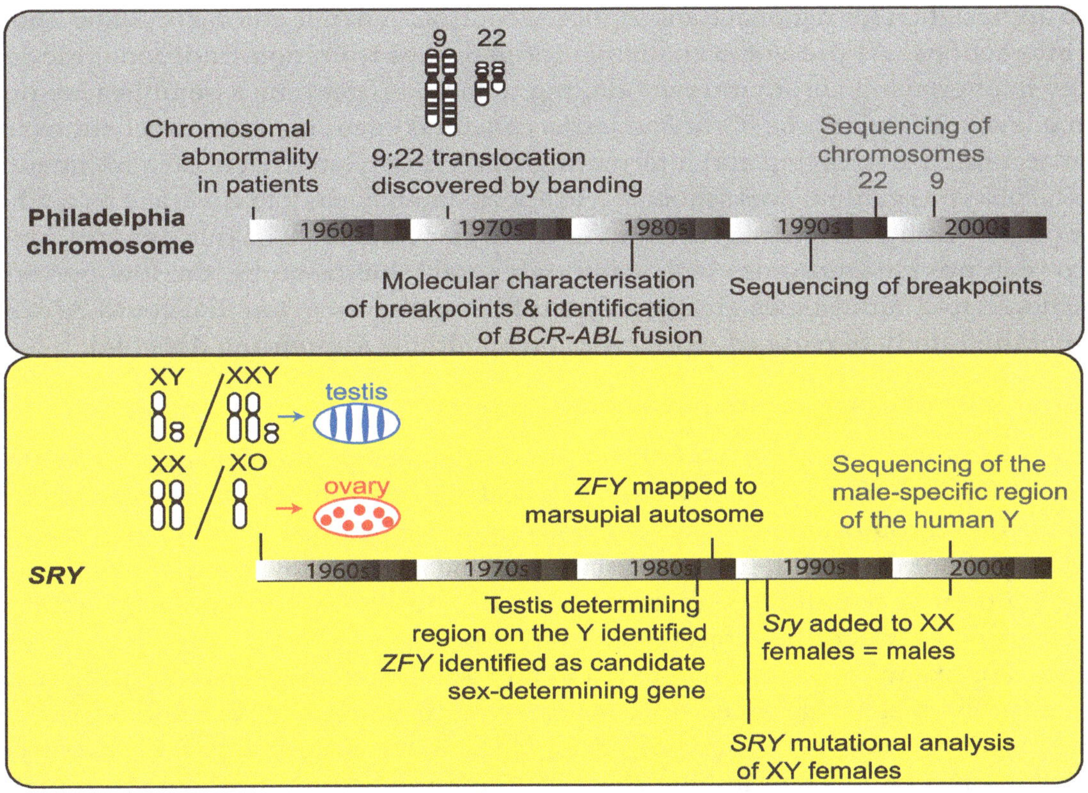

Figure 2. The incremental advances made through combined cytogenetic and genomic information in the discovery of the Philadelphia chromosome causing chronic myelogenous leukemia [29–33] and the discovery of the sex-determining gene *SRY* [34–38].

2. Development of Genome Sequencing from Cytogenetics

The proposal to sequence the human genome led to the start of the genomics era. When we consider the human genome project, it was approached from an understanding that the position of the sequence on the chromosome was important [39]. Indeed, the forerunner of the Human Genome Project was a series of meetings of an international Human Gene Mapping consortium, which met annually to put together increasingly detailed physical maps of all the human chromosomes, combining data from linkage analysis, somatic cell genetics and radiation hybrid analysis, and in situ hybridisation. This consortium was organised into separate committees for each human chromosome, as well as committees for the mitochondrial genome. Moreover, the consortium included a comparative gene mapping committee, which started out largely focused on mouse but grew to encompass many other mammals, birds and fishes. Slowly, the physical maps developed by these working groups expanded and were filled in with other markers. Sequencing crept in to offer the ultimate detail of individual genes (or at least exomes). The advent of large insert clones like BACs (bacterial artificial chromosomes) greatly aided the extension of DNA sequence to encompass larger genomic intervals beyond individual genes [39].

Physical BAC or yeast artificial chromosome (YAC) maps of each chromosome were constructed and sequenced, resulting in a chromosome-based genome assembly and enabling the integration of gene and genetic mapping data accumulated over many years and by many different researchers [40]. The subsequent ENCODE (Encyclopedia of DNA elements) project saw the integration of sequence data with information on chromatin states, which provided an exceptional insight into dynamic gene regulation [41]. However, just as genome sequences for other species were needed to help interpret the human genome [42], comparative data from a broad range of species are required to fully understand the role many chromatin modifications play in genome function. The interpretation of the ENCODE data for the human genome was only made possible by the chromosome-based genome assembly, affording an appreciation for regional transcriptional control, dynamic chromatin states and long-range interlocus interactions. At present, a challenge for genomes from non-traditional model species is the difficulty in overlaying chromatin remodelling data when genome assemblies are not yet at the chromosome level. The platypus (*Ornithorhynchus anatinus*) genome is an excellent example of the difficulty in accurately overlaying and interpreting DNA methylation data on a fragmented genome assembly. The platypus genome was sequenced to approximately six-fold coverage by a whole genome shotgun approach using Sanger sequencing, and only around 21% of this genome assembly was anchored to platypus chromosomes [43]. Although a valuable resource, the low percentage of the genome anchored to chromosomes greatly reduced the number of genes that could be examined in a recent comparative study of reduced representation bisulphite sequencing data [44].

3. The Integral Role of Cytogenetics in Genome Projects

With increasingly cost-effective high throughput sequencing, most recently assembled genomes feature short contigs and often lack even a basic physical map or chromosome number and morphology information. While chromosome-level assemblies might not be feasible for all genomes targeted for sequencing, they should be well represented across all lineages to allow comparative genome biology studies that, by their very nature, rely on knowing the position of orthologous sequences among genomes. All genome sequencing projects should incorporate some level of cytogenetic analysis from the very start. For example, a logical first step in any whole genome sequence project would be to ensure that the individual being sequenced is not carrying chromosomal aberrations, particularly if the genome assembly is to be used as a reference for population-level sequencing. Basic karyotyping would ensure the ploidy level of the species, the absence of aneuploidy and confirm the genetic sex of the individual when cytogenetically distinguishable sex chromosomes are present, a particularly important consideration in species subject to environmental sex reversal (e.g., *Pogona vitticeps* [45]). Karyotyping will also determine if there are large heterochromatic chromosomes or regions. The flow sorting of chromosomes can be used for gross assessment of aneuploidy. Flow cytometry with appropriate standards is a reliable and fast method for estimating genome size [46], an important consideration in determining the amount of sequencing required to achieve the desired level of genome assembly.

A whole genome sequence is much more informative if it is assigned and oriented onto chromosomes, and is far more intuitive to visualise as chromosomes than unconnected and unordered scaffolds. When whole genome sequences fall short of this 'chromosome level assembly', their use for critical aspects of evolutionary and applied biology is significantly limited. Assigning sequence contigs to chromosomes has most often been achieved by integrating sequence data with molecular cytogenetic mapping data. This can be achieved by determining the location of a large-insert clone by fluorescence in situ hybridisation (FISH) on metaphase chromosomes or even extended chromatin fibres (fibre FISH), facilitating physical fine mapping of contigs. In Sanger sequenced genomes, this was accomplished by assigning BAC clones corresponding to individual, large sequence scaffolds. For example, the opossum genome, with a scaffold N50 (a measurement of assembly quality where 50% of scaffolds are this size or larger) of 59.8 Mb and 97% of the sequence contained in 216 scaffolds, was anchored onto the eight opossum autosomes and the X chromosome by mapping 415 BAC clones [47,48]. However, the proportion of the genome assembly assigned to chromosomes is dependent on the quality of the genome (i.e., N50 size and number of scaffolds). The platypus genome is a prime example, where the high repeat content resulted in a scaffold N50 of 957 kb, and thus only about 21% of the genome was chromosome-anchored [43]. An excellent example of where a cytogenetics approach vastly improved the accuracy of genome assembly is the tomato genome. The tomato genome, sequenced by a combination of Sanger and next generation sequencing technologies [49], benefitted greatly from the physical assignment of sequence scaffolds of BACs by FISH and confirmation by optical mapping [50]. The original tomato assembly was ordered based on a high-density linkage map. Differences in arrangement between the linkage and cytogenetic/optical maps were detected for one-third of these scaffolds, mainly in pericentric regions where a reduced level of recombination renders linkage mapping less reliable [50]. The benefits gained from assigning even a portion of the sequence to chromosomes are immense, as highlighted by the chromosomics successes listed in Table S1.

Many genomes sequenced over the past decade have used a 'shotgun' approach based on short read sequence technologies to produce a series of scaffolds, often several hundred per chromosome, which are neither anchored to, nor ordered on, the chromosomes. Anchoring every scaffold to a chromosome would be a labour-intensive task, particularly if the assembly has a higher number of scaffolds; by combining computational approaches to merge scaffolds with either cytogenetic mapping and/or PCR-based scaffold verification, chromosome-level assemblies are a more achievable exercise [9,51]. In addition, the development of universal BAC clone probe sets that can be used in a high-throughput, cross-species, multiple hybridisation approach are speeding up the process of developing cytogenetic maps [9]. The advances in sequencing technology that are producing more contiguous genome assemblies, such as the contact sequencing approach of HiC-seq (e.g., Dovetail) [17], linked-read sequencing approach (10X Genomics) [12,16], long read technologies like PacBio [13] and Oxford Nanopore [15], and optical mapping (BioNano) [12], combined with high-throughput cytogenetic methods, will place chromosome-level assemblies within reach for many species.

4. The Big Questions in Genome Biology Requiring a Chromosomics Approach

Despite the meticulous approach taken for the human genome, gaps remained in the genome sequence when the 'finished' euchromatic sequence of the human genome was published in 2004 [52]. These 'black holes' of the genome corresponded to the most repetitive regions such as, but not limited to, the critically important centromeres [53], nucleolar organiser regions (NORs) [54] and the Y chromosome [36]. Repetitive regions are some of the most rapidly evolving sequences and therefore, are among the most interesting regions of the genome. By employing a chromosomics approach, these hot spots of evolution are beginning to lose their black hole status in the human genome, as well as in other species. We discuss the fundamental questions arising from these evolutionary dynamic regions in relation to two overarching themes associated with genome evolution: genome plasticity and sex chromosome evolution.

4.1. *Genome Plasticity and Chromosome Evolution*

Why do species have specific karyotypes? Why do chromosome numbers vary greatly within some groups, but are largely the same in others? Why are some of the regions of the genome so well conserved? Why are genomes so extensively changed among closely related species and others strongly conserved? Why do some chromosome rearrangements appear to lead to speciation, yet others are tolerated within a species or population of species? Is the underlying mechanism responsible for chromosomal speciation the same as that leading to chromosomal rearrangements in a disease context (i.e., cancer)? These are fundamental questions regarding genome plasticity that remain unanswered, and a chromosomics approach is essential for major breakthroughs. The answers to these questions will have wide-ranging impacts in the field of biology. Unlocking the genomic basis of speciation is a biological research priority, fuelled by the ongoing debate on species concepts and facilitated by the availability of an unprecedentedly large number of genomic resources.

The concept of chromosomal speciation, at one stage considered a major contributor in separating populations that differ by a structural rearrangement, was virtually abandoned in favour of theories of a gradual accumulation of mutations in 'speciation genes' (e.g., Reference [55]). The implementation of the most recent 'suppressed recombination model' [56,57] has now fuelled the field using a combination of sequence and cytogenetics [58–60]. In this context, chromosome rearrangements could have a minimal influence on fitness, but would suppress recombination, leading to the reduction of gene flow across genomic regions and to the accumulation of incompatibilities.

Understanding chromosomal speciation is also critical to determining the mechanism(s) underlying genome adaptation to environmental factors and how biodiversity is generated and transmitted to subsequent generations. With so many threatened species across the globe, understanding why some structural variants are tolerated within a population while others lead to reproductive isolation could prove important for the management of breeding programs for species conservation programs. In a disease context, a greater knowledge of the drivers of genome instability will aid research into human and animal diseases, particularly cancers.

Regions of genome instability can have dramatic effects for an organism. Despite being the subject of many studies using a range of species, the underlying molecular mechanisms resulting in genome restructuring/reshuffling are relatively poorly understood. For example, it remains unclear if chromosomal changes associated with speciation arise because there is an adaptive value to a specific chromosomal configuration, and what causes the genomic instability in the first place. The combined use of comparative genomics and cytogenetics of both closely and distantly related mammalian species has been extremely useful in defining models that explain genome structure and evolution [61–67]. Such reconstructions have revealed that the genomic regions implicated in structural evolutionary changes disrupt genomic synteny (evolutionary breakpoint regions, EBRs) and are clustered in regions more prone to breaking and reorganisation [61–63]. In searching for the origin (and consequences) of this evolutionary instability, approaches based purely on genome sequence have only revealed that EBRs are enriched for repetitive sequences, including segmental duplications and transposable elements, which provide the templates for non-allelic homologous recombination, resulting in inversions and additional structural changes [68,69]. Likewise, repetitive sequences in centromeric regions have been implicated in illegitimate recombination events forming Robertsonian fusions [62,70,71]. EBRs also typically occur in gene-dense regions, enriched with genes involved in adaptive processes, where changes to gene expression caused by a chromosomal rearrangement may provide a selective advantage [60,63,72]. Consequently, given the diversity of factors associated with EBRs, it is unlikely that the sequence composition of genomes is solely responsible for genomic instability during evolution and speciation.

As chromosomes are more than just DNA sequence, a more comprehensive approach that incorporates global genomic information on recombination rates, chromatin accessibility, gene function data and nuclear architecture is providing more insight into the factors underpinning genome instability. Of course, chromosome-level assemblies are an essential resource for accurate interpretation of the combination of all these data because, without such assemblies, we have a very limited (if any) understanding of the extent of the genomic restructuring that may have occurred between species, or between normal and disease states, that facilitated changes in global genomic features.

Furthermore, it has become clear that the interplay between the organisation of the genome and nuclear architecture is central to genome function [73]. We have seen a rapid evolution of methods by which to analyse genome organisation and nuclear architecture, moving from cytogenetic approaches, providing a direct measurement within individual cells of distances between loci, to chromosome conformation capture approaches (3C, 4C, 5C and Hi-C), which infer the contact among loci, typically in populations of cells rather than single cells [74,75]. However, comparison of the results obtained from chromosome conformation capture methods and FISH analyses demonstrates that care needs to be taken when interpreting data obtained solely by one method, suggesting that the use of a combined molecular and cytogenetic approach will lead to more accurate 3D models of genome organisation [76,77].

A new model for genome rearrangements, referred to as the integrative breakage model, has recently been proposed, bringing together all of these features [64]. It posits that genome reshuffling permissiveness is influenced by (i) the physical interaction of genomic regions inside the nucleus, (ii) the accessibility of chromatin states and (iii) the maintenance of essential genes and/or their association with long-range cis-regulatory elements (Figure 3). An initial test of the integrative breakage model using rodents has supported this model [65]. EBRs were found to not only coincide with regions enriched for repetitive sequences and genes, especially genes involved in reproduction and pheromone detection, but possessed the characteristics of open, actively transcribed chromatin. The challenge remains to use a broader spectrum of species to fully test this model and dissect the underlying mechanism for chromosomal rearrangements. Such studies will now be possible with the ability to achieve chromosome-level assemblies and obtain information on chromatin modifications and nuclear architecture for non-traditional model species. A similar approach could be extended to intraspecific comparisons, such as normal versus disease samples or samples across a population where structural variants are known.

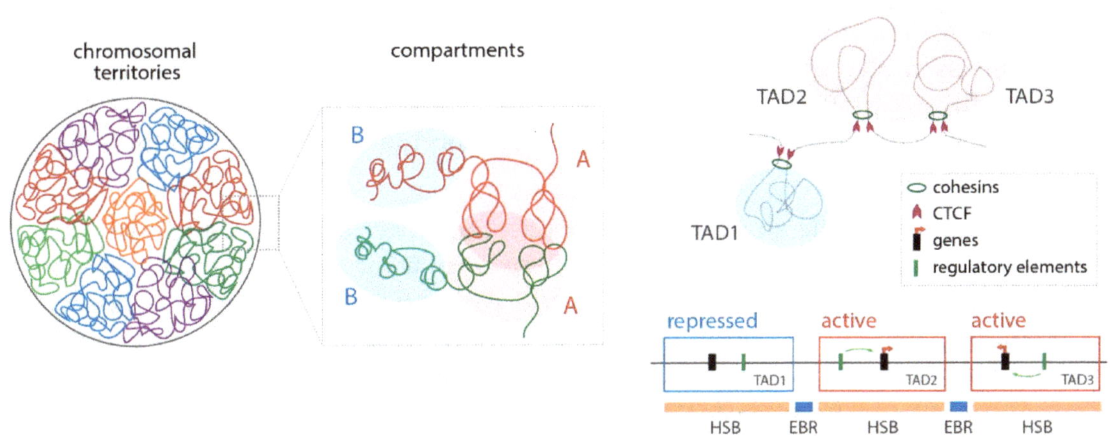

Figure 3. The integrative breakage model, a multilayer framework for the study of genome evolution that takes into account the high-level structural organisation of genomes and the functional constraints that accompany genome reshuffling [64]. Genomes are compartmentalised into different levels of organisation that include: (i) chromosomal territories, (ii) 'open' (termed 'A')/'closed' (termed 'B') compartments inside chromosomal territories, (iii) topologically associated domains (TADs) and (iv) looping interactions. TADs, which are delimited by insulating factors such as CTCF and cohesins, harbour looping topologies that permit long-range interactions between target genes and their distal enhancers, thus providing 'regulatory neighbourhoods' within homologous syntenic blocks (HSBs). In this context, the integrative breakage model proposes that genomic regions involved in evolutionary reshuffling (evolutionary breakpoint regions, EBRs) which will likely be fixed within populations are (i) those that contain open chromatin DNA configurations and epigenetic features that could promote DNA accessibility and therefore genomic instability, and (ii) that do not disturb essential genes and/or gene expression.

4.2. Sex Chromosome Evolution: Genetics and Epigenetics

Sex chromosomes represent one of the most dynamic parts of any genome, as they are highly variable in morphology and sequence content across the plant and animal kingdoms. The special evolutionary forces experienced by sex chromosomes have rendered them highly complex entities within the genome; thus, it remains a challenge for evolutionary biologists to disentangle the varied mechanisms involved in their evolution. There are still many fundamental questions that remain unanswered because of the genomic black hole status of sex chromosomes. Why do sex chromosomes evolve and degenerate in some species but not in others? Why do sex chromosomes have a propensity to accumulate repetitive sequences that, in most cases, are species-specific? How do sex chromosomes drive speciation and hybrid incompatibilities? Why do sex chromosomes vary within a species? Why do some species have complete dosage compensation mechanisms while others do not? The complexity of sex chromosome origin, evolution and gene organisation is multilayered and cannot be understood by studying a single aspect of its biology alone. Therefore, a chromosomics approach, taking into account all aspects of cellular and molecular biology, will be essential to answering these questions.

Many genome projects were undertaken without considering the sex chromosomes, with most projects intentionally choosing to sequence the homogametic sex in order to obtain higher sequence coverage and better assembly of the X or Z chromosome, completely neglecting to obtain sequence for the Y or W, thus ignoring the complexities of sex-delimited sex chromosome variation. Simple karyotyping with basic banding analysis, or painting one sex chromosome onto the other, can be very informative about the DNA content of the sex chromosomes. Such experiments can provide valuable information to support the adoption of appropriate sequencing technologies to obtain sequences from those unique but difficult to sequence regions of the genome. In a recent review, Tomaszkiewicz et al. [78] highlighted the need to sequence sex chromosomes, and elegantly described challenges and opportunities for combining new and emerging technologies to sequence these difficult regions of the genome. Only a chromosomics approach, combining cytogenetics and appropriate sequencing platform(s), can answer the fundamental questions regarding sex chromosome evolution. As an example, a human Y chromosome of African origin was recently assembled by flow sorting nine million Y chromosomes and sequencing using the Oxford Nanopore MinION platform, resulting in a Y chromosome assembly with an N50 of 1.46Mb [79], yet this method was much more time- and cost-efficient than that used to obtain the original human Y chromosome sequence [36,80].

Determining the epigenetic status of the sex chromosomes and the genes they contain is also extremely valuable. For example, the Chinese half-smooth tongue sole (*Cynoglossus semilaevis*) is a species with genetic sex determination (ZZ males and ZW females), but with a temperature override mechanism, where exposure of developing embryos to high temperatures causes genetic ZW females to develop as males (sex reversal). The sex determining gene *dmrt1* [81] is epigenetically silenced by DNA methylation in ZW females, but not in sex-reversed ZW males, where *dmrt1* expression is upregulated, leading to initiation of the male development pathway [82].

Dosage compensation, a mechanism equalizing the expression of genes on the sex chromosomes between males and females, is epigenetically controlled. A comparison of the gene content of the X chromosomes of eutherians and marsupials would suggest that a dosage compensation mechanism, in the form of X chromosome inactivation, may be shared between these two mammalian groups, as the X chromosome of marsupials is homologous to approximately two-thirds of the X of eutherians [83]. However, epigenetic analyses point to an independent evolutionary origin of X chromosome inactivation in marsupials and eutherians [84]. Similarly, there are striking differences in the extent and mechanisms of dosage compensation between more divergent taxa. For example, *Drosophila melanogaster* increases X chromosome transcription by the binding of Male Specific Lethal (MSL) complex to the single X chromosome in males to achieve dosage compensation [85]. In contrast, many species, including insects, fishes, birds, reptiles and platypus, have incomplete dosage compensation [86]. Reports of incomplete dosage compensation have most often relied purely on a sequence-based approach to measure the average transcriptional output of the X or Z chromosome between males and females for a population of cells, which does not afford an understanding of the mechanisms that impact differential transcription. However, examination of individual cells and measures of gene copies from the two Z or X chromosomes using a technique detecting nascent transcription (RNA-FISH) provides information on a single cell basis. For example, in the homogametic sex of chicken (*Gallus gallus*) and platypus, RNA-FISH detected a portion of cells expressing a gene from one copy of the X/Z, while a portion was expressed from both copies, which explains the incomplete dosage compensation pattern observed by transcriptome approaches measuring population of cells [87–89].

Sex chromosomes also have an impact beyond simply facilitating sex determination. In *Drosophila*, for example, polymorphisms in repetitive sequences on the Y chromosome influence gene expression of genes across the genome, particularly those involved in chromosome organisation and chromatin assembly [90,91]. Essentially, the polymorphic Y chromosome is a source of epigenetic variation in *Drosophila*. This epigenetic variation has implications for speciation. Engineered species hybrids showed either reduced fertility or rescued fertility depending on the origin of the Y chromosome and grandparental genetic background of the hybrid, suggesting that the Y chromosome may contribute to reproductive isolation [92]. The regulatory effect of the Y chromosome on gene expression is not limited to *Drosophila*, but has been demonstrated, at least for immune-related genes, in humans and mice [93,94]. This regulatory role for the Y highlights the importance of ascertaining not only the DNA sequence, but also the epigenetic status of sex chromosomes, in addition to, and in the context of, the rest of the genome.

5. Challenges Ahead for Chromosomics

Chromosomics approaches can have far greater success for answering fundamental biological questions than either genomics or cytogenetics approaches alone; this begs the question: why haven't these two fields merged more extensively? We have identified three major challenges that may be preventing a closer union of these fields.

The biggest challenge for chromosomics is the dwindling number of researchers worldwide with expertise in cytogenetics. Rejuvenating the training of cytogeneticists is essential if the potential of chromosomics as a field is to reach its full potential. At a "Cytogenetics in the Genomics Era" workshop held in 2017 at the University of Canberra, we identified a need to renew excitement in chromosomes among undergraduate and graduate students worldwide. Genetics and genomics courses are often taught by those who have little experience in or appreciation for chromosomes, perhaps leading to anxiety in students around the study of chromosome biology. The origin of this mismatch can also be found in the backgrounds of leaders in the fields of DNA sequencing and of cytogenetics. Genome sequencing researchers often have backgrounds in biochemistry, and their training may be entirely devoid of exposure to genetics. In contrast, those who gravitate to chromosome work often have a background of zoology or botany, and may perhaps have had little exposure to biochemistry. At a time when we are more aware than ever before of the important role genome organisation and nuclear architecture plays in genome function, it is imperative that students gain an understanding of, and appreciation for, the basics of chromosome biology from their first introduction to the world of genetics and genomics. This early introduction needs to be followed by reinforcement throughout their studies. Some good courses in integrated cell and molecular biology would be a step in the right direction.

Furthermore, graduate students have been attracted to the rapidly advancing world of genome sequencing, where mountains of data are now being rapidly and cheaply generated, as opposed to cytogenetics projects, where data are accumulated more slowly. We need to instil in students the incredible experience of observing the amazing structure of chromosomes under a microscope, and the importance of understanding chromosome structure and continuing to develop more high-throughput approaches to cytogenetics to keep pace with the rapidly advancing world of genome technology. The genomics field over the past decade has made major technological advances in obtaining genome sequences faster and more cheaply, driven mostly by the large community in this field. Advancements of a similar level in cytogenetics will require a larger community of researchers to drive the need for the technology. By increasing the training of researchers in cytogenetics, we not only increase the uptake of chromosomics, but generate a potential pool of people able to develop technologies to achieve cytogenetic analyses fast and cheaper.

A challenge for cytogeneticists is that chromosome work, which used to be the cheapest aspect of a genome project, is now the most expensive. New sequencing technology has brought down the cost of sequencing by six orders of magnitude, and the speed at which data are generated has dramatically increased. Technical innovation in cell biology, in contrast, has greatly magnified costs. High throughput really does not apply to chromosome observation or experimentation.

The second most challenging skills area for chromosomics is the need for sophisticated bioinformatics. Whole genomes can now be rapidly sequenced, but assembling the sequence is much slower, and more labour- and computationally-intensive. Likewise, overlaying genome sequence with 3D chromatin structure data often presents a computational challenge [95]. More importantly, the incorporation of cytogenetic information with genomic data is not commonly attempted. For chromosomics approaches to be more readily applied in the future, we require more bioinformaticians to be trained generally in genome assembly, as well as with an appreciation for cytogenetics.

Another challenge is ensuring samples, whether from wild species, laboratory species or clinical specimens, are collected appropriately for cytogenetic analysis. The collection of samples for DNA analysis is now routine, but the collection of material for the combination of cytogenetic and genomic analysis is not. The special requirements of samples collected for cytogenetic analysis need to be disseminated more widely. This is a relatively easy challenge to address by making a 'field guide to chromosomics' available to field researchers and a similar one to those working in a clinical setting, detailing how samples should be collected and stored for the implementation of chromosomic techniques. With the appropriate samples, a chromosomics approach could be employed to study structural and epigenetic variation at population or biogeography levels, where there is the potential to uncover the underlying genetic or epigenetic basis for adaptation to a particular environment. These data could have an impact on the conservation and management of threatened species and lead to a greater understanding of the factors underlying disease phenotypes.

6. Future Opportunities

6.1. *Sequencing Genome Black Holes*

Resolving highly repetitive regions of genomes, i.e., the black holes, is now possible, and we will soon have the capability to explore these previously under-represented regions of the genome to more fully understand their evolution and function. An important step towards understanding the evolution and function of repetitive regions has been the development of tools able to analyse and visualise these regions of the genome, such as RepeatExplorer [96]. Furthermore, sequencing of highly repetitive regions is now possible with long-read sequencing technology. For example, the centromeric sequence of the human Y chromosome has recently been assembled [97], demonstrating that nanopore long technology has the potential to fill in genome black holes. Likewise, combinations of different approaches where individual sex chromosomes are sequenced are proving successful in resolving at least the non-repetitive regions of sex chromosomes and identifying candidate sex-determining genes [98].

6.2. *Spatial Chromosome Organisation*

The importance of the territorial organisation of chromosomes (chromosome territories) in plant and animal cells was proposed over a century ago by several cytologists (reviewed in Reference [3]). We are currently at a stage where the capacity to study the changes in spatial organisation in a population of cells, or even a single cell, is possible. The recent advances in chromatin analysis, coupled with next generation sequencing (e.g., Hi-C, Chromatin Interaction Analysis by Paired-End Tag Sequencing (ChIA-PET)) and 3D and 4D FISH, live-cell and super-resolution microscopy, provide opportunities to garner a more comprehensive understanding of chromosomal activities within the nucleus contributing to gene regulation, expression and ultimate phenotypic outcomes of an individual [22–25,99]. Such a chromosomics approach is already underway for the human genome with the launch of the 4D Nucleome project to understand how the changes in chromosome dynamics contribute to gene regulation in different cell types and biological states [99]. We will gain unprecedented insight into the role of the spatial organisation of chromosomes in genome evolution by extending this same approach to many more species.

7. Concluding Remarks

We are on the verge of an exciting new revolution in biology, with a change from thinking of genomes as one-dimensional entities to defining the ways every component of the genome is packaged and changes through space and time. Chromosomics is the best path forward, providing one comprehensive analysis to answer complex questions in evolution and disease contexts. However, this can only be achieved if genomicists, cytogeneticists, cell biologists and bioinformaticians commit to forming a closer union for advancing this new era in genome biology.

Author Contributions: The concept and idea of this review was developed during the workshop "Cytogenetics in the Genomics Era" organised by T.E. and J.E.D. at the Institute for Applied Ecology, University of Canberra, February 2017. J.E.D. and T.E. prepared the first draft after S.P., R.O., A.R-H., M.B.C., M.D.B.E., K.F., J.A.M.G., D.G., F.G., L.K., I.M., M.R., K.S. and E.W. contributed to drafting the manuscript outline. JED, S.P., R.O, A.R-H., M.B.C., M.D.B.E., K.F., J.A.M.G., D.G., F.G., L.K., I.M., M.R., K.S., E.W. and T.E. contributed to writing and revising drafts of the manuscripts. J.E.D., S.P. and A.R.-H. prepared figures. All authors approved the final version.

Acknowledgments: We thank Arthur Georges, Craig Moritz, Stephen Sarre who contributed to discussions as part of the "Cytogenetics in the Genomics Era"workshop.

References

1. Claussen, U. Chromosomics. *Cytogenet. Genome Res.* **2005**, *111*, 101–106. [CrossRef] [PubMed]
2. Winkler, H. *Verbreitung und Ursache der Parthenogenesis im Pflanzen-und Tierreiche*; Verlag von Gustav Fischer: Jena, Germany, 1920.
3. Traut, W.; Vogel, H.; Glöckner, G.; Hartmann, E.; Heckel, D.G. High-throughput sequencing of a single chromosome: A moth W chromosome. *Chromosom. Res.* **2013**, *21*, 491–505. [CrossRef] [PubMed]
4. Seifertova, E.; Zimmerman, L.B.; Gilchrist, M.J.; Macha, J.; Kubickova, S.; Cernohorska, H.; Zarsky, V.; Owens, N.D.L.; Sesay, A.K.; Tlapakova, T.; et al. Efficient high-throughput sequencing of a laser microdissected chromosome arm. *BMC Genom.* **2013**, *14*, 1. [CrossRef] [PubMed]
5. Ma, L.; Li, W.; Song, Q. Chromosome-range whole-genome high-throughput experimental haplotyping by single-chromosome microdissection. *Methods Mol. Biol.* **2017**, *1551*, 161–169.
6. Makunin, A.I.; Kichigin, I.G.; Larkin, D.M.; O'Brien, P.C.M.; Ferguson-Smith, M.A.; Yang, F.; Proskuryakova, A.A.; Vorobieva, N.V.; Chernyaeva, E.N.; O'Brien, S.J.; et al. Contrasting origin of B chromosomes in two cervids (Siberian roe deer and grey brocket deer) unravelled by chromosome-specific DNA sequencing. *BMC Genom.* **2016**, *17*, 618. [CrossRef] [PubMed]
7. Murchison, E.P.; Schulz-Trieglaff, O.B.; Ning, Z.; Alexandrov, L.B.; Bauer, M.J.; Fu, B.; Hims, M.; Ding, Z.; Ivakhno, S.; Stewart, C.; et al. Genome sequencing and analysis of the Tasmanian devil and its transmissible cancer. *Cell* **2012**, *148*, 780–791. [CrossRef] [PubMed]
8. Skinner, B.M.; Sargent, C.A.; Churcher, C.; Hunt, T.; Herrero, J.; Loveland, J.E.; Dunn, M.; Louzada, S.; Fu, B.; Chow, W.; et al. The pig X and Y Chromosomes: Structure, sequence, and evolution. *Genome Res.* **2016**, *26*, 130–139. [CrossRef] [PubMed]
9. Damas, J.; O'Connor, R.; Farré, M.; Lenis, V.P.E.; Martell, H.J.; Mandawala, A.; Fowler, K.; Joseph, S.; Swain, M.T.; Griffin, D.K.; et al. Upgrading short read animal genome assemblies to chromosome level using comparative genomics and a universal probe set. *Genome Res.* **2017**, *27*, 875–884. [CrossRef] [PubMed]

10. Beliveau, B.J.; Boettiger, A.N.; Avendaño, M.S.; Jungmann, R.; McCole, R.B.; Joyce, E.F.; Kim-Kiselak, C.; Bantignies, F.; Fonseka, C.Y.; Erceg, J.; et al. Single-molecule super-resolution imaging of chromosomes and in situ haplotype visualization using Oligopaint FISH probes. *Nat. Commun.* **2015**, *6*, 7147. [CrossRef]
11. Lakadamyali, M.; Cosma, M.P. Advanced microscopy methods for visualizing chromatin structure. *FEBS Lett.* **2015**, *589*, 3023–3030. [CrossRef]
12. Mostovoy, Y.; Levy-Sakin, M.; Lam, J.; Lam, E.T.; Hastie, A.R.; Marks, P.; Lee, J.; Chu, C.; Lin, C.; Džakula, Ž.; et al. A hybrid approach for de novo human genome sequence assembly and phasing. *Nat. Methods* **2016**, *13*, 12–17. [CrossRef] [PubMed]
13. Rhoads, A.; Au, K.F. PacBio Sequencing and Its Applications. *Genom. Proteom. Bioinform.* **2015**, *13*, 278–289. [CrossRef] [PubMed]
14. Magi, A.; Semeraro, R.; Mingrino, A.; Giusti, B.; D'Aurizio, R. Nanopore sequencing data analysis: State of the art, applications and challenges. *Brief. Bioinform.* **2017**, *19*, 1256–1272. [CrossRef] [PubMed]
15. Lu, H.; Giordano, F.; Ning, Z. Oxford Nanopore MinION Sequencing and Genome Assembly. *Genom. Proteom. Bioinform.* **2016**, *14*, 265–279. [CrossRef] [PubMed]
16. Zheng, G.X.Y.; Lau, B.T.; Schnall-Levin, M.; Jarosz, M.; Bell, J.M.; Hindson, C.M.; Kyriazopoulou-Panagiotopoulou, S.; Masquelier, D.A.; Merrill, L.; Terry, J.M.; et al. Haplotyping germline and cancer genomes with high-throughput linked-read sequencing. *Nat. Biotechnol.* **2016**, *34*, 303–311. [CrossRef] [PubMed]
17. Putnam, N.H.; Connell, B.O.; Stites, J.C.; Rice, B.J.; Hartley, P.D.; Sugnet, C.W.; Haussler, D.; Rokhsar, D.S. Chromosome-scale shotgun assembly using an in vitro method for long-range linkage. *Genome Res.* **2016**, *26*, 342–350. [CrossRef] [PubMed]
18. Dudchenko, O.; Batra, S.S.; Omer, A.D.; Nyquist, S.K.; Hoeger, M.; Durand, N.C.; Shamim, M.S.; Machol, I.; Lander, E.S.; Aiden, A.P. De novo assembly of the *Aedes aegypti* genome using Hi-C yields chromosome-length scaffolds. *Science* **2017**, *10*, 92–95. [CrossRef] [PubMed]
19. Fullwood, M.J.; Liu, M.H.; Pan, Y.F.; Liu, J.; Xu, H.; Mohamed, Y.B.; Orlov, Y.L.; Velkov, S.; Ho, A.; Mei, P.H.; et al. An oestrogen-receptor-α-bound human chromatin interactome. *Nature* **2009**, *462*, 58–64. [CrossRef] [PubMed]
20. Cremer, T.; Cremer, C. Rise, fall and resurrection of chromosome territories: A historical perspective. Part II. Fall and resurrection of chromosome territories during the 1950s to 1980s. Part III. Chromosome territories and the functional nuclear architecture: Experiments and models from the 1990s to the present. *Eur. J. Histochem.* **2006**, *50*, 223–272.
21. Gibcus, J.H.; Samejima, K.; Goloborodko, A.; Samejima, I.; Naumova, N.; Nuebler, J.; Kanemaki, M.T.; Xie, L.; Paulson, J.R.; Earnshaw, W.C.; et al. A pathway for mitotic chromosome formation. *Science* **2018**, *359*, eaao6135. [CrossRef]
22. Naumova, N.; Imakaev, M.; Fudenberg, G.; Zhan, Y.; Lajoie, B.R.; Mirny, L.A.; Dekker, J. Organization of the mitotic chromosome. *Science* **2013**, *342*, 948–953. [CrossRef]
23. Patel, L.; Kang, R.; Rosenberg, S.C.; Qiu, Y.; Raviram, R.; Chee, S.; Hu, R.; Ren, B.; Cole, F.; Corbett, K.D. Dynamic reorganization of the genome shapes the recombination landscape in meiotic prophase. *Nat. Struct. Mol. Biol.* **2019**, *26*, 164–174. [CrossRef] [PubMed]
24. Alavattam, K.G.; Maezawa, S.; Sakashita, A.; Khoury, H.; Barski, A.; Kaplan, N.; Namekawa, S.H. Attenuated chromatin compartmentalization in meiosis and its maturation in sperm development. *Nat. Struct. Mol. Biol.* **2019**, *26*, 175–184. [CrossRef] [PubMed]
25. Vara, C.; Paytuví-Gallart, A.; Cuartero, Y.; Le Dily, F.; Garcia, F.; Salvà-Castro, J.; Gómez-H, L.; Julià, E.; Moutinho, C.; Aiese Cigliano, R.; et al. Three-dimensional genomic structure and cohesin occupancy correlate with transcriptional activity during spermatogenesis. *Cell Rep.* **2019**, *28*, 352–367. [CrossRef] [PubMed]
26. Poonperm, R.; Takata, H.; Hamano, T.; Matsuda, A.; Uchiyama, S.; Hiraoka, Y.; Fukui, K. Chromosome scaffold is a double-stranded assembly of scaffold proteins. *Sci. Rep.* **2015**, *5*, 11916. [CrossRef] [PubMed]
27. Wako, T.; Fukuda, M.; Furushima-Shimogawara, R.; Belyaev, N.D.; Turner, B.M.; Fukui, K. Comparative analysis of topographic distribution of acetylated histone H4 by using confocal microscopy and a deconvolution system. *Anal. Chim. Acta* **1998**, *365*, 9–17. [CrossRef]
28. Nir, G.; Farabella, I.; Pérez Estrada, C.; Ebeling, C.G.; Beliveau, B.J.; Sasaki, H.M.; Lee, S.H.; Nguyen, S.C.; McCole, R.B.; Chattoraj, S.; et al. Walking along chromosomes with super-resolution imaging, contact maps, and integrative modeling. *PLoS Genet.* **2018**, *14*, 1–35. [CrossRef] [PubMed]

29. Nowell, P.C.; Hungerford, D.A. Chromosome studies on normal and leukemic human leukocytes. *J. Natl. Cancer Inst.* **1960**, *25*, 85–109.
30. Rowley, J.D. A new consistent chromosomal abnormality in chronic myelogenous leukaemia identified by quinacrine fluorescence and Giemsa staining. *Nature* **1973**, *243*, 290. [CrossRef]
31. Shtivelman, E.; Lifshitz, B.; Gale, R.P.; Canaani, E. Fused transcript of abl and bcr genes in chronic myelogenous leukaemia. *Nature* **1985**, *315*, 550–554. [CrossRef]
32. Humphray, S.J.; Oliver, K.; Hunt, A.R.; Plumb, R.W.; Loveland, J.E.; Howe, K.L.; Andrews, T.D.; Searle, S.; Hunt, S.E.; Scott, C.E.; et al. DNA sequence and analysis of human chromosome 9. *Nature* **2004**, *429*, 369–374. [CrossRef] [PubMed]
33. Dunham, I.; Shimizu, N.; Roe, B.A.; Chissoe, S.; Dunham, I.; Hunt, A.R.; Collins, J.E.; Bruskiewich, R.; Beare, D.M.; Clamp, M.; et al. The DNA sequence of human chromosome 22. *Nature* **1999**, *402*, 489–495. [CrossRef] [PubMed]
34. Foster, J.W.; Graves, J.A.M. An SRY-related sequence on the marsupial X chromosome: Implications for the evolution of the mammalian testis-determining gene. *Proc. Natl. Acad. Sci. USA* **1994**, *91*, 1927–1931. [CrossRef] [PubMed]
35. Sinclair, A.H.; Berta, P.; Palmer, M.S.; Hawkins, J.R.; Griffiths, B.L.; Smith, M.J.; Foster, J.W.; Frischauf, A.M.; Lovell-Badge, R.; Goodfellow, P.N. A gene from the human sex-determining region encodes a protein with homology to a conserved DNA-binding motif. *Nature* **1990**, *346*, 240–244. [CrossRef] [PubMed]
36. Skaletsky, H.; Kuroda-Kawaguchi, T.; Minx, P.J.; Cordum, H.S.; Hillier, L.; Brown, L.G.; Repping, S.; Pyntikova, T.; Ali, J.; Bieri, T.; et al. The male-specific region of the human Y chromosome is a mosaic of discrete sequence classes. *Nature* **2003**, *423*, 825–837. [CrossRef] [PubMed]
37. Koopman, P.; Gubbay, J.; Vivian, N.; Goodfellow, P.; Lovell-Badge, R. Male development of chromosomally female mice transgenic for Sry. *Nature* **1991**, *351*, 117–121. [CrossRef] [PubMed]
38. Page, D.C.; Mosher, R.; Simpson, E.M.; Fisher, E.M.C.; Mardon, G.; Pollack, J.; McGillivray, B.; de la Chapelle, A.; Brown, L.G. The sex-determining region of the human Y chromosome encodes a finger protein. *Cell* **1987**, *27*, 67–82. [CrossRef]
39. McPherson, J.D.; Marra, M.; Hillier, L.D.; Waterston, R.H.; Chinwalla, A.; Wallis, J.; Sekhon, M.; Wylie, K.; Mardis, E.R.; Wilson, R.K.; et al. A physical map of the human genome. *Nature* **2001**, *409*, 934–941.
40. International Human Genome Sequencing Consortium Initial sequencing and analysis of the human genome. *Nature* **2001**, *409*, 860–921. [CrossRef] [PubMed]
41. Bernstein, B.E.; Birney, E.; Dunham, I.; Green, E.D.; Gunter, C.; Snyder, M. An integrated encyclopedia of DNA elements in the human genome. *Nature* **2012**, *489*, 57–74.
42. Green, P. 2x genomes—Does depth matter? *Genome Res.* **2007**, *17*, 1547–1549. [CrossRef] [PubMed]
43. Warren, W.C.; Hillier, L.W.; Marshall Graves, J.A.; Birney, E.; Ponting, C.P.; Grützner, F.; Belov, K.; Miller, W.; Clarke, L.; Chinwalla, A.T.; et al. Genome analysis of the platypus reveals unique signatures of evolution. *Nature* **2008**, *453*, 175–183. [CrossRef] [PubMed]
44. Waters, S.A.; Livernois, A.M.; Patel, H.; O'Meally, D.; Craig, J.M.; Graves, J.A.M.; Suter, C.M.; Waters, P.D. Landscape of DNA methylation on the marsupial X. *Mol. Biol. Evol.* **2018**, *35*, 431–439. [CrossRef]
45. Georges, A.; Li, Q.; Lian, J.; Meally, D.O.; Deakin, J.; Wang, Z.; Zhang, P.; Fujita, M.; Patel, H.R.; Holleley, C.E.; et al. High-coverage sequencing and annotated assembly of the genome of the Australian dragon lizard *Pogona vitticeps*. *Gigascience* **2015**, *4*, 45. [CrossRef] [PubMed]
46. Doležel, J.; Greilhuber, J.; Suda, J. Estimation of nuclear DNA content in plants using flow cytometry. *Nat. Protoc.* **2007**, *2*, 2233–2244. [CrossRef] [PubMed]
47. Duke, S.E.; Samollow, P.B.; Mauceli, E.; Lindblad-Toh, K.; Breen, M. Integrated cytogenetic BAC map of the genome of the gray, short-tailed opossum, *Monodelphis domestica*. *Chromosom. Res.* **2007**, *15*, 361–370. [CrossRef] [PubMed]
48. Mikkelsen, T.S.; Wakefield, M.J.; Aken, B.; Amemiya, C.T.; Chang, J.L.; Duke, S.; Garber, M.; Gentles, A.J.; Goodstadt, L.; Heger, A.; et al. Genome of the marsupial *Monodelphis domestica* reveals innovation in non-coding sequences. *Nature* **2007**, *447*, 167–177. [CrossRef] [PubMed]
49. The Tomato Genome Consortium. The tomato genome sequence provides insights into fleshy fruit evolution. *Nature* **2012**, *485*, 635–641. [CrossRef]

50. Shearer, L.A.; Anderson, L.K.; de Jong, H.; Smit, S.; Goicoechea, J.L.; Roe, B.A.; Hua, A.; Giovannoni, J.J.; Stack, S.M. Fluorescence in situ hybridization and optical mapping to correct scaffold arrangement in the tomato genome. *G3* **2014**, *4*, 1395–1405. [CrossRef]
51. Deakin, J.E.; Edwards, M.J.; Patel, H.; O'Meally, D.; Lian, J.; Stenhouse, R.; Ryan, S.; Livernois, A.M.; Azad, B.; Holleley, C.E.; et al. Anchoring genome sequence to chromosomes of the central bearded dragon (*Pogona vitticeps*) enables reconstruction of ancestral squamate macrochromosomes and identifies sequence content of the Z chromosome. *BMC Genom.* **2016**, *17*, 447. [CrossRef]
52. Collins, F.S.; Lander, E.S.; Rogers, J.; Waterson, R.H. Finishing the euchromatic sequence of the human genome. *Nature* **2004**, *431*, 931–945.
53. Miga, K.H. The promises and challenges of genomic studies of human centromeres. *Prog. Mol. Subcell. Biol.* **2017**, *56*, 285–304. [PubMed]
54. McStay, B. Nucleolar organizer regions: Genomic 'dark matter' requiring illumination. *Genes Dev.* **2016**, *30*, 1598–1610. [CrossRef] [PubMed]
55. Presgraves, D.C.; Balagopalan, L.; Abmayr, S.M.; Orr, H.A. Adaptive evolution drives divergence of a hybrid inviability gene between two species of *Drosophila*. *Nature* **2003**, *423*, 715–719. [CrossRef] [PubMed]
56. Noor, M.A.F.; Grams, K.L.; Bertucci, L.A.; Reiland, J. Chromosomal inversions and the reproductive isolation of species. *Proc. Natl. Acad. Sci. USA* **2001**, *98*, 12084–12088. [CrossRef] [PubMed]
57. Rieseberg, L.H. Chromosomal rearrangements and speciation. *Trends Ecol. Evol.* **2001**, *16*, 351–358. [CrossRef]
58. Dumas, D.; Britton-Davidian, J. Chromosomal rearrangements and evolution of recombination: Comparison of chiasma distribution patterns in standard and Robertsonian populations of the house mouse. *Genetics* **2002**, *162*, 1355–1366. [PubMed]
59. Farré, M.; Micheletti, D.; Ruiz-Herrera, A. Recombination rates and genomic shuffling in human and chimpanzee—A new twist in the chromosomal speciation theory. *Mol. Biol. Evol.* **2013**, *30*, 853–864. [CrossRef] [PubMed]
60. Ullastres, A.; Farré, M.; Capilla, L.; Ruiz-Herrera, A. Unraveling the effect of genomic structural changes in the rhesus macaque—implications for the adaptive role of inversions. *BMC Genom.* **2014**, *15*, 530. [CrossRef] [PubMed]
61. Murphy, W.J.; Larkin, D.M.; Everts-van der Wind, A.; Bourque, G.; Tesler, G.; Auvil, L.; Beever, J.E.; Chowdhary, B.P.; Galibert, F.; Gatzke, L.; et al. Dynamics of mammalian chromosome evolution inferred from multispecies comparative maps. *Science* **2005**, *309*, 613–617. [CrossRef] [PubMed]
62. Ruiz-Herrera, A.; Castresana, J.; Robinson, T.J. Is mammalian chromosomal evolution driven by regions of genome fragility? *Genome Biol.* **2006**, *7*, R115. [CrossRef] [PubMed]
63. Larkin, D.M.; Pape, G.; Donthu, R.; Auvil, L.; Welge, M.; Lewin, H.A. Breakpoint regions and homologous synteny blocks in chromosomes have different evolutionary histories. *Genome Res.* **2009**, *19*, 770–777. [CrossRef] [PubMed]
64. Farré, M.; Robinson, T.J.; Ruiz-Herrera, A. An Integrative Breakage Model of genome architecture, reshuffling and evolution: The Integrative Breakage Model of genome evolution, a novel multidisciplinary hypothesis for the study of genome plasticity. *BioEssays* **2015**, *37*, 479–488. [CrossRef] [PubMed]
65. Capilla, L.; Sánchez-Guillén, R.A.; Farré, M.; Paytuví-Gallart, A.; Malinverni, R.; Ventura, J.; Larkin, D.M.; Ruiz-Herrera, A. Mammalian comparative genomics reveals genetic and epigenetic features associated with genome reshuffling in rodentia. *Genome Biol. Evol.* **2016**, *8*, 3703–3717. [CrossRef] [PubMed]
66. Froenicke, L.; Graphodatsky, A.; Müller, S.; Lyons, L.A.; Robinson, T.J.; Volleth, M.; Yang, F.; Wienberg, J. Are molecular cytogenetics and bioinformatics suggesting diverging models of ancestral mammalian genomes ? *Genome Res.* **2006**, *16*, 306–310. [CrossRef]
67. Kim, J.; Farre, M.; Auvil, L.; Capitanu, B.; Larkin, D.M.; Ma, J.; Lewin, H.A. Reconstruction and evolutionary history of eutherian chromosomes. *Proc. Natl. Acad. Sci. USA* **2017**, *114*, E5379–E5388. [CrossRef] [PubMed]
68. Bailey, J.A.; Baertsch, R.; Kent, W.J.; Haussler, D.; Eichler, E.E. Hotspots of mammalian chromosomal evolution. *Genome Biol.* **2004**, *5*, R23. [CrossRef]
69. Batzer, M.A. The impact of retrotransposons on human genome evolution. *Nat. Rev. Genet.* **2009**, *10*, 691–703.
70. Slijepcevic, P. Telomeres and mechanisms of Robertsonian fusion. *Chromosoma* **1998**, *107*, 136–140. [CrossRef]
71. Garagna, S.; Page, J.; Fernandez-Donoso, R.; Zuccotti, M.; Searle, J.B. The Robertsonian phenomenon in the house mouse: Mutation, meiosis and speciation. *Chromosoma* **2014**, *123*, 529–544. [CrossRef]

72. Lemaitre, C.; Zaghloul, L.; Sagot, M.-F.; Gautier, C.; Arneodo, A.; Tannier, E.; Audit, B. Analysis of fine-scale mammalian evolutionary breakpoints provides new insight into their relation to genome organisation. *BMC Genom.* **2009**, *10*, 335. [CrossRef] [PubMed]
73. Van Bortle, K.; Corces, V.G. Nuclear organization and genome function. *Annu. Rev. Cell Dev. Biol.* **2012**, *28*, 163–187. [CrossRef] [PubMed]
74. Giorgetti, L.; Heard, E. Closing the loop: 3C versus DNA FISH. *Genome Biol.* **2016**, *17*, 215. [CrossRef] [PubMed]
75. Davies, J.O.J.; Oudelaar, A.M.; Higgs, D.R.; Hughes, J.R. How best to identify chromosomal interactions: A comparison of approaches. *Nat. Methods* **2017**, *14*, 125–134. [CrossRef] [PubMed]
76. Williamson, I.; Berlivet, S.; Eskeland, R.; Boyle, S.; Illingworth, R.S.; Paquette, D.; Dostie, J.; Bickmore, W.A. Spatial genome organization: Contrasting views from chromosome conformation capture and fluorescence in situ hybridization. *Genes Dev.* **2014**, *28*, 2778–2791. [CrossRef]
77. Denker, A.; De Laat, W. The second decade of 3C technologies: Detailed insights into nuclear organization. *Genes Dev.* **2016**, *30*, 1357–1382. [CrossRef]
78. Tomaszkiewicz, M.; Rangavittal, S.; Cechova, M.; Sanchez, C.; Fescemyer, H.W.; Harris, R.; Ye, D.; Brien, C.M.O.; Chikhi, R.; Ryder, O.A.; et al. A time- and cost-effective strategy to sequence mammalian Y Chromosomes: An application to the de novo assembly of gorilla Y. *Genome Res.* **2016**, *26*, 530–540. [CrossRef]
79. Kuderna, L.F.K.; Lizano, E.; Julià, E.; Gomez-Garrido, J.; Serres-Armero, A.; Kuhlwilm, M.; Alandes, R.A.; Alvarez-Estape, M.; Juan, D.; Simon, H.; et al. Selective single molecule sequencing and assembly of a human Y chromosome of African origin. *Nat. Commun.* **2019**, *10*, 4. [CrossRef]
80. Hughes, J.F.; Skaletsky, H.; Pyntikova, T.; Graves, T.A.; van Daalen, S.K.M.; Minx, P.J.; Fulton, R.S.; McGrath, S.D.; Locke, D.P.; Friedman, C.; et al. Chimpanzee and human Y chromosomes are remarkably divergent in structure and gene content. *Nature* **2010**, *463*, 536–539. [CrossRef]
81. Cui, Z.; Liu, Y.; Wang, W.; Wang, Q.; Zhang, N.; Lin, F.; Wang, N.; Shao, C.; Dong, Z.; Li, Y.; et al. Genome editing reveals dmrt1 as an essential male sex-determining gene in Chinese tongue sole (*Cynoglossus semilaevis*). *Sci. Rep.* **2017**, *7*, 42213. [CrossRef]
82. Shao, C.; Li, Q.; Chen, S.; Zhang, P.; Lian, J.; Hu, Q.; Sun, B.; Jin, L.; Liu, S.; Wang, Z.; et al. Epigenetic modification and inheritance in sexual reversal of fish. *Genome Res.* **2014**, *24*, 604–615. [CrossRef] [PubMed]
83. Graves, J.A.M. The origin and function of the mammalian Y chromosome and Y-borne genes—An evolving understanding. *BioEssays* **1995**, *17*, 311–320. [CrossRef] [PubMed]
84. Deakin, J.E. Marsupial X chromosome inactivation: Past, present and future. *Aust. J. Zool.* **2013**, *61*, 13–23. [CrossRef]
85. Straub, T.; Becker, P.B. Dosage compensation: The beginning and end of generalization. *Nat. Rev. Genet.* **2007**, *8*, 47–57. [CrossRef] [PubMed]
86. Gu, L.; Walters, J.R. Evolution of sex chromosome dosage compensation in animals: A beautiful theory, undermined by facts and bedeviled by details. *Genome Biol. Evol.* **2017**, *9*, 2461–2476. [CrossRef] [PubMed]
87. Deakin, J.E.; Hore, T.A.; Koina, E.; Graves, J.A.M. The status of dosage compensation in the multiple X chromosomes of the platypus. *PLoS Genet.* **2008**, *4*, e1000140. [CrossRef] [PubMed]
88. Livernois, A.M.; Waters, S.A.; Deakin, J.E.; Graves, J.A.M.; Waters, P.D. Independent evolution of transcriptional inactivation on sex chromosomes in birds and mammals. *PLoS Genet.* **2013**, *9*, e1003635. [CrossRef] [PubMed]
89. Julien, P.; Brawand, D.; Soumillon, M.; Necsulea, A.; Liechti, A.; Schütz, F.; Daish, T.; Grützner, F.; Kaessmann, H. Mechanisms and evolutionary patterns of mammalian and avian dosage compensation. *PLoS Biol.* **2012**, *10*, e1001328. [CrossRef]
90. Lemos, B.; Branco, A.T.; Hartl, D.L. Epigenetic effects of polymorphic Y chromosomes modulate chromatin components, immune response, and sexual conflict. *Proc. Natl. Acad. Sci. USA* **2010**, *107*, 15826–15831. [CrossRef]
91. Sackton, T.B.; Hartl, D.L. GBE Meta-Analysis Reveals that genes regulated by the preferentially localized to repressive chromatin. *Genome Biol. Evol.* **2013**, *5*, 255–266. [CrossRef]
92. Araripe, L.O.; Tao, Y.; Lemos, B. Interspecific Y chromosome variation is sufficient to rescue hybrid male sterility and is influenced by the grandparental origin of the chromosomes. *Heredity* **2016**, *116*, 516–522. [CrossRef] [PubMed]

93. Case, L.K.; Wall, E.H.; Dragon, J.A.; Saligrama, N.; Krementsov, D.N.; Moussawi, M.; Zachary, J.F.; Huber, S.A.; Blankenhorn, E.P.; Teuscher, C. The Y chromosome as a regulatory element shaping immune cell transcriptomes and susceptibility to autoimmune disease. *Genome Res.* **2013**, *23*, 1474–1485. [CrossRef] [PubMed]
94. Krementsov, D.N.; Case, L.K.; Dienz, O.; Raza, A.; Fang, Q.; Ather, J.L. Genetic variation in chromosome Y regulates susceptibility to influenza A virus infection. *Proc. Natl. Acad. Sci. USA* **2017**, *114*, 3491–3496. [CrossRef] [PubMed]
95. Yardımcı, G.G.; Noble, W.S. Software tools for visualizing Hi-C data. *Genome Biol.* **2017**, *18*, 26. [CrossRef] [PubMed]
96. Novák, P.; Neumann, P.; Macas, J. Graph-based clustering and characterization of repetitive sequences in next-generation sequencing data. *BMC Bioinform.* **2010**, *11*, 378. [CrossRef] [PubMed]
97. Jain, M.; Olsen, H.E.; Turner, D.J.; Stoddart, D.; Bulazel, K.V.; Paten, B.; Haussler, D.; Willard, H.; Akeson, M.; Miga, K.H. Linear assembly of a human Y centromere using nanopore long reads. *Nat. Biotechnol.* **2018**, *36*, 321–323. [CrossRef] [PubMed]
98. Tomaszkiewicz, M.; Medvedev, P.; Makova, K.D. Y and W Chromosome assemblies: Approaches and discoveries. *Trends Genet.* **2017**, *33*, 266–282. [CrossRef] [PubMed]
99. Dekker, J.; Belmont, A.S.; Guttman, M.; Leshyk, V.O.; Lis, J.T.; Lomvardas, S.; Mirny, L.A.; O'Shea, C.C.; Park, P.J.; Ren, B.; et al. The 4D nucleome project. *Nature* **2017**, *549*, 219–226. [CrossRef]

6

Complex Structure of *Lasiopodomys mandarinus vinogradovi* Sex Chromosomes, Sex Determination and Intraspecific Autosomal Polymorphism

Svetlana A. Romanenko [1,*], Antonina V. Smorkatcheva [2], Yulia M. Kovalskaya [3], Dmitry Yu. Prokopov [1], Natalya A. Lemskaya [1], Olga L. Gladkikh [1], Ivan A. Polikarpov [2], Natalia A. Serdyukova [1], Vladimir A. Trifonov [1,4], Anna S. Molodtseva [1], Patricia C. M. O'Brien [5], Feodor N. Golenishchev [6], Malcolm A. Ferguson-Smith [5] and Alexander S. Graphodatsky [1]

1 Institute of Molecular and Cellular Biology, Siberian Branch of Russian Academy of Sciences, Novosibirsk 630090, Russia; dprokopov@mcb.nsc.ru (D.Y.P.); lemnat@mcb.nsc.ru (N.A.L.); olga_gladkikh@mcb.nsc.ru (O.L.G.); serd@mcb.nsc.ru (N.A.S.); vlad@mcb.nsc.ru (V.A.T.); rada@mcb.nsc.ru (A.S.M.); graf@mcb.nsc.ru (A.S.G.)

2 Department of Vertebrate Zoology, St. Petersburg State University, St. Petersburg 199034, Russia; tonyas1965@mail.ru (A.V.S.); ivanApolikarpov@gmail.com (I.A.P.)

3 Severtzov Institute of Ecology and Evolution, Russian Academy of Sciences, Moscow 119071, Russia; sicistam@yandex.ru

4 Department of Natural Science, Novosibirsk State University, Novosibirsk 630090, Russia

5 Cambridge Resource Centre for Comparative Genomics, Department of Veterinary Medicine, University of Cambridge, Madingley Road, Cambridge CB3 OES, UK; allsorter@gmail.com (P.C.M.O.B); maf12@cam.ac.uk (M.A.F.-S.)

6 Zoological Institute, Russian Academy of Sciences, Saint-Petersburg 199034, Russia; f_gol@mail.ru

* Correspondence: rosa@mcb.nsc.ru

Abstract: The mandarin vole, *Lasiopodomys mandarinus*, is one of the most intriguing species among mammals with non-XX/XY sex chromosome system. It combines polymorphism in diploid chromosome numbers, variation in the morphology of autosomes, heteromorphism of X chromosomes, and several sex chromosome systems the origin of which remains unexplained. Here we elucidate the sex determination system in *Lasiopodomys mandarinus vinogradovi* using extensive karyotyping, crossbreeding experiments, molecular cytogenetic methods, and single chromosome DNA sequencing. Among 205 karyotyped voles, one male and three female combinations of sex chromosomes were revealed. The chromosome segregation pattern and karyomorph-related reproductive performances suggested an aberrant sex determination with almost half of the females carrying neo-X/neo-Y combination. The comparative chromosome painting strongly supported this proposition and revealed the mandarin vole sex chromosome systems originated due to at least two *de novo* autosomal translocations onto the ancestral X chromosome. The polymorphism in autosome 2 was not related to sex chromosome variability and was proved to result from pericentric inversions. Sequencing of microdissection derived of sex chromosomes allowed the determination of the coordinates for syntenic regions but did not reveal any Y-specific sequences. Several possible sex determination mechanisms as well as interpopulation karyological differences are discussed.

Keywords: aberrant sex determination; chromosome painting; comparative cytogenetics; genome architecture; mandarin vole; microdissection; high-throughput sequencing; rearrangements; rodents; sex chromosomes

1. Introduction

Most therian mammals have a conventional XX/XY sex chromosome system with the Y-borne testis-determining *SRY* gene. Nevertheless, several dozen species with nonstandard systems of chromosomal sex determination have been described among mammals [1]. There are species with isomorphic sex chromosomes in males and females (three species of *Ellobius* genus), with the absence of the regular Y chromosome (e.g., *Dicrostonyx torquatus*) or the *SRY* gene (e.g., *Ellobius lutescens, Tokudaia*), with the Y chromosome in females (e.g., *Myopus schisticolor*), with heteromorphism of the X chromosomes or multiple sex chromosomes (see more examples in [2]). Most species of mammals with aberrant sex chromosome systems belong to the subfamily Arvicolinae (Myomorpha, Rodentia). One such example is the mandarin vole, *Lasiopodomys mandarinus*.

The first karyotype descriptions of *L. mandarinus* made in the 1970s and further works showed the variability of chromosomal numbers among and within populations of this species. In the mandarin voles from Mongolia and Buryatia (*Lasiopodomys mandarinus vinogradovi*) the diploid chromosome number (2n) is 47–48 [3], whereas Chinese populations display 2n = 49–52 (*L. m. mandarinus*, Henan province [4–6]), 2n = 48–50 (*Lasiopodomys mandarinus mandarinus*, Shandong province [7]), or 2n = 47–50 (*Lasiopodomys mandarinus faeceus*, Jiangsu province [8]). Comparative cytogenetic studies made with G-banding and routine staining indicated intrapopulation variability in morphology of some chromosome pairs in karyotypes of *L. mandarinus*, specifically, two pairs of autosomes (No. 1 and No. 2) and sex chromosomes. Each of the studied populations is characterized by large heteromorphic X chromosomes that differ both in shape and size. Wang et al. [7] suggested that the unusual X chromosome variability in *L. mandarinus* originated through translocation of autosomes onto sex chromosomes. The autosomal polymorphism is subspecies-specific, not associated with sex and sex chromosomes, and caused by presumed inversions based on G-banding analysis [7,9]).

Sex chromosome systems of *L. m. vinogradovi* have been investigated first with G-banding and routine staining [3] and recently by cross-species chromosome painting [10]. Using the last method, Gladkikh et al. [10] demonstrated the origin of neo-X chromosomes by at least two independent autosome-sex chromosome translocation events. The complex of sex chromosomes in the only female (2n = 47) studied by these authors consisted of one metacentric chromosome (neo-X1), one submetacentric chromosome (neo-X2), and one small acrocentric (neo-X3). But at least two other sex chromosome systems exist in *L. m. vinogradovi*. In some females (2n = 47) the system is represented by the neo-X2 and two small acrocentrics. A male combination (2n = 48) is represented by the neo-X1 plus three small acrocentrics, one of which is considered to be the Y chromosome [3]. Both of these karyomorphs were described based on the examination of a relatively small sample with traditional methods unable to determine homology among small acrocentrics [3].

All studied males from the Chinese population had an unpaired acrocentric chromosome that could be a Y chromosome [7,8,11]. Analysis of the synaptonemal complex of *L. mandarinus* from China showed that there was indeed a chromosome that could pair with the X chromosome [12,13]. It was also shown that the sex chromosomes of the male *L. m. vinogradovi* pair and recombine at pachytene [14]. Studies on *L. m. mandarinus* demonstrated that sex determination in the subspecies is independent of *SRY* or *R-spondin 1* [13]. Chen et al. [15] also excluded the *Sall 4* gene as a potential testis-determining factor in this subspecies. Up to now, all attempts to find a chromosome carrying any Y chromosome-specific genes or regions (*SRY*, *Rbm*-gene family, PAR) in *L. mandarinus* using molecular approaches failed [5,15,16]. Thus, the question about the presence of a Y chromosome in karyotypes of male mandarin voles is actually controversial.

Despite the unusual heteromorphism of X chromosomes and failure to detect any testis-determining gene, the mandarin vole was, by default, considered as a species with a standard, XY males/non-Y females, sex determination system. Within the framework of this hypothesis, the absence of several predicted sex chromosome combinations (specifically, neo-X1/neo-X1 females expected in the progeny of males and females carrying neo-X1 chromosome, and neo-X2/Y males, expected in the progeny of males and females carrying neo-X2 chromosome) needs explanation. The failure to reveal these combinations may be either the consequence of small sample size or low viability of their carriers. In the latter case, the reduced fertility is expected for the neo-X2 females because three-quarters of their offspring from crossing with neo-X1/Y males (Y/0, neo-X1/0, and neo-X2/Y) should be nonviable. Also, under conventional sex determination with normal meiotic chromosome segregation, female carriers of a single neo-X2 should deliver only daughters. These predictions can be tested by the crossbreeding experiments.

To elucidate the sex determination system in *L. mandarinus vinogradovi,* we carried out a comprehensive study which combined several different conventional and molecular cytogenetic methods, single chromosome DNA sequencing, and breeding experiments revealing the chromosome segregation pattern as well as the reproductive performance of different karyomorphs. Comparative molecular cytogenetic research methods have been applied to achieve a more detailed description of the karyotype of this species and a deeper study of the autosomal polymorphism.

2. Materials and Methods

2.1. Ethics Statement

All applicable international, national, and/or institutional guidelines for the care and use of animals were followed. All experiments were approved by the Ethics Committee on Animal and Human Research of the Institute of Molecular and Cellular Biology, Siberian Branch of the Russian Academy of Sciences (IMCB, SB RAS), Russia (order No. 32 of 5 May 2017). This article does not contain any studies with human participants performed by any of the authors.

2.2. Specimens Sampled

In total, the karyotypes of 205 voles (163 females and 42 males) were examined with conventional cytogenetic methods. Of them, 27 individuals were captured in Selenginskii and Dzhidinskii districts of Buryatia in 2002–2017. The rest were captive-born descendants of these voles. Twelve animals (7 voles from the same laboratory colony and 5 voles captured in Selenginskii districts of Buryatia in 2017) were chosen for molecular cytogenetic study and chromosome sequencing.

2.3. Chromosome Preparation and Chromosome Staining

For karyotyping, chromosome suspensions were obtained from bone marrow and/or spleen by a standard method with preliminary colchicination of animals [17]. For some individuals, short-term culture of bone marrow was used. For molecular cytogenetic study, metaphase chromosome spreads were prepared from primary fibroblast cultures as described previously [18,19]. The fibroblast cell lines were derived from biopsies of skin, lung, and tail tissues in the case of laboratory animals and from finger biopsy in the case of wild animals as described previously [10]. All cell lines were deposited in the IMCB, SB RAS, cell bank ("The general collection of cell cultures", No. 0310-2016-0002). Cell cultures and chromosome suspensions were obtained in the Laboratory of animal cytogenetics, the IMCB, SB RAS, Russia.

G-banding was performed on chromosomes of all animals prior to fluorescence in situ hybridization, using the standard trypsin/Giemsa treatment procedure [20]. C-banding has followed the classical method [21] or the method with some modifications [10,21].

2.4. Crossbreeding Experiments

We sexed 327 offspring delivered by 38 females and surviving to at least 24 days of age. The dams were karyotyped, and offspring sex ratios in the pooled progeny obtained from dams of each karyomorph were compared. For each female karyomorph, the observed ratio of male to female offspring in the pooled progeny was compared with an even sex ratio with Chi-square goodness-of-fit test or, in case of a small sample, with Fisher's exact test. Female offspring ($n = 64$) born to 19 of the same dams were karyotyped, and the proportions of daughters holding different karyomorphs in the pooled progeny were calculated and compared between karyomorphs using 2 X 3 Fisher's exact test.

2.5. Female Reproductive Success in Relation to Karyomorphs

Thirty-two virgin females older than 70 days were paired with unrelated unfamiliar males. All pairs were maintained under standard conditions (see [22] for details). The females were weighed weekly until the detection of pregnancy, after which the nests were checked every two days until delivery, and then again once a week. Thus, the litter sizes were determined no later than the second day after birth. The number of surviving offspring was determined at weaning (on Day 24 after birth). The pairs were monitored for three months. The dams were karyotyped immediately or within a few months after the end of this experiment. We estimated the effects of the dam's karyomorph on the following characteristics of reproductive success over a three-month period: Proportion of females who gave birth (Fisher's exact test), number of litters, total number of the delivered offspring, and total number of the weaned offspring; the last three parameters were determined and compared for those females who gave birth (Student's *t*-test).

All tests were two-tailed and the α level of significance was 0.05.

2.6. Microdissection and Probe Amplification

G-banding by trypsin using Giemsa (GTG-banding) was performed before microdissection to accurately identify chromosomes. Glass needle-based microdissection was performed as described earlier [23]. One copy of each sex chromosome was collected. Chromosome-specific libraries were obtained using whole genome amplification (WGA) kits (Sigma). After amplification, DNA was purified using nucleic acid purification kits for DNA (BioSilica). DNA libraries were labeled using WGA. Chromosome-specific probes were obtained for chromosomes neo-X1 (probe L2), neo-X2 (probe L3) from LMAN19f, neo-X2 (probe L33) and neo-Y (probes L11 and L13 (distal part)) from LMAN14f, neo-X1 (probe L8) and neo-Y (probe L5) from LMAN15m, and neo-X2 (probe L31) from LMAN5f. Here, LMAN is an individual of *L. m. vinogradovi* (f, female; m, male).

2.7. Fluorescence in Situ Hybridization (FISH)

The sets of flow-sorted chromosomes of field vole (*Microtus agrestis*, MAG) and Arctic lemming (*Dicrostonyx torquatus*, DTO) painting probes were described previously [10,24–27]. The telomeric DNA probe was generated by PCR using the oligonucleotides $(TTAGGG)_5$ and $(CCCTAA)_5$ [28]. Clones of human ribosomal DNA (rDNA) containing partial 18S, full 5.8S, and a part of the 28S ribosomal genes and two internal transcribed spacers were obtained as described in Maden et al. [29]. FISH was performed following previously published protocols [30,31]. Images were captured using VideoTest-FISH software (Imicrotec) with a JenOptic charge-coupled device (CCD) camera mounted on an Olympus BX53 microscope. Hybridization signals were assigned to specific chromosome regions defined by G-banding patterns previously photographed and captured by the CCD camera. All images were processed using Corel Paint Shop Pro X3 (Jasc Software).

2.8. Sequencing

Libraries for sequencing were prepared according to the TruSeq Nano Library Preparation Kit (Illumina). Size selection was performed using the Pippin Prep. Quantification of the libraries before sequencing was performed using real-time PCR with SYBR GREEN. Then, 300-base pair paired-end reads were generated on Illumina MiSeq using the Illumina MiSeq Reagent Kit v3, according to the manufacturer's instructions. Raw reads were deposited in the Sequence Read Archive of the National Center for Biotechnology Information under accession PRJNA613194.

2.9. Bioinformatic Analysis

The reads obtained by sequencing were used in the DOPseq_analyzer pipeline (https://github.com/ilyakichigin/DOPseq_analyzer) to search for syntenic regions in the mouse genome assembly GRCm38. The operation of this pipeline was reported by Makunin et al. [32]. It can be briefly described as follows. First, the cutadapt 1.18 tool [33] removes the sequences of Illumina adapters and primers used for amplification. The purified reads are aligned on the mouse genome GRCm38 (to search for target regions) and the human genome GRCh38 (to remove contamination reads) using Burrows-Wheeler Aligner 0.7.17 [34], low-quality alignments (alignment length <20, mapping quality <20) were discarded. Then, by calculation, the density of alignment and the identification of target regions occurs using DNAcopy package [35]. The resulting coordinates are then checked manually in the UCSC (University of California, Santa Cruz) genome browser (https://genome.ucsc.edu).

The obtained coordinates for syntenic blocks are slightly different between libraries since they contain different amounts and diversity of the target DNA. To establish more accurate averaged boundaries of the evolutionary breakpoints, the reads obtained for all libraries were combined and reused in DOPseq_analyzer.

3. Results

3.1. Sex Chromosome Combinations Revealed by Extensive Karyotyping

The karyotypes of the studied individuals included, in addition to 22 pairs of autosomes common to males and females, four combinations of large heteromorphic sex chromosomes and small acrocentric chromosomes unidentifiable with conventional cytogenetic methods.

All studied males (42 individuals) had 2n = 48 and an identical system of sex chromosomes: Neo-X1 and three small acrocentrics (karyomorph I, thereafter KI). Males with neo-X2 were not found. Among 163 females, three sex chromosome combinations were revealed. Karyomorph II (KII, 47% of the studied females) corresponded to the sex chromosome system described by Gladkikh et al. [10] and had 2n = 47 with neo-X1, neo-X2, and neo-X3. Karyomorph III (KIII, also 47% of the studied females) had 2n = 47 with neo-X2 and two small acrocentrics. Karyomorph IV (KIV, 6% of females) had 2n = 48 with neo-X1, neo-X1, and two small acrocentrics.

3.2. Hybridization Experiment

Males were presented in the pooled progeny of all female karyomorphs. Sex ratio was female-biased in offspring of KII and KIII females. To the contrary, only sons were born to the few breeding KIV females (Table 1). The differences in the offspring sex ratio between KIV and the other two karyomorphs were significant (Fisher's exact test: KII vs. KIV, $p = 0.001$; KIII vs. KIV, $p = 0.002$). Sex ratio in the pooled sample, including the offspring of all females, was significantly female-biased (40% of males, $\chi^2 = 12.96$, df (degrees of freedom) = 1, $p < 0.001$).

Table 1. Comparison of sex ratio in the progeny born to the female carriers of different karyomorphs

Dam's Karyomorph (number of dams)	Number of Sons/Number of Progeny (%)	Deviation from the Expected = 0.5
II (17)	54/149 (36.2)	$\chi^2 = 11.2$; $p < 0.001$
III (18)	70/171 (40.9)	$\chi^2 = 5.66$; $p = 0.017$
IV (3)	7/7 (100)	Fisher's exact test: $p = 0.070$

Karyotyping of 19 dams and their 64 daughters showed that KII and KIII females produced mainly KIII and KII daughters, respectively. As one can expect, KIV females were not found among the daughters of KIII dams. This variant was very rare in the progeny of KII females. There was a significant difference between the two most common karyomorphs in the proportions of KII:KIII:KIV daughters in progeny (KII dams: 9:25:3; KIII dams: 21:6:0; $p < 0.001$).

3.3. *Female Reproductive Success Related to Their Karyomorphs*

Of the 32 females participating in the experiment, 13 (41%) belonged to KII, 15 (47%) to KIII, and four (13%) to KIV. Female carriers of the two most common karyomorphs did not differ in any measure of reproductive success (Table S1). At the same time, none of the rare KIV females produced offspring during a three-month period. This karyomorph significantly differed from the other two in the proportion of carriers that gave birth (Fisher's exact test for KII vs. KIV: $p = 0.002$; KIII vs. KIV: $p = 0.009$) (Table S1).

3.4. *Comparative Molecular Cytogenetic Investigation of Different L. m. vinogradovi Karyomorphs*

Comparative chromosome painting with two sets of painting probes was used for the analysis of karyotypes of 12 animals (Table 2). Since the set of *M. agrestis* probes showed almost complete identity of the autosomal sets in various individuals of *L. m. vinogradovi*, only partial localization of the *D. torquatus* probes was carried out on the chromosomes of most individuals. Application of comparative chromosome painting allowed us to establish that the acrocentric chromosomes participating in formation of complex sex chromosome systems in *L. m. vinogradovi* are homologous to MAG13/X/13 (designated here as neo-Y) and MAG17/19 (designated here as neo-X3, according to [10]) (Figures 1 and 2).

Table 2. The list of investigated individuals of *Lasiopodomys mandarinus vinogradovi* with abbreviated names, diploid numbers (2n), origin, systems of sex chromosomes, and types of autosome LMAN2; f, female; m, male; 2a, the acrocentric with the order of syntenic blocks MAG1/5; 2b, the acrocentric with the order of syntenic blocks MAG1/5/1/5/1/5; 2c, the submetacentric with the order of syntenic blocks MAG5/1/5/1. See comments in the text.

Abbreviation	2n	Origin	Complex of Sex Chromosomes	Type of Autosome LMAN2
LMAN0f	47	laboratory colony	neo-X1/neo-X2/neo-X3	2b2b
LMAN1f	47	laboratory colony	neo-X1/neo-X2/neo-X3	2b2b
LMAN2f	47	laboratory colony	neo-X1/neo-X2/neo-X3	2a2b
LMAN3f	47	laboratory colony	neo-X1/neo-X2/neo-X3	2a2b
LMAN5f	47	laboratory colony	neo-X2/neo-X3/neo-Y	2a2b
LMAN6f	47	laboratory colony	neo-X2/neo-X3/neo-Y4	2a2a
LMAN10m	48	laboratory colony	neo-X1/neo-Y/neo-X3/neo-X3	2a2a
LMAN14f	47	wild	neo-X2/neo-X3/neo-Y	2a2c
LMAN15m	48	wild	neo-X1/neo-Y/neo-X3/neo-X3	2a2c
LMAN16f	47	wild	neo-X1/neo-X2/neo-X3	2a2a
LMAN17f	47	wild	neo-X1/neo-X2/neo-X3	2a2c
LMAN19f	47	wild	neo-X1/neo-X2/neo-X3	2c2c

Sex chromosomes of KI (males) were represented by the largest metacentric (neo-X1) and a small-sized acrocentric (putative neo-Y) (Figure 1a). The autosomes homologous to MAG17/19 (neo-X3) should also be included in the complex of male sex chromosomes as they were present in single copy in KII. The MAGX probe hybridized to the p-arm of the neo-X1 chromosome and to the interstitial part of neo-Y chromosome (Figure 2e). MAG13 labeled the q-arms of neo-X and neo-Y (Figure 1a). The neo-X1 chromosome had three C-positive blocks on the q-arm (Figure 3a,b).

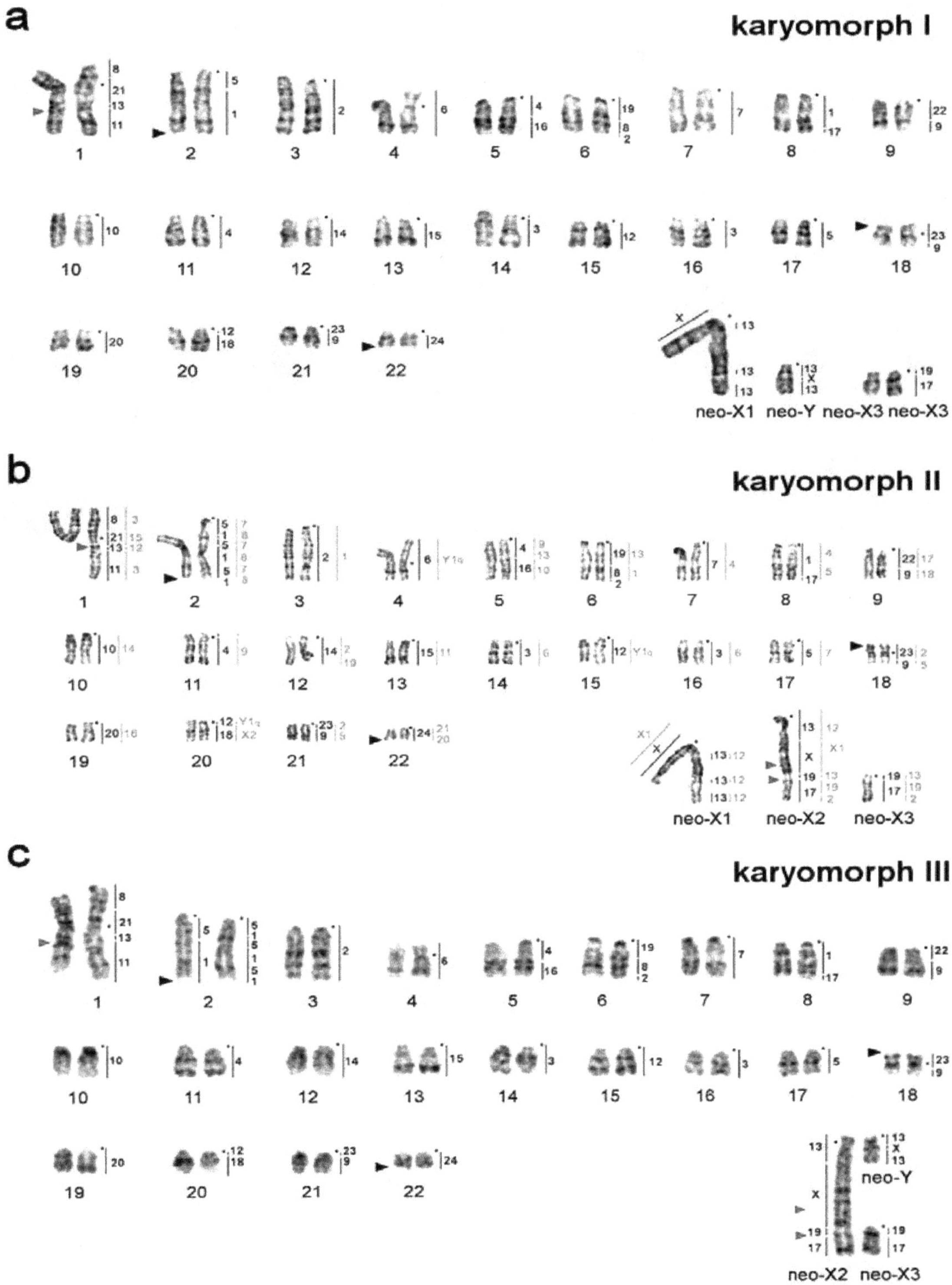

Figure 1. GTG-banded karyotype of *L. m. vinogradovi*: (**a**) LMAN10m, (**b**) LMAN0f [10], (**c**) LMAN5f. Black dots mark the position of centromeres. Vertical black bars mark the localization of *Microtus agrestis* (MAG) chromosome painting probes, vertical grey bars mark the localization of *Dicrostonyx torquatus* (DTO) painting probes. Numbers along the vertical lines correspond to chromosome numbers of *M. agrestis* and *D. torquatus*. Black triangles indicate sites of localization of ribosomal DNA clusters. Grey triangles indicate localization of the largest interstitial telomeric block.

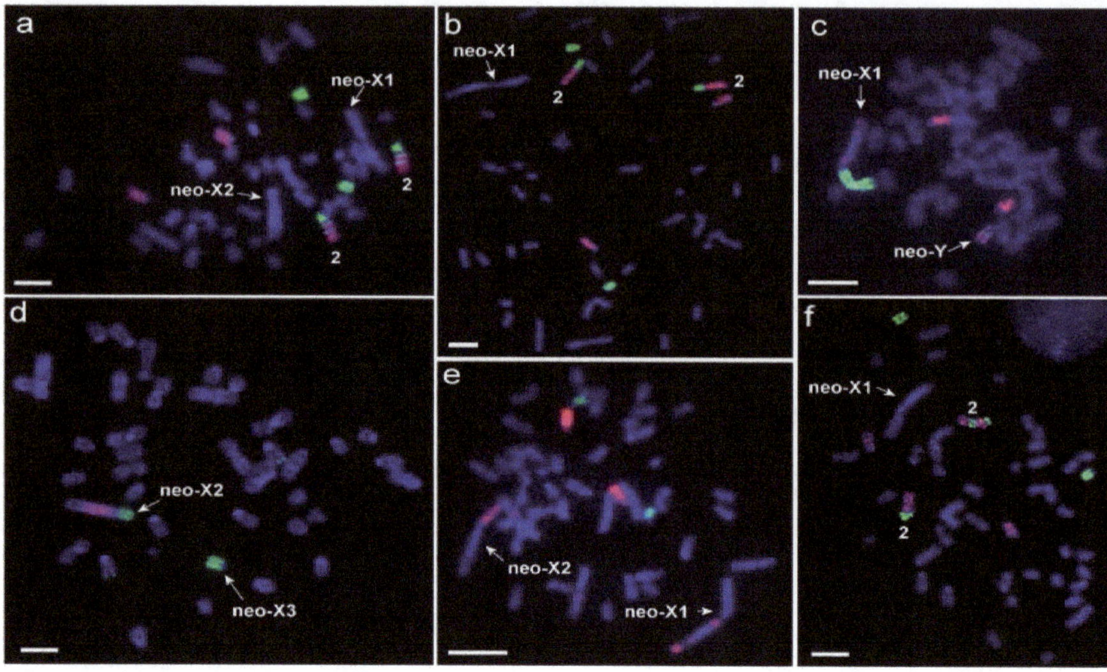

Figure 2. Examples of fluorescent in situ hybridization: (**a**) MAG5 (green) and MAG1 (red) onto LMAN1f, (**b**) MAG5 (green) and MAG1 (red) onto LMAN10m, (**c**) MAGX (green) and MAG13+14 (red) onto LMAN10m, (**d**) MAG17 (green) and MAGX (red) onto LMAN6f, (**e**) MAG23 (green) and MAG13+14 (red) onto LMAN1f, (**f**) MAG1 (red) and MAG5 (green) onto LMAN15m. Scale bar is 10 μm.

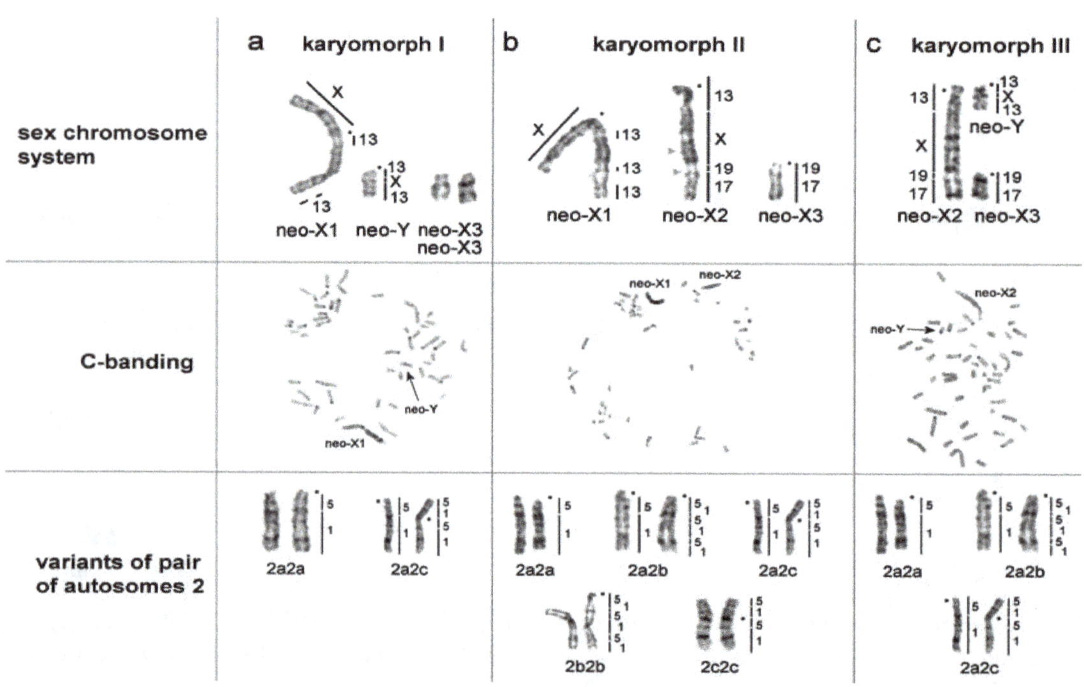

Figure 3. Polymorphic sex chromosomes and a pair of autosomes 2 in karyotypes of *Lasiopodomys mandarinus vinogradovi*: (a) Males (karyomorph I), (b) females, karyomorph II, (c) females, karyomorph III. From top to bottom: Sex chromosome system in the karyomorph, C-banding of metaphase chromosomes (chromosomes of LMAN10m (a), LMAN19f (b), and LMAN14f (c) are given as an example for each karyomorph), a pair of autosomes 2 of different individuals with localization of *Microtus agrestis* samples. Black arrows marked possible neo-Y.

Sex chromosomes of KIII (females) were represented by the large neo-X2, one small unpaired acrocentric corresponding to neo-X3, and another small unpaired acrocentric homologous to MAG13/X/13 (neo-Y) (Figure 1c). As in the case of KII the neo-X2 chromosome had a block of grey heterochromatin in the area homologous to MAGX (Figure 3b,c)

KIV (females) have not been found among the animals analyzed by molecular cytogenetic methods but the structure of their karyotype can be unequivocally reconstructed based on the analysis of other karyomorphs. Their sex chromosome complex can be described as a pair of neo-X1 chromosomes and two acrocentrics homologous to MAG17/19 (neo-X3).

MAGY probe labeled heterochromatic, C-positive parts of neo-X1 and neo-X2 chromosomes of male and female *L. m. vinogradovi*.

The field vole and the Arctic lemming painting probes showed the conservatism of autosomal sets in the species and the same localization of the probes as was shown previously for LMAN0f with the only exception the chromosome 2 [10]. The chromosome is designated here as LMAN2 without an additional letter. We described four variants of the distribution of syntenic blocks homologous to MAG1 and MAG5 onto LMAN2 due to inversions: Both homologs are acrocentrics with the order of probes MAG1/5 (LMAN6f, 10m, 16f) or MAG1/5/1/5/1/5 (LMAN0f, 1f), or MAG1/5 and MAG1/5/1/5/1/5 (LMAN2f, 3f, 5f). Both homologs are submetacentric with the order of probes MAG5/1/5/1 (LMAN19f). The LMAN2 is heteromorphic, and this heteromorphism is formed by an acrocentric chromosome homologous to MAG1/5 and a submetacentric chromosome with a derived structure MAG5/1/5/1 (LMAN14f, 15m, 17f) (Figures 1 and 2, Table 2).

Noncentromeric interstitial telomeric sequences (ITS) were localized on the autosome 1 and on the neo-X2 chromosome of all studied specimens. Each individual carried three rDNA clusters located on chromosomes 2, 18, and 22 (Figure 2). The size of rDNA clusters was different on homologs of autosome 2.

In general, no correlation was observed between sex chromosome systems and the morphology of chromosome pair 2, both among wild-caught and captive-born animals (Table 2).

3.5. Sequencing and Bioinformatic Analysis

Eight sex chromosome libraries were sequenced on the Miseq Illumina platform (Table S2). All sequencing and alignment statistics are presented in Table S3. Based on the low-coverage sequencing data obtained, the coordinates of large syntenic regions and the boundaries of evolutionary rearrangements relative to the mouse (*Mus musculus*, MMU) genome were identified (Table S2).

False-positive regions were detected on the chromosome neo-Y where the painting probe of MAGX chromosome labeled a small interstitial region (Figure 2a,c). The sequencing data did not show homology of the chromosome to the MMUX but indicated the synteny of the neo-Y chromosome with a region of the MMU18 only.

4. Discussion

The fascinating sex chromosome polymorphism of the mandarin vole raises three closely related questions: (1) How is such an unusual system maintained in a population? (2) How did the system evolve, that is, what chromosome rearrangements led to the polymorphism of sex chromosomes? (3) What is the specific (molecular) mechanism of sex determination in this species?

Up to now, the investigators mainly addressed the latter two problems. Here, on the contrary, we found it useful to focus on the first two questions, and our results provided the premise to approach solving the third, and perhaps, the main question for many evolutionary biologists.

4.1. How Is Such an Unusual System Maintained in Populations of L. m. vinogradovi?

If one assumes the presence of a single neo-Y chromosome to be both necessary and sufficient to initiate testis development, the described variation of sex chromosome combination is hard to explain.

Why some of the expected karyomorphic variants are missing, and where do the females with a single X chromosome (KIII) come from?

Theoretically, the offspring resulting from a cross between the females with the neo-X1 chromosome and males also carrying the neo-X1 chromosome should include neo-X1/neo-X1 females. These females have not been revealed in previous studies, but here we did find this rather infrequent karyomorph (KIV) due to comprehensive sampling. At the same time, no males with neo-X2/neo-Y, which are expected to occur among both KII and KIII females' progeny, were detected among more than 40 karyotyped male voles. From the "standard sex determination" point of view, their absence may be explained by the lethality of male embryos with neo-X2/neo-Y combination. However, this means that the KIII females produce three-quarters of nonviable embryos (neo-Y/0, neo-X1/0, and neo-X2/neo-Y), and KII females produce one-quarter of the nonviable embryos (neo-X2/neo-Y). According to this hypothesis, only KIV females are lucky to have no costs associated with nonviable offspring and, therefore, are predicted to have the highest reproductive success. This prediction does not seem to be supported by the results of our experiments. Further and most important, under the standard sex determination and normal sex chromosome segregation, KIII dams should produce only KII daughters. In fact, they also produce KIII daughters and KI sons. Finally, the finding that KII dams delivered a large proportion of KIII daughters also needs explanation. Within the framework of the standard sex determination hypothesis, the observed patterns require both a nondisjunction of Y chromosome in the second division and nonviability of most karyomorphs. This scenario appears to be very unlikely. On the other hand, the chromosome segregation pattern inferred from our results appears to be in good agreement with the "XY female hypothesis" [36–38]. This hypothesis requires neither chromosome nondisjunction nor the lethality of large proportions of offspring; the only one nonviable combination (neo-Y/neo-Y) is implied. The unequal proportions of different karyomorphs in progeny of each type of females may be explained by two phenomena.

The first one is apparent lower viability of neo-X1/neo-X1 females due to, for example, a violation in gene dosage compensation, or some other unknown causes. Unfortunately, the mechanisms of gene dosage compensation in this species are unexplored. The second plausible phenomenon is that the relative success of male neo-X1 and neo-Y spermatozoa depends on the karyomorph of a fertilized female. From our chromosome segregation data, it looks like male gametes carrying neo-Y are favored in neo-X1/neo-X2 females, resulting in higher than expected proportions of both sons (neo-X1/neo-Y) and neo-X2/neo-Y daughters (each about 40% vs. the expected 25%). To the contrast, an excess of sons (40% vs. 33% expected) and neo-X1/neo-X2 females (40% vs. 33% expected) in a progeny of neo-X2/neo-Y dams and predominance of sons in a progeny of neo-X1/neo-X1 (100% in our small sample) dams suggest the spermatozoa with neo-X1 chromosome to be favored in these crossbreeding combinations.

The cytogenetic mechanisms underlying these phenomena are unclear but it is noteworthy that they all reduce a sex ratio distortion. Thus, in terms of ultimate causes, these mechanisms might be selected as they increased the investment to sons which have higher reproductive value than daughters in a population with a female-biased primary sex ratio [39].

Our proposition that neo-X2/neo-Y mandarin voles are actually present but have a female phenotype received strong support from the results of fluorescence in situ hybridization and the comparative molecular sex chromosome investigation.

4.2. Complex Systems of Sex Chromosomes in L. m. vinogradovi and Their Origin

The comparative chromosome painting convincingly showed that at least two autosomal translocations on sex chromosomes took place in the evolution of the mandarin vole karyotypes forming the neo-X and neo-Y chromosomes in *L. m. vinogradovi*.

Comparing our sex chromosome sequencing data to previous comparative chromosome painting data [26] confirms that MAG13=MMU18, MAG17=MMU13/15, and MAG19=MMU15. However, our neo-Y sequencing data did not show homology of this chromosome to the mouse X chromosome, and this contrasts to the clear detection of X chromosome signal on the neo-Y chromosome by FISH. Therefore, it is likely that the FISH-signal represents shared repetitive sequences that are not included in the bioinformatic analyses. The fact that an unpaired chromosome with a small interstitial block of heterochromatin was detected by C-banding in karyotypes of both males and females further confirmed the presence of a block of repeated sequences on the chromosome (Figure 3a,c).

The regular Y chromosome has not been revealed in comparative chromosome painting experiments based on localization of the *M. agrestis* Y chromosome probe. Previously Zhao et al. [40] also failed to find a regular Y by FISH experiments with localization of partial human and whole mouse Y. Detection of signals from the MAGY probe in the heterochromatic, C-positive, parts of neo-X1 and neo-X2 chromosome of *L. m. vinogradovi* males and females may be caused by repeated sequences. As the Y chromosome of *M. agrestis* carries a huge block of heterochromatin, it can be assumed that both these arvicoline species have similar repeated sequences on their sex chromosomes. This assumption does not exclude the presence of sequences responsible for masculinization function in this area. This phenomenon requires a thorough study.

By analyzing the low-coverage chromosome sequencing data, we also failed to identify any Y chromosome-specific genes or regions. This may indicate either the elimination of the regular Y chromosome in this species or the insufficiency of our approach of aligning the reads of chromosomes on the mouse genome to search for the Y chromosome due to its rapid evolution and complex repetitive structure. To date, there are no Y chromosome assemblies for the representatives of Cricetidae family; it is possible that the forthcoming release of such genomic assemblies will allow us to answer the question about the presence of a regular Y chromosome using bioinformatic comparative genomic analysis methods.

It should be noted here that *L. mandarinus* is one of the most unusual species in terms of the synaptic behavior of its sex chromosomes. The nature of the XY pairing observed in this species differs markedly from that revealed in all other arvicolines. It was proposed based on the recombination pattern detected in pachytene that the XY synapsis in *L. m. vinogradovi* is a derivative condition resulting from de novo translocated autosomal material [14]. The FISH results obtained here completely confirm the suggestion. Preservation of ITS at the confluence sites of ancestral autosomes and sex chromosomes also indicates that the translocation has occurred recently.

4.3. *Possible Mechanisms of Sex Determination in L. m. vinogradovi*

The pattern of association between phenotypic sex and sex chromosome combinations found in *L. m. vinogradovi* is similar to that in the wood lemming (*Myopus schisticolor*), collared lemmings (genus *Dicrostonyx*), and the African pygmy mouse (*Mus minutoides*) [41–44]. It suggests an X-linked mutation (in *L. m. vinogradovi*, the mutation on the neo-X2 chromosome) to prevent masculinization of neo-X2/neo-Y individuals, but exact genetic bases of the male-to-female sex reversal are unknown. In the case of the mandarin vole, the problem of sex determination mechanism is additionally complicated by the failure to reveal any Y-specific genes or regions. The following scenarios can be suggested.

(1) The neo-X2 chromosome contains an epistatic locus (B) suppressing the dominant male development trigger (A) located on the neo-Y (Figure 4a).

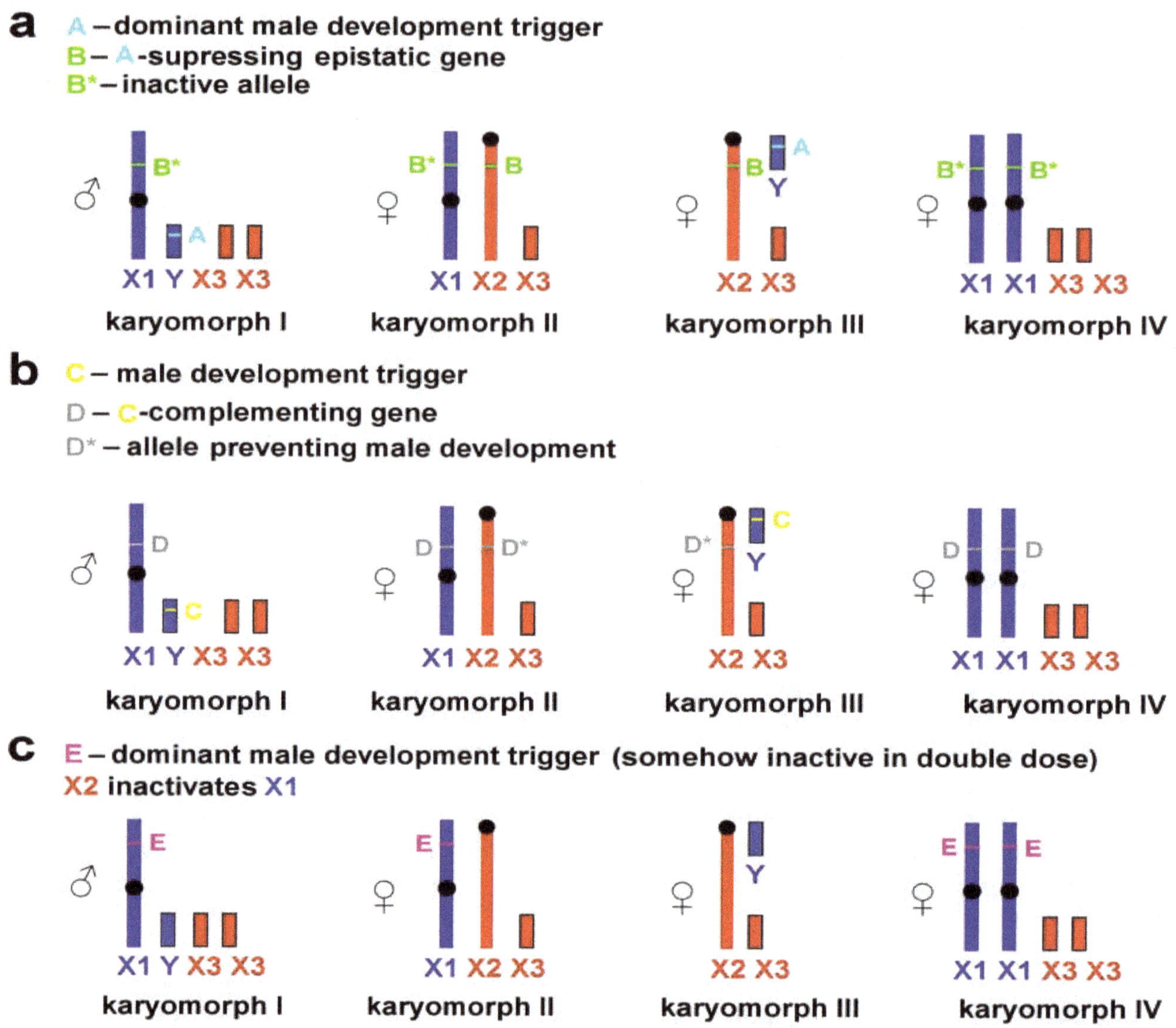

Figure 4. Possible mechanisms of sex determination in *Lasiopodomys mandarinus vinogradovi*: (**a**) a mechanism suggesting the presence of an epistatic locus on neo-X2 (X2) chromosome suppressing the dominant male development trigger on neo-Y (Y) chromosome, (**b**) a mechanism suggesting the presence of a locus on neo-X1 (X1) chromosome complementing the dominant male development trigger on neo-Y, (**c**) a mechanism suggesting the presence of male development trigger on neo-X1 chromosome and nonrandom inactivation of neo-X chromosomes. See comments in the text.

(2) The neo-X1 chromosome contains a locus (D) complementing the male development trigger (C), whereas this locus is absent from or is inactive on the homologous neo-X2 (Figure 4b). In this case, the sex-determination system is similar to that described for *Myopus schisticolor* [41] where a deletion differentiates two types of X chromosomes [43]; the same has been suggested for *Dicrostonyx torquatus* [42]. Zhu et al. [8] proposed a possible role of deletions in the formation of sex chromosomes in *L. m. faeceus*.

(3) Taking into account the uncertainty of the location of male development gene(s), it cannot be ruled out that they are associated not with neo-Y but with neo-X1 chromosome only (Figure 4c). According to this scenario, the neo-X2 chromosome either is capable of inactivating neo-X1 or carries a trigger-suppressing gene resulting in female phenotype of neo-X1/neo-X2 carriers Although nonrandom inactivation of the X chromosome has not been described for the mandarin voles, it has been identified in experiments on interspecific crosses of several arvicoline species [45].

This scenario, however, appears to be the least plausible as it requires the dominant male development trigger to be somehow inactive in double doses to produce neo-X1/neo-X1 females.

Whichever of the proposed scenarios is true, we believe that it should be the same in *L. m. vinogradovi* and *L. m. faeceus*. This assumption is based on the fact that the studied sample of voles from this Chinese population (Henan province) was represented by the same combinations of large X chromosomes and small acrocentrics, and in approximately the same proportions. In this population, a male karyomorph, KI, and female karyomorphs, KII and KIII, were common while KIV females were found as a rare variant [5]. In our opinion, information on the karyotypes of the mandarin voles from another Chinese population, Shandong province, deserves special attention. According to Wang et al. [7], females with a single X chromosome (KIII), common in other populations, were not found here, whereas an additional male karyomorph corresponding KIII has been reported. Thus, we assume that the aberrant sex determination system, in which some of the carriers of the Y chromosome display a female phenotype, has either not yet appeared or has already disappeared in this population. In our opinion, a detailed study of the sex chromosomes of voles from Shandong could shed light on the evolution of molecular mechanisms of sexual determination in this species.

4.4. Chromosomal Differences of L. mandarinus from Different Populations

Comparison of *L. mandarinus* showed clear differences in the karyotypes of individuals from different populations. So, the mandarin voles from Mongolia and Buryatia (*L. m. vinogradovi*), having diploid chromosome number 2n = 47–48, carried three pairs of metacentric chromosomes corresponding to pairs 1, 4, and 18 described in this work ([3], present data). The sizes of two types of chromosomes bearing regions homologous to MAGX, were approximately the same. All animals analyzed in the current work had stable and identical pairs of chromosome 1 (LMAN1), presented by two metacentric chromosomes. Both homologs carried interstitial telomeric sequences (ITS) in q-arms, separating synteny MAG8/21 and MAG11/13 (Figure 2). As previously suggested, that the association MAG11/13 is ancestral for the subgenus *Lasiopodomys*, the presence of ITS showed that the first pair of metacentric chromosomes was formed by the evolutionary recent fusion of the two ancestral pairs of chromosomes [10].

LMAN2 is homologous to MAG1 and MAG5. This fusion is characteristic of *L. m. vinogradovi* and it has never been found in any other arvicolines. The pair was polymorphic in the animal analyzed here due to multiple para- and pericentric inversions. It is important to notice here that both homologs of LMAN2 carry big clusters of ribosomal genes in the distal part of q-arms. It is possible that the fusion MAG1/5 is also characteristic of mandarin voles from other populations, but molecular cytogenetic methods must be used to verify this assumption.

In the karyotypes of all the studied individuals from China, there are only two stable pairs of bi-armed autosomes (corresponding to LMAN1 and 4 described in this work). Polymorphism of LMAN1 revealed in the Chinese population is not characteristic for individuals from the Buryatia. Zhang and Zhu [46] postulated that a Robertsonian fission is the main reason for the polymorphism of chromosome 1.

L. m. faeceus inhabiting the Jiangsu province in China has 2n = 47–50. The pair of largest autosomes is formed by two submetacentrics [8]. In the absence of molecular cytogenetic studies, it is difficult to state unequivocally, but it seems that sex chromosome systems are similar to those described in this work. Nevertheless, the relative sizes and ratio of lengths of arms of the submeta- and metacentric chromosomes attributed to the sex chromosomes are different, which may indicate a different accumulation of repeated sequences, or the presence of intrachromosomal rearrangements. It cannot be ruled out that other pairs of autosomes participated in the translocation of X chromosomes to autosomes. However, there is a polymorphism in one pair of autosomes in *L. m. faeceus*, apparently smaller than a pair of chromosomes 2 in *L. m. vinogradovi*. So, the autosomal polymorphism, previously described only for pairs of chromosomes 1 and 2, possibly affects other pairs of autosomes in karyotypes of the mandarin voles from different populations.

It is shown that *L. m. mandarinus* from Henan province (China) has a diploid number 2n = 49–52 [4–6], while the same subspecies in the Shandong province (China) has 2n = 48–50 [7]. Based on G-banding comparison we propose that the pair of chromosomes 1 in [7] is homologous to our LMAN1q. We should also note that the number of chromosomes bearing nucleolus organizer regions identified by Wang et al. [7] and in this work is different (4 vs. 3). Moreover, clusters of ribosomal genes are located on a pair of chromosomes 1 [7] (which corresponds to localization on LMAN1q). The morphology of sex chromosomes in *L. m. mandarinus* is similar to that described for *L. m. faeceus*. Surprisingly, among *L. m. mandarinus*, some males with a sex chromosome system morphologically similar to the karyomorphs III (females) described in this work were found [7], whereas among *L. m. faeceus*, females were discovered whose sex chromosome system was represented by two large submetacentric chromosomes [5].

It is known that the number of rDNA clusters and their localization can vary on various chromosomes even between closely related species [47]. This instability can be caused by a clustered structure of ribosomal genes that facilitate translocations by illegitimate recombination between nonhomologous chromosomes. Among mammals, multiple cases of interspecific variation in localization of rDNA clusters were described including presence on sex chromosomes [48]. Two of the three pairs of chromosomes carrying clusters of ribosomal genes in *L. m. vinogradovi* had stable morphology and localization of probes. At the moment, it remains unclear whether the polymorphism of the pair of autosomes 2 in *L. m. vinogradovi* and pair of autosomes 1 in *L. m. mandarinus* is associated with the location of the cluster of rDNA on them. The reasons for the significant polymorphism of these particular pairs of autosomes remain unclear.

In addition to the differences described above, individuals from different populations exhibit a different amount and distribution of heterochromatin ([5,7]; this work).

Thus, the karyotypes of the mandarin vole from all currently studied geographical populations are significantly different. In order to give a taxonomic assessment of these differences, it is necessary to study the karyotypes of *L. mandarinus* from different populations by molecular cytogenetic methods as well as applying molecular genetic data for the establishment of phylogenetic relationships between populations.

5. Conclusions

Euchromatic parts of mammalian sex chromosomes are highly conserved. Only rare cases of their involvement in rearrangements have been described in myomorph rodents, bats, carnivores, primates, and cetartiodactyls. Mandarin voles undoubtedly represent a unique species even among myomorphs and arvicolines. Their karyotypic features, such as the presence of different polymorphic pairs of autosomes and nonstandard sex chromosome systems, indicate significant plasticity of their genome, as well as ongoing processes of karyotypic evolution within the species. Such a diversity of sex chromosome systems as found in mandarin voles (within the same species) seems unique and has not been described yet in any other mammalian species. Such factors as modifications of the epigenetic state of DNA and accumulation of a large number of repeats may be required to trigger evolutionary plasticity [49].

Supplementary Materials
Table S1: Comparison between KII and KIII females by the reproductive success over a three-month period for the two most common female karyomorphs. Table S2: Chromosome-specific microdissected probes of *L. m. vinogradovi* and syntenic regions in mouse genome assembly GRCm38. Table S3: Sequencing and alignment statistics.

Author Contributions: Conceptualization, A.S.G. and A.V.S.; methodology, A.S.M., A.V.S., Y.M.K., S.A.R., O.L.G., N.A.L., and P.C.M.O.; software, D.Y.P.; validation, A.V.S., Y.M.K., S.A.R., and V.A.T.; formal analysis, S.A.R., A.V.S., Y.M.K., N.A.L., O.L.G., D.Y.P., and V.A.T.; investigation, A.V.S., Y.M.K., S.A.R., N.A.L., O.L.G., P.C.M.O., V.A.T., and N.A.S.; resources, A.V.S., Y.M.K., I.A.P., and F.N.G.; data curation, A.V.S. and S.A.R.; writing—original draft preparation, S.A.R. and A.V.S.; writing—review and editing, V.A.T., D.Y.P., M.A.F.-S., and P.C.M.O.; visualization, S.A.R. and A.V.S.; supervision, A.S.G.; project administration, A.S.G. and A.V.S.; funding acquisition, A.S.G. and A.V.S. All authors have read and agreed to the published version of the manuscript.

Acknowledgments: The authors gratefully acknowledge the resources provided by the "Molecular and Cellular Biology" core facility of the IMCB, SB RAS.

References

1. Bakloushinskaya, I.Y. Evolution of sex determination in mammals. *Biol. Bull.* **2009**, *36*, 167–174. [CrossRef]
2. Romanenko, S.A.; Volobouev, V.T. Non-sciuromorph rodent karyotypes in evolution. *Cytogenet. Genome Res.* **2012**, *137*, 233–245. [CrossRef] [PubMed]
3. Kovalskaya, Y.M.; Orlov, V.N. Unusual sex chromosomes and intrapopulation chromosomal polymorphism in the Chinese vole. *Tsitologiia* **1974**, *16*, 497.
4. Zhu, B.C.; Wang, H.Y.; Qu, A. Studies on fertility of XO females in the *Microtus mandarinus*. *Acta Zool. Sin.* **1998**, *44*, 209–212.
5. Zhu, B.; Gao, H.; Wang, H.; Gao, J.; Zhang, Y.; Dong, Y.; Hou, J.; Nan, X. The origin of the genetical diversity of *Microtus mandarinus* chromosomes. *Hereditas* **2003**, *139*, 90–95. [CrossRef]
6. Zhu, B.; Liu, J.; Xu, Y.; Zhang, Y.; Wang, T. Cytogenetic studies of brown field-mouse. *Yi Chuan Xue Bao* **1993**, *20*, 135–140.
7. Wang, J.X.; Zhao, X.F.; Deng, Y.; Qi, H.Y.; Wang, Z.J. Chromosomal polymorphism of mandarin vole, *Microtus mandarinus* (Rodentia). *Hereditas* **2003**, *138*, 47–53. [CrossRef]
8. Zhu, B.; Dong, Y.; Gao, J.; Li, P.; Pang, Y.; Liu, H.; Chen, H. Numerical and structural variations of the X chromosomes and no. 2 autosomes in mandarin vole, *Microtus mandarinus* (Rodentia). *Hereditas* **2006**, *143*, 130–137. [CrossRef]
9. Kovalskaya, Y.M.; Smorkatcheva, A.V. Overview of geographical variability of karyotype of Chinese vole *Lasiopodomys mandarinus* Milne-Edwards 1871 (Rodentia, Arvicolinae)–Another group of close species? In Proceedings of the Teriofauna of Russia and adjacent territories; Nauka: Moscow, 2011; p. 227.
10. Gladkikh, O.L.; Romanenko, S.A.; Lemskaya, N.A.; Serdyukova, N.A.; O'Brien, P.C.; Kovalskaya, J.M.; Smorkatcheva, A.V.; Golenishchev, F.N.; Perelman, P.L.; Trifonov, V.A. Rapid karyotype evolution in *Lasiopodomys* involved at least two autosome–sex chromosome translocations. *PLoS ONE* **2016**, *11*, e0167653. [CrossRef]
11. Liu, H.; Yan, N.; Zhu, B. Two new karyotypes and bandings in *Microtus mandarinus faeceus* (Rodentia): Two new karyotypes and bandings in *M. m. faeceus*. *Hereditas* **2010**, *147*, 123–126. [CrossRef]
12. Zhang, Z.; Gu, W.; Zhu, B.; Wang, T. Study on synaptonemal complexes in spermatocytes of *Microtus mandarinus*. *J. Shaanxi Norm. Univ. Nat. Sci. Ed.* **1999**, *27*, 93–96.
13. Yan, N.; Zhu, B.C.; Wang, Y.F. Recent advances on sex determining mechanisms of *Microtus mandarinus*. *Yi Chuan Hered.* **2009**, *31*, 587–594. [CrossRef] [PubMed]
14. Borodin, P.M.; Basheva, E.A.; Torgasheva, A.A.; Dashkevich, O.A.; Golenishchev, F.N.; Kartavtseva, I.V.; Mekada, K.; Dumont, B.L. Multiple independent evolutionary losses of XY pairing at meiosis in the grey voles. *Chromosome Res.* **2012**, *20*, 259–268. [CrossRef] [PubMed]
15. Chen, Y.; Ming, Q.; Zhu, B. Exclusion of Sall 4 as the sex-determining gene in the Mandarin vole *Microtus mandarinus mandarinus*. *Hereditas* **2011**, *148*, 93–97. [CrossRef]
16. Chen, Y.; Dong, Y.; Xiang, X.; Zhang, X.; Zhu, B. Sex determination of *Microtus mandarinus mandarinus* is independent of Sry gene. *Mamm. Genome* **2008**, *19*, 61–68. [CrossRef]
17. Graphodatsky, A.S.; Radjabli, S.I. *Chromosomes of Agricultural and Laboratory Mammals; An atlas*; Nauka: Novosibirsk, 1989; p. 128.
18. Stanyon, R.; Galleni, L. A rapid fibroblast culture technique for high resolution karyotypes. *Ital. J. Zool.* **1991**, *58*, 81–83. [CrossRef]
19. Romanenko, S.A.; Perelman, P.L.; Trifonov, V.A.; Serdyukova, N.A.; Li, T.; Fu, B.; O'Brien, P.C.; Ng, B.L.; Nie, W.; Liehr, T. A first generation comparative chromosome map between Guinea pig (*Cavia porcellus*) and humans. *PLoS ONE* **2015**, *10*, e0127937. [CrossRef]
20. Seabright, M. A rapid banding technique for human chromosomes. *Lancet* **1971**, *298*, 971–972. [CrossRef]
21. Sumner, A.T. A simple technique for demonstrating centromeric heterochromatin. *Exp. Cell Res.* **1972**, *75*, 304–306. [CrossRef]
22. Smorkatcheva, A.V. Parental care in the captive mandarin vole, *Lasiopodomys mandarinus*. *Can. J. Zool.* **2003**, *81*, 1339–1345. [CrossRef]

23. Yang, F.; Trifonov, V.A.; Ng, B.L.; Kosyakova, N.; Carter, N.P. Generation of paint probes from flow-sorted and microdissected chromosomes. In *Fluorescence in Situ Hybridization (FISH)-Application Guide*; Springer: Berlin, Heidelberg, 2017; pp. 67–79.
24. Telenius, H.; Ponder, B.A.; Tunnacliffe, A.; Pelmear, A.H.; Carter, N.P.; Ferguson-Smith, M.A.; Behmel, A.; Nordenskjöld, M.; Pfragner, R. Cytogenetic analysis by chromosome painting using DOP-PCR amplified flow-sorted chromosomes. *Genes. Chromosomes Cancer* **1992**, *4*, 257–263. [CrossRef]
25. Yang, F.; Carter, N.P.; Shi, L.; Ferguson-Smith, M.A. A comparative study of karyotypes of muntjacs by chromosome painting. *Chromosoma* **1995**, *103*, 642–652. [CrossRef]
26. Sitnikova, N.A.; Romanenko, S.A.; O'Brien, P.C.M.; Perelman, P.L.; Fu, B.; Rubtsova, N.V.; Serdukova, N.A.; Golenishchev, F.N.; Trifonov, V.A.; Ferguson-Smith, M.A.; et al. Chromosomal evolution of Arvicolinae (Cricetidae, Rodentia). I. The genome homology of tundra vole, field vole, mouse and golden hamster revealed by comparative chromosome painting. *Chromosome Res.* **2007**, *15*, 447–456. [CrossRef] [PubMed]
27. Romanenko, S.A.; Lemskaya, N.A.; Trifonov, V.A.; Serdukova, N.; O'Brien, P.C.M.; Bulatova, N.S.; Golenishchev, F.N.; Ferguson-Smith, M.A.; Yang, F.; Graphodatsky, A.S. Genome-wide comparative chromosome maps of *Arvicola amphibius, Dicrostonyx torquatus*, and *Myodes rutilus*. *Chromosome Res.* **2016**, *24*, 145–159. [CrossRef]
28. Ijdo, J.W.; Wells, R.A.; Baldini, A.; Reeders, S.T. Improved telomere detection using a telomere repeat probe (TTAGGG)n generated by PCR. *Nucleic Acids Res.* **1991**, *19*, 4780. [CrossRef] [PubMed]
29. Maden, B.E.; Dent, C.L.; Farrell, T.E.; Garde, J.; McCallum, F.S.; Wakeman, J.A. Clones of human ribosomal DNA containing the complete 18 S-rRNA and 28 S-rRNA genes. Characterization, a detailed map of the human ribosomal transcription unit and diversity among clones. *Biochem. J.* **1987**, *246*, 519–527. [CrossRef] [PubMed]
30. Yang, F.; O'Brien, P.C.M.; Milne, B.S.; Graphodatsky, A.S.; Solanky, N.; Trifonov, V.; Rens, W.; Sargan, D.; Ferguson-Smith, M.A. A Complete comparative chromosome map for the dog, red fox, and human and its integration with canine genetic maps. *Genomics* **1999**, *62*, 189–202. [CrossRef] [PubMed]
31. Graphodatsky, A.S.; Sablina, O.V.; Meyer, M.N.; Malikov, V.G.; Isakova, E.A.; Trifonov, V.A.; Polyakov, A.V.; Lushnikova, T.P.; Vorobieva, N.V.; Serdyukova, N.A.; et al. Comparative cytogenetics of hamsters of the genus *Calomyscus*. *Cytogenet. Genome Res.* **2000**, *88*, 296–304. [CrossRef]
32. Makunin, A.I.; Kichigin, I.G.; Larkin, D.M.; O'Brien, P.C.; Ferguson-Smith, M.A.; Yang, F.; Proskuryakova, A.A.; Vorobieva, N.V.; Chernyaeva, E.N.; O'Brien, S.J. Contrasting origin of B chromosomes in two cervids (Siberian roe deer and grey brocket deer) unravelled by chromosome-specific DNA sequencing. *BMC Genomics* **2016**, *17*, 618. [CrossRef]
33. Martin, M. Cutadapt removes adapter sequences from high-throughput sequencing reads. *EMBnet.journal* **2011**, *17*, 10–12. [CrossRef]
34. Li, H.; Durbin, R. Fast and accurate short read alignment with Burrows-Wheeler Transform. *Bioinformatics* **2009**, *25*, 1754–251760. [CrossRef] [PubMed]
35. Olshen, A.B.; Venkatraman, E.S.; Lucito, R.; Wigler, M. Circular binary segmentation for the analysis of array-based DNA copy number data. *Biostatistics* **2004**, *5*, 557–572. [CrossRef] [PubMed]
36. Bull, J.J.; Bulmer, M.G. The evolution of XY females in mammals. *Heredity* **1981**, *47*, 347–365. [CrossRef] [PubMed]
37. Fredga, K. Aberrant sex chromosome mechanisms in mammals. *Diffentiation* **1983**, *23*, S23–S30.
38. Bianchi, N.O. *Akodon* sex reversed females: The never ending story. *Cytogenet. Genome Res.* **2002**, *96*, 60–65. [CrossRef]
39. Fisher, R.A. *The Genetical Theory of Natural Selection*; Clarendon Press: Oxford, UK, 1930.
40. Zhao, S.; Shetty, J.; Hou, L.; Delcher, A.; Zhu, B.; Osoegawa, K.; de Jong, P.; Nierman, W.C.; Strausberg, R.L.; Fraser, C.M. Human, mouse, and rat genome large-scale rearrangements: Stability versus speciation. *Genome Res.* **2004**, *14*, 1851–1860. [CrossRef]
41. Fredga, K.; Groop, A.; Winking, H.; Frank, F. A hypothesis explaining the exceptional sex ratio in the wood lemming (*Myopus schisticolor*). *Hereditas* **1977**, *85*, 101–104. [CrossRef]
42. Gileva, E.A.; Chebotar, N.A. Fertile XO males and females in the varying lemming, *Dicrostonyx torquatus* Pall.(1779). *Heredity* **1979**, *42*, 67. [CrossRef]

43. Liu, W.-S.; Eriksson, L.; Fredga, K. XY sex reversal in the wood lemming is associated with deletion of Xp21-23 as revealed by chromosome microdissection and fluorescence in situ hybridization. *Chromosome Res.* **1998**, *6*, 379–383. [CrossRef]
44. Veyrunes, F.; Chevret, P.; Catalan, J.; Castiglia, R.; Watson, J.; Dobigny, G.; Robinson, T.J.; Britton-Davidian, J. A novel sex determination system in a close relative of the house mouse. *Proc. R. Soc. Lond. B Biol. Sci.* **2009**, *277*, 1049–1056. [CrossRef]
45. Zakian, S.M.; Kulbakina, N.A.; Meyer, M.N.; Semenova, L.A.; Bochkarev, M.N.; Radjabli, S.I.; Serov, O.L. Non-random inactivation of the X-chromosome in interspecific hirbid voles. *Genet Res.* **1987**, *50*, 23–27. [CrossRef] [PubMed]
46. Zhang, M.; Zhu, B. Relation between polymorphism of chromosome No. 1 and chromosome number in *Microtus mandarinus*. *J. Shaanxi Norm. Univ. Nat. Sci. Ed.* **2000**, *28*, 83–86.
47. Hsu, T.C.; Spirito, S.E.; Pardue, M.L. Distribution of 18+28S ribosomal genes in mammalian genomes. *Chromosoma* **1975**, *53*, 25–36. [CrossRef] [PubMed]
48. Proskuryakova, A.; Kulemzina, A.; Perelman, P.; Serdukova, N.; Ryder, O.; Graphodatsky, A. The Case of X and Y Localization of Nucleolus Organizer Regions (NORs) in *Tragulus javanicus* (Cetartiodactyla, Mammalia). *Genes* **2018**, *9*, 312. [CrossRef] [PubMed]
49. Cazaux, B.; Catalan, J.; Veyrunes, F.; Douzery, E.J.; Britton-Davidian, J. Are ribosomal DNA clusters rearrangement hotspots? A case study in the genus *Mus* (Rodentia, Muridae). *BMC Evol. Biol.* **2011**, *11*, 124. [CrossRef]

7

Spatial and Temporal Dynamics of Contact Zones between Chromosomal Races of House Mice, *Mus musculus domesticus*

Joaquim T. Tapisso [1,†], Sofia I. Gabriel [1,2,†], Ana Mota Cerveira [1,2], Janice Britton-Davidian [3,‡], Guila Ganem [3], Jeremy B. Searle [4], Maria da Graça Ramalhinho [1] and Maria da Luz Mathias [1,*]

1 CESAM – Centro de Estudos do Ambiente e do Mar, Faculdade de Ciências, Universidade de Lisboa, 1749-016 Lisboa, Portugal; jstapisso@fc.ul.pt (J.T.T.); sigabriel@fc.ul.pt (S.I.G.); ana.cerveira@gmail.com (A.M.C.); gramalhinho@fc.ul.pt (M.d.G.R.)

2 Departamento de Biologia, Universidade de Aveiro, Campus Universitário de Santiago, 3810-193 Aveiro, Portugal

3 ISEM – Institut des Sciences de l'Evolution de Montpellier, Université de Montpellier, CNRS, EPHE, IRD, 34095 Montpellier, France; mlmathias@fc.ul.p (J.B.-D.); Guila.Ganem@umontpellier.fr (G.G.)

4 Department of Ecology and Evolutionary Biology, Corson Hall, Cornell University, Ithaca, NY 14853, USA; jeremy.searle@cornell.edu

* Correspondence: mlmathias@fc.ul.pt

† These authors contributed equally to this work.

‡ In memoriam.

Abstract: Analysis of contact zones between parapatric chromosomal races can help our understanding of chromosomal divergence and its influence on the speciation process. Monitoring the position and any movement of contact zones can allow particular insights. This study investigates the present (2012–2014) and past (1998–2002) distribution of two parapatric house mouse chromosomal races—PEDC (Estreito da Calheta) and PADC (Achadas da Cruz)—on Madeira Island, aiming to identify changes in the location and width of their contact. We also extended the 1998–2002 sampling area into the range of another chromosomal race—PLDB (Lugar de Baixo). Clinal analysis indicates no major geographic alterations in the distribution and chromosomal characteristics of the PEDC and PADC races but exhibited a significant shift in position of the Rb (7.15) fusion, resulting in the narrowing of the contact zone over a 10+ year period. We discuss how this long-lasting contact zone highlights the role of landscape on mouse movements, in turn influencing the chromosomal characteristics of populations. The expansion of the sampling area revealed new chromosomal features in the north and a new contact zone in the southern range involving the PEDC and PLDB races. We discuss how different interacting mechanisms (landscape resistance, behaviour, chromosomal incompatibilities, meiotic drive) may help to explain the pattern of chromosomal variation at these contacts between chromosomal races.

Keywords: Robertsonian fusions; chromosomal evolution; distribution; clinal analysis

---

1. Introduction

Since the first study by Gropp et al. [1], numerous chromosomal races of the western house mouse (*Mus musculus domesticus*), resulting from centric or Robertsonian (Rb) fusions of pairs of acrocentric chromosomes, have been identified in Europe and North Africa [2,3]. Robertsonian races are defined by presence of metacentric chromosomes and chromosome numbers that can be reduced as low as 2n = 22, in contrast to the all-acrocentric 2n = 40 standard race [2]. Further chromosomal variability arises from the occurrence of whole-arm reciprocal translocations (WARTs) and through hybridisation [4,5]. Areas of contact between different chromosomal races have been extensively studied in the house mouse providing insights into the role of chromosomal variation on population divergence and speciation [3,5–9]. Specifically, it has been argued that chromosomal rearrangements may promote hybrid unfitness leading to reduced interracial gene flow close to the mutation breakpoint, as well as recombination suppression in the same genomic region. This can result in genetic divergence, ultimately leading to 'parapatric speciation', i.e., the situation where, in a geographical context, two neighbouring forms/races become separate species whilst in contact and hybridising [8,10–12]. Additionally, behavioural factors, such as divergence of mate recognition systems, environmental factors, such as barriers to dispersal, and stochastic factors, resulting from source/sink demographic population systems, may also contribute to decreased gene flow, thus enhancing reproductive isolation between the hybridising races [5,13–22].

The majority of chromosomal races in the house mouse are grouped into metacentric 'systems' occupying defined geographical areas. These enable contact and hybridisation between chromosomal races with different sets of metacentric chromosomes, but also between those metacentric races and the standard all-acrocentric race that surrounds the 'system' [2]. At present, about 20 hybrid zones have been described, which are discrete contacts between chromosomal races as well as some polymorphic areas referred to as unconfirmed hybrid zones, e.g. [3].

The Madeira 'system', characterising the western house mouse populations on the island of Madeira, involves an extraordinary chromosomal variability and includes six metacentric races with diploid numbers ranging from 22 to 38 [4,23]. So far, a single contact zone between two neighbouring parapatric races—the PEDC race (Estreito da Calheta) and the PADC race (Achadas da Cruz) [17]—has been found in a previous survey (1998–2002). The two races share seven Rb fusions (2n = 24) and differ by the occurrence of Rb (6.7) in the PEDC race and Rb (7.15) in PADC [23]. A notable characteristic of this zone is the staggered clines of the diagnostic fusions due to the presence of chromosome 7 in an acrocentric state, e.g. [5]. Human activities influencing the wider environment were proposed to be the main factors regulating the structure and location of this zone [17]. These two races occur in the westernmost part of Madeira, the PEDC race in the southwest plus in an apparently isolated area in the north coast, and the PADC race in a more north-westward area, located between the main distributional area of PEDC and its isolated population in the north (Figure 1).

In the present study, based on a recent survey (2012-2014), we reanalyse the distribution of the PADC and PEDC races and the extent and structure of the previously identified contact zone on the south coast of Madeira. We also extend the sampling area northwards to investigate the northern contact between the PEDC and PADC races. Furthermore, we include a third race (Lugar de Baixo, PLDB) in the analysis, occurring eastward of the southern range of PEDC and separated from this race by a 2 km-wide area of natural vegetation [17], that may represent a natural barrier for active dispersal of house mice. The PLDB race also shares a total of seven fusions (2n = 24) with the PEDC race, again only differing in one, Rb (15.18) vs. (5.18), respectively [23].

Most of the studies carried out on the structure of contact zones involving Rb races of *M. m. domesticus* have described the contact of two freely interacting populations, such as PADC and PEDC. Few studies have considered the influence of physical barriers on hybridising forms (e.g. [24] and references therein). Here we aimed to investigate: a) the structure of the contact zone between the PEDC and PADC races, by comparing the results from two surveys conducted over ten years apart, b) the chromosomal features of the northern populations of mice, that could not be analysed in the 1998–2002 survey because of unsuccessful trapping, and c) the effect of both environmental and physical barriers on mouse dispersal and the influence they may have on the chromosomal characteristics of the PEDC-PADC and PEDC-PLDB contact zones. Previously it was hypothesised that PEDC and PADC occupied habitats differed in quality, providing a basis for source-sink dynamics to emerge [17]. Spatial differentiation in habitat quality is the major factor stabilising source-sink systems, although other factors may also affect movements of individuals among populations, e.g. [25–27]. By attaining the above goals we expect to answer the following questions: i) has the contact zone between the PEDC and PADC races moved or changed in width?, ii) do populations of the PADC race and the northern isolated PEDC population interact?, iii) can source-sink dynamics explain the distribution of PEDC and PLDB and the movement of mice between these populations? iv) is there a natural barrier that could block active dispersal of house mice in this system (therefore influencing chromosomal characteristics)?

By answering these questions, we hope to better understand to what extent the chromosomal races studied are physically isolated and the potential impact of interaction or hybridisation between the races on further reproductive isolation and divergence.

2. Material and Methods

2.1. Study Area and Sampling of Mice

A total of five field trips to Madeira Island, each lasting approximately three weeks, were carried out between 2012 and 2014. Sampling of house mice took place across the geographical range of three metacentric races, Lugar de Baixo (hereafter PLDB), Estreito da Calheta (hereafter PEDC) and Achadas da Cruz (hereafter PADC) [17,23]. The races were named following Ramalhinho et al. [28] and Piálek et al. [2] and are all characterised by eight Rb fusions differing from each other by single fusions [17,23]. PLDB differs from PEDC by the presence of Rb (15.18) and absence of Rb (5.18) fusions, while PEDC differs from PADC by the presence of Rb (6.7) and absence of Rb (7.15) fusions. The combination of chromosomal fusions displayed by each race is as follows:

PLDB: Rb (2.4) (3.14) (6.7) (8.11) (9.12) (10.16) (13.17) (15.18)

PEDC: Rb (2.4) (3.14) (5.18) (6.7) (8.11) (9.12) (10.16) (13.17)

PADC: Rb (2.4) (3.14) (5.18) (7.15) (8.11) (9.12) (10.16) (13.17)

Based on the habitat preferences of the house mouse in Madeira [17], sampling sites were mainly located in human dwellings and human-modified habitats (farms, cultivated and fallow fields). Altogether, we sampled 65 sites along a south-northwest-east transect c. 83 km long, between the villages of Ponta do Sol and Chão da Ribeira (Figure 1; Table S1). The sites were numbered sequentially from site 1 at the southern end, to site 65 at the north-eastern end of the transect. Sites 6 to 50 correspond approximately to the sites previously surveyed by Nunes et al. [17] over the period 1998–2002. The distance between sampling sites ranged between c. 300 m and 8 km, according to terrain (Figure 1). All animals were captured in Sherman live-traps baited with sardine paste. Traps were set for one to three nights depending on trapping success. Mice were transported to animal facilities in Funchal ("Estação de Biologia Marinha") for karyotyping and data gathering. All applicable international, national, and/or institutional guidelines for the care and use of animals were followed. Capturing and testing were conducted by researchers certified by Ministério da Agricultura, do Mar do Ambiente e do Ordenamento do Território (01/2014/CAPT) for Portugal.

2.2. Chromosome Analysis

Karyotypes were prepared from bone marrow cells, following standard protocols used in earlier studies [4], namely the 'air-drying' procedure [29] and G-banding of chromosomes [30]. A minimum of three metaphases per individual were analysed, making sure that all homologous pairs of chromosomes were identified in each. Metaphases were observed under an Olympus microscope and karyotyped using Leica Chantal software. Chromosomes were identified following Cowell [31] at the Faculty of Sciences in Lisbon and at the Institute of Evolutionary Sciences, Montpellier. Chromosomal analyses were performed using an Olympus BX41 microscope with an attached Leica DC 250 camera equipped with a Leica CW4000 Karyo image analysis system.

The distribution of metacentric races was recorded following the methodology described in Nunes et al. [17]. Accordingly, the sampling sites were distributed along the main road running around the western region of Madeira Island (ER222 and ER101, Figure 1), because most houses were there and it likely acts as the dispersal route for mice. Excluding two sampling points, located 1250 m and 1750 m from the road, all other sites were located c. 2m to 865 m away from the road. In order to fully compare the results from both surveys (1998–2002 vs. 2012-2014), the exact same methodology was employed. As such, the positions of sampling sites off the main road were orthogonally projected onto it, see [17]. Sites located less than 200 m apart after projection were pooled and treated as a single site in all analyses. For each site, we: i) determined the diploid number of all animals, ii) identified all fusions, and iii) determined the percentage of acrocentric and metacentric chromosomes for diagnostic chromosomes and fusions.

2.3. Temporal Analysis

For the temporal analysis, we compared our new dataset with the one obtained between 1998 and 2002 (Table S2). Previously sampled sites were re-projected orthogonally along the main road to avoid errors when comparing distances among sites between the two periods. The 1998–2002 dataset includes the results reported by Nunes et al. [17] and unpublished data (Table S2). Temporal comparisons included: i) the chromosomal characteristics of sites and groups of sites for the whole transect, ii) the percentage of acrocentrics and metacentrics in sites and groups of sites also for the whole extension of the transect, and iii) the clinal pattern for diagnostic fusions of PEDC (Rb 6.7) and PADC (Rb 7.15) along the previously sampled transect analysed by Nunes et al. [17], i.e., sites 6 to 50 (km 10.1 to 58.17). The software package C-Fit8 was used to compute eight regression cline models (Table S3). Model selection was then performed based on the Akaike's information criterion [32] (Table S3). Logistic or scaled logistic functions were selected to best fit the metacentric clines:

$$f(x) = (e^{s(x-c)}/1 + e^{s(x-c)}) \times h, \quad (1)$$

where x is the geographic distance, s and c are the maximum slope and the centre of the cline, respectively, and h is the height. The width (expressed as an inverse of the slope) and cline centres were compared between the pair of fusions Rb (6.7) and Rb (7.15) for both time periods. To check for differences between clines, we assumed that twice the difference in $\log_e$ likelihood values between the constrained and unconstrained model (on cline centre for coincidence and on cline width for concordance) followed a χ^2 distribution with a *d.f.* equal to the difference of the parameters estimated, see [33].

3. Results

3.1. Current Chromosomal Variation in Western Madeira

After projecting all sampling sites on the road (Figure 1), 29 sites were pooled (in combinations of 2 to 4 sites). Pooling of sites restricted chromosomal analyses to a total of 48 sites. A total of 449 mice were trapped along the transect, 363 of which were successfully karyotyped. The number of karyotyped animals per site ranged from 1 to 31. Each site was allocated to a zone/race on the basis of frequency of the race-specific Rb fusions, i.e., Rb (15.18) vs. (5.18) and Rb (6.7) vs. (7.15) (Figure 1).

Fusion Rb (15.18) was identified along the first six sites of the transect (site 1 to 6 at km 0 to 10.1). From sites 1 to 4 (km 0 to 8.4) all mice carried this fusion in a homozygous state, and in sites 5 and 6 (km 8.48 and 10.1) in a heterozygous state as a result of hybridisation between PEDC and PLDB.

Fusion Rb (5.18), in a homozygous state, first occurred in a mouse at site 3 (km 7.16) and again at sites 5 and 6 (km 8.48 and 10.1), both in a homozygous state (5 out of 21 mice) and a heterozygous state (10 out of 21 mice). These heterozygous Rb (5.18) mice were the same that also carried the fusion Rb (15.18) in a heterozygous state mentioned above, again indicating the occurrence of hybridisation. From site 7 to site 48 (km 12 to 49.54) all mice carried the fusion Rb (5.18) in a homozygous state (211 mice). Exceptionally, at site 32 (km 32.95) chromosomes 5 and 18 were acrocentric in one of the sampled mice. From sites 49 to 57 (km 57.83 to 77.36) the fusion occurred in both homozygous (49 out of 58 mice) and heterozygous states (9 out of 58 mice). Along the last sites on the transect (site 58 to 65 at km 79.12 to 82.78) fusion Rb (5.18) was again homozygous (i.e., fixed) in all mice sampled (58 individuals) (Figure 2; Figure 3).

Fusion Rb (6.7) was homozygous in most animals collected from sites 1 to 18 (km 0 to 16.63) (96 out of 97 mice), with the exception of one heterozygous mouse for Rb (6.7) at site 6 (km 10.1). From site 19 (km 19.85) northwards the frequency of occurrence of this fusion gradually decreased, being absent between sites 35 to 55 (km 33.47 to 72.72) but occurring again between sites 56 and 65 (km 75.59 to 82.78) (51 out of 67 mice). In this section of the transect the fusion occurred both in a homozygous or heterozygous state.

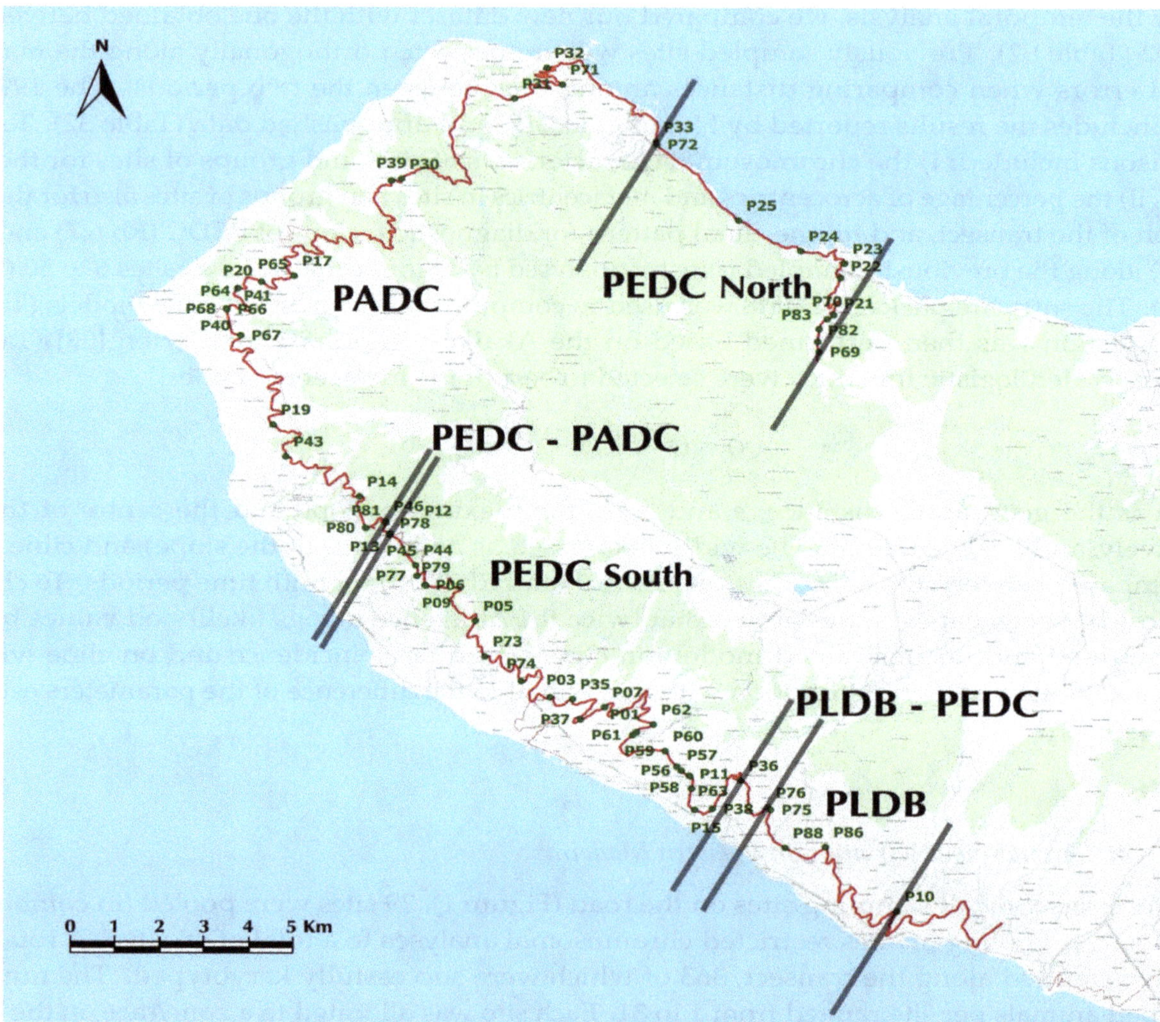

Figure 1. Location of sampling sites (dark green dots) orthogonally projected on the roads ER222 and ER101 (red line) of Madeira. See Table S1 for details on sampling sites. Black lines represent the limits of different chromosomal zones, PLDB represents the Lugar de Baixo race, PLDB-PEDC the contact zone between the Lugar de Baixo and Estreito da Calheta races, PEDC the Estreito da Calheta race, PEDC-PADC the contact zone between the Estreito da Calheta and Achadas da Cruz races and PADC the Achadas da Cruz race.

Fusion Rb (7.15) occurred in mice sampled from sites 33 to 53 (km 33 to 68.39), usually in a heterozygous state. Mice with acrocentric chromosomes 6, 7, and 15 co-occurred with mice carrying both Rb (6.7) and Rb (7.15) fusions from site 25 (km 28.6) until the end of the transect (km 82.78).

Based on the above description, it was possible to classify different segments of the transect according to the relative frequency of the four diagnostic Rb fusions (Figures 1 and 2). Animals assigned to the PLDB race occurred in the first segment on the south coast of Madeira (sites 1 to 6 at km

0 to 10.1). The identification of PLDB-PEDC hybrids in the last two sites of this segment (sites 5 to 6 at km 8.48 to 10.1) confirmed a hybrid zone between the two races. The next segment is characterised by animals carrying the PEDC diagnostic fusions in a homozygous state (sites 7 to 18 at km 12 to 16.63), followed by a longer segment where animals carried the fusion Rb (6.7) in a heterozygous state (sites 19 to 34 at km 19.85 to 33.42). This segment is followed by a very small area, where PEDC and PADC overlap, although no hybrids between them were identified (sites 33 to 34 at km 33 to 33.42). The next segment occupies the entire north-west corner of Madeira Island (sites 35 to 53 at km 33.47 to 68.39), where individuals were either assigned to PADC, due to the presence of Rb (7.15), or carried the acrocentrics 6, 7, and 15. The last segment of the transect (sites 56 to 65 at km 75.59 to 82.78) was characterised by mice with Rb (6.7), either in a homozygous or heterozygous state. A few animals along this last section of the transect had acrocentrics 6, 7, and 15 (16 mice).

3.2. *Temporal Variation in the Distribution of Madeira Rb Races*

The distribution of races PLDB, PEDC and PADC shows no significant differences between the two periods analysed (1998–2002 versus 2012-2014). Animals assigned to the race PLDB were found until km 10.1 (Figures 2a and 3a). In this section of the transect, the main difference between temporal periods is the detection of hybrids between races PLDB and PEDC in the 2012-2014 sampling (Figure 2a).

The next segment of the transect is characterised by mice assigned to race PEDC, with fusion Rb (6.7) in a homozygous state (Figures 2b and 3b), in the period 1998–2002 between km 11.7 and 13.9 and in the period 2012-2014 between km 12 and 16.6.

The following segment, starting at km 16.9 in 1998–2002 and at km 19.8 in 2012-2014, was characterised by the presence of homozygous and heterozygous mice for Rb (6.7), and mice carrying acrocentric chromosomes 6 and 7. This segment ended at km 31.9 in both periods analysed. This was followed by a very short segment defined by the presence of mice carrying fusions Rb (6.7) or (7.15) either in homozygous or heterozygous states and mice acrocentric for the chromosomes involved in both fusions.

Despite the substantial sampling effort in this segment, hybrids between races PEDC and PADC were exclusively found in the period 1998–2002. This segment was located between km 32.2 and 33.5 for the period 1998–2002 and between km 33.0 and 33.4 for the period 2012-2014.

Adjacent to this short region, a longer segment, ranging from km 34.2 to km 63.2 in 1998–2002 and from km 34.2 to km 68.4 in 2012-2014, is characterised by the presence of mice homozygous or heterozygous for the fusion Rb (7.15) as well as mice acrocentric for chromosomes 7 and 15.

The last segment of the transect was characterised by the presence of mice either homozygous or heterozygous for Rb (6.7), and mice carrying acrocentric chromosomes 6 and 7. For the period 1998–2002 this segment was located between km 77.2 and km 81.7, while for the period 2012-2014 it was located between km 72.7 and km 82.8.

Following the above results we assessed the present extent and location of the different chromosomal zones (Figure 1): zone **PLDB**; zone **PLDB–PEDC,** a contact zone between the PLDB and PEDC races; zone **PEDC South**, including mice either homozygous or heterozygous for Rb (6.7) and mice carrying acrocentric 6 and 7; zone **PEDC–PADC,** a contact zone between the PEDC and PADC races; zone **PADC,** with mice either homozygous or heterozygous for Rb (7.15) plus mice acrocentric for 7 and 15; and zone **PEDC North,** with identical characteristics to zone PEDC South.

Cline model estimates are summarised in Table 1. Tests of concordance and coincidence of the fusion Rb (6.7) revealed similarities between the two periods analysed (Figure 4). The results of maximum likelihood ratio tests revealed no significant differences in the cline width between years 1998–2002 and 2012-2014 (12.48 vs. 10.61, respectively; maximum likelihood difference of 1.43, $p = 0.232$), nor between cline centres (19.07 vs. 18.26; maximum likelihood difference of 1.70, $p = 0.192$). Cline analysis of the Rb (7.15) fusion also showed a similarity between cline width in the two periods analysed (2.09 vs. 1.96; maximum likelihood difference of 0.03, $p = 0.862$) but a significant difference between cline centres (24.71 vs. 23.89; maximum likelihood difference of 9.04, $p = 0.003$) was observed.

Table 1. Maximum likelihood estimates of cline centre position and cline width for Rb 6.7 (logistic regression model) and Rb 7.15 (scaled logistic regression model) and respective confidence intervals.

Clines	Centre	Confidence Interval	Width	Confidence Interval
Rb 6.7 1998–2002	19.07	18.37 - 19.76	12.48	10.70 - 14.95
Rb 6.7 2012–2014	18.26	17.22 - 19.30	10.61	8.73 - 13.52
Rb 7.15 1998–2002	24.71	24.16 - 25.26	2.09	1.49 - 3.52
Rb 7.15 2012–2014	23.89	23.45 - 24.33	1.96	1.19 - 5.52

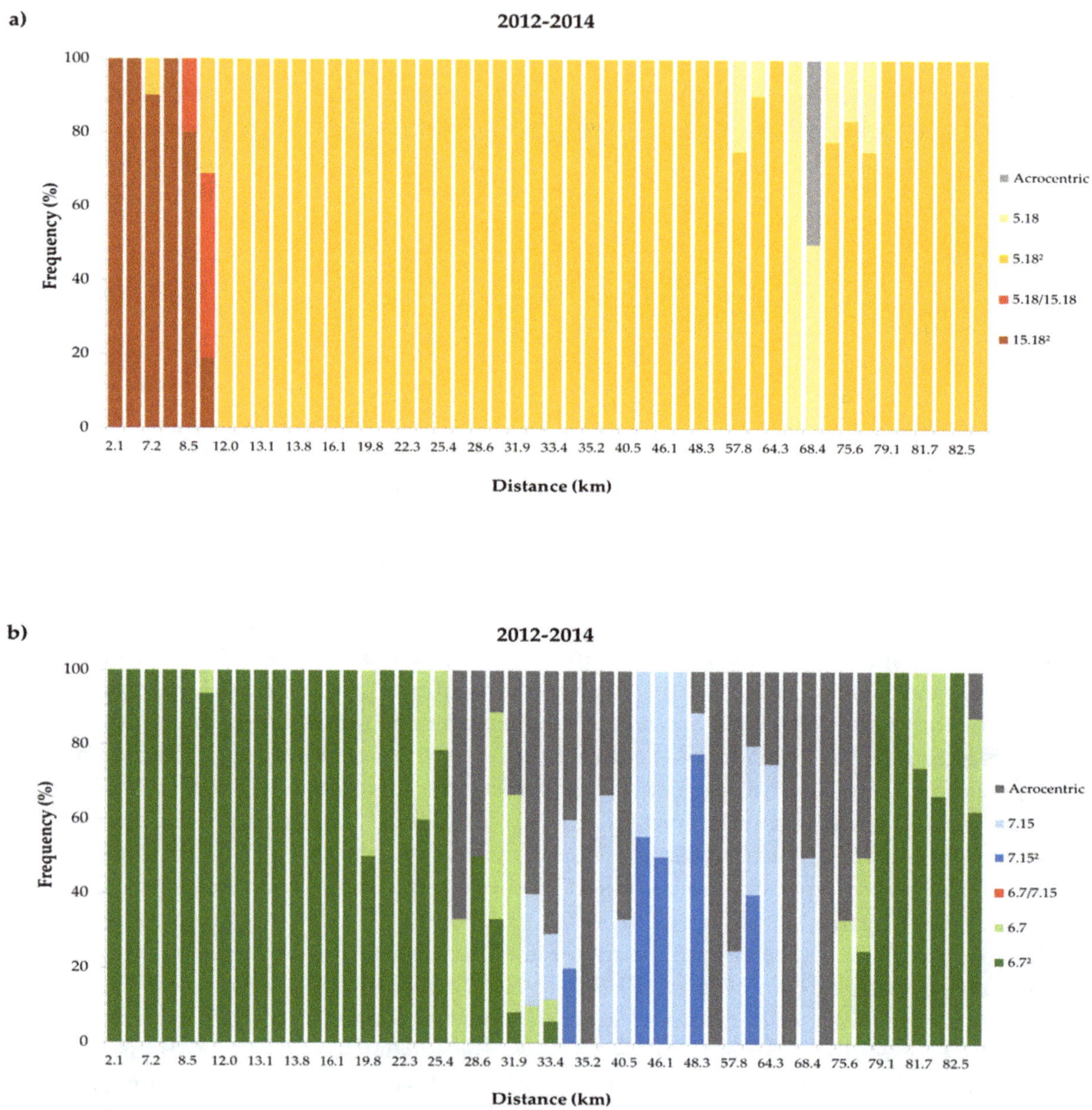

Figure 2. Frequency of diagnostic Rb fusions for the period 2012-2014. a) the upper panel shows frequencies of fusions Rb (15.18) and Rb (5.18) in homozygous [$(15.18)^2$; $(5.18)^2$] and heterozygous [(5.18)] state, frequency of hybrid mice carrying both fusions [(5.18)/(15.18)] and frequencies of mice

carrying none of the fusions (Acrocentric). b) the lower panel shows frequencies of fusions Rb (6.7) and Rb (7.15) in homozygous [$(6.7)^2$; $(7.15)^2$] and heterozygous [(6.7); (7.15)] states, frequency of hybrid mice carrying both fusions [(6.7)/(7.15)] and frequencies of mice carrying none of the fusions (Acrocentric).

a)

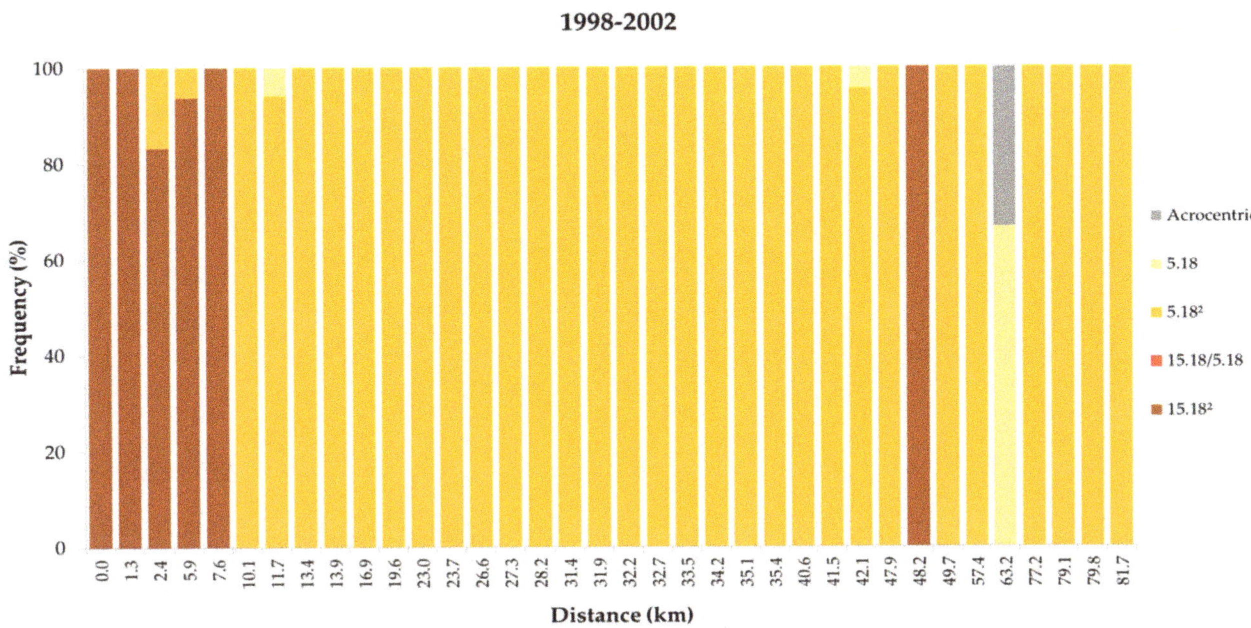

b)

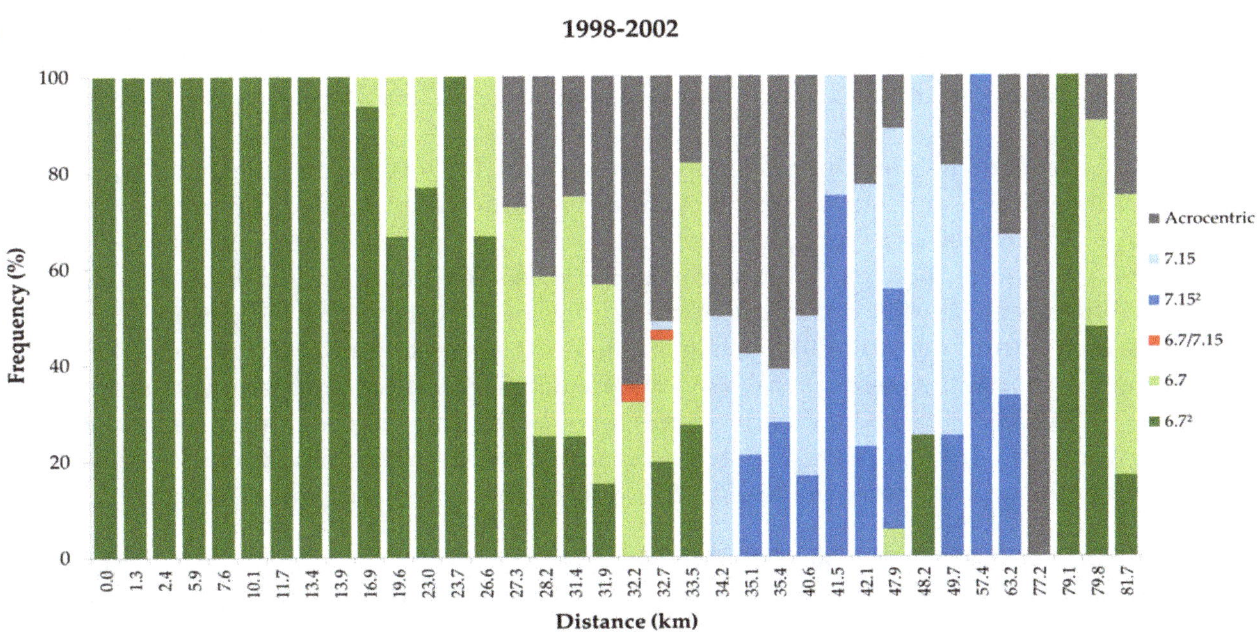

Figure 3. Frequency of diagnostic chromosomal fusions for the period 1998–2002. a) the upper panel shows frequencies of fusions Rb (15.18) and Rb (5.18) in homozygous [$(15.18)^2$; $(5.18)^2$] and heterozygous [(5.18)] states, frequency of hybrid mice carrying both fusions [(5.18)/(15.18)] and frequencies of mice carrying none of the fusions (Acrocentric). b) the lower panel shows frequencies of fusions Rb (6.7) and Rb (7.15) in homozygous [$(6.7)^2$; $(7.15)^2$] and heterozygous [(6.7); (7.15)] states, frequency of hybrid mice carrying both fusions [(6.7)/(7.15)] and frequencies of mice carrying none of the fusions (Acrocentric).

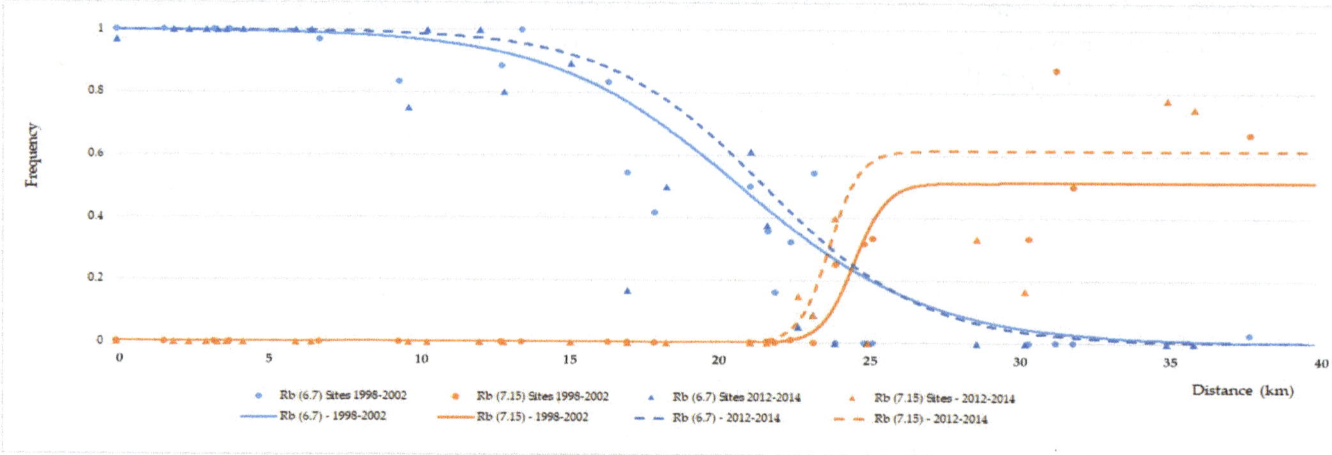

Figure 4. Comparison of clinal patterns for fusions Rb (6.7) (blue) and Rb (7.15) (orange) between the periods 1998–2002 (circles and full line) and 2012-2014 (triangles and dashed line).

4. Discussion

4.1. Current Distribution of Estreito da Calheta (PEDC) and Achadas da Cruz (PADC) Races

The possible involvement of chromosomal rearrangements in species formation has engendered much interest. Several models of speciation have been proposed, emphasising the role of hybrid zones between chromosomal races, e.g. [34,35]. The western house mouse, due to the accumulation and fixation of Robertsonian fusions, has been considered an excellent model to study how chromosomal rearrangements may be implicated in population divergence, race formation, and speciation, including the analysis of chromosomal hybrid zones [9]. Here, we compared the structure, position and width of the hybrid zone between two chromosomal races (PEDC and PADC) on Madeira Island at two time periods 10 years apart, and expanded the sampling efforts at both ends of the previously analysed transect.

Our results confirm the previous findings by Nunes et al. [17] of four chromosomal zones occurring along the south-western and northward distribution of the PEDC-PADC contact. As described in the previous study, the frequency of diagnostic fusions Rb (6.7) and Rb (7.15) varied along the transect. One of the main differences between the two time periods was the broadening of the PADC chromosomal zone, i.e., a wider geographic detection of mice assigned to this race, reflecting the increase of the sampling area during the 2012-2014 survey in the north-western part of the island. The increase of the transect along the north-western coast towards the northern side of the island also allowed another contact zone – PEDC North – to be identified between PADC and the previously described (and very geographically circumscribed) PEDC isolate on the north coast, characterised by the presence of mice mostly homozygous or heterozygous for Rb (6.7), with a few occasional mice exclusively acrocentric for these chromosomes.

Our 2012-2014 survey, particularly along the northern side of Madeira also revealed the occurrence of fusion Rb (1.15) whose distribution had not previously been identified on the island. This fusion, always found in a heterozygous state, appears to be more frequent in the PEDC North zone, having been detected in a total of 12 mice from 6 sampling sites. However, it was also very occasionally found outside this range, namely in the PEDC South zone (in two mice in a single location) and in the northernmost end of the PADC zone in a single mouse, heterozygous for Rb (7.15). In nature, chromosome 1 has been involved in several fusions, however Rb (1.15) is not a common one [2,36]. In Madeira, we suggest that Rb (1.15) may result from a recent mutation event, considering the fusion's somewhat restricted geographic distribution and its exclusive heterozygous state. This fusion seems to have emerged multiple times in the house mouse in situations where chromosomes 1 and 15 are available to fuse but, so far, it has never been found in a fixed state. A similar situation has been suggested by

Sans-Fuentes et al. [37] for the appearance of a rare fusion, Rb (7.17) in the polymorphic Barcelona 'system', and by Adolph & Klein [38] for several rare Rb fusions in mice from southern Germany.

Our results suggest that some temporal change has occurred in the spatial distribution of the diagnostic fusions in the PEDC-PADC parapatric area, resulting in the narrowing of the contact zone between them. Comparisons involving Rb (6.7) indicate that the general clinal pattern for this fusion did not change significantly between 1998–2002 and 2012–2014, i.e., no significant variation was observed in either the cline centre or width. However, for Rb (7.15), a statistically significant shift in the cline centre was detected towards the PEDC-PADC contact zone. This encompasses the main area of polymorphism, exhibiting an increased number of homozygous mice for the Rb (7.15) fusion over the last decade. Nevertheless, despite the change of the cline centre of Rb (7.15), the absence of a major geographical shift of this zone seems to exclude a possible numerical imbalance between the races, possibly related with a lower density of mice at the hybrid zone [17], allowing each population to evolve independently and 7.15 to increase towards fixation. A similar situation has been already described by Castiglia & Capanna [14] in a contact zone between two chromosomal races in Italy. Considering the observed narrowing of the contact zone after about 10 years, it would be interesting to address the dynamics of both chromosomal races and their hybrid zone after a longer timeframe.

4.2. *Dynamics of the Contact Zone Between Estreito da Calheta (PEDC) and Achadas da Cruz (PADC) Races*

Contact or hybrid zones are narrow regions where two more or less homogeneous parental forms meet and interbreed [39,40]. These zones are maintained by the influence of diverse factors, e.g., a balance between incompatibility of chromosomal races and dispersal [3,8,10,41–44]. This definition clearly fits the hybrid zone between PEDC and PADC chromosomal races. Two main mechanisms may determine patterns of gene flow between the hybridising taxa: i) 'suppressed recombination', contributing to the accumulation of genetic incompatibilities between parental types, and ii) 'hybrid dysfunction', suggesting that hybrids or heterozygotes are less fit, and thus are selected against [8,44,45]. The role of both mechanisms has been discussed for different geographic metacentric 'systems' of the house mouse [8,44,46,47].

A single study on the fertility of hybrids between PEDC and PADC discussed the more traditional concept model of 'hybrid dysfunction'. Results revealed that hybrids between these two races obtained under experimental conditions exhibited moderate subfertility (approximately a 50% decrease), suggesting that, under natural conditions, this level of underdominance could contribute to limit gene flow between parental populations [48].

A tension zone may move if there is a difference in population size (affecting dispersal rate) for the two races that are hybridising [42]. Nunes et al. [17], on the basis of the results of the 1998–2002 survey and on the distribution and abundance of crops and agricultural areas over the studied transect, suggested that potential habitats for mice are of better quality, more abundant and more evenly distributed over the area occupied by PEDC, while potential habitats for PADC are of poorer quality, fewer and more dispersed. This environmental contrast defines a source-sink continuum and could result in an asymmetrical dispersal rate into the contact zone between the two races [49,50]. In a source-sink system the contribution of sinks is indeed relevant for the persistence of individuals but a stable equilibrium between populations can be maintained in invariant environments [26]. According to this hypothesis, PEDC mice are expected to move more into/across the hybrid zone when compared with PADC mice. Despite a slight (non-significant) shift of the Rb (6.7) cline in that direction, further contributing to the observed narrowing of the contact zone, it was the Rb (7.15) fusion that significantly increased in frequency in the contact zone. Also, this increase may also be favoured by sexual selection as predicted in Nunes et al. [20], where patterns of preference of both males and females in the contact zone would favour Rb (7.15) individuals and thus contribute to the observed cline shift. Genomic analysis would be valuable in assessing the level and direction of gene flow across the hybrid zone, helping clarify the putative contribution of environmental quality on the dynamics of mouse movement across this area.

The relative stability of the zone may also be explained by the fact that environmental conditions in this area did not change significantly. In fact, during the period of approximately 10 years, between the past and present survey, no marked changes were noticed. The types of crops and land use were almost the same in 2012-2014 as before, with no notable reduction in agricultural areas associated with urbanisation, suggesting environmentally stable conditions. Hence, we did not expect demographical changes both within and at the border of the hybrid zone.

The role of commensalism of mice in combination with a fragmented landscape matrix and the extreme topography of Madeira has restricted the movement of mice and contributed to population isolation and fixation of fusions, namely the fixation of Rb (7.15) in the northernmost part of the transect [17,23].

The maintenance of the contact zone, although narrower, can also be explained by the persistence of an 'acrocentric peak', where mice carry acrocentric chromosomes 6, 7, and 15, limiting the production of hybrids (only two PEDC-PADC hybrids were identified in 1998–2002 and none in 2012-2014). There is an expectation that individuals that are acrocentric for chromosomes 6, 7, and 15 will be favoured because they cannot produce PEDC x PADC F_1 hybrids and therefore will have more grand-offspring than other karyotypic categories. This situation is described by Nunes et al. [46 and references therein] and Gündüz et al. [5]. These authors suggest that the comparatively lower fitness of PEDC x PADC F_1 hybrids, characterised by a chain-of-four configuration at meiosis I (6-6.7-7.15-15), will favour individuals homozygous for chromosomes 6, 7, and 15 thus leading to a high frequency of acrocentrics in the contact zone. On the other hand, meiotic drive may help to explain the slight but significant narrowing of the hybrid zone by maintaining metacentrics involving the fusions Rb (6.7) and Rb (7.15) in the area [51]. Natural variation in centromere strength has been found in wild populations of the house mouse, including in Madeira, where relatively stronger centromeres of metacentric chromosomes are preferentially retained in the egg in opposition to centromeres of acrocentrics, more likely to be segregated to the polar body [51]. As such, in these populations, where metacentrics tend to naturally accumulate, hybrid zones characterised by an 'acrocentric peak' may persist but become reduced in width, as observed in our study, reflecting the balancing forces in action as discussed above.

Thus, it seems reasonable to propose that multiple selective pressures (e.g., hybrid underdominance coincident with a peak of acrocentrics, meiotic drive, differences in habitat quality, habitat fragmentation) may be acting together resulting in the observed changes of the position and structure of the PEDC-PADC contact zone.

4.3. Contact Zone Between Races Estreito da Calheta (PEDC) and Lugar de Baixo (PLDB)

In the present study, we also analysed to what extent landscape discontinuities constrain mouse migration and the karyotypic structure of the contact zone between PEDC and PLDB. These two races have very limited distributions, with the PLDB occupying a restricted area east of PEDC which in turn is parapatric with PADC as discussed above.

The chromosomal analysis of mice across the distributional areas of the two races shows a contact zone between them encompassing approximately 3 km, i.e., from km 7.2 to km 10.1 along the transect, with all hybrid individuals detected between both races located between km 8.48 and 10.1.

It is worth pointing out that similarly to metacentrics Rb (6.7) and Rb (7.15), there was a possible role of WARTs in generating the diagnostic metacentrics Rb (5.18) and Rb (15.18) [4]. As such, it was not expected that chromosome 18, in particular, could occur here in an acrocentric form unless by migration of acrocentric-bearing mice, as inferred for the PEDC-PADC contact zone [5] or for contact zones between chromosomal races in Italy [7,14] and Spain [52].

The offspring resulting from the reproduction between mice of the PEDC and the PLDB races will most likely exhibit hybrid unfitness. Because races have metacentrics with monobrachial homology, the F_1 between Rb (5.18) and Rb (15.18) form a chain-of-four at meiosis I, therefore, as for PEDC x PADC offspring, underdominance in hybrids can be anticipated [46]. Nevertheless, a surprisingly large number of hybrids were detected between both races across a much wider area when compared to the PEDC-PADC contact zone. Assuming that the 2 km-wide band of non-commensal habitat is not an unbridgeable obstacle for active dispersal between the two races, in particular for PEDC mice, in addition to the possibility of passive dispersal movements, one can ask what mechanisms could have hampered the movement of PLDB mice across the valley. It would be interesting to further understand whether there is a sex-biased dispersion (expectedly by young males, [53]) into the PLDB race acting as a vehicle of gene flow contributing to the width of the contact zone between both races. Although the area inhabited by the PLDB race is occupied by sparse commensal habitat mostly surrounded by forests and woodlands, potentially offering fewer opportunities for mice to find shelter and food [54], whether this area can function as a sink as opposed to the PEDC source-area should be better investigated. In support of this hypothesis, a previous study showed that the amount of energy spent for maintenance is different between races and slightly lower in PLDB (energy intake: PLDB 28.96 ± 7.59 KJ day^{-1}; PEDC 32.80 ± 2.64 KJ day^{-1}) [55]. These differences in energy balance may reflect typically lower food availability for PLDB mice as a proxy for lower habitat quality [55,56]. This is again consistent with previous findings by Nunes et al. [17] regarding the PADC and PEDC races. Future studies will allow us to focus on a better understanding of the formation, maintenance, and evolution of this contact zone.

Most of the previously described hybrid zones in house mice involve a metacentric race and the all-acrocentric standard race [3]. The present data on the house mouse metacentric 'system' in Madeira reinforce the knowledge on the contact zones between different metacentric races. Further behavioural experiments coupled with genome-wide analysis will allow a better understanding of the underlying mechanisms maintaining the equilibrium of contact zones between Rb races. As such, it will be possible to assess levels of gene flow, its preferential direction (if any), identification of loci under selection, loci presenting clines across the contact zone, and its potential role in survival and reproduction [57,58].

Supplementary Materials

Table S1: Description of localities sampled between 2012 and 2014 along a transect, including location name, identification code (ID), site number, GPS coordinates (Latitude and Longitude), distance along the transect (km) and number of house mice karyotyped (N), Table S2: Description of sites sampled between 1998 and 2002 including location name, identification code (ID), site number, GPS coordinates (Latitude and Longitude), distance along the transect (km) and number of animals karyotyped (N), Table S3: Maximum likelihood estimates and Akaike values computed on CFit8 considering eight cline models.

Author Contributions: Conceptualization, J.T.T., S.I.G, A.M.C., J.B.S. and M.d.L.M.; Data curation, J.T.T., S.I.G. and A.M.C.; Formal analysis, J.T.T. and A.M.C.; Funding acquisition, S.I.G, A.M.C., G.G., J.B.S. and M.d.L.M.; Investigation, J.T.T., S.I.G, A.M.C., J.B.-D., G.G., M.d.G.R., J.B.S. and M.d.L.M.; Methodology, J.T.T., S.I.G, A.M.C., J.B.-D. and M.d.G.R.; Project administration, J.T.T., S.I.G, A.M.C. and M.d.L.M.; Resources, M.d.L.M.; Supervision, S.I.G; M.d.L.M.; Validation, J.T.T., S.I.G., A.M.C., J.B.-D., M.d.G.R. and J.B.S.; Visualization, J.T.T., A.M.C. and M.d.L.M.; Writing—Original draft, J.T.T. and M.d.L.M.; Writing—Review & editing, J.T.T., S.I.G, A.M.C., G.G., J.B.S. and M.d.L.M. All authors have read and agreed to the published version of the manuscript.

Acknowledgments: We here pay tribute to our colleague Janice Britton-Davidian - for her dynamism and dedicated scientific contribution that was absolutely central to our studies of the Madeira Rb system - and for her unforgettable friendship. We would like to thank Manuel Biscoito, Director of Estação de Biologia Marinha do Funchal for all the support in Madeira, in particular for the mice maintenance and laboratory facilities, Tomé Neves for his precious help with the Madeira map and statistical analysis, and Sophie von Merten for additional statistical advice. Finally, we would like to thank the three anonymous reviewers whose comments greatly improved the final version of the manuscript.

References

1. Gropp, A.; Tettenborn, U.; von Lehemann, E. Chromosomenuntersuchungen bei der Tabakmaus (*Mus poschiavinus*) und bei den Hybriden mit der Laboratoriummaus. *Experientia* **1969**, *25*, 875–876. [CrossRef] [PubMed]
2. Piálek, J.; Hauffe, H.C.; Searle, J.B. Chromosomal variation in the house mouse: A review. *Biol. J. Linn. Soc.* **2005**, *84*, 535–563. [CrossRef]

3. Hauffe, H.C.; Giménez, M.D.; Searle, J.B. Chromosomal hybrid zones in the house mouse. In *Evolution of the House Mouse*; Macholán, M., Baird, S.J., Munclinger, P., Piálek, J., Eds.; Cambridge University Press: Cambridge, UK, 2012; pp. 407–430.
4. Britton-Davidian, J.; Catalan, J.; Ramalhinho, M.G.; Auffray, J.C.; Nunes, A.C.; Gazave, E.; Searle, J.B.; Mathias, M.L. Chromosomal phylogeny of Robertsonian races of the house mouse on the island of Madeira: Testing between alternative mutational processes. *Genet. Res.* **2005**, *86*, 171–183. [CrossRef] [PubMed]
5. Gündüz, İ.; Pollock, C.L.; Giménez, M.D.; Förster, D.W.; White, T.A.; Sans-Fuentes, M.A.; Hauffe, H.C.; Ventura, J.; López-Fuster, M.J.; Searle, J.B. Staggered chromosomal hybrid zones in the house mouse: Relevance to reticulate evolution and speciation. *Genes* **2010**, *1*, 193–209. [CrossRef]
6. Brünner, H.; Hausser, J. Genetic and karyotypic structure of a hybrid zone between the chromosomal races Cordon and Valais in the common shrew *Sorex araneus*. *Hereditas* **1996**, *125*, 147–158. [CrossRef]
7. Franchini, P.; Castiglia, R.; Capanna, E. Reproductive isolation between chromosomal races of the house mouse *Mus musculus domesticus* in a parapatric contact area revealed by an analysis of multiple unlinked loci. *J. Evol. Biol.* **2008**, *21*, 502–513. [CrossRef]
8. Franchini, P.; Colangelo, P.; Solano, E.; Capanna, E.; Verheyen, E.; Castiglia, R. Reduced gene flow at pericentromeric loci in a hybrid zone involving chromosomal races of the house mouse *Mus musculus domesticus*. *Evolution* **2010**, *64*, 2020–2032.
9. Giménez, M.D.; Förster, D.W.; Jones, E.P.; Jóhannesdóttir, F.; Gabriel, S.I.; Panithanarak, T.; Scascitelli, M.; Merico, V.; Garagna, S.; Searle, J.B.; et al. A half century of studies on a chromosomal hybrid zone of the house mouse. *J. Hered.* **2017**, *108*, 25–35. [CrossRef]
10. White, M.J.D. Chain processes in chromosomal speciation. *Syst. Zool.* **1978**, *27*, 285–298. [CrossRef]
11. Rieseberg, L.H. Chromosomal rearrangements and speciation. *Trends Ecol. Evol.* **2001**, *16*, 351–358. [CrossRef]
12. Panithanarak, T.; Hauffe, H.C.; Dallas, J.F.; Glover, A.; Ward, R.G.; Searle, J.B. Linkage-dependent gene flow in a house mouse chromosomal hybrid zone. *Evolution* **2004**, *58*, 184–192. [CrossRef] [PubMed]
13. Ganem, G.; Searle, J.B. Behavioural discrimination among chromosomal races of the house mouse (*Mus musculus domesticus*). *J. Evol. Biol.* **1996**, *9*, 817–831. [CrossRef]
14. Castiglia, R.; Capanna, E. Contact zones between chromosomal races of *Mus musculus domesticus*. 1. Temporal analysis of a hybrid zone between the CD chromosomal race (2n = 22) and populations with the standard karyotype. *Heredity* **1999**, *83*, 319–326. [CrossRef]
15. Chatti, N.; Ganem, G.; Benzekri, K.; Catalan, J.; Britton-Davidian, J.; Saïd, K. Microgeographical distribution of two chromosomal races of house mice in Tunisia: Pattern and origin of habitat partitioning. *Proc. R. Soc. Lond. B* **1999**, *266*, 1561–1569. [CrossRef]
16. Ratkiewicz, M.; Banaszek, A.; Jadwiszczak, W.; Chetnicki, W.; Fedyk, S. Genetic diversity, stability of population structure and barriers to gene flow in a hybrid zone between two *Sorex araneus* chromosome races. *Mammalia* **2003**, *68*, 275–283. [CrossRef]
17. Nunes, A.C.; Britton-Davidian, J.; Catalan, J.; Ramalhinho, M.G.; Capela, R.; Mathias, M.L.; Ganem, G. Influence of physical environmental characteristics and anthropogenic factors on the position and structure of a contact zone between two chromosomal races of the house mouse on the island of Madeira (North Atlantic, Portugal). *J. Biogeogr.* **2005**, *32*, 2123–2134. [CrossRef]
18. Ganem, G.; Litel, C.; Lenormand, T. Variation in mate preference across a house mouse hybrid zone. *Heredity* **2008**, *100*, 594–601. [CrossRef] [PubMed]
19. Nosil, P. Ernst Mayr and the integration of geographic and ecological factors in speciation. *Biol. J. Linn. Soc.* **2008**, *95*, 26–46. [CrossRef]
20. Nunes, A.C.; Mathias, M.L.; Ganem, G. Odor preference in house mice: Influences of habitat heterogeneity and chromosomal incompatibility. *Behav. Ecol.* **2009**, *20*, 1252–1261. [CrossRef]
21. Hiadlovská, Z.; Strnadová, M.; Macholán, M.; Bímová, B.V. Is water really a barrier for the house mouse? A comparative study of two mouse subspecies. *Folia Zool.* **2012**, *61*, 319–329. [CrossRef]
22. Wierzbicki, H.; Moska, M.; Strzala, T.; Macierzynska, A. Do aquatic barriers reduce male-mediated gene flow in a hybrid zone of the common shrew (*Sorex araneus*)? *Hereditas* **2011**, *148*, 114–117. [CrossRef] [PubMed]
23. Britton-Davidian, J.; Catalan, J.; Ramalhinho, M.G.; Ganem, G.; Auffray, J.C.; Capela, R.; Biscoito, M.; Searle, J.B.; Mathias, M.L. Rapid chromosomal evolution in island mice. *Nature* **2000**, *403*, 158. [CrossRef]

24. Dureje, L.; Macholán, M.; Baird, S.J.E.; Piálek, J. The mouse hybrid zone in central Europe: From morphology to molecules. *Folia Zool.* **2012**, *61*, 308–318. [CrossRef]
25. Chambers, L.K.; Singleton, G.R.; van Wensveen, M. Spatial heterogeneity in wild populations of house mice (*Mus domesticus*) on the Darling Downs, south-eastern Queensland. *Wildl. Res.* **1996**, *23*, 23–38. [CrossRef]
26. Dias, P.C.; Verheyen, G.R.; Raymond, M. Source–sink populations in Mediterranean blue tits: Evidence using single-locus minisatellite probes. *J. Evol. Biol.* **1996**, *9*, 965–978. [CrossRef]
27. Hauffe, H.C.; Panithanarak, T.; Dallas, J.F.; Piálek, J.; Gündüz, İ.; Searle, J.B. The tobacco mouse and its relatives: A 'tail' of coat colours, chromosomes, hybridization and speciation. *Cytogenet. Genome Res.* **2004**, *105*, 395–405. [CrossRef]
28. Ramalhinho, M.G.; Braz, C.; Catalan, J.; Mathias, M.L.; Britton-Davidian, J. AgNOR variability among Robertsonian races of the house mouse from the island of Madeira: Implications for patterns of Rb fusion formation and genetic differentiation. *Biol. J. Linn. Soc.* **2005**, *84*, 585–591. [CrossRef]
29. Lee, M.R.; Elder, F.F.B. Yeast stimulation of bone marrow mitosis for cytogenetic preparations. *Cytogenet. Cell Genet.* **1980**, *26*, 36–40. [CrossRef]
30. Seabright, M. A rapid banding technique for human chromosomes. *Lancet* **1971**, *2*, 971–972. [CrossRef]
31. Cowell, J.K. A photographic representation of the variability of G-banded structure of the chromosomes of the mouse karyotype. *Chromosoma* **1984**, *89*, 294–320. [CrossRef]
32. Akaike, H. A new look at the statistical model identification. *IEEE Trans. Autom. Contr.* **1974**, *19*, 716–723. [CrossRef]
33. Fel-Clair, F.; Lenormand, T.; Catalan, J.; Grobert, J.; Orth, A.; Boursot, P.; Viroux, M.C.; Britton-Davidian, J. Genomic incompatibilities in the hybrid zone between house mice in Denmark: Evidence from steep and non-coincident chromosomal clines for Robertsonian fusions. *Genet. Res.* **1996**, *67*, 123–134. [CrossRef] [PubMed]
34. Sites, J.W.; Moritz, C. Chromosomal evolution and speciation revisited. *Syst. Zool.* **1987**, *36*, 153–174. [CrossRef]
35. Bidau, C.J.; Giménez, M.D.; Palmer, C.L.; Searle, J.B. The effects of Robertsonian fusions on chiasma frequency and distribution in the house mouse (*Mus musculus domesticus*) from a hybrid zone in northern Scotland. *Heredity* **2001**, *87*, 305–313. [CrossRef]
36. Gazave, E.; Catalan, J.; Ramalhinho, M.G.; Mathias, M.L.; Nunes, A.C.; Dumas, D.; Britton-Davidian, J.; Auffray, J.C. The non-random occurrence of Robertsonian fusion in the house mouse. *Genet. Res.* **2003**, *81*, 33–42. [CrossRef]
37. Sans-Fuentes, M.A.; Muñoz-Muñoz, F.; Ventura, J.; López-Fuster, M.J. Rb (7.17) a rare Robertsonian fusion in wild populations of the house mouse. *Genet. Res.* **2007**, *89*, 207–213. [CrossRef] [PubMed]
38. Adolph, S.; Klein, J. Genetic variation of wild mouse populations in southern Germany. I. Cytogenetic study. *Genet. Res.* **1983**, *41*, 117–134. [CrossRef] [PubMed]
39. Jiggins, C.D.; Mallet, J. Bimodal hybrid zones and speciation. *Trends Ecol. Evol.* **2000**, *15*, 250–255. [CrossRef]
40. Castiglia, R.; Annesi, F.; Capanna, E. Contact zones between chromosomal races of *Mus musculus domesticus*. 3. Molecular and chromosomal evidence of restricted gene flow between the CD race (2n = 22) and the ACR race (2n = 24). *Heredity* **2002**, *89*, 219–224. [CrossRef] [PubMed]
41. Barton, N.H. The dynamics of hybrid zones. *Heredity* **1979**, *43*, 341–359. [CrossRef]
42. Barton, N.H.; Hewitt, G.M. Analysis of hybrid zones. *Annu. Rev. Ecol. Syst.* **1985**, *16*, 113–148. [CrossRef]
43. Barton, N.H.; Hewitt, G.M. Adaptation, speciation and hybrid zones. *Nature* **1989**, *341*, 497–503. [CrossRef] [PubMed]
44. Navarro, A.; Barton, N.H. Chromosomal speciation and molecular divergence-accelerated evolution in rearranged chromosomes. *Science* **2003**, *300*, 321–324. [CrossRef] [PubMed]
45. Capilla, L.; Medarde, N.; Alemany-Schmidt, A.; Oliver-Bonet, M.; Ventura, J.; Ruiz-Herrera, A. Genetic recombination variation in wild Robertsonian mice: On the role of chromosomal fusions and Prdm9 allelic background. *Proc. R. Soc. B* **2014**, *281*, 20140297. [CrossRef]
46. Dumas, D.; Britton-Davidian, J. Chromosomal rearrangements and evolution of recombination: Comparison of chiasma distribution patterns in standard and Robertsonian populations of the house mouse. *Genetics* **2002**, *162*, 1355–1366.
47. Giménez, M.D.; White, T.A.; Hauffe, H.C.; Panithanarak, T.; Searle, J.B. Understanding the basis of

diminished gene flow between hybridizing chromosome races of the house mouse. *Evolution* **2013**, *67*, 1446–1462. [CrossRef] [PubMed]

48. Nunes, A.C.; Catalan, J.; Lopez, J.; Ramalhinho, M.G.; Mathias, M.L.; Britton-Davidian, J. Fertility assessment in hybrids between monobrachially homologous Rb races of the house mouse from the island of Madeira: Implications for modes of chromosomal evolution. *Heredity* **2011**, *106*, 348–356. [CrossRef]
49. Donahue, M.J.; Holyoak, M.; Feng, C. Patterns of dispersal and dynamics among habitat patches varying in quality. *Am. Nat.* **2003**, *162*, 302–317. [CrossRef] [PubMed]
50. Johnson, D.M. Source-sink dynamics in a temporally heterogeneous environment. *Ecology* **2004**, *85*, 2037–2045. [CrossRef]
51. Chmátal, L.; Gabriel, S.I.; Mitsainas, G.P.; Martínez-Vargas, J.; Ventura, J.; Searle, J.B.; Schultz, R.M.; Lampson, M.A. Centromere strength provides the cell biological basis for meiotic drive and karyotype evolution in mice. *Curr. Biol.* **2014**, *24*, 2295–2300. [CrossRef] [PubMed]
52. Medarde, N.; López-Fuster, M.J.; Muñoz-Muñoz, F.; Ventura, J. Spatio-temporal variation in the structure of a chromosomal polymorphism zone in the house mouse. *Heredity* **2012**, *109*, 78–89. [CrossRef] [PubMed]
53. Pocock, M.J.; Hauffe, H.C.; Searle, J.B. Dispersal in house mice. *Biol. J. Linn. Soc.* **2005**, *84*, 565–583. [CrossRef]
54. Górecki, A.; Meczeva, R.; Pis, T.; Gerasimov, S.; Walkowa, W. Geographical variation of thermoregulation in wild populations of *Mus musculus* and *Mus spretus*. *Acta Theriol.* **1990**, *35*, 209–214. [CrossRef]
55. Mathias, M.L.; Nunes, A.C.; Marques, C.C.; Auffray, J.C.; Britton-Davidian, J.; Ganem, G.; Gündüz, İ.; Ramalhinho, M.G.; Searle, J.B.; Speakman, J. Effects of climate on oxygen consumption and energy intake of chromosomally divergent populations of the house mouse (*Mus musculus domesticus*) from the island of Madeira (North Atlantic, Portugal). *Funct. Ecol.* **2006**, *20*, 330–339. [CrossRef]
56. McNab, B.K. Minimizing energy expenditure facilitates vertebrate persistence on oceanic islands. *Ecol. Lett.* **2002**, *5*, 693–704. [CrossRef]
57. Teeter, K.C.; Payseur, B.A.; Harris, L.W.; Bakewell, M.A.; Thibodeau, L.M.; O'Brien, J.E.; Krenz, J.G.; Sans-Fuentes, M.A.; Nachman, M.W.; Tucker, P.K. Genome-wide patterns of gene flow across a house mouse hybrid zone. *Genome Res.* **2008**, *18*, 67–76. [CrossRef]
58. Turner, L.M.; Harr, B. Genome-wide mapping in a house mouse hybrid zone reveals hybrid sterility loci and Dobzhansky-Muller interactions. *eLife* **2014**, *3*, e02504. [CrossRef] [PubMed]

8

Taxonomic Diversity not Associated with Gross Karyotype Differentiation: The Case of Bighead Carps, Genus *Hypophthalmichthys* (Teleostei, Cypriniformes, Xenocyprididae)

Alexandr Sember [1,*], Šárka Pelikánová [1], Marcelo de Bello Cioffi [2], Vendula Šlechtová [1], Terumi Hatanaka [2], Hiep Do Doan [3], Martin Knytl [4] and Petr Ráb [1]

1 Laboratory of Fish Genetics, Institute of Animal Physiology and Genetics, Czech Academy of Sciences, Rumburská 89, 277-21 Liběchov, Czech Republic; pelikanova@iapg.cas.cz (Š.P.); v.slechtova@iapg.cas.cz (V.Š.); rab@iapg.cas.cz (P.R.)
2 Departamento de Genética e Evolução, Universidade Federal de São Carlos, Rod. Washington Luiz km 235 cep, São Carlos 13565-905, Brazil; mbcioffi@ufscar.br (M.d.B.C.); hterumi@yahoo.com.br (T.H.)
3 Research Institute of Aquaculture No. 1, Dinh Bang, Tu Son, Bac Ninh 16000, Vietnam; xinh@yahoo.com
4 Department of Cell Biology, Faculty of Science, Charles University, Viničná 7, 2-128-43 Prague, Czech Republic; knytlma@natur.cuni.cz
* Correspondence: sember@iapg.cas.cz

Abstract: The bighead carps of the genus *Hypophthalmichthys* (*H. molitrix* and *H. nobilis*) are important aquaculture species. They were subjected to extensive multidisciplinary research, but with cytogenetics confined to conventional protocols only. Here, we employed Giemsa-/C-/CMA_3- stainings and chromosomal mapping of multigene families and telomeric repeats. Both species shared (i) a diploid chromosome number $2n = 48$ and the karyotype structure, (ii) low amount of constitutive heterochromatin, (iii) the absence of interstitial telomeric sites (ITSs), (iv) a single pair of 5S rDNA loci adjacent to one major rDNA cluster, and (v) a single pair of co-localized U1/U2 snDNA tandem repeats. Both species, on the other hand, differed in (i) the presence/absence of remarkable interstitial block of constitutive heterochromatin on the largest acrocentric pair 11 and (ii) the number of major (CMA_3-positive) rDNA sites. Additionally, we applied here, for the first time, the conventional cytogenetics in *H. harmandi*, a species considered extinct in the wild and/or extensively cross-hybridized with *H. molitrix*. Its $2n$ and karyotype description match those found in the previous two species, while silver staining showed differences in distribution of major rDNA. The bighead carps thus represent another case of taxonomic diversity not associated with gross karyotype differentiation, where 2n and karyotype structure cannot help in distinguishing between genomes of closely related species. On the other hand, we demonstrated that two cytogenetic characters (distribution of constitutive heterochromatin and major rDNA) may be useful for diagnosis of pure species. The universality of these markers must be further verified by analyzing other pure populations of bighead carps.

Keywords: comparative fish cytogenetics; cytotaxonomy; chromosome banding; East Asian cypriniform fishes; FISH; rDNA; snDNA

---

1. Introduction

The bighead carps of the genus *Hypophthalmichthys* (Bleeker, 1860) represent a small, well-defined group of morphologically highly distinct and ecologically unique cyprinoid fishes [1] formerly recognized as cyprinid subfamily Hypophthalmichthiynae. Recent formal taxonomy includes this genus into family Xenocyprididae (sensu [2]) and, at the same time, the genus is a member of monophyletic clade harboring several East Asian morphologically distinctly differentiated genera [3]. Collectively, the bighead carps once consisted of monotypic genus *Aristichthys* (Oshima, 1919) with species *Aristichthys nobilis* (Richardson, 1844) (bighead carp) and genus *Hypophthalmichthys* (Bleeker, 1860) with two recognized species: silver carp, *Hypophthalmichthys molitrix* (Valenciennes, in Cuvier & Valenciennes, 1844) and Harmand's silver carp (or large-scaled silver carp), *Hypophthalmichthys harmandi* (Sauvage, 1884). However, Howes [1] synonymized the genus *Aristichthys* with *Hypophthalmichthys* based on morphological characteristics—a taxonomic action not always accepted [4]. The systematic status of *H. harmandi* is not well understood at present, and while some authors [4] recognized it as a species distinct from *H. molitrix*, others [5] consider it as subspecies of silver carp only; nevertheless, both species differ in a number of morphological, physiological and reproductive characters (for details, see Supplementary File 1: Text S1).

In their native range (from Amur R. in the north to the Red R. basin in Vietnam and Hainan Island in the south) and elsewhere in temperate regions in Eurasia, they are highly economically important fishes as objects of both lacustrine and riverine fishery and aquaculture [6]. However, bighead carps have been introduced and/or stocked into rivers and lakes outside their native range such as, e.g., in North America (see [7] and references therein), India [8], South Africa [9], and elsewhere in a number of countries [10], where they consequently became invasive aliens which degraded aquatic ecosystems, changing significantly the food webs (see, e.g., in [10–14]). Bighead carps have been and still are objects of intense investigation in various types of studies; for instance, search on 25 April 2020 shows 1155 records on Web of Science and ~19,200 records on Google Scholar when using the term 'Hypophthalmichthys'. Similarly, the chromosomes of bighead and silver carp have been studied by relatively high number of authors (reviewed in Table 1), although mostly just at the level of conventionally Giemsa-stained chromosomes.

Table 1. Summary of reported data on diploid chromosome number (2*n*), numbers of chromosomes in particular morphological categories (m—metacentric, sm—submetacentric, st—subtelocentric, a—acrocentric) and number of chromosome arms (NF value).

Species	2*n*	Karyotype Composition				NF	References
		m	sm	st	a		
H. nobilis	48	20	16		12	84	[15–18]
	48	- 26 -		20	2	74	[19]
	48	18	30			96	[20,21]
	48	14	24		10	86	[22,23]
	48	6	36	6		96	[24,25]
	48	26	20	2		96	[26]
	48	- 24 -		24		72	[27]
H. molitrix	48	10	- 26 -		12	84	[28]
	48	20	12	6	10	82	[29]
	48	- 20 -		- 28 -		68	[19]
	48	22	14		12	84	[15,16]
	48	14	24		10	86	[22,23]
	48	24	16	8		96	[30]
	48	20	24		4	96	[31]
	48	18	22	8		88	[32]
	48	- 24 -		24		72	[27]
	48	18	22	8		92	[33]

Note: During the search for data on cytogenetics of bighead carps, we found also eight other studies (published between years 1976–1985) but we did not include them in this summary because they provided 2*n* only and/or were found methodically very problematic. Their list is available upon request from the corresponding author.

All those studies identically reported $2n = 48$ but differed markedly in the karyotype description, evidently due to the low quality of chromosome preparations, except the reports of Liu [26,30] where mitotic chromosomes from the leukocyte cultures were successfully prepared. Only a few of those studies tried to investigate some other chromosomal characteristics using silver staining of nucleolar organizer regions (NORs; Ag-NOR technique) [27], C-banding [27,34], G-banding [35], or BrdU replication banding [33], all with very ambiguous and not reliable results except the one of Almeida-Toledo et al. [27] who evidenced multiple NOR regions on chromosomes of both bighead carp species. However, the chromosomes of *H. harmandi* have not been studied as yet.

Aiming to more deeply examine the karyotype organization in *H. molitrix* and *H. nobilis*, we combined conventional cytogenetics (Giemsa-, C-, and CMA_3- stainings) with the chromosomal mapping of 5S and 18S rDNA, U1 and U2 snDNA, and $(TTAGGG)_n$ tandem repeats. In addition, we have undertaken Giemsa karyotyping and Ag-NOR analysis in a third species, *H. harmandi*, which is considered extinct in the wild and/or extensively cross-hybridized with *H. molitrix*. We analyzed individuals of *H. harmandi* from a unique gene pool strain, not hybridized with silver carp.

2. Material and Methods

2.1. Sampling

We analyzed four juveniles of *H. molitrix* and five juveniles of *H. nobilis* originated from Fishery Farm, Pohořelice, Czech Republic. The geographical origin of the stock of the former is unknown (original brood fishes were imported from Hungary), while the stock of the latter has been derived from imports from U.S.S.R., which have originated in Amur R. Nine juveniles of *H. harmandi* belonged to a pure line maintained at the Research Institute of Aquaculture No. 1, Dinh Bang, Tu Son, Bac Ninh, Vietnam; it originates from Red River in Vietnam and has been derived from the wild population in the late 1950s, i.e., before silver carp introductions from China. These fishes were imported into the Laboratory of Fish Genetics in 1991. Individuals of *H. molitrix* and *H. nobilis* used for the cytogenetic analysis were tested biochemically to confirm the species identity according to the method of Šlechtová et al. [36], who found species-specific alleles in eight allozyme loci. As the analyzed fishes were juveniles, the sex could not be determined. Samples came from the Czech Republic (Petr Ráb) and Vietnam (Hiep Do Doan) in accordance with the national legislation of the countries concerned. To prevent fish suffering, all handling of fish by collaborators followed European standards in agreement with §17 of the Act No. 246/1992 coll. The procedures involving fish were also supervised by the Institutional Animal Care and Use Committee of the Institute of Animal Physiology and Genetics CAS, v.v.i., the supervisor´s permit number CZ 02361 certified and issued by the Ministry of Agriculture of the Czech Republic. All fishes were euthanized using 2-phenoxyethanol (Sigma-Aldrich, St. Louis, MO, USA) before being dissected.

2.2. Chromosome Preparation and Conventional Cytogenetics

Chromosome preparations were produced using leukocyte cultures in the case of juveniles of *H. molitrix* and *H. nobilis* [37,38], while those of *H. harmandi* were achieved by a direct preparation from the cephalic kidney [39,40]. The quality of chromosomal spreading was enhanced by a dropping method described by Bertollo et al. [40]. Chromosomes were stained with 5% Giemsa solution (pH 6.8) (Merck, Darmstadt, Germany) for a conventional cytogenetic analysis or kept unstained for other methods. For sequential stainings, selected Giemsa-stained slides were distained in a cold fixation with methanol: acetic acid 3:1 (v/v) before the application of other technique. For FISH, slides were dehydrated in an ethanol series (70, 80, and 96%, 3 min each) and stored at −20 °C.

Constitutive heterochromatin was visualized by C-banding according to Haaf and Schmid [41]; chromosomes were counterstained with 4′,6-diamidino-2-phenolindole (DAPI) (Sigma-Aldrich). Fluorescence staining was done by GC-specific fluorochrome Chromomycin A_3 (CMA_3) and AT-specific fluorochrome DAPI (both Sigma-Aldrich), following Mayr et al. [42] and Sola et al. [43]. The banding protocols were performed either separately or sequentially on the metaphases previously treated by other method(s). In *H. harmandi,* only silver-nitrate impregnation of NORs (i.e, Ag-NOR staining) was performed, according to Howell and Black [44].

2.3. DNA Isolation and Preparation of FISH Probes

Total genomic DNA was extracted from fin and blood tissue using the Qiagen DNeasy Blood & Tissue Kit (Qiagen, Hilden, Germany). 5S and 28S rDNA fragments were obtained by polymerase chain reaction (PCR) with primers and thermal profiles described in Sember et al. [45]. Amplification of 18S rDNA and U1 snDNA was done by PCR with the primers 18SF (5′-CCGAGGACCTCACTAAACCA-3′) and 18SR (5′-CCGCTTTGGTGACTCTTGAT-3′) [46]; U1F (5′-GCAGTCGAGATTCCCACATT-3′) and U1R (5′-CTTACCTGGCAGGGGAGATA-3′) [47], using the thermal profiles described in Yano et al. [48] and Silva et al. [47], respectively. The obtained PCR products were purified using NucleoSpin Gel and PCR Clean-up (Macherey-Nagel GmbH, Düren, Germany) according to manufacturer's instructions. The subsequent procedures involving cloning of the purified products and a plasmid isolation, sequencing (in both strands) of selected positive clones, assembly of chromatograms from obtained sequences and sequence alignment followed essentially the same workflow as described in Sember et al. [49]. Some portion of obtained products was sequenced (in both strands) by Macrogen company (Netherlands). The content of resulting consensus sequences was verified using NCBI BLAST/N analysis [50] and selected clones were used for a FISH probe preparation. For the chromosomal mapping of U2 snDNA, we used the probe obtained previously from a botiid fish *Leptobotia elongata* (for details, see Sember et al. [49]). Furthermore, the FISH results from the mapping of *Hypophthalmichthys*-derived 28S rDNA probe were verified by 28S rDNA probes generated from the nemacheilid loach *Schistura corica* [45] and botiid loach *Botia almorhae* [49].

DNA probes were labeled mostly by PCR, either with biotin-16-dUTP or with digoxigenin-11-dUTP (both Roche, Mannheim, Germany). Due to its long size, the 18S rDNA probe was generated in two steps: (i) non-labeling PCR amplification from a verified 18S rDNA clone and (ii) nick translation (2 h) of the amplified 18S rDNA product using Nick Translation Mix (Abbott Molecular, Illinois, USA). A portion of U1 and U2 snDNA probes was also labeled by Nick Translation Mix (Abbott Molecular); the template DNA was in this case the entire plasmid DNA containing U1 or U2 snDNA insert. A dual-color FISH for each slide involved 200 ng of each probe and 25 μg of sonicated salmon sperm DNA (Sigma-Aldrich). The final hybridization mixtures were prepared according to Sember et al. [45].

2.4. FISH Analysis

Dual-color FISH experiments were conducted essentially according to Sember et al. [45]. Briefly, chromosome preparations were thermally aged (overnight at 37 °C and 1 h at 60 °C), then pre-treated in RNase A (200 μg/mL in 2× SSC, 60–90 min, 37 °C) (Sigma-Aldrich) and pepsin (50 μg/mL in 10 mM HCl, 3 min, 37 °C), and finally denatured in 75% formamide in 2× SSC (pH 7.0) (Sigma-Aldrich) for 3 min at 72 °C. Probes were denatured at 86 °C for 6 min, cooled on ice, and dropped on the chromosome slides. Hybridization took place in a moist chamber at 37 °C overnight. A post-hybridization washing was done under high stringency, i.e., two times in 50% formamide/2× SSC (42 °C, 10 min) and three times in 1× SSC (42 °C, 7 min). Prior to the probe detection, 3% bovine serum albumin (BSA) (Vector Labs, Burlington, Canada) in 0.01% Tween 20/ 4× SSC was applied to the slides to block unspecific binding of antibodies. Hybridization signals were detected by Anti-Digoxigenin-FITC (Roche; dilution 1:10 in 0.5% BSA/PBS) and Streptavidin-Cy3 (Invitrogen Life Technologies, San Diego, CA, USA; dilution 1:100 in 10% NGS (normal goat serum)/PBS). Experiments with altered labeling (e.g., biotin for 18S and digoxigenin for 5S rDNA) were included to verify the observed patterns. All FISH images presented here have a unified system of pseudocolored signals—red for the 18S rDNA and U2 snDNA probes, and green for the 5S rDNA and U1 snDNA probes. Finally, all FISH slides were mounted in antifade containing 1.5 μg/mL DAPI (Cambio, Cambridge, UK).

Telomeric $(TTAGGG)_n$ repeats were detected by FISH using a commercial telomere PNA (peptide nucleic acid) probe directly labeled with Cy3 (DAKO, Glostrup, Denmark) according to the manufacturer's instructions, with a single modification concerning the prolonged hybridization time (1.5 h).

2.5. Microscopic Analyses and Image Processing

Giemsa-stained chromosomes and FISH images were inspected using a Provis AX70 Olympus microscope equipped with a standard fluorescence filter set. FISH images were captured under immersion objective 100× with a black and white CCD camera (DP30W Olympus) for each fluorescent dye separately using DP Manager imaging software (Olympus). The same software was used to superimpose the digital images with the pseudocolors. Karyotypes from Giemsa-stained chromosomes were arranged in Ikaros (Metasystems) software. Final images were optimized and arranged using Adobe Photoshop, version CS6.

At least 15 metaphases per individual and method were analyzed, some of them sequentially. Chromosomes were classified according to Levan et al. [51], but modified as m—metacentric, sm—submetacentric, st—subtelocentric, and a—acrocentric, where st and a chromosomes were scored as uniarmed to calculate NF value (Nombre Fondamental, number of chromosome arms sensu Matthey [52]). Chromosome pairs were arranged according to their size in each chromosome category.

3. Results

3.1. Karyotypes and Chromosome Banding Characteristics

Analyzed fishes of all three species possessed invariably a $2n = 48$ (Figure 1a,c,e), confirming thus previous reports (Table 1). Besides, they also possessed the same karyotype compositions: four pairs of m, 12 pairs of sm, and eight pairs of st-a chromosomes (Figure 1). Chromosomes of *H. molitrix* and *H. nobilis* displayed a very low content of constitutive heterochromatin concentrated in the pericentromeric chromosome regions, except for significantly heterochromatinized short (*p*) arms of the largest st chromosome pair in *H. molitrix* and additional interstitial block of heterochromatin on this pair in *H. nobilis* only (Figure 1b,d). CMA_3 fluorescence revealed six positive signals in the karyotype of *H. molitrix* (*p*-arms of the largest and middle-sized st chromosome pairs; Figure 2a), while it displayed altogether 10 signals in *H. nobilis* (all in *p*-arms of st chromosome pairs including the largest st element; Figure 2b). In the karyotype of *H. harmandi*, four Ag-positive signals in the *p*-arms in st chromosome pairs (likely Nos. 17 and 18) were observed (Figure 1f).

3.2. Sequence Analysis of Repetitive DNA Fragments

PCR amplification resulted consistently in approximately 150 bp (U1 snDNA), 200 bp (5S rDNA), 300 bp (28S rDNA), and 1800 bp (18S rDNA) long fragments. Searches with the BLAST/N program at NCBI yielded the following results; 18S rDNA (*H. molitrix*)—sequenced 1380 bp long part showed 96–99% identity with 18S rDNA fragments of many fish species; 28S rDNA (both from *H. molitrix* and *H. nobilis*) displayed high similarity results (96–98% identity) with 28S rDNA sequences of many teleosts; 5S rDNA (both from *H. molitrix* and *H. nobilis*): 176–178 nt of our sequenced fragment was subjected to BLAST/N and showed 87–88% identity with sequence of 5S rDNA and non-transcribed spacer of *Megalobrama amblycephala* (Sequence ID: KT824058.1), *Cyprinus carpio* (Sequence ID: LN598602.1) and *Danio rerio* (Sequence ID: AF213516.1), and further 97% identity was shown in 104–114 nt long part of our PCR fragment with the coding region of 5S rDNA of many fishes. Finally, 123 nt of our U1 snDNA fragment showed 97% identity with the predicted U1 snRNA gene region of many fish species. Sequences for 18S rDNA and U1 snDNA (from *H. molitrix*) and for 5S rDNA (from both *H. molitrix* and *H. nobilis*) were deposited in GenBank under the accession numbers MT165584-MT165587. We have not investigated U2 snDNA genes from *Hypophthalmichthys* as the U2 snDNA probe from *Leptobotia elongata* has proven to be fully sufficient for FISH.

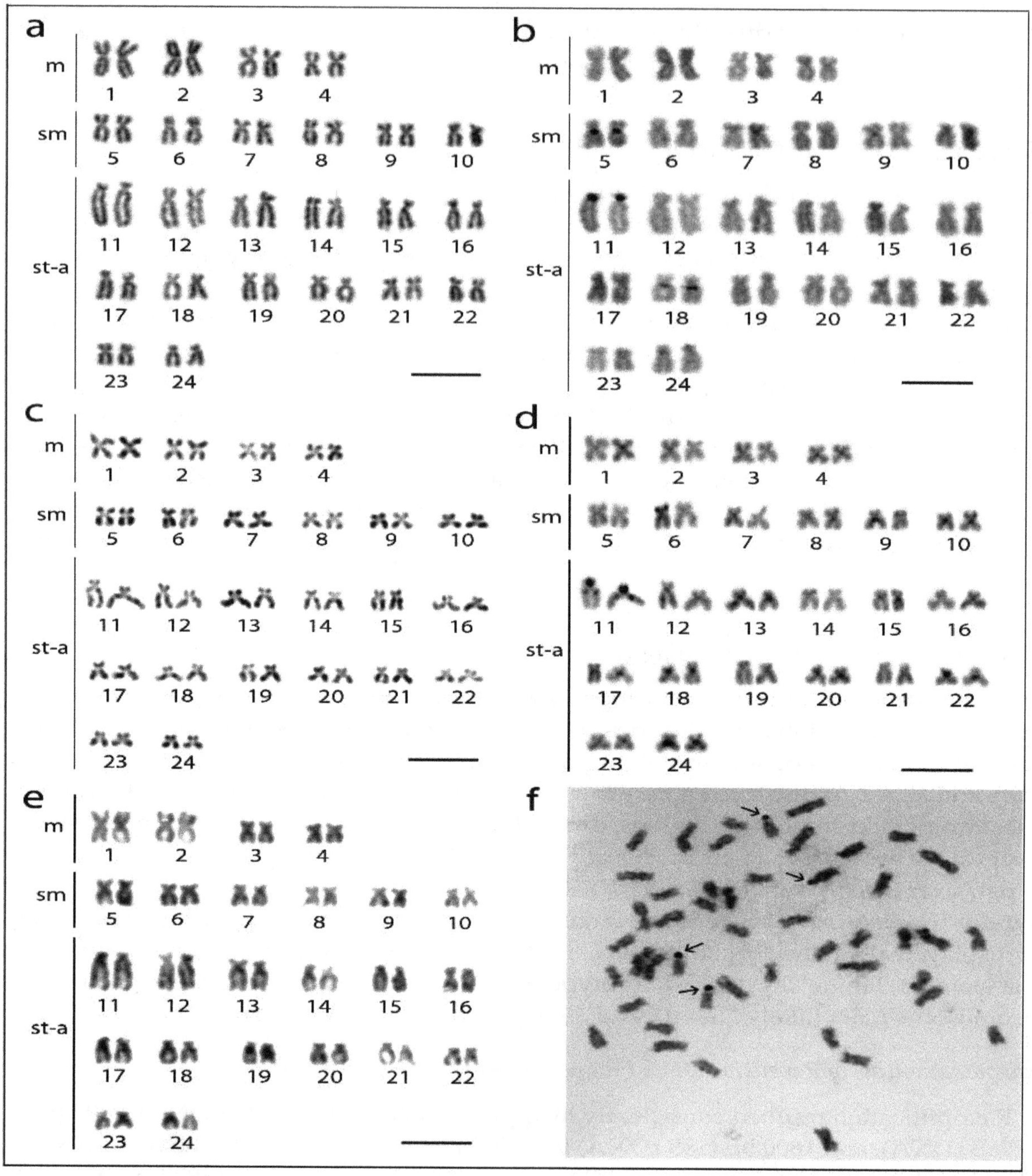

Figure 1. Karyotypes of three *Hypophthalmichthys* species arranged from mitotic metaphases after Giemsa staining, C-banding or Ag-NOR staining. (**a**,**b**) *H. molitrix* (individual HM3), (**c**,**d**) *H. nobilis* (individual HN4), and (**e**) *H. harmandi* (individual HH1). (**a**,**c**,**e**) Giemsa staining; (**b**,**d**) C-banding. Note two distinct blocks of constitutive heterochromatin on pair No. 11 in *H. nobilis* (**d**). (**f**) Ag-NOR staining in *H. harmandi* (individual HH3). The metaphase is incomplete (one chromosome missing; $2n = 47$), but the most representative one regarding the spreading quality and the signal strength and it is also to higher extent sufficient enough to present required features (i.e., note a lack of Ag-NOR signal on the largest acrocentric chromosome pair No. 11). Scale bar = 10 μm.

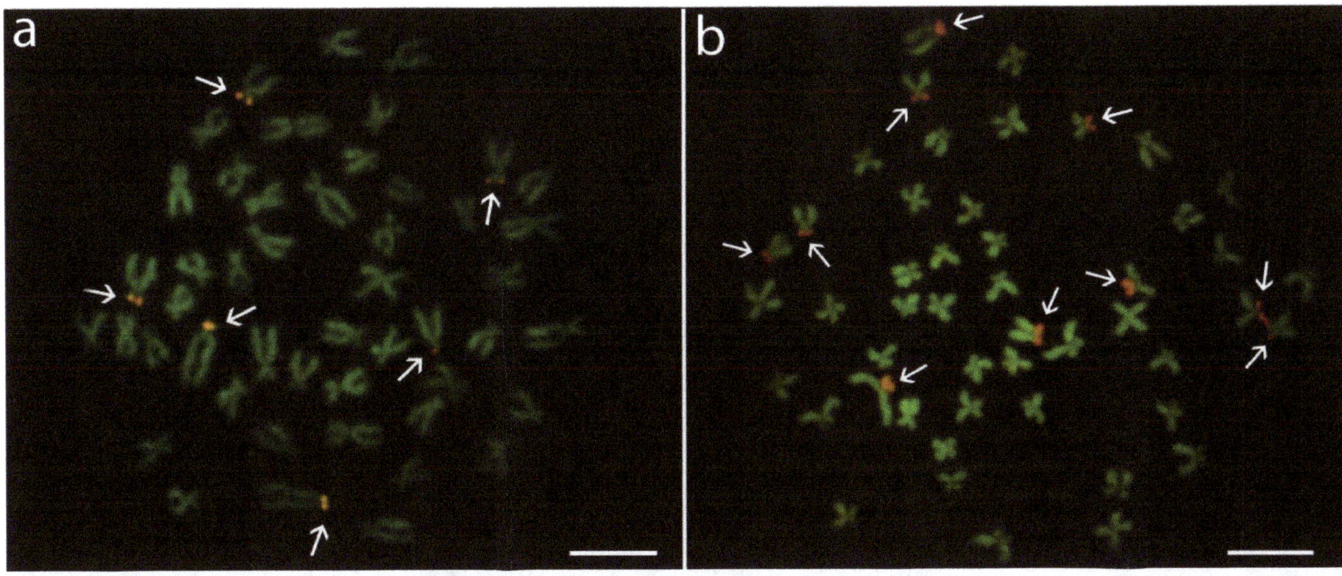

Figure 2. Mitotic metaphases of *Hypophthalmichthys molitrix* and *H. nobilis* after CMA_3/DAPI staining. (**a**) *H. molitrix*, individual HM3, (**b**) *H. nobilis*, individual HN4. For better contrast, images were pseudocolored in red (for CMA_3) and green (for DAPI). Arrows indicate CMA_3-positive sites. Scale bar = 10 μm.

3.3. Hybridization Patterns of Repetitive DNA Probes

5S rDNA probe mapped consistently to the proximal region of the largest acrocentric pair No. 11 in both species (Figure 3a,b). On the same chromosome pair, adjacent to 5S rDNA cluster, tandem arrays of 18S rDNA were found to cover the entire *p*-arms (Figure 3a,b). Additional 18S rDNA loci resided in the terminal part of *p*-arms or encompassed entire *p*-arms of several chromosomes. The complete number of 18S rDNA signals was eight in *H. molitrix* (chromosome pairs 11, 14, 20, and 21) and ten in *H. nobilis* (chromosome pairs 11, 14, 15, 20, and 21) (Figure 3a,b). On the other hand, 28S rDNA probes (generated from herein studied species or utilized from other cypriniforms formerly analyzed by us [45,49]) did not generate any hybridization signals, suggesting that a 300 bp long probe is too short to visualize small rDNA clusters present in *Hypophthalmichthys*, while 1800 bp of 18S rDNA can produce signals of sufficient intensity. Although all the 18S rDNA sites corresponded with CMA_3-positive signals, some 18S rDNA clusters in *H. molitrix* were not revealed by this GC-specific fluorochrome (compare Figures 2a and 3a), again probably reflecting small size (i.e., relatively low copy number of tandem arrays) of major rDNA cistrons.

U1 and U2 snDNA probes co-localized in both species in a pericentromeric region of small st chromosome pair (No. 7) (Figure 3c,d). Neither the co-localization between snDNA and rDNA (Figure 4a,b and Figure 5), nor intraspecific variability in the number of hybridization signals of any multigene family were observed among analyzed individuals of both species. Telomere FISH marked only ends of all chromosomes, with no additional interstitial sites (Figure 4c,d).

As we analyzed not sexed juvenile individuals, we could not directly assess possible sex-related differences in the karyotypes and in patterns of analyzed cytogenetic markers. Nonetheless, we did not observe any type of within-species polymorphism in our sampling, and it has been formerly shown that both *Hypophthalmichthys* species display a sex ratio around 1:1 due to genetic sex determination governed most likely by a homomorphic (i.e., cytologically indistinguishable) XX/XY sex chromosome system [53].

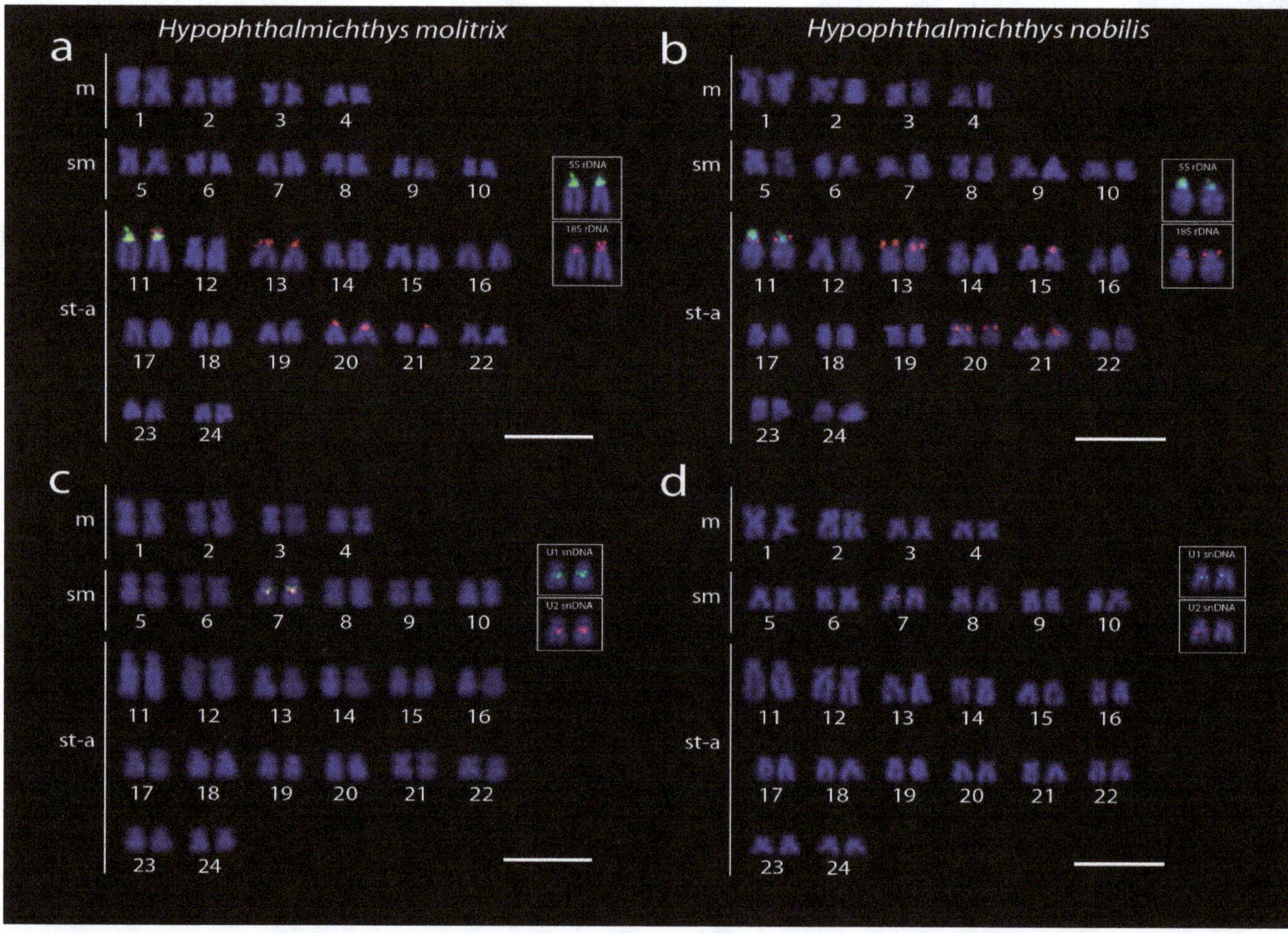

Figure 3. Karyotypes of *Hypophthalmichthys molitrix* and *H. nobilis* arranged after 5S/18S rDNA and U1/U2 snDNA FISH. (**a**,**b**) 18S rDNA (red) and 5S rDNA (green) probes. Insets show separately 5S and 18S rDNA signals on the largest acrocentric pair. Note the adjacent position of 5S and 18S rDNA signals on chromosome pair No. 11 in both species. (**c**,**d**) U1 (green) and U2 (red) snDNA probes mapped on mitotic chromosomes of (**c**) *H. molitrix* and (**d**) *H. nobilis*. Note the co-localization of a single pair of U1 and U2 snDNA signals in small sm chromosome pair No. 7. Insets show separate hybridization signals for each individual probe. Chromosomes were counterstained with DAPI (blue). Identification codes of individuals: (**a**) *H, molitrix* HM2, (**b**) *H. nobilis* HN1, (**c**) *H. molitrix* HM4, and (**d**) *H. nobilis* HN3. Scale bar = 10 μm.

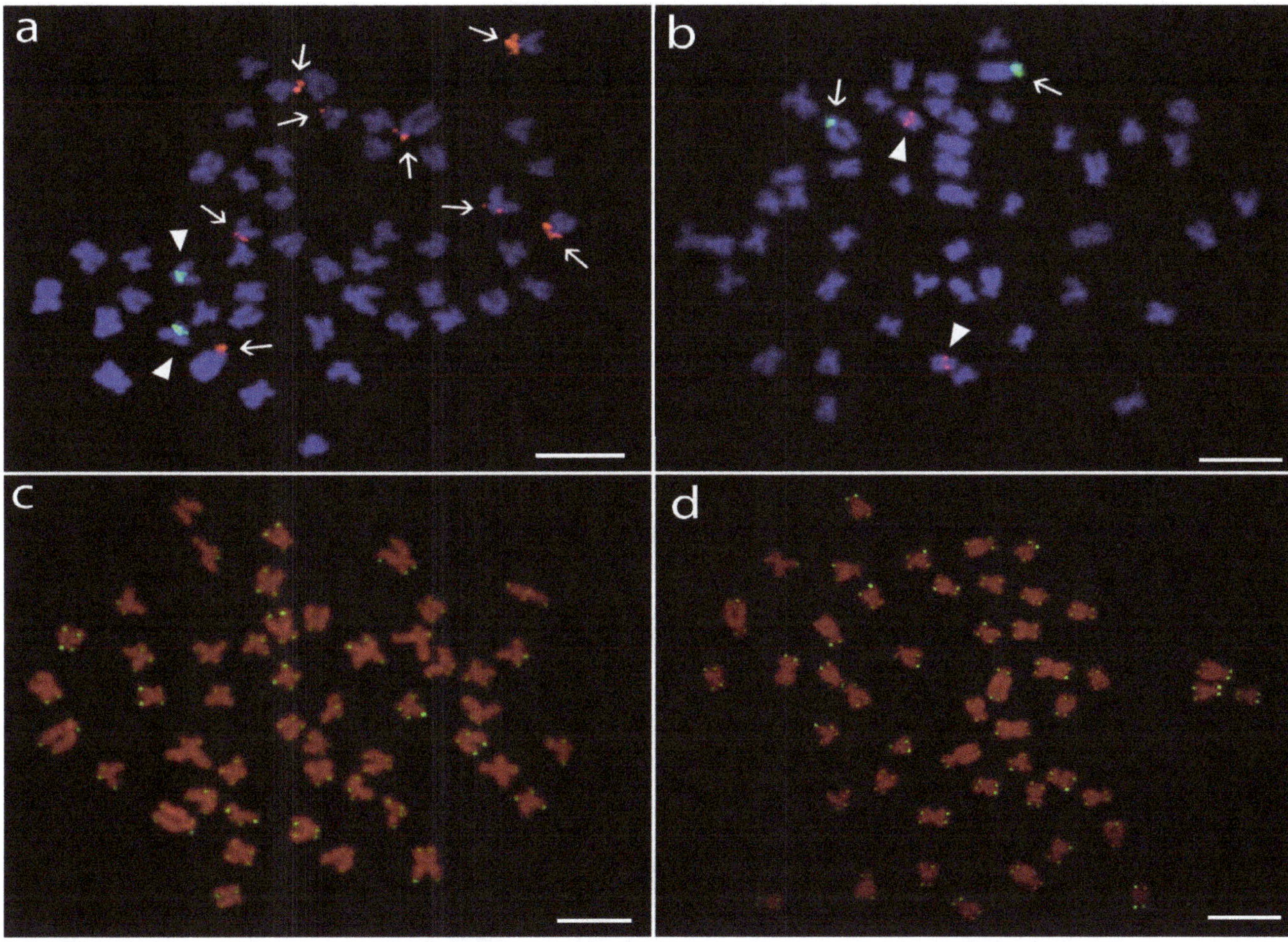

Figure 4. Mitotic metaphases of *Hypophthalmichthys molitrix* and *H. nobilis* after different cytogenetic treatments. (**a**,**c**) *H. molitrix* and individual HM4 in both methods; (**b**,**d**) *H. nobilis*, individuals HN4, and HN3, respectively. Images (**a**,**b**) clarify an independent location of distinct cytogenetic markers. (**a**) FISH with U1 snDNA (green, arrowheads) and 18S rDNA (red, arrows) probes. (**b**) FISH with U2 snDNA (red, arrowheads) and 5S rDNA (green, arows) probes. Chromosomes were counterstained with DAPI (blue). (**c**,**d**) PNA FISH with telomeric probe; for better contrast, pictures were pseudocolored in green (telomeric repeat probe) and red (DAPI). Scale bar = 10 μm.

4. Discussion

The chromosomes of the two species of bighead carps, *H. molitrix* and *H. nobilis*, were extensively studied (Table 1), evidently due to their high aquacultural value. On the other hand, 2*n* and karyotype of the third species of the genus, *H. harmandi*, is reported in our study for the first time. Our current assessment of the karyotype structure and the hybridization patterns of selected multigene families in *H. molitrix* and *H. nobilis* is summarized in Figure 5. Our study confirmed 2*n* = 48 for these two species and revealed the same chromosome count for *H. harmandi*. The karyotype structures in *H. molitrix* and *H. nobilis*, however, differed markedly among various studies. The reason for these discrepancies might be linked with the following facts; (i) chromosomes of cypriniform fishes generally exhibit very small size when compared to other teleosts (see, e.g., in [45,54–56]); (ii) furthermore, cyprinoid chromosomes also exhibit a gradual decrease in size, with the centromere positions ranging stepwise from median to nearly terminal, making it difficult to assess the chromosomal categories with accuracy; and (iii) inspection of published chromosome pictures showed that previous reports were based on highly condensed chromosomes which also made it impossible to describe the karyotype accurately. However, careful analysis of a number of metaphase cells with less condensed chromosomes demonstrated that karyotypes of all three species of bighead carps at the level of conventionally Giemsa-stained chromosomes are in fact identical.

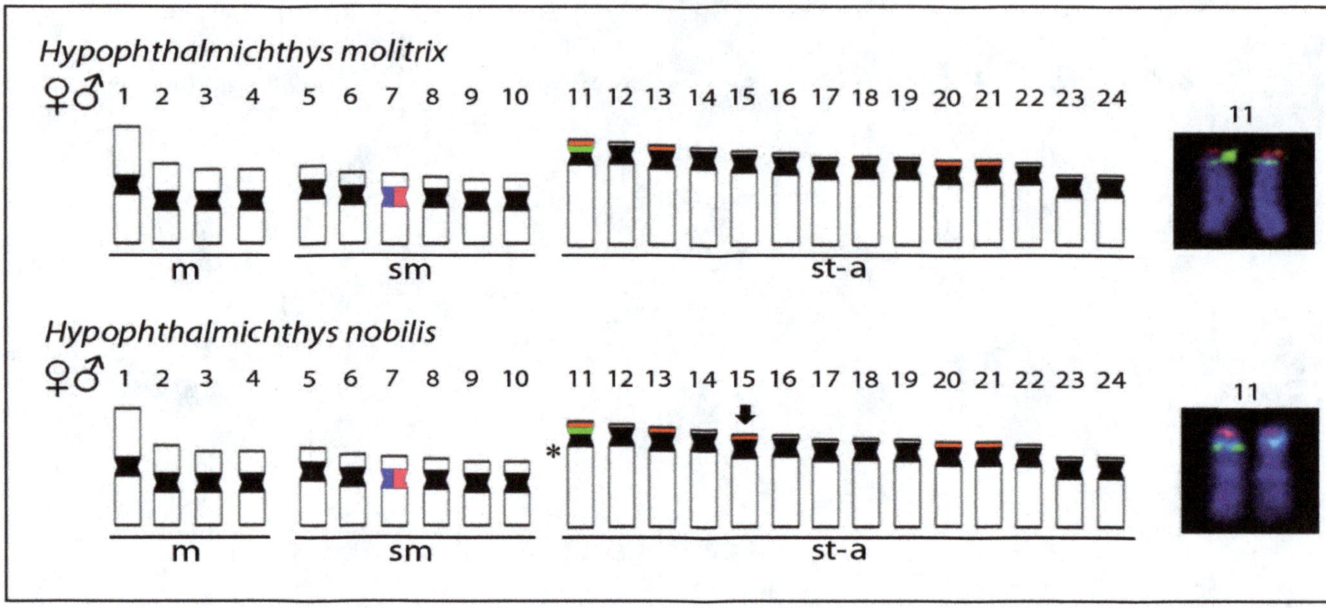

Figure 5. Representative idiograms of two *Hypophthalmichthys* species highlighting the distribution of analyzed multigene families. 18S (red) and 5S (green) rDNA sites and U1 (blue) and U2 (pink) snDNA sites on the chromosomes of *H. molitrix* and *H. nobilis*. Note the co-localization of snDNA sites on the chromosome pair 7 and the adjacent arrangement of 5S and 18S rDNA sites on the chromosome pair 11. Moreover, notice the additional 18S rDNA site on chromosome pair 15 in *H. nobilis* (marked by arrow) in comparison to the karyotype of *H. molitrix*. Finally, an asterisk denotes the location of the differential interstitial C-band, which is present in *H. nobilis* but absent in *H. molitrix*. Insets with the chromosome pair 11 (right) display the chromosomes dissected from prometaphase plates after rDNA FISH, where the adjacent arrangement of both rDNA classes is clearly visible.

We thus show that potential interspecific hybrids between *H. harmandi* and *H. molitrix* cannot be revealed after basic karyotype analysis alone. Nonetheless, we observed that karyotypes of *H. molitrix* and *H. nobilis* differ in two other cytogenetic characters; one of them displays distinctive pattern also in *H. harmandi*. First, the presence of an additional interstitial C-band on the largest acrocentric pair in all individuals of *H. nobilis* clearly distinguishes this species from *H. molitrix*, at least in our sampled populations. This additional location of constitutive heterochromatin in *H. nobilis* might potentially emerge after a pericentric inversion which did not affect the general morphology of the chromosome but relocated part of the heterochromatic block from the *p*-arm to the proximal region of the long (*q*) arm. Our data, however, cannot rule out the involvement of other mechanisms such as centromere repositioning [57,58]. For the third species, *H. harmandi*, data from C-banding are not available; therefore, we cannot confirm if this method alone may provide enough information to discriminate the karyotypes of all three species. Nonetheless, even if we could do that, we would have to take into account that constitutive heterochromatin might display a polymorphic distribution among populations of diverse taxa (including teleosts; exemplified in [59–62]) and thus this feature might limit the resolution power of C-banding for interspecific diagnosis. Second, we found a difference in the number and position of major rDNA sites—four loci in *H. harmandi*, eight in *H. molitrix* and ten in *H. nobilis*. Our results are partially not consistent with those of Almeida-Toledo et al. [27] who also reported four pairs bearing Ag-NORs in *H. molitrix*, but only three pairs in *H. nobilis* (in contrast to five pairs revealed by us via FISH). Our view on this discrepancy is that either (i) the Ag-NOR method detected only clusters active in preceding interphase, while our FISH analysis showed all major rDNA sites irrespective of their transcriptional activity, or (ii) Almeida-Toledo et al. [27] examined the hybridized individuals which remained undetected due to lack of testing for genome admixtures, i.e., the step that we included in our present study. In either case, both studies collectively suggest that the patterns of major rDNA distribution might be stable at least in *H. molitrix* and that it differs from the one found in *H. nobilis*, strengthening the possibility that this marker may be useful also in species diagnosis in other *Hypophthalmichthys* populations.

Major (45S; NOR-forming) and minor (5S; located outside NOR) rDNA clusters are by far the most utilized cytotaxonomic markers in fishes [63–65]. Major rDNA is usually visualized by 18S or 28S rDNA probes. Despite the ever-growing number of studies showing lability of their site number and patterns of distribution in fish genomes (with many cases documenting intra- and inter-populational variability) (see, e.g., in [66–68]) and even their vulnerability to change rapidly under different environmental conditions [69] or hybridization [70], certain arrangements of rDNA classes can help to clarify a presence of species complexes or cryptic species (see, e.g., in [71–73]), to uncover the genome composition in hybrid specimens [74,75], to confirm the ploidy level, and to deduce the mechanism of polyploidy [76–79]. It has been repeatedly documented that even closely related species may possess dramatically different number of rDNA loci [45,71,80]. A difference in number of 5S rDNA clusters between emerald and darter goby (two vs. 42) [81] may serve as an illustrative example. Besides the difference in number and position of positive signals, also the linkage between 45S and 5S or them with other multigene families may represent a valuable cytotaxonomic determiner (see, e.g., in [82–85]).

Among Cypriniformes, many studies have been conducted on polyploid species and especially on those of high aquacultural importance, such as genera *Cyprinus* and *Carassius* (see, e.g., in [55,74,86,87]) or on unisexually reproducing taxa such as *Squalius*, *Cobitis*, and *Misgurnus* and on species closely related to them [45,77,79,88–92]. Some reports revealed amplified number of either 5S or 18S rDNA signals [45,55], different types of inter-individual/inter-populational polymorphisms in number and location of rDNAs [70,88–91,93] or high interspecific variability in this character [91,92,94], while still other studies found rather standard patterns, with just one locus of one or both rDNA classes per haploid genome [45,77,95] or only a slight elevation in number of sites [56,95,96]. Among two *Hypophthalmichthys* species analyzed herein, a single pair of 5S rDNA loci occupied apparently homeologous chromosomes and were found adjacent to one of the multiple 18S rDNA clusters. Similar links between 5S and 18S rDNA sites provided valuable cytotaxonomic markers in some cyprinids (see, e.g., in [86,92,96]). In our study, however, as this arrangement is shared by both species, it cannot be considered as useful cytotaxonomic determiner. Nonetheless, Ag-NOR analysis in *H. harmandi* clearly showed that NORs are not present on this largest acrocentric pair, hence potential hybrids containing the *H. harmandi* genome could be identified this way. What is further evident is the interspecific difference in the number of 18S rDNA sites, which could be helpful as a cytotaxonomic marker, but its intraspecific stability must be further verified in other pure populations of both species. In this sense, it may be difficult to discriminate all 18S rDNA loci due to their tiny size, therefore the analysis should be treated with caution.

Genes for small nuclear RNA (snRNA) are yet readily used for chromosome mapping in fishes, though studies employing U2 snDNA as a cytogenetic marker are steadily growing in the last years ([84,97–101], to name a few). On the other hand, U1 snDNA has been so far chromosomally mapped only in a cichlid *Oreochromis niloticus* [102], several South American characiforms of the genera *Astyanax* [47] and *Triportheus* [85], and further in African characiform representative *Hepsetus odoe* [103], one species from Gadiformes [104] and one taxon (suspected species complex) belonging to Mugiliformes [73]. Among cypriniforms, only a single recent work mapped U2 snDNA sites, namely in diploid and tetraploid loaches of the family Botiidae [49], therefore our present study is the first one showing the position of U1 snDNA on cypriniform chromosomes. In botiids, perhaps surprisingly, the mapping of U2 snDNA showed mostly a conserved single pair of U2 snDNA signals irrespective of the ploidy level. What is more, the location of U2 snRNA arrays in the pericentromeric/interstitial region as revealed in botiids was also found herein in both *Hypophthalmichthys* species and, interestingly, the same or similar pattern has been encountered in approximately half of fish species inspected for U2 snDNA distribution to date (see [84,101] and examples listed in Yano et al. [99]). It seems that a strong selective pressure operates to maintain such a location for this gene. Moreover, in botiids [49] as well as in two herein studied *Hypophthalmichthys* species and in some other fish species [47,101,105,106] snDNA clusters are located on rather small-sized chromosomes. It is tempting to hypothesize that this location may facilitate more efficient expression as small chromosomes tend to occupy rather interior, transcriptionally active part of the interphase nucleus (see, e.g., in [107]). What is less conserved, is the so far known association of U2 snDNA with other multigene families. Several combinations of syntenic/adjacent or intermingled arrangements can be found among fishes such as between 5S rDNA and U1 snDNA [47,85], 5S rDNA and U2 snDNA [84,98,100,105], 18S rDNA with U2 snDNA [99,106,108], 5S and 18S rDNA together with U2 snDNA [99] and even with several histone genes [109]; further U1 and U2 snDNA [97,104] or U1 and U2 snDNA together with 5S rDNA [110]. Therefore, these arrangements may potentially serve as useful cytotaxonomic markers. In our study, both investigated *Hypophthalmichthys* species shared the co-localization of U1 and U2 snDNA cistrons along with an independent location of these sites with respect to rDNA classes.

FISH aimed to map the vertebrate telomeric $(TTAGGG)_n$ repeat motif showed signals only in their usual location at termini of all chromosomes. No interstitial telomeric sequences (ITSs), which might point to previous structural chromosomal rearrangements (see, e.g., in [111]), were detected, neither in *H. molitrix* nor in *H. nobilis*. More importantly, this type of analysis did not reveal any differences between analyzed species that would be helpful in their discrimination.

Recently, all three herein studied species are included in the genus *Hypophthalmichthys* (Bleeker, 1860) [1] but Kottelat [4] noted that not all authors agree with synonymization of the genus *Aristichthys* (Oshima, 1919). From the cytotaxonomic view, it is not possible to contribute to this problem due to lack of significant karyotype differences. Table 2 further summarizes all available data for members of the monophyletic East Asian clade of the family Xenocyprididae (sensu Tan and Ambruster [2]). Though the quality of such data was affected by the facts discussed above (i.e., the characteristics of cyprinoid chromosomes), their critical assessment demonstrates that these species possess (i) the same $2n = 48$; (ii) very similar karyotype structures; and, where studied [112,113], also (iii) multiple NOR sites, supporting thus molecular phylogeny of the clade [3].

Table 2. Review of reported cytogenetic data for members of the monophyletic clade of several East Asian morphologically distinct genera.

Species	2*n*	Karyotype Composition				NF	References
		m	sm	st	a		
Elopichthys bambusa	48	10	24	12	2	82	[112,114]
Luciobrama microcephalus	48	12	22	12	2	82	[114]
Ochetobius elongates	48	10	16	22		74	[114]
Squaliobarbus curriculus	48	14	30		4	92	[114]
Culter oxycephaloides	48	20	24		4	92	[115]
Xenocypris macrolepis	48	20	26	2		94	[112,115]
Xenocypris davidi	48	18	26	4		92	[112,115]
Xenocypris fangi	48	16	28	4		92	[112]
Xenocypris sechuanensis	48	18	26	4		92	[112]
Megalobrama amblycephala	48	18	26	4		92	[115–117]
Megalobrama terminalis	48	18	22	8		88	[115]
Ctenopharyngodon idella	48	18	24	6		90	[115]
	48	18	30			96	[118]
	48	18	22	8		88	[116]
Mylopharyngodon piceus	48	14	- 34 -			96	[25]
	48	16	28	4		92	[119]

The stability of 2n (with either 48 or 50 chromosomes) is widely documented for majority of non-polyploid cyprinoids [91,92,95,120] as well as in other related cypriniforms (see, e.g., in [45,89]), indicating its high conservatism. These signs of the so-called karyotype stasis, in which identical or almost identical karyotypes are maintained within a certain taxonomic group even over considerable long evolutionary time, are observable also in other teleost lineages such as in the pikes of the genus *Esox* [121,122], several lineages of salmonid fishes with A-type karyotype [123,124], and further especially in knifefishes of the family Notopteridae (see [125] and references therein) and many percomorph groups [126–130].

Karyotype stasis has been also documented in diverse clades across the tree of life (e.g., typically in birds [131] and in feline lineages [132]). The underlying evolutionary mechanisms for this mode of karyotype evolution have not been identified so far but they may be at least partially linked with the functional arrangement of chromatin within the interphase nucleus and the degree of tolerance to its change [133,134]. Nonetheless, it is highly probable that such a high degree of karyotype similarity may significantly contribute to the rate of interspecific hybridization [135], which has been repeatedly documented among many cyprinids [19,55,70,75] as well as between the *Hypophthalmichthys* species [18–21,36].

5. Conclusions

Our cytogenetic study of all three species of the genus *Hypophthalmichthys* documented that their karyotype macrostructure, i.e., the number of chromosomes in respective morphological categories, is identical, therefore these characteristics alone may not help in the identification of pure species and interspecific hybridizations. A brief overview of available cytogenetic data of other members of the monophyletic clade of East Asian fishes, to which *Hypophthalmichthys* belongs, shows identical $2n = 48$, very similar karyotypes and, in a subset of analyzed species, also multiple NOR sites, supporting thus the molecular phylogeny of the clade. The bighead carps thus belong to the teleost lineages where the taxonomic diversity is not associated with extensive karyotype repatterning.

However, an important difference has been unraveled in the present study between *H. molitrix* and *H. nobilis* as the latter species exhibits additional interstitial band of constitutive heterochromatin on the largest acrocentric pair 11. Lack of data for *H. harmandi* did not allow us to assess the usefulness of this marker in this practically extinct species.

On the other hand, a combined set of FISH and Ag-NOR results showed that the karyotypes of all three species differ among each other in the number and position of major rDNA sites—four in *H. harmandi*, eight in *H. molitrix*, and ten in *H. nobilis*. Particularly important is the absence of major rDNA on the largest pair 11 in the karyotype of *H. harmandi*, which may distinguish this species from the other two. Therefore, the combination of both cytogenetic methods may be useful for the species diagnosis inside *Hypophthalmichthys*. Testing of their universality across different pure *Hypophthalmichthys* populations together with concomitant generation of another cytogenetic markers (such as, e.g., species-specific satellite DNA classes) is an inevitable further research step.

Author Contributions: H.D.D. and P.R. managed the fish material. A.S. and P.R. conceived and designed the experiments. A.S., Š.P., M.d.B.C., T.H. and M.K. conducted the experiments. A.S., M.d.B.C., M.K., Š.P., V.Š., P.R. analyzed the data. M.K. and P.R. contributed reagents, materials and analytical equipment. A.S. and P.R. wrote the draft of the manuscript. All authors have read and agreed to the published version of the manuscript.

Acknowledgments: The authors are grateful to the fishery manager Vladimír Chytka, Fishery Farm, Pohořelice, Czech Republic, for providing bigheads carp individuals for this study and information about stock origins. Thanks are due to also Věra Šlechtová who biochemically analyzed individuals of *H. molitrix* and *H. nobilis* to confirm pure species. Zuzana Majtánová is deeply acknowledged for help in arranging *H. harmandi* karyotype from ancient photographs.

References

1. Howes, G.J. Anatomy and phylogeny of the Chinese major carps *Ctenopharyngodon* Steind., 1866 and *Hypophthalmichthys* Blkr., 1860. *Bull. Br. Mus. (Nat. Hist.) Zool* **1981**, *41*, 1–52.
2. Tan, M.; Armbruster, J.W. Phylogenetic classification of extant genera of fishes of the order Cypriniformes (Teleostei. Ostariophysi). *Zootaxa* **2018**, *4476*, 6–39. [CrossRef] [PubMed]
3. Tao, W.; Zou, M.; Wang, X.; Gan, X.; Mayden, R.L.; He, S. Phylogenomic analysis resolves the formerly intractable adaptive diversification of the endemic clade of East Asian Cyprinidae (Cypriniformes). *PLoS ONE* **2010**, *5*, e13508. [CrossRef] [PubMed]
4. Kottelat, M. The fishes of the inland waters of Southeast Asia. In *A Catalogue and Core Bibliography of the Fishes Known to Occur in Freshwaters, Mangroves and Estuaries*; Raffles Bull. Zool: Kent Ridge, Singapore, 2013; pp. 1–663.
5. Yen, M.D. Species composition and distribution of the freshwater fish fauna of the North of Vietnam. *Hydrobiologia* **1985**, *121*, 281–286. [CrossRef]
6. FAO Statistics Food and Agriculture. Available online: http://www.fao.org/fishery/culturedspecies/Hypophthalmichthys_molitrix/en (accessed on 29 January 2020).
7. Stepien, C.A.; Snyder, M.R.; Elz, A.E. Invasion genetics of the silver carp *Hypophthalmichthys molitrix* across North America: Differentiation of fronts, introgression, and eDNA metabarcode detection. *PLoS ONE* **2019**, *14*, e0203012. [CrossRef]
8. Sing, A.K.; Kumar, D.; Srivastava, S.C.; Ansan, A.; Jena, T.K.; Sarkar, U.K. Invasion and impacts of alien fish species in the Ganga River, India. *Aquat. Ecosyst. Health Manag.* **2013**, *16*, 408–414. [CrossRef]
9. Luebcker, N.; Zengeva, T.A.; Dabrowski, J.; Robertson, J.P. Predicting the potential distribution of invasive silver carp *Hypophthalmichthys molitrix* in South Africa. *Afr. J. Aquat. Sci.* **2014**, *39*, 157–165. [CrossRef]
10. Garvey, J.E. Bigheads of the genus *Hypophthalmichthys*. In *A Handbook of Global Freshwater Invasive Species*; Francis, R.A., Ed.; Earthscan: New York, NY, USA, 2012; pp. 235–245.
11. Hayer, C.A.; Breeggemann, J.J.; Jason, J.; Klumb, R.A.; Graeb, B.D.S.; Bertrand, K.N. Population characteristics of bighead and silver carp on the northwestern front of their North American invasion. *Aquat. Invasions* **2014**, *9*, 289–303. [CrossRef]
12. Farrington, H.L.; Edwards, C.E.; Bartron, M.; Lance, R.F. Phylogeography and population genetics of introduced silver carp (*Hypophthalmichthys molitrix*) and bighead carp (*H. nobilis*) in North America. *Biol. Invasions* **2017**, *19*, 2789–2811. [CrossRef]
13. DeBoer, J.A.; Anderson, A.M.; Casper, A.F. Multi-trophic response to invasive silver carp (*Hypophthalmichthys molitrix*) in large floodplain river. *Freshw. Biol.* **2018**, *63*, 597–611. [CrossRef]
14. Tristano, E.P.; Coulter, A.A.; Newton, T.J.; Garvey, J.E. Invasive silver carp may compete with unionid mussels for algae: First experimental evidence. *Aquat. Conserv.* **2019**, *29*, 1749–1757. [CrossRef]
15. Marián, T.; Krasznai, Z. Karyological investigation on Ctenopharyngodon idella and Hypophthalmichthys nobilis and their cross-breeding. *Aquac. Hung. (Szarvas)* **1978**, *1*, 44–50.
16. Marián, T.; Krasznai, Z. Comparative karyological studies on Chinese carps. *Aquaculture* **1979**, *13*, 325–336. [CrossRef]
17. Marián, T.; Krasznai, Z. Zylogische Untersuchungen bei der Familie Cyprinidae (Pisces). *Biol. Zent.* **1987**, *97*, 205–214.
18. Marián, T.; Krasznai, Z.; Oláh, J. Characteristic karyological, biochemical and morphological markers of silver carp (*Hypophthalmichthys molitrix*), bighead carp (*Aristicthys nobilis*) and their hybrids. *Aquac. Hung.* **1986**, *5*, 15–30.
19. Vasiliev, V.P.; Makeeva, A.P.; Ryabov, I.N. The study of chromosome complexes in cyprinid fish and their hybrids. *Genetika* **1978**, *14*, 1450–1460. (In Russian)
20. Beck, M.L.; Biggers, C.J.; Dupree, H.K. Karyological analysis of *Ctenopharyngodon idella, Aristichthys nobilis* and their F1 hybrid. *Trans. Am. Fish. Soc.* **1980**, *109*, 433–438. [CrossRef]
21. Beck, M.L.; Biggers, C.J. Chromosomal investigation of Ctenopharyngodon idella × Aristichthys nobilis hybrids. *Experientia* **1982**, *38*, 319. [CrossRef]
22. Zan, R.G.; Song, Z. Analysis and comparison between karyotypes of Cyprinus carpio and Carassius auratus and Aristichthys nobilis and Hypophthalmichthys molitrix. *Acta Genet. Sin.* **1980**, *7*, 72–77.

23. Zan, R.G.; Song, Z.; Liu, W.G. Studies on karyotypes and nuclear DNA contents of some cyprinoid fishes, with notes on fish polyploids in China. In *Indo-Pacific Fish Biology: Proceedings of the Second International Conference on Indo-Pacific Fish*; Uyeno, T., Arai, R., Taniuchi, T., Matsuura, K., Eds.; Ichthyological Society of Japan: Tokyo, Japan, 1986; pp. 877–885.
24. Zhou, T. Analysis of karyotype in bighead carp. *Freshw. Fish. (Shashi)* **1980**, *4*, 3–7. (In Chinese)
25. Zhou, T.; Lin, J.; Yang, Y. The karyotype of black carp [Mylopharyngodon piceus (Rich.)]. *J. Wuhan Univ.* **1980**, *4*, 112–116. (In Chinese with English Abstract)
26. Liu, L.Y. Karyotype analysis of *Aristichthys nobilis* by the leucocyte culture method. *J. Beijing Norm. Univ. (Nat. Sci.)* **1981**, *3*, 79–83. (In Chinese with English Abstract)
27. Almeida-Toledo, L.F.; Bigoni, A.P.V.; Bernardino, G.; De Almeida Filho, S. Chromosomal location of NORs and C bands in F1 hybrids of bighead carp and silver carp reared in Brazil. *Aquaculture* **1995**, *135*, 277–284. [CrossRef]
28. Anonymous. A preliminary report on the chromosomes number of 52 species of fishes of the Pearl River system in Guandong Province. *J. Wuhan Univ.* **1983**, *3*, 123–125.
29. Manna, G.K.; Khuda-Bukhsh, A.R. Karyomorphology of cyprinid fishes and cytological evaluation of the family. *Nucleus* **1978**, *43*, 119–127.
30. Liu, L.Y. Karyotype analysis of *Hypophthalmichthys molitrix*. *Acta Genet. Sin.* **1981**, *8*, 251–255. (In Chinese with English Abstract)
31. Su, Z.; Xu, K.; Chen, S.; Bai, G. Studies on triploid silver carp and its karyotype. *Chin. Zool. Res. (Suppl.) Kunming* **1984**, *5*, 15–19. (In Chinese with English Abstract)
32. Yu, X.L.; Zhou, T.; Li, Z.C.; Li, K.; Zhou, M. On the karyosystematics of cyprinid fishes and a summary of fish chromosome studies in China. *Genetica* **1987**, *72*, 225–236. [CrossRef]
33. Zhou, M.; Li, Y.C.; Zhou, T.; Yu, X.J. A BrdU-BSG method for G-banding in fish chromosomes and an idiogram of G-banded karyotype of silver carp. *Acta Genet. Sin.* **1989**, *16*, 184–187.
34. Ren, X.H.; Cui, J.; Yu, Q. Chromosomal heterochromatin differentiation of cyprinid fishes. *Genetica* **1992**, *87*, 47–51. [CrossRef]
35. Khuda-Bukhsh, A.R.; Chakrabarti, C. Induction of serial bands akin to G-type on metaphase chromosomes of two species of fish, *Rita rita* (Bagridae) and *Hypophthalmichthys molitrix* (Cyprinidae). *Indian J. Exp. Biol.* **1996**, *34*, 1271–1273.
36. Šlechtová, V.; Šlechta, V.; Hiep, D.D.; Valenta, M. Biochemical genetic comparison of bighead (*Aristichthys nobilis*) and silver carp (*Hypophthalmichthys molitrix*) and their hybrids in Czechoslovakia. *J. Fish Biol.* **1991**, *39*, 349–357. [CrossRef]
37. Fujiwara, A.; Nishida-Umehara, C.; Sakamoto, T.; Okamoto, N.; Nakayama, I.; Abe, S. Improved fish lymphocyte culture for chromosome preparation. *Genetica* **2001**, *111*, 77–89. [CrossRef] [PubMed]
38. Symonová, R.; Flajšhans, M.; Sember, A.; Havelka, M.; Gela, D.; Kořínková, T.; Rodina, M.; Rábová, M.; Ráb, P. Molecular cytogenetics in artificial hybrid and highly polyploid sturgeons: An evolutionary story narrated by repetitive sequences. *Cytogenet. Genome Res.* **2013**, *141*, 153–162. [CrossRef]
39. Ráb, P.; Roth, P. Cold-blooded vertebrates. In *Methods of Chromosome Analysis*; Balicek, P., Forejt, J., Rubeš, J., Eds.; Cytogenetická Sekce Československé Biologické Společnosti pri. CSAV: Brno, Czech Republic, 1988; pp. 115–124.
40. Bertollo, L.A.C.; Cioffi, M.B.; Moreira-Filho, O. Direct chromosome preparation from freshwater teleost fishes. In *Fish Cytogenetic Techniques (Ray-Fin Fishes and Chondrichthyans)*, 2nd ed.; Ozouf-Costaz, C., Pisano, E., Foresti, F., Toledo, L.F.A., Eds.; CRC Press: Enfield, CT, USA, 2015; Volume 1, pp. 21–26.
41. Haaf, T.; Schmid, M. An early stage of ZZ/ZW sex chromosomes differentiation in *Poecilia sphenops* var. *melanistica (Poeciliidae, Cyprinodontiformes)*. *Chromosoma* **1984**, *89*, 37–41. [CrossRef]
42. Mayr, B.; Ráb, P.; Kalat, M. Localisation of NORs and counterstain-enhanced fluorescence studies in *Perca fluviatilis* (Pisces, Percidae). *Genetica* **1985**, *67*, 51–56. [CrossRef]
43. Sola, L.; Rossi, A.R.; Iaselli, V.; Rasch, E.M.; Monaco, P.J. Cytogenetics of bisexual/unisexual species of *Poecilia*. II. Analysis of heterochromatin and nucleolar organizer regions in *Poecilia mexicana mexicana* by C-banding and DAPI, quinacrine, chromomycin A_3, and silver staining. *Cytogenet. Cell Genet* **1992**, *60*, 229–235. [CrossRef]
44. Howell, W.M.; Black, D.A. Controlled silver-staining of nucleolus organizer regions with a protective colloidal developer: A 1-step method. *Experientia* **1980**, *36*, 1014–1015. [CrossRef]

45. Sember, A.; Bohlen, J.; Šlechtová, V.; Altmanová, M.; Symonová, R.; Ráb, P. Karyotype differentiation in 19 species of river loach fishes (Nemacheilidae, Teleostei): Extensive variability associated with rDNA and heterochromatin distribution and its phylogenetic and ecological interpretation. *BMC Evol. Biol.* **2015**, *15*, 251. [CrossRef]
46. Cioffi, M.B.; Martins, C.; Centofante, L.; Jacobina, U.; Bertollo, L.A.C. Chromosomal variability among allopatric populations of Erythrinidae fish *Hoplias malabaricus*: Mapping of three classes of repetitive DNAs. *Cytogenet. Genome Res.* **2009**, *125*, 132–141. [CrossRef]
47. Silva, D.M.Z.A.; Utsunomia, R.; Pansonato-Alves, J.C.; Oliveira, C.; Foresti, F. Chromosomal mapping of repetitive DNA sequences in five species of *Astyanax* (Characiformes, Characidae) reveals independent location of U1 and U2 snRNA sites and association of U1 snRNA and 5S rDNA. *Cytogenet. Genome Res.* **2015**, *146*, 144–152. [CrossRef] [PubMed]
48. Yano, C.F.; Bertollo, L.A.C.; Cioffi, M.B. Fish-FISH: Molecular cytogenetics in fish species. In *Fluorescence In Situ Hybridization (FISH)—Application Guide*, 2nd ed.; Liehr, T., Ed.; Springer: Berlin, Germany, 2017; pp. 429–443.
49. Sember, A.; Bohlen, J.; Šlechtová, V.; Altmanová, M.; Pelikánová, Š.; Ráb, P. Dynamics of tandemly repeated DNA sequences during evolution of diploid and tetraploid botiid loaches (Teleostei: Cobitoidea: Botiidae). *PLoS ONE* **2018**, *13*, e0195054. [CrossRef] [PubMed]
50. Altschul, S.F.; Gish, W.; Miller, W.; Myers, E.W.; Lipman, D.J. Basic local alignment search tool. *J. Mol. Biol.* **1990**, *215*, 403–410. Available online: http://blast.ncbi.nlm.nih.gov/blast (accessed on 20 March 2018). [CrossRef]
51. Levan, A.; Fredga, K.; Sandberg, A.A. Nomenclature for centromeric position on chromosomes. *Hereditas* **1964**, *52*, 201–220. [CrossRef]
52. Matthey, R. L′ evolution de la formule chromosomiale chez les vertebrees. *Experientia* **1945**, *1*, 78–86. [CrossRef]
53. Liu, H.; Pang, M.; Yu, X.; Zhou, Y.; Tong, J.; Fu, B. Sex-specific markers developed by next-generation sequencing confirmed an XX/XY sex determination system in bighead carp (*Hypophthalmichehys nobilis*) and silver carp (*Hypophthalmichthys molitrix*). *DNA Res.* **2018**, *25*, 257–264. [CrossRef]
54. Ráb, P.; Collares-Pereira, M.J. Chromosomes of European cyprinid fishes (Cyprinidae, Cypriniformes). A review. *Folia Zool.* **1995**, *44*, 193–214.
55. Knytl, M.; Kalous, L.; Rylková, K.; Choleva, L.; Merilä, J.; Ráb, P. Morphologically indistinguishable hybrid *Carassius* female with 156 chromosomes: A threat for the threatened crucian carp, *C. carassius*, L. *PLoS ONE* **2018**, *13*, e0190924. [CrossRef]
56. Saenjundaeng, P.; Oliveira, E.A.; Tanomtong, A.; Supiwong, W.; Phimphan, S.; Collares-Pereira, M.J.; Sember, A.; Yano, C.F.; Hatanaka, T.; Bertollo, L.A.C.; et al. Chromosomes of Asian cyprinid fishes: Cytogenetic analysis of two representatives of small paleotetraploid tribe Probarbini. *Mol. Cytogenet.* **2018**, *11*, 51. [CrossRef]
57. Rocchi, M.; Archidiacono, N.; Schempp, W.; Capozzi, O.; Stanyon, R. Centromere repositioning in mammals. *Heredity* **2012**, *108*, 59–67. [CrossRef]
58. Schubert, I. What is behind "centromere repositioning"? *Chromosoma* **2018**, *127*, 229–234. [CrossRef] [PubMed]
59. Souza, I.L.; Moreira-Filho, O. Constitutive heterochromatin and Ag-NOR polymorphisms in the small characid fish *Astyanax scabripinnis* (Jenyns, 1842). *Comp. Gen. Pharm.* **2007**, *72*, 63–69. [CrossRef]
60. Hashimoto, D.T.; Porto-Foresti, F. Chromosome polymorphism of heterochromatin and nucleolar regions in two populations of the fish *Astyanax bockmanni* (Teleostei: Characiformes). *Neotrop. Ichthyol.* **2010**, *8*, 861–866. [CrossRef]
61. Baumgärtner, L.; Paiz, L.M.; Zawadzki, C.H.; Margarido, V.P.; Castro, A.L.B.P. Heterochromatin polymorphism and physical mapping of 5S and 18S ribosomal DNA in four populations of *Hypostomus strigaticeps* (Regan, 1907) from the Paraná River basin, Brazil: Evolutionary and environmental correlation. *Zebrafish* **2014**, *11*, 479–487. [CrossRef] [PubMed]
62. Utsunomia, R.; Pansonato-Alves, J.C.; Costa-Silva, G.J.; Mendonça, F.F.; Scacchetti, P.C.; Oliveira, C.; Foresti, F. Molecular and cytogenetic analyses of cryptic species within the *Synbranchus marmoratus* Bloch, 1795 (Synbranchiformes: Synbranchidae) grouping: Species delimitations, karyotypic evolution and intraspecific diversification. *Neotrop. Ichthyol.* **2014**, *12*, 903–911. [CrossRef]

63. Cioffi, M.B.; Bertollo, L.A.C. Chromosomal distribution and evolution of repetitive DNAs in fish. *Genome Dyn.* **2012**, *7*, 197–221. [CrossRef]
64. Gornung, E. Twenty years of physical mapping of major ribosomal RNA genes across the teleosts: A review of research. *Cytogenet. Genome Res.* **2013**, *141*, 90–102. [CrossRef]
65. Sochorová, J.; Garcia, S.; Gálvez, F.; Symonová, R.; Kovařík, A. Evolutionary trends in animal ribosomal DNA loci: Introduction to a new online database. *Chromosoma* **2018**, *127*, 141–150. [CrossRef]
66. Castro, J.; Rodriguez, S.; Pardo, B.G.; Sanchez, L.; Martinez, P. Population analysis of an unusual NOR-site polymorphism in brown trout (*Salmo trutta* L.). *Heredity* **2001**, *86*, 291–302. [CrossRef]
67. Piscor, D.; Alves, A.L.; Parise-Maltempi, P.P. Chromosomal microstructure diversity in three *Astyanax* (Characiformes, Characidae) species: Comparative analysis of the chromosomal locations of the 18S and 5S rDNAs. *Zebrafish* **2015**, *12*, 81–90. [CrossRef]
68. Traldi, J.B.; Vicari, M.R.; Martinez, J.D.F.; Blanco, D.R.; Lui, R.L.; Moreira-Filho, O. Chromosome analyses of *Apareiodon argenteus* and *Apareiodon davisi* (Characiformes, Parodontidae): An extensive chromosomal polymorphism of 45S and 5S ribosomal DNAs. *Zebrafish* **2016**, *13*, 19–25. [CrossRef]
69. Da Silva, A.F.; Feldberg, E.; Moura Carvalho, N.D.; Hernández Rangel, S.M.; Schneider, C.H.; Carvalho-Zilse, G.A.; Fonsêca da Silva, V.; Gross, M.C. Effects of environmental pollution on the rDNAomics of Amazonian fish. *Environ. Pollut.* **2019**, *252*, 180–187. [CrossRef] [PubMed]
70. Pereira, C.S.A.; Aboim, M.A.; Ráb, P.; Collares-Pereira, M.J. Introgressive hybridization as a promoter of genome reshuffling in natural homoploid fish hybrids (Cyprinidae, Leuciscinae). *Heredity* **2014**, *112*, 343–350. [CrossRef] [PubMed]
71. Cioffi, M.B.; Molina, W.F.; Artoni, R.F.; Bertollo, L.A.C. Chromosomes as tools for discovering biodiversity—The case of Erythrinidae fish family. *Recent Trends Cytogenet. Stud. Methodol. Appl.* **2012**, 125–146. [CrossRef]
72. Prizon, A.C.; Bruschi, D.P.; Borin-Carvalho, L.A.; Cius, A.; Barbosa, L.M.; Ruiz, H.B.; Zawadzki, C.H.; Fenocchio, A.S.; Portela-Castro, A.L.B. Hidden diversity in the populations of the armored catfish *Ancistrus* Kner, 1854 (Loricariidae, Hypostominae) from the Paraná River Basin revealed by molecular and cytogenetic data. *Front. Genet.* **2017**, *8*, 185. [CrossRef] [PubMed]
73. Nirchio, M.; Paim, F.G.; Milana, V.; Rossi, A.R.; Oliveira, C. Identification of a new mullet species complex based on an integrative molecular and cytogenetic investigation of *Mugil hospes* (Mugilidae: Mugiliformes). *Front. Genet.* **2018**, *9*, 17. [CrossRef]
74. Zhu, H.P.; Gui, J.F. Identification of genome organization in the unusual allotetraploid form of *Carassius auratus gibelio*. *Aquaculture* **2007**, *265*, 109–117. [CrossRef]
75. Zhang, C.; Ye, L.; Chen, Y.; Xiao, J.; Wu, Y.; Tao, M.; Xiao, Y.; Liu, S. The chromosomal constitution of fish hybrid lineage revealed by 5S rDNA FISH. *BMC Genet.* **2015**, *16*, 140. [CrossRef]
76. Zhu, H.P.; Ma, D.M.; Gui, J.F. Triploid origin of the gibel carp as revealed by 5S rDNA localization and chromosome painting. *Chromosome Res.* **2006**, *14*, 767–776. [CrossRef]
77. Li, Y.J.; Tian, Y.; Zhang, M.Z.; Tian, P.P.; Yu, Z.; Abe, S.; Arai, K. Chromosome banding and FISH with rDNA probe in the diploid and tetraploid loach *Misgurnus anguillicaudatus*. *Ichthyol. Res.* **2010**, *57*, 358–366. [CrossRef]
78. Da Silva, M.; Matoso, D.A.; Ludwig, L.A.M.; Gomes, E.; Almeida, M.C.; Vicari, M.R.; Artoni, R.F. Natural triploidy in *Rhamdia quelen* identified by cytogenetic monitoring in Iguaçu basin, southern Brazil. *Environ. Biol. Fishes* **2011**, *91*, 361–366. [CrossRef]
79. Spóz, A.; Boroń, A.; Ocalewicz, K.; Kirtiklis, L. Polymorphism of the rDNA chromosomal regions in the weatherfish *Misgurnus fossilis* (Teleostei: Cobitidae). *Folia Biol.* **2017**, *65*, 63–70. [CrossRef]
80. Symonová, R.; Majtánová, Z.; Sember, A.; Staaks, G.B.; Bohlen, J.; Freyhof, J.; Rábová, M.; Ráb, P. Genome differentiation in a species pair of coregonine fishes: An extremely rapid speciation driven by stress-activated retrotransposons mediating extensive ribosomal DNA multiplications. *BMC Evol. Biol.* **2013**, *13*, 42. [CrossRef] [PubMed]
81. Lima-Filho, P.A.; Bertollo, L.A.C.; Cioffi, M.B.; Costa, G.W.W.F.; Molina, W.F. Karyotype divergence and spreading of 5S rDNA sequences between genomes of two species: Darter and emerald gobies (*Ctenogobius*, Gobiidae). *Cytogenet. Genome Res.* **2014**, *142*, 197–203. [CrossRef] [PubMed]

82. Mazzei, F.; Ghigliotti, L.; Bonillo, C.; Coutanceau, J.-P.; Ozouf-Costaz, C.; Pisano, E. Chromosomal patterns of major and 5S ribosomal DNA in six icefish species (Perciformes, Notothenioidei, Channichthyidae). *Polar Biol.* **2004**, *28*, 47–55. [CrossRef]
83. Ziemniczak, K.; Barros, A.V.; Rosa, K.O.; Nogaroto, V.; Almeida, M.C.; Cestari, M.M.; Moreira-Filho, O.; Artoni, R.F.; Vicari, M.R. Comparative cytogenetics of Loricariidae (Actinopterygii: Siluriformes): Emphasis in Neoplecostominae and Hypoptopomatinae. *Ital. J. Zool.* **2012**, *79*, 492–501. [CrossRef]
84. Piscor, D.; Fernandes, C.A.; Parise-Maltempi, P.P. Conserved number of U2 snDNA sites in *Piabina argentea*, *Piabarchus stramineus* and two *Bryconamericus* species (Characidae, Stevardiinae). *Neotrop. Ichthyol.* **2018**, *16*, e170066. [CrossRef]
85. Yano, C.F.; Merlo, M.A.; Portela-Bens, S.; Cioffi, M.B.; Bertollo, L.A.C.; Santos-Júnior, C.D.; Rebordinos, L.; Esteban, M.A.; Albert, J.S. Evolutionary dynamics of multigene families in *Triportheus* (Characiformes, Triportheidae): A transposon mediated mechanism? *Front. Mar. Sci.* **2020**, *7*, 6. [CrossRef]
86. Inafuku, J.; Nabeyama, M.; Kikuma, Y.; Saitoh, J.; Kubota, S.; Kohno, S.I. Chromosomal location and nucleotide sequences of 5S ribosomal DNA of two cyprinid species (Osteichthyes, Pisces). *Chromosome Res.* **2000**, *8*, 193–199. [CrossRef]
87. He, W.; Qin, Q.; Liu, S.; Li, T.; Wang, J.; Xiao, J.; Xie, L.; Zhang, C.; Liu, Y. Organization and variation analysis of 5S rDNA in different ploidy-level hybrids of red crucian carp × topmouth culter. *PLoS ONE* **2012**, *7*, e38976. [CrossRef]
88. Rábová, M.; Ráb, P.; Ozouf-Costaz, C. Extensive polymorphism and chromosomal characteristics of ribosomal DNA in a loach fish, *Cobitis vardarensis* (Ostariophysi, Cobitidae) detected by different banding techniques and fluorescence in situ hybridization (FISH). *Genetica* **2001**, *111*, 413–422. [CrossRef] [PubMed]
89. Boroń, A.; Ozouf-Costaz, C.; Coutanceau, J.-P.; Woroniecka, K. Gene mapping of 28S and 5S rDNA sites in the spined loach *Cobitis taenia* (Pisces, Cobitidae) from a diploid population and a diploid-tetraploid population. *Genetica* **2006**, *128*, 71–79. [CrossRef] [PubMed]
90. Gromicho, M.; Coutanceau, J.-P.; Ozouf-Costaz, C.; Collares-Pereira, M.J. Contrast between extensive variation of 28S rDNA and stability of 5S rDNA and telomeric repeats in the diploid-polyploid *Squalius alburnoides* complex and in its maternal ancestor *Squalius pyrenaicus* (Teleostei, Cyprinidae). *Chromosome Res.* **2006**, *14*, 297–306. [CrossRef] [PubMed]
91. Pereira, C.S.A.; Ráb, P.; Collares-Pereira, M.J. Chromosomes of European cyprinid fishes: Comparative cytogenetics and chromosomal characteristics of ribosomal DNAs in nine Iberian chondrostomine species (Leuciscinae). *Genetica* **2012**, *140*, 485–495. [CrossRef] [PubMed]
92. Rossi, A.R.; Milana, V.; Hett, A.K.; Tancioni, L. Molecular cytogenetic analysis of the Appenine endemic cyprinid fish *Squalius lucumonis* and three other Italian leuciscines using chromosome banding and FISH with rDNA probes. *Genetica* **2012**, *140*, 469–476. [CrossRef]
93. Singh, M.; Kumar, R.; Nagpure, N.S.; Kushwaha, B.; Mani, I.; Lakra, W.S. Extensive NOR site polymorphism in geographically isolated populations of Golden mahseer, *Tor putitora*. *Genome* **2009**, *52*, 783–789. [CrossRef]
94. Libertini, A.; Sola, L.; Rampin, M.; Rossi, A.R.; Iijima, K.; Ueda, T. Classical and molecular cytogenetic characterization of allochthonous European bitterling *Rhodeus amarus* (Cyprinidae, Acheilognathinae) from Northern Italy. *Genes Genet. Syst.* **2008**, *83*, 417–422. [CrossRef]
95. Han, C.C.; Yen, T.B.; Chen, N.C.; Tseng, M.C. Comparative studies of 5S rDNA profiles and Cyt *b* sequences in two *Onychostoma* species (Cyprinidae). *Int. J. Mol. Sci.* **2015**, *16*, 29663–29672. [CrossRef]
96. Kirtiklis, L.; Porycka, K.; Boroń, A.; Coutanceau, J.-P.; Dettai, A. Use of the chromosomal co-location of the minor 5S and the major 28S rDNA as a cytogenetic marker within the genus *Leuciscus* (Pisces, Cyprinidae). *Folia Biol.* **2010**, *58*, 245–249. [CrossRef]
97. Merlo, M.A.; Cross, I.; Rodríguez-Rúa, A.; Manchado, M.; Rebordinos, L. First approach to studying the genetics of the meagre (*Argyrosomus regius*; Asso, 1801) using three multigene families. *Aquac. Res.* **2013**, *44*, 974–984. [CrossRef]
98. Scacchetti, P.C.; Utsunomia, R.; Pansonato-Alves, J.C. Repetitive DNA sequences and evolution of ZZ/ZW sex chromosomes in *Characidium* (Teleostei: Characiformes). *PLoS ONE* **2015**, *10*, e0137231. [CrossRef] [PubMed]
99. Yano, C.F.; Bertollo, L.A.C.; Rebordinos, L.; Merlo, M.A.; Liehr, T.; Portela-Bens, S.; Cioffi, M.B. Evolutionary dynamics of rDNAs and U2 small nuclear DNAs in *Triportheus* (Characiformes, Triportheidae): High variability and particular syntenic organization. *Zebrafish* **2017**, *14*, 146–154. [CrossRef] [PubMed]

100. Araya-Jaime, C.; Mateussi, N.T.B.; Utsunomia, R.; Costa-Silva, G.J.; Oliveira, C.; Foresti, F. ZZ/Z0: The new system of sex chromosomes in *Eigenmannia* aff. *trilineata* (Teleostei: Gymnotiformes: Sternopygidae) characterized by molecular cytogenetics and DNA barcoding. *Zebrafish* **2017**, *14*, 464–470. [CrossRef]
101. Ponzio, J.C.; Piscor, D.; Parise-Maltempi, P.P. Chromosomal locations of U2 snDNA clusters in *Megaleporinus*, *Leporinus* and *Schizodon* (Characiformes: Anostomidae). *Biologia* **2018**, *73*, 295–298. [CrossRef]
102. Cabral-de-Mello, D.C.; Valente, G.T.; Nakajima, R.T.; Martins, C. Genomic organization and comparative chromosome mapping of the U1 snRNA gene in cichlid fish, with an emphasis in *Oreochromis niloticus*. *Chromosome Res.* **2012**, *20*, 279–292. [CrossRef]
103. Carvalho, P.C.; Oliveira, E.A.; Bertollo, L.A.C.; Yano, C.F.; Al-Rikabi, A.B.H.; Cioffi, M.B. First chromosomal analysis in Hepsetidae (Actinopterygii, Characiformes): Insights into relationship between African and Neotropical fish groups. *Front. Genet.* **2017**, *8*, 203. [CrossRef] [PubMed]
104. García-Souto, D.; Troncoso, T.; Pérez, M.; Pasantes, J.J. Molecular cytogenetic analysis of the european hake *Merluccius merluccius* (Merlucciidae, Gadiformes): U1 and U2 snRNA gene clusters map to the same location. *PLoS ONE* **2015**, *10*, e0146150. [CrossRef]
105. Piscor, D.; Paiz, L.M.; Baumgärtner, L.; Cerqueira, F.J.; Fernandes, C.A.; Lui, R.L.; Parise-Maltempi, P.P.; Margarido, V.P. Chromosomal mapping of repetitive sequences in *Hyphessobrycon eques* (Characiformes, Characidae): A special case of the spreading of 5S rDNA clusters in a genome. *Genetica* **2020**, *148*, 25–32. [CrossRef]
106. Usso, M.C.; dos Santos, A.R.; Gouveia, J.G.; Frantine-Silva, W.; Araya-Jaime, C.; Oliveira, M.L.M.; Foresti, F.; Giuliano-Caetano, L.; Dias, A.L. Genetic and chromosomal differentiation of *Rhamdia quelen* (Siluriformes, Heptapteridae) revealed by repetitive molecular markers and DNA barcoding. *Zebrafish* **2019**, *16*, 87–97. [CrossRef]
107. Cremer, M.; Von Hase, J.; Volm, T.; Brero, A.; Kreth, G.; Walter, J.; Fischer, C.; Solovei, I.; Cremer, C.; Cremer, T. Non-random radial higher-order chromatin arrangements in nuclei of diploid human cells. *Chromosome Res.* **2001**, *9*, 541–567. [CrossRef]
108. Úbeda-Manzanaro, M.; Merlo, M.A.; Palazón, J.L.; Cross, I.; Sarasquete, C.; Rebordinos, L. Chromosomal mapping of the major and minor ribosomal genes, $(GATA)_n$ and U2 snRNA gene by double-colour FISH in species of the Batrachoididae family. *Genetica* **2010**, *138*, 787–794. [CrossRef] [PubMed]
109. Malimpensa, G.C.; Traldi, J.B.; Toyama, D.; Henrique-Silva, F.; Vicari, M.R.; Moreira-Filho, O. Chromosomal mapping of repeat DNA in *Bergiaria westermanni* (Pimelodidae, Siluriformes): Localization of 45S rDNA in B chromosomes. *Cytogenet. Genome Res.* **2018**, *154*, 99–106. [CrossRef] [PubMed]
110. Manchado, M.; Zuasti, E.; Cross, I.; Merlo, A.; Infante, C.; Rebordinos, L. Molecular characterization and chromosomal mapping of the 5S rRNA gene in *Solea senegalensis*: A new linkage to the U1, U2, and U5 small nuclear RNA genes. *Genome* **2006**, *49*, 79–86. [CrossRef] [PubMed]
111. Ruiz-Herrera, A.; Nergadze, S.G.; Santagostino, M.; Giulotto, E. Telomeric repeats far from the ends: Mechanisms of origin and role in evolution. *Cytogenet. Genome Res.* **2009**, *122*, 219–228. [CrossRef]
112. Zhou, M.; Kang, Y.; Li, Y.C.; Zhou, T. Studies on silver-stained karyotypes of 7 species in Cyprinidae (Pisces). *Zool. Res* **1988**, *9*, 225–229. (In Chinese)
113. Ren, X.H.; Yu, X.J. Characterization of nucleolar organizer regions of twelve species of Chinese cyprinid fishes. *Caryologia* **1993**, *46*, 201–207. [CrossRef]
114. Li, Y.C.; Li, K.; Hong, Y.H.; Gui, J.F.; Zhou, T. Studies on karyotypes of Chinese cyprinid fishes. VII. Karyotypic analyses of seven species in the subfamily Leuciscinae with a consideration for the phylogenetic relationships of some cyprinid fishes concerned. *Acta Genet. Sin.* **1985**, *12*, 367–372. (In Chinese)
115. Li, K.; Li, Y.C.; Zhou, M.; Zhou, T. Studies on the karyotypes of Chinese cyprinid fishes. II. Karyotypes of four species of Xenocyprininae. *Acta Zool. Sin.* **1983**, *29*, 207–213. (In Chinese)
116. Zan, R.G.; Song, Z. Analysis and comparison between the karyotypes of *Ctenopharyngodon idella* and *Megalobrama amblycephala*. *Acta Genet. Sin.* **1979**, *6*, 205–210. (In Chinese)
117. Lu, R.H.; Li, Y.J.; Xu, K.S. A chromosome study of *Megalobrama amblycephala*. *Oceanol. Limnol. Sin.* **1984**, *15*, 487–492. (In Chinese)
118. Liu, Y.U. On the karyotype of the grass-carp Ctenopharyngodon Idella. *Acta Zool. Sin* **1980**, *26*, 14–16. (In Chinese)
119. Lou, Y.; Zhang, K.; Wu, Y.; Wang, Y. Studies on karyotype of black carp (*Mylopharyngodon piceus*). *J. Fish. China* **1983**, *7*, 77–81. (In Chinese)

120. Arai, R. *Fish Karyotypes: A Check List*, 1st ed.; Springer: Tokyo, Japan, 2011.
121. Ráb, P.; Crossman, E.J. Chromosomal NOR phenotypes in North American pikes and pickerels, genus *Esox* with notes on Umbridae (Euteleostei: Esocae). *Can. J. Zool.* **1994**, *72*, 1951–1956. [CrossRef]
122. Symonová, R.; Ocalewicz, K.; Kirtiklis, L.; Delmastro, G.B.; Pelikánová, Š.; Garcia, S.; Kovařík, A. Higher-order organisation of extremely amplified, potentially functional and massively methylated 5S rDNA in European pikes (*Esox* sp.). *BMC Genom.* **2017**, *18*, 391. [CrossRef] [PubMed]
123. Phillips, R.B.; Ráb, P. Chromosome evolution in the Salmonidae (Pisces): An update. *Biol. Rev. Camb. Philos. Soc.* **2001**, *76*, 1–25. [CrossRef] [PubMed]
124. Dion-Côté, A.-M.; Symonová, R.; Lamaze, F.; Pelikánová, Š.; Ráb, P.; Bernatchez, L. Standing chromosomal variation in Lake Whitefish species pairs: The role of historical contingency and relevance for speciation. *Mol. Ecol.* **2017**, *26*, 178–192. [CrossRef]
125. Barby, F.F.; Bertollo, L.A.C.; Oliveira, E.A.; Yano, C.F.; Hatanaka, T.; Ráb, P.; Sember, A.; Ezaz, T.; Artoni, R.F.; Liehr, T.; et al. Emerging patterns of genome organization in Notopteridae species (Teleostei, Osteoglossiformes) as revealed by Zoo-FISH and Comparative Genomic Hybridization (CGH). *Sci. Rep.* **2019**, *9*, 1112. [CrossRef]
126. Molina, W.F. Chromosomal changes and stasis in marine fish groups. In *Fish Cytogenetics*, 1st ed.; Pisano, E., Ozouf-Costaz, C., Foresti, F., Kapoor, B.G., Eds.; Science Publishers: Enfield, CT, USA, 2007; pp. 69–110.
127. Motta-Neto, C.C.; Cioffi, M.B.; Bertollo, L.A.C.; Molina, W.F. Extensive chromosomal homologies and evidence of karyotypic stasis in Atlantic grunts of the genus *Haemulon* (Perciformes). *J. Exp. Mar. Biol. Ecol.* **2011**, *401*, 75–79. [CrossRef]
128. Motta-Neto, C.C.; Cioffi, M.B.; Bertollo, L.A.C.; Molina, W.F. Molecular cytogenetic analysis of Haemulidae fish (Perciformes): Evidence of evolutionary conservation. *J. Exp. Mar. Biol. Ecol.* **2011**, *407*, 97–100. [CrossRef]
129. Motta-Neto, C.C.; Lima-Filho, P.A.; Araújo, W.C.; Bertollo, L.A.C.; Molina, W.F. Differentiated evolutionary pathways in Haemulidae (Perciformes): Karyotype stasis versus morphological differentiation. *Rev. Fish Biol. Fish.* **2012**, *22*, 457–465. [CrossRef]
130. Motta-Neto, C.C.; Cioffi, M.B.; Costa, G.W.W.F.; Amorim, K.D.J.; Bertollo, L.A.C.; Artoni, R.F.; Molina, W.F. Overview on karyotype stasis in Atlantic grunts (Eupercaria, Haemulidae) and the evolutionary extensions for other marine fish groups. *Front. Mar. Sci.* **2019**, *6*, 628. [CrossRef]
131. Ellegren, H. Evolutionary stasis: The stable chromosomes of birds. *Trends Ecol. Evol.* **2010**, *25*, 283–291. [CrossRef]
132. Tian, Y.; Nie, W.; Wang, J.; Ferguson-Smith, M.A.; Yang, F. Chromosome evolution in bears: Reconstructing phylogenetic relationships by cross-species chromosome painting. *Chromosome Res.* **2004**, *12*, 55–63. [CrossRef]
133. Razin, S.V.; Gavrilov, A.A.; Vassetzky, Y.S.; Ulianov, S.V. Topologically-associating domains: Gene warehouses adapted to serve transcriptional regulation. *Transcription* **2016**, *7*, 84–90. [CrossRef]
134. Rosin, L.F.; Crocker, O.; Isenhart, R.L.; Nguyen, S.C.; Xu, Z.; Joyce, E.F. Chromosome territory formation attenuates the translocation potential of cells. *eLife* **2019**, *8*, e49553. [CrossRef]
135. Alves, M.J.; Coelho, M.M.; Collares-Pereira, M.J. Evolution in action through hybridization and polyploid in an Iberain freshwater fish: A genetic review. *Genetica* **2001**, *111*, 375–385. [CrossRef]

9

Comparative Chromosome Mapping of Musk Ox and the X Chromosome among some Bovidae Species

Anastasia A. Proskuryakova [1,*], Anastasia I. Kulemzina [1], Polina L. Perelman [1,2], Dmitry V. Yudkin [3], Natalya A. Lemskaya [1,3], Innokentii M. Okhlopkov [4], Egor V. Kirillin [4], Marta Farré [5,6], Denis M. Larkin [5], Melody E. Roelke-Parker [7], Stephen J. O'Brien [8,9], Mitchell Bush [10] and Alexander S. Graphodatsky [1,2]

1 Institute of Molecular and Cellular Biology, SB RAS, 630090 Novosibirsk, Russia; zakal@mcb.nsc.ru (A.I.K.); polina.perelman@gmail.com (P.L.P.); lemnat@mcb.nsc.ru (N.A.L.); graf@mcb.nsc.ru (A.S.G.)
2 Novosibirsk State University, 630090 Novosibirsk, Russia
3 State Research Center of Virology and Biotechnology "Vector", Federal Service for Surveillance on Consumer Rights Protection and Human Well-being (FBRI SRC VB "Vector", Rospotrebnadzor), 630559 Koltsovo, Novosibirsk Region, Russia; yudkin_dv@vector.nsc.ru
4 Institute for Biological Problems of Cryolithozone Siberian Branch of RAS, 677980 Yakutsk, Russia; imo-ibpc@yandex.ru (I.M.O.); e.kir@mail.ru (E.V.K.)
5 The Royal Veterinary College, Royal College Street, University of London, London NW1 0TU, UK; mfarrebelmonte@gmail.com (M.F.); dlarkin@rvc.ac.uk (D.M.L.)
6 School of Biosciences, University of Kent, Canterbury CT2 7NJ, UK
7 Frederick National Laboratory of Cancer Research, Leidos Biomedical Research, Inc., Frederick, MD 21701, USA; melody.roelke-parker@nih.gov
8 Theodosius Dobzhansky Center for Genome Bioinformatics, Saint-Petersburg State University, Sredniy Av. 41A, 199034 Saint-Petersburg, Russia; lgdchief@gmail.com
9 Oceanographic Center, Nova Southeastern University, Fort Lauderdale 3301 College Ave, Fort Lauderdale, FL 33314, USA
10 The Center for Species Survival, Department of Reproductive Sciences, Smithsonian Conservation Biology Institute, Smithsonian's National Zoological Park, 1500 Remount Road, Front Royal, VA 22630, USA; giraffedoc@hotmail.com
* Correspondence: andrena@mcb.nsc.ru

Abstract: Bovidae, the largest family in Pecora infraorder, are characterized by a striking variability in diploid number of chromosomes between species and among individuals within a species. The bovid X chromosome is also remarkably variable, with several morphological types in the family. Here we built a detailed chromosome map of musk ox (*Ovibos moschatus*), a relic species originating from Pleistocene megafauna, with dromedary and human probes using chromosome painting. We trace chromosomal rearrangements during Bovidae evolution by comparing species already studied by chromosome painting. The musk ox karyotype differs from the ancestral pecoran karyotype by six fusions, one fission, and three inversions. We discuss changes in pecoran ancestral karyotype in the light of new painting data. Variations in the X chromosome structure of four bovid species nilgai bull (*Boselaphus tragocamelus*), saola (*Pseudoryx nghetinhensis*), gaur (*Bos gaurus*), and Kirk's Dikdik (*Madoqua kirkii*) were further analyzed using 26 cattle BAC-clones. We found the duplication on the X in saola. We show main rearrangements leading to the formation of four types of bovid X: Bovinae type with derived cattle subtype formed by centromere reposition and Antilopinae type with Caprini subtype formed by inversion in XSB1.

Keywords: BAC-clones; chromosome painting; Kirk's Dikdik; musk ox; saola; nilgai bull; gaur

1. Introduction

Cetartiodactyla is a large mammalian order, including camels, whales, pigs, hippos, and ruminants—the suborder of animals with divided stomach. Bovidae is the most specious family in Ruminantia comprising 143 species with 50 genera [1]. Bovidae is generally subdivided into 2 subfamilies: Bovinae (the bovids: tribes Bovini, Tragelaphini and Boselaphini), represented by nilgai, four-horned antelope, wild cattle, bison, Asian buffalo, African buffalo, and kudu; and Antilopinae (antelopes and caprini: Neotragini, Aepycerotini, Cephalophini, Oreotragini, Hippotragini, Alcelaphini, Caprini, Antilopini, and Reduncini tribes), represented by antelopes, gazelles, goats and their relatives [2]. Bovids include several domesticated species (cattle, goat, and sheep) with high economic significance. Although recent advances have been made in the genomic inference of these species, phylogenetic relationships of species within the family are complex and remain somewhat unresolved [3].

Accumulated cytogenetic data for the Bovidae family allow tracing the trends in evolution of karyotypes of the Bovinae [4–7] and Antilopinae [7–11] subfamilies. Bovid karyotypes are characterized by tandem and Robertsonian translocations of acrocentric chromosomes, producing wide variation in (2n) chromosome numbers [12,13]. Previously, some 43 bovid species have been studied by comparative chromosome painting, mostly with cattle painting probes [14]. Due to its economic importance, the cattle genome has been widely studied, identifying interchromosome rearrangements between species but resolving few intrachromosomal rearrangements. The use of high-resolution dromedary probes [5] and BAC mapping [9,15] has provided increased precision with respect to syntenic segments orientation relative to centromeres and inversions. Proposed ancestral Bovidae and pecoran karyotypes were imputed based on genomic [16] and cytogenetic data [5,16,17]. Although the diploid chromosome number of Bovidae species ranges from 30 to 60, the number of autosomal arms in karyotypes is stable at 58 for most karyotyped species [18]. Comparative chromosome painting data have been applied to phylogenetic analyses of relationships between tribes. For example, a clear marker association was detected for Tragelaphini: the translocation of cattle chromosomes 1;29 [8], which is shared by all members of this group. Because large chromosomal rearrangements by and large correspond in a parsimony sense to morphology-based phylogenies for the group, it seems that chromosomal rearrangements played an important role in the speciation of the Bovidae family [12].

The X chromosome in the Bovidae family presents a special case in Bovidae evolution. displaying marked karyotypic variability between species, in contrast to highly conserved X chromosome morphology seen for the majority of eutherian mammals [19]. Surprisingly, unlike conserved X chromosome of the majority of eutherian mammals noted by Ohno [19], bovid chromosome X displayed variability. Several types of the X chromosome were identified: the cattle type appeared to be submetacentric, while the tragelaphines and antilopinae was acrocentric [20].

Cytogenetic maps of *Bos taurus*, *Bubalus bubalis*, and *Ovis aries* X chromosomes, when compared with that of *Homo sapiens*, revealed the conservation of pseudoautosomal region position in the Antilopinae subfamily, which was detected using a molecular cytogenetics approach [21,22]. In Ruminants and, especially, in the Bovidae family, a substantial number of intrachromosomal rearrangements, including inversions and centromere repositions, have been identified previously by using comparative bacterial artificial chromosome (BAC) mapping [15]. The cited research, however, described species from only three tribes: Bovini, Hippotragini, and Caprini. Moreover, several independent bovid lineages show the presence of compound sex chromosomes resulting from gonosome and autosome fusions [8,14,20,21,23]. Here, we extend the list of 18 species studied by detailed X chromosome BAC mapping to include species from three tribes: nilgai bull (Boselaphini), saola, gaur (Bovini), and Kirk's Dikdik (Antilopini) to reveal intrachromosomal rearrangements that occurred on the X chromosome in Bovidae family.

Among species of Bovidae family, musk ox (*Ovibos moschatus*), deserves special attention. This species belongs to a basal branch of the Caprini tribe [3], and along with reindeer, they are the only ungulates of the Arctic to survive the late Pleistocene extinction associated the most recent retreat from glaciation [24]. There are classic cytogenetic [25–27] and molecular cytogenetic data on chromosome fusions [14] in musk ox karyotype. Moreover, there is a huge pool of data describing cetariodactyl karyotypes using chromosome painting [14,28], especially using human and dromedary probes [5,29,30]. Here, we used the combination of these human and dromedary painting probes to establish a detailed comparative chromosome map for musk ox to interpret the descent of this species' genome organization in an evolutionary context.

2. Material and Methods

2.1. Species

The list of studied species, diploid chromosome number, and the source of cell lines are presented in Table 1. All cell lines belong to the cell culture collection of general biological purpose (No. 0310-2016-0002) of the Institute of Molecular and Cellular Biology (IMCB) of the Siberian Branch, Russian Academy of Sciences (SB RAS).

Table 1. The list of Bovidae cell lines used in this study.

Scientific Name, Abbreviation	Common Name	Subfamily	Diploid Number	Sample/Cell Line Source/Acknowledgment
Ovibos moschatus, OMO	Musk ox	Antilopinae (Caprini)	48, XX	Allaikhovsky District, Sakha Republic, Yakutia, Russia. IMCB SB RAS, Novosibirsk
Ovis aries musimon, OAR	Sheep	Antilopinae (Caprini)	54, XX	Melody Roelke and June Bellizzi, Catoctin Wildlife Preserve and Zoo, Maryland, USA; Laboratory of Genomic Diversity, NCI-Frederick, MD, USA
Madoqua kirkii, MKI	Kirk's Dikdik	Antilopinae (Antilopini)	48, XY	Mitchell Bush, Conservation and Research Center, National Zoological Park, Virginia, USA; Laboratory of Genomic Diversity, NCI-Frederick, MD, USA
Bos taurus, BTA	Cattle	Bovinae (Bovini)	60, XX	IMCB SB RAS, Novosibirsk.
Bos gaurus, BGA	Gaur	Bovinae (Bovini)	58, XX	Doug Armstrong, Henry Doorly Zoo, Omaha, NE, USA; Laboratory of Genomic Diversity, NCI-Frederick, MD, USA
Pseudoryx nghetinhensis, PNG	Saola	Bovinae (Bovini)	50, XX	[5]
Boselaphus tragocamelus, BTR	Nilgai bull	Bovinae (Boselaphini)	44, X+14, X+14	Melody Roelke and June Bellizzi, Catoctin Wildlife Preserve and Zoo, Maryland, USA; Laboratory of Genomic Diversity, NCI-Frederick, MD, USA

2.2. Chromosome Preparation

Metaphase chromosomes were obtained from fibroblast cell lines. Briefly, cells were incubated at 37 °C in 5% CO2 in medium αMEM (Gibco), supplemented with 15% fetal bovine serum (Gibco), 5% AmnioMAX-II complete (Gibco) and antibiotics (ampicillin 100 μg/mL, penicillin 100 μg/mL, amphotericin B 2.5 μg/mL). Metaphases were obtained by adding colcemid (0.02 mg/L) and ethidium bromide (1.5 mg/mL) to actively dividing culture for 3–4 hours. Hypotonic treatment was performed with 3 mM KCl, 0.7 mM sodium citrate for 20 min at 37 °C and followed by fixation with 3:1 methanol:glacial acetic acid (Carnoy's) fixative. Metaphase chromosome preparations were made from a suspension of fixed fibroblasts, as described previously [31,32]. G-banding on metaphase chromosomes prior to fluorescence in situ hybridization (FISH) was performed using the standard procedure [33]. Heterochromatin analysis was performed by the Combined Method of Heterogeneous Heterochromatin Detection (CDAG) [34]. AT- and GC-enriched repetitive sequences were detected by DAPI (40-6-diamidino-2-phenylindol) and CMA3 (chromomycin A3) fluorescent dyes following formamide denaturation and renaturation in hot salt solution.

2.3. FISH Probes

The protocol for the selection of BAC-clones was reported previously [15]. Briefly, we selected 26 BAC clones highly conserved among Cetartiodactyla from bovine CHORI-240 library using bioinformatic tools. BAC DNA was isolated using the Plasmid DNA Isolation Kit (BioSilica, Novosibirsk, Russia) and amplified with GenomePlex Whole Genome Amplification kit (Sigma-Aldrich Co., St. Louis, MO, USA). Labeling of BAC DNA was performed using GenomePlex WGA Reamplification Kit (Sigma-Aldrich Co., St. Louis, MO, USA) by incorporating biotin-16-dUTP or digoxigenin-dUTP (Roche, Basel, Switzerland). The list of BAC-clones is shown in Table 2. Plasmid containing ribosomal

DNA [35] was amplified and labeled as described above. Telomere repeats were synthesized and labeled in non-template PCR using primers (TTAGGG)5 and (CCCTAA)5 [36]. Human and dromedary chromosome-specific probes were described previously [6,32] and were labeled by DOP-PCR [37] with biotin-16-dUTP or digoxigenin-dUTP (Roche, Basel, Switzerland).

2.4. *FISH Procedure*

Dual-color FISH experiments on G-banded metaphase chromosomes were conducted as described by Yang and Graphodatsky [32]. Tripsin-treated chromosomes were immobilized in 0.5% formaldehyde in PBS followed by formamid denaturing and overnight probe hybridization at 40 °C. Digoxigenin-labeled probes were detected using anti-digoxigenin-CyTM3 (Jackson ImmunoResearch Laboratories, Inc., West Grove, PA, whereas biotin-labeled probes were identified with avidin-FITC (Vector Laboratories) and anti-avidin FITC (Vector Laboratories, Inc., Burlingame, CA, USA). Images were captured and processed using VideoTesT 2.0 Image Analysis System (Zenit, St. Petersburg, Russia) and a Baumer Optronics CCD camera mounted on a BX53 microscope (Olympus, Shinjuku, Japan).

3. Results

3.1. *Comparative Chromosome Map of Musk Ox, Dromedary, and Human*

The musk ox karyotype includes six submetacentric and 17 acrocentric autosomes and one sex chromosomal pair (2N = 48) (Figure 1). The fundamental number of autosomal arms in musk ox is 58, which in general is characteristic for karyotypes of the Bovidae family [18]. To establish the genome-wide chromosome comparative map of the musk ox, human and dromedary painting probes were used. The chromosome map (Figure 1) and additional comparison with cattle chromosomes (the reference karyotype for ruminants) and with pecoran ancestral karyotype (PAK) are summarized in Table 2. The painting probes from 22 human (HSA) and 35 dromedary (CDR) autosomal paints revealed 51 and 61 conserved segments on the musk ox karyotype, respectively.

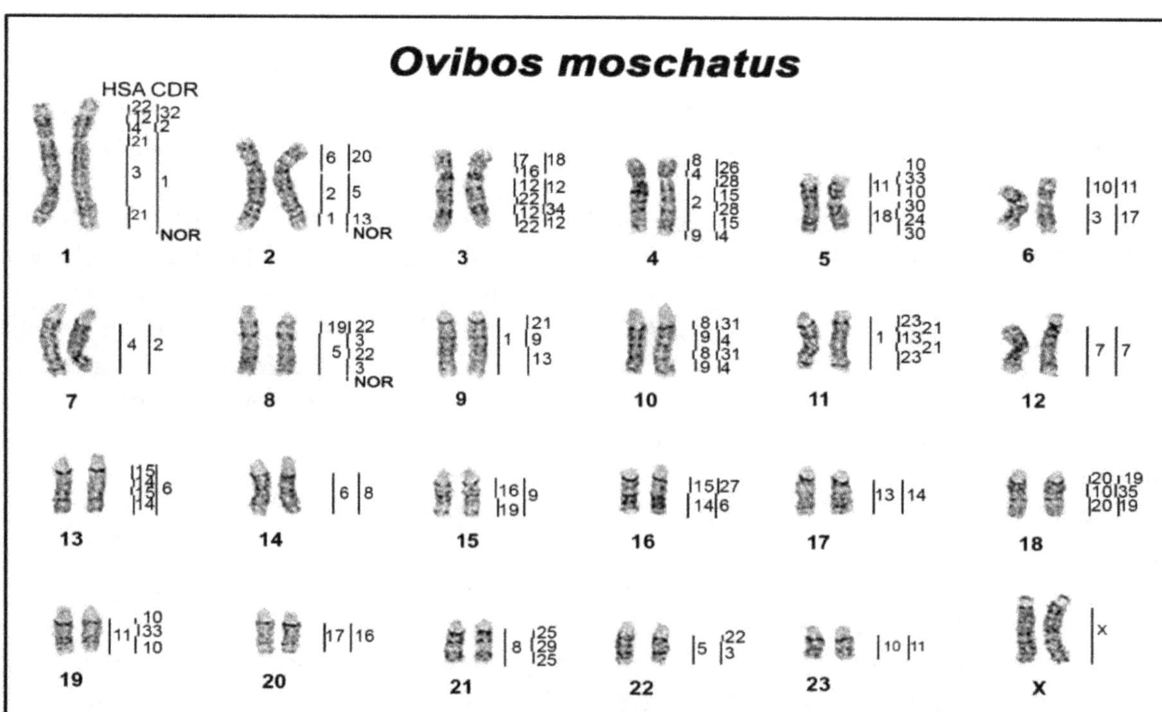

Figure 1. Comparative chromosome painting map of musk ox with homologies to human (HSA) and dromedary (CDR). Nucleolar organizing regions (NOR) show the localization of the nucleolar organizing region.

Table 2. Correspondence between conserved chromosomal segments in musk ox (OMO), human (HSA), dromedary (CDR), cattle (BTA) and Pecoran ancestral karyotype (PAK) [5] revealed by chromosome painting. The order of conservative segments is started from centromere.

OMO	HSA	CDR	BTA	PAK
1p	22q′/12q″/4pq	32/2	17	N1
1q	21/3/21	1	1	A2
2p	6p	20	23	R
2q	2q″/1	5/13	2	B2
3p	16p/7	18	25	T
3q	12pq′/22q″/12pq′/22q″	12/34/12	5	C2
4p	4/8p″	26	27	V
4q	2pq/9	28/15/28/15/4	11	C1
5p	11	10/33/10	29	W
5q	18	30/24/30	24	S
6p	10q	11	28	U
6q	3	17	22	Q
7	4pq	2	6	F
8	19p/5	22/3/22/3	7	E
9	1	21/9/13	3	A1
10	8p′/9/8p′/9	31/4/31/4	8	B1
11	1	23/21/13/21/23	16	K
12	7	7	4	D
13	15/14/15/14	6	10	G
14	6q	8	9	H1
15	16q/19p	9	18	M
16	15/14	27/6	21	P
17	13	14	12	I
18	20/10p/20	19/35/19	13	J
19	11	10/33/10	15	L
20	17	16	19	N2
21	8q	25/29/25	14	H2
22	5	22/3	20	O
23	10q	11	26	U
X	X	X	X	X

We studied the distribution of repeated sequences in the musk ox karyotype using several methods. FISH analysis revealed the localization of telomere repeat and ribosomal DNA sequences. Six nucleolar organizing regions (NORs) with telomeric localization on OMO 1, OMO 2 and OMO 8 chromosome pairs were identified (Figure 1). Telomeric repeats are situated on terminal regions of chromosomes. The CDAG differential staining revealed centromeric and telomeric GC-enriched heterochromatin with prominent AT- and GC-enriched pericentromeric blocks of heterochromatin on acrocentric chromosomes (Figure 2). Smaller blocks of heterochromatin were observed on meta- and submetacentric chromosomes, except chromosome 6, where a large block of heterochromatin was identified. Enlarged telomeric blocks were observed on at least 3 pairs of autosomes, while only two pairs of small acrocentrics appeared to have repeated sequences distributed over the whole chromosome.

3.2. *Mapping of the X Chromosome in Bovidae*

To investigate the order of conserved syntenic segments on X chromosomes in the Bovidae family, 26 BAC-clones were localized using FISH on X chromosomes of four species (nilgai bull, saola, gaur, and Kirk's Dikdik) in a series of pairwise FISH experiments (Table 3). In all Bovinae species, similar order of BAC-clones was observed. The same order was observed earlier in Antilopinae subfamily, except for the marker inversion in Caprini [15]. In total, comparative analysis of BAC order revealed identical syntenic blocks: X Syntenic Block 1 (13 BACs, XSB1), X Syntenic Block 2 (7 BACs, XSB2), and X Syntenic Block 3 (6 BACs, XSB3) [15]. Two types of chromosome X changes were identified in a course of Bovidae evolution: a centromere reposition, and inversions of an entire syntenic block. Interestingly, a segmental duplication in XSB3 containing CH-108D16 was detected on saola X (Figure 3).

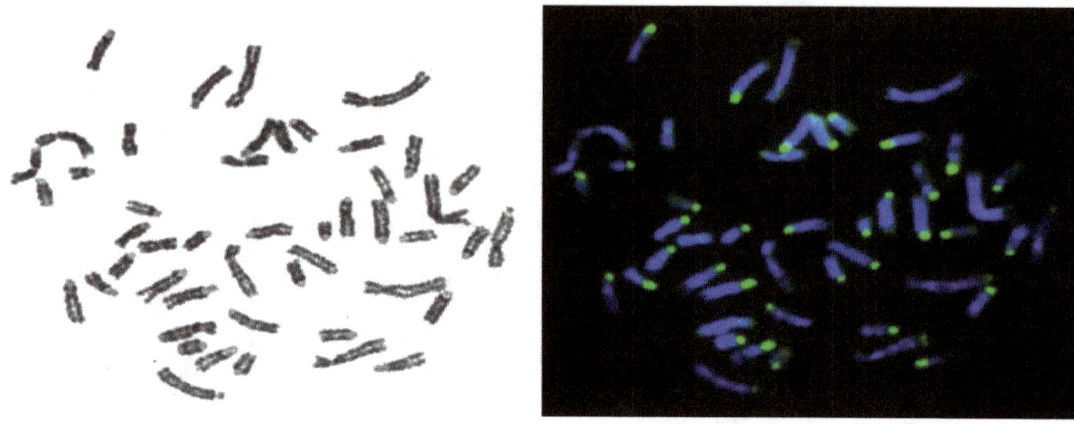

Figure 2. Chromomycin A3-DAPI after G-banding (CDAG) staining performed on metaphase chromosomes of musk ox: GTG-banding (**left**) and CMA3/DAPI-staining after denaturation and renaturation procedure (**right**).

Table 3. The order of 26 CHORI-240 BACs on Bovidae X chromosomes. The color of the cells corresponds to a given conserved syntenic segment. To display the complete scheme of evolution in Bovidae, the X chromosome maps published previously are also presented [15]. The region duplicated in saola and inverted in Caprini is labeled with a lighter colour.

Syntenic Block	X BAC's Order in Bovinae Subfamily		X BAC's Order in Antilopinae Subfamily	
	In most Bovinae	In saola	Caprini tribe	Hippotragini and Antilopini tribe
X syntenic block 1 (XSB1)	CH240-514O22	CH240-514O22	CH240-66H2	CH240-66H2
	CH240-287O21	CH240-287O21	CH240-155A13	CH240-155A13
	CH240-128C9	CH240-128C9	CH240-90L14	CH240-90L14
	CH240-106A3	CH240-106A3	CH240-373L23	CH240-373L23
	CH240-229I15	CH240-229I15	CH240-62M10	CH240-62M10
	CH240-103E10	CH240-103E10	CH240-122P17	CH240-122P17
	CH240-386M8	CH240-386M8	CH240-252G15	CH240-252G15
X syntenic block 2 (XSB2)	CH240-108D16	CH240-108D16	CH240-375C5	CH240-375C5
	CH240-54D24	CH240-54D24	CH240-130I15	CH240-130I15
	CH240-93K24	CH240-93K24 CH240-108D16	CH240-118P13	CH240-118P13
	CH240-122N13	CH240-122N13	CH240-25P8	CH240-25P8
	CH240-195J23	CH240-195J23	CH240-14O10	CH240-14O10
	CH240-316D2	CH240-316D2	CH240-214A3	CH240-214A3
X syntenic block 3 (XSB3)	CH240-214A3	CH240-214A3	CH240-386M8	CH240-386M8
	CH240-14O10	CH240-14O10	CH240-103E10	CH240-103E10
	CH240-25P8	CH240-25P8	CH240-128C9	CH240-229I15
	CH240-118P13	CH240-118P13	CH240-106A3	CH240-106A3
	CH240-130I15	CH240-130I15	CH240-229I15	CH240-128C9
	CH240-375C5	CH240-375C5	CH240-287O21	CH240-287O21
	CH240-252G15	CH240-252G15	CH240-514O22	CH240-514O22
	CH240-122P17	CH240-122P17	CH240-316D2	CH240-316D2
	CH240-62M10	CH240-62M10	CH240-195J23	CH240-195J23
	CH240-373L23	CH240-373L23	CH240-122N13	CH240-122N13
	CH240-90L14	CH240-90L14	CH240-93K2	CH240-93K2
	CH240-155A13	CH240-155A13	CH240-54D24	CH240-54D24
	CH240-66H2	CH240-66H2	CH240-108D16	CH240-108D16

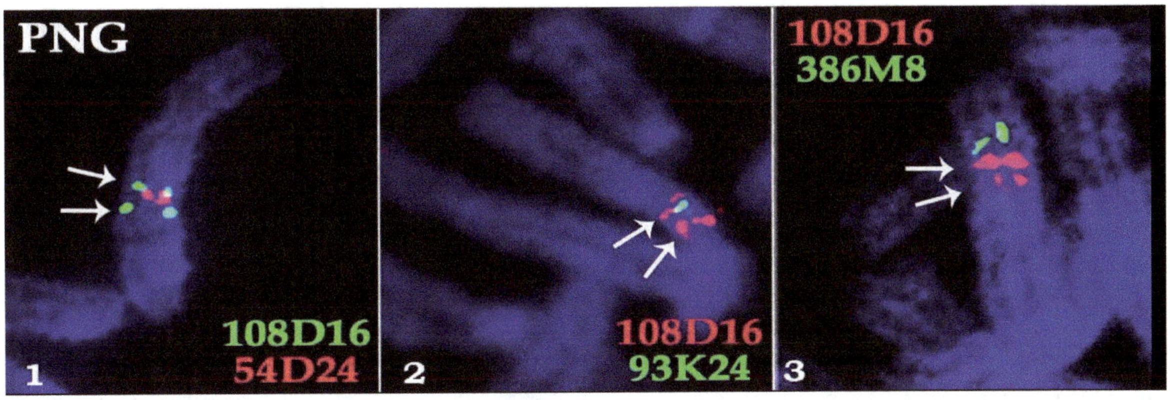

Figure 3. The duplication of the X chromosome segment in *Pseudoryx nghetinhensis* (PNG) shown by dual color FISH of cattle BAC-clones (pink and green) from CHORI-240 library. Three FISH experiments illustrate the revealed order of BACs on the X chromosome: 386M8, 108D16, 54D24, 93K24, 108D16. White arrows indicate the duplicated region corresponding to 108D16: 1, 2 –54D24 and 93K24 are between duplicated regions, 3 –386M8 is outside of duplicated region.

4. Discussion

4.1. Evolution of Musk Ox and Bovid Karyotypes

Prior G-banding karyotypes of musk ox [25,26] revealed five fusions that formed submetacentric chromosomes described using cattle microdissected chromosomes [14]. Here, a complete high-resolution comparative map for musk ox karyotype was obtained using human and camel chromosome specific probes and compared to cattle karyotype (Table 2). Our results are in agreement with previous publications showing the origin of musk ox submetacentric chromosomes [14].

We show the presence of prominent heterochromatin blocks at centromeric positions in the musk ox karyotype. Many bovid species are characterized by prominent pericentromeric blocks of heterochromatin and their karyotype evolution is marked by frequent occurrence of Robertsonian fusions [2,18]. Several hypotheses point to the role that repetitive sequences may have in driving chromosome evolution in bovids by increasing the occurrence of Robertsonian translocations due to the physical proximity of centromeres of acrocentric chromosomes during meiosis [38]. As shown previously, repetitive sequences were involved in formation of Robertsonian translocations in mice [38]. Therefore, the presence of heterochromatic blocks on acrocentric chromosomes in bovid species may contribute to the high variability of bovid karyotypes, including the occurrence of cytotypes in many species, high frequency of Robertsonian fusions, and autosome to the X chromosome translocations.

Overall, the obtained comparative map indicates that musk ox karyotype is nearly homologous to the pecoran ancestral karyotype [5] (Table 2). The comparison of the ancestral elements of the musk ox with other pecoran species demonstrates the rearrangements that formed its karyotype, but also more events occurring in different lineages (Figure 5). The musk ox karyotype has evolved from PAK through six fusions (CDR 1 + 2/32, CDR 20 + 5/13, CDR 18 + 12/34/12, CDR 26 + 28/15/28/15/4, CDR 33/10 + 24/30, and CDR 4+17), one fission (CDR 11), and three inversions (on HSA 21/3/21, CDR 22/3/22/3, HSA 12pq'/22q"12pq'/22q") (Figure 4). It is characterized by inversions on ancestral elements A2, C2, E, and a split of U.

The musk ox is a representative of subfamily Antilopinae, Caprini tribe. Both comparative linkage and FISH maps showed one major distinction between ovine (Antilopinae, Caprini) and bovine (Bovinae) karyotypes. This difference resulted from a translocation involving segments homologous to BTA 9 and BTA 14 [4,7,39]. However, this association is not observed in musk ox. Presumably, this is determined by the basal position of Ovibovina [40], suggesting that the BTA 9/14 translocation occurred only in Caprina subtribe.

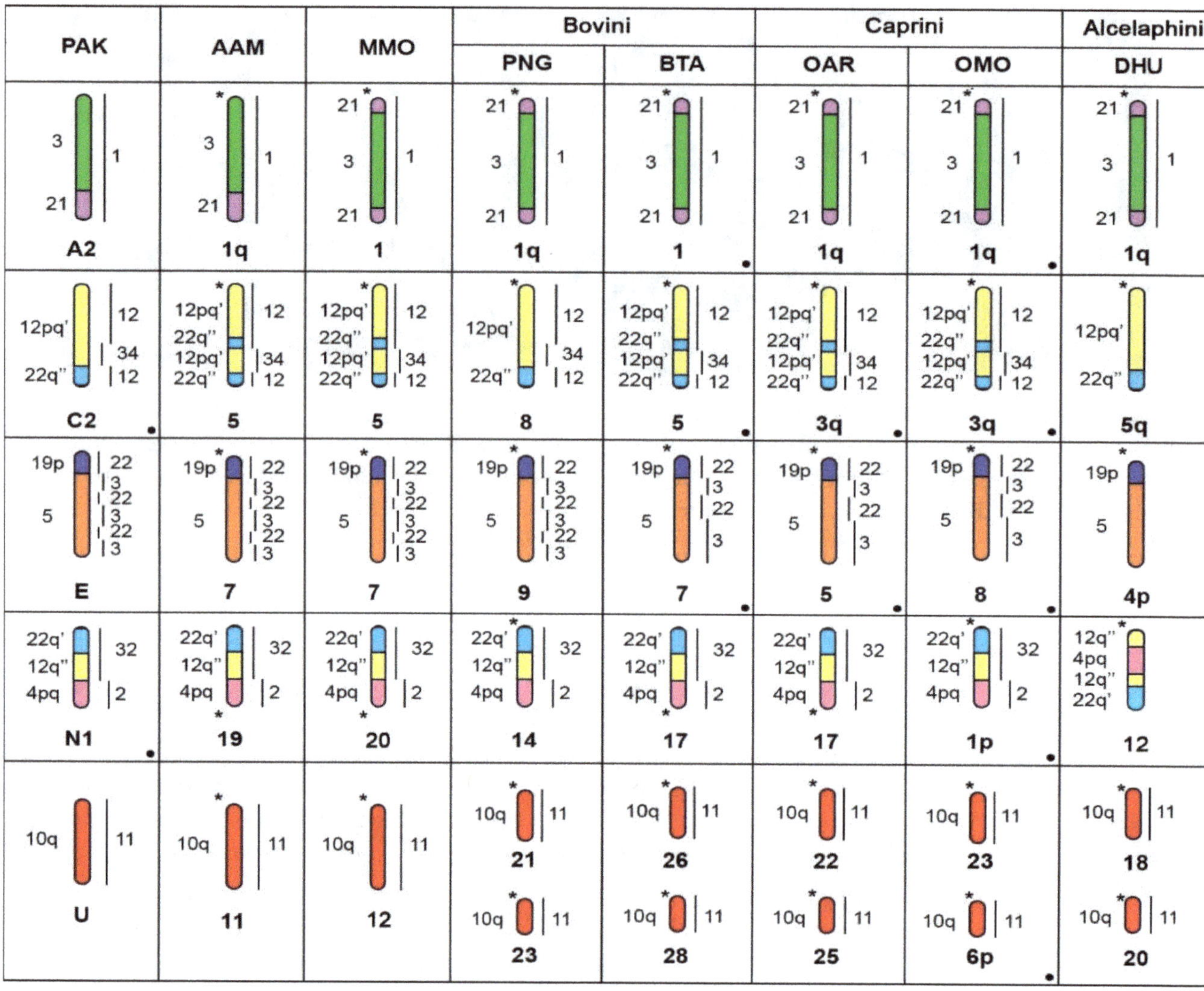

Figure 4. A scheme depicting chromosome homologies of pecoran species to the ancestral karyotype chromosomes (PAK) [5] with human homologies on the left and dromedary on the right. Presented species include AAM (*Antilocapra americana*) [5], MMO (*Moschus moschiferus*) [30], and representatives of different Bovidae tribes: PNG (*Pseudoryx nghetinhensis*) [5], BTA (*Bos taurus*) [6] (Bovini), OAR (*Ovis aries*) [10], OMO (*Ovibos moschatus*) (Caprini), DHU (*Damaliscus hunteri*) [11] (Alcelaphini). Centromere positions are shown by an asterisk. New data obtained in this study are marked by a black circle in cell corners.

The ancestral form of PAC A2 is similar to pronghorn (AAM) 1q [5]. However, other bovid species and Moschidae (Siberian musk deer) [30] showed an inversion changing the order of homologous segments into HSA 21/3/21 (Figure 4). Therefore, this inversion likely represents a cytogenetic marker for at least Bovidae+Moschidae. The verification of this human syntenic association is needed in Cervidae where the fission of the synteny was shown for *Muntiacus muntjac* [41]. Although Cervidae have been well studied with bovid [42] and muntjac [41] probes, comparison to human probes is still unknown, hindering the deduction of ancestral rearrangements.

The ancestral pecoran synteny PAK C2 represents an interesting case. An inversion in HSA 12pq′/22q″ occurred independently in different phylogenetic lineages in Pecora (Antilocapridae, Moschidae, Bovina, Caprina) (Figure 4) [16]. Therefore, there is a hot spot of chromosome evolution in the region homologous to HSA12pq′/22q″ in Pecora. Additional investigation is required to verify if this inversion occurred in the same region, with an in-depth analysis of the DNA sequence surrounding this region needed to elucidate the genomic elements causing repeated rearrangements. Contrary to what has been previously suggested [5], this ancestral PAK chromosome C2 was composed of HSA 12/22, and not of HSA 12/22/12/22, because the outgroup and many species from basal lineages have the HSA 12/22 association (whales, Java mouse deer, giraffe, okapi, saola, hirola) [5,11,30,43]. Similarly, the independent inversions in PAK chromosome E (CDR 22/3/22/3/22/3) in BTA 7 (Bovina), OAR 5, OMO 8 (Caprini) (CDR 22/3/22/3), and giraffe (Giraffidae) [30], while ancestral conditions were retained in Java mouse deer, pronghorn, Siberian musk deer and saola, mark another hot spot of chromosome evolution that requires further study.

During FISH experiments on the localization of CDR 22 on OMO chromosomes, an additional small region of homology on OMO 22 was detected. This region was also detected in other Bovidae species: cow, sheep (Figure 5), and saola [5]. The sequence homologies of HSA5 = BTA20 are confirmed by ENSEMBLE genome browser data, also blast data of alpaca RH markers from the chromosome homologous to CDR22 show homology with BTA20 (unpublished data). The order of conservative segments on BTA7 is HSA19p/5 and CDR22/3/22/3. These data differ from research published previously reporting HSA5/19p/5 and CDR3/22/3/22 [6].

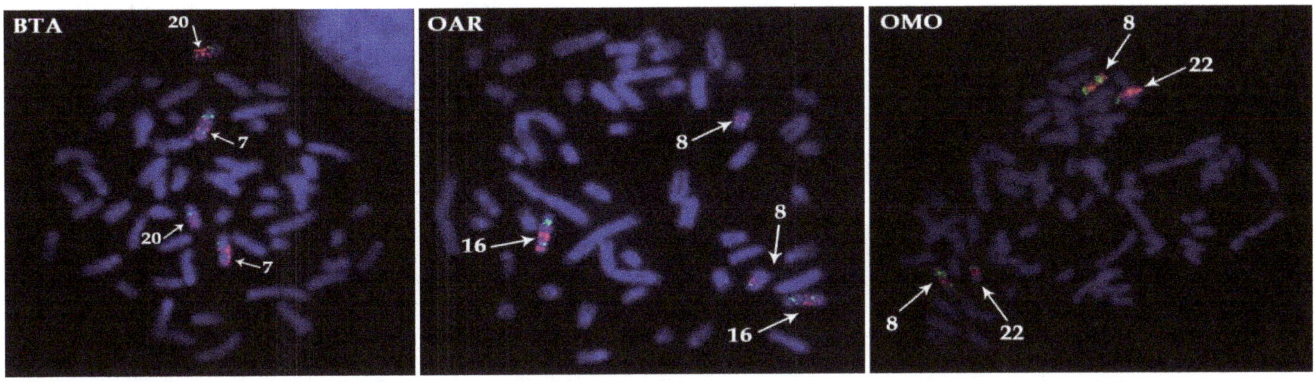

Figure 5. Localization of CDR 22 (green) and CDR 3 (pink) on BTA (*Bos taurus*), OAR (*Ovis aries*), and OMO (*Ovibos moschatus*) metaphase chromosomes by FISH showing additional previously unreported by painting fragment homologous to CDR22. White arrows indicate chromosomes with specific signal.

The split of the ancestral element PAK U is shown in musk ox and also in PNG, BTA, OAR, and DHU karyotypes, thus suggesting that this fission is a marker for the Bovidae lineage [5].

On another ancestral chromosome, PAK N1 centromere reposition events occurred independently on homologous chromosomes in several bovid lineages (Pseudoryina, Ovibovina, Alcelaphini) (Figure 5). Further refinement of the ancestral chromosome N1 was achieved (Figure 4), showing that the order of human syntenic regions on the PAK ancestral chromosome N1 is HSA 22′/12″/4pq and not 12′/22″/4pq, as was reported earlier [5]. OMO 1p (HSA 22q/12pq″/4pq) retained the ancestral order of conserved segments N1 but was then tandemly fused with PAK A2 based on the position of the ancestral centromere.

4.2. Bovine X Chromosome Evolution

The family Bovidae includes two major branches: Bovinae and Antilopinae [40]. Earlier cytogenetic studies identified three types of morphological diversity of the X chromosome in Bovidae: an antilopinae type (acrocentric), a tragelaphines type (acrocentric), and a cattle type (submetacentric) [20]. Tragelaphines chromosome X was likely formed from the ancestral pecoran X by two inversions. This type is ancestral in Bovinae and presented in nilgai, saola (Figure 6), and domestic river buffalo [44]. The X chromosome in nilgai and saola are marked by several morphological features. The first one is an autosomal translocation onto chromosome X in nilgai karyotype. In Tragelaphini and Boselaphini tribes, independent autosomal translocations were observed [9,23,45,46]. Such rearrangements have an impact on the behavior of chromosomes in meiosis, manifested as a lowering of synapsis in the pseudoautosomal region [45]. The second one is a segmental duplication of sequences homologous to CH-108D16 in saola (PNG) (Table 3). Unfortunately, we cannot determine whether this segmental duplication is characteristic for the entire species due to the lack of information of other individuals. However, the duplicated region contains genes responsible for intrauterine development and may have an adaptive value.

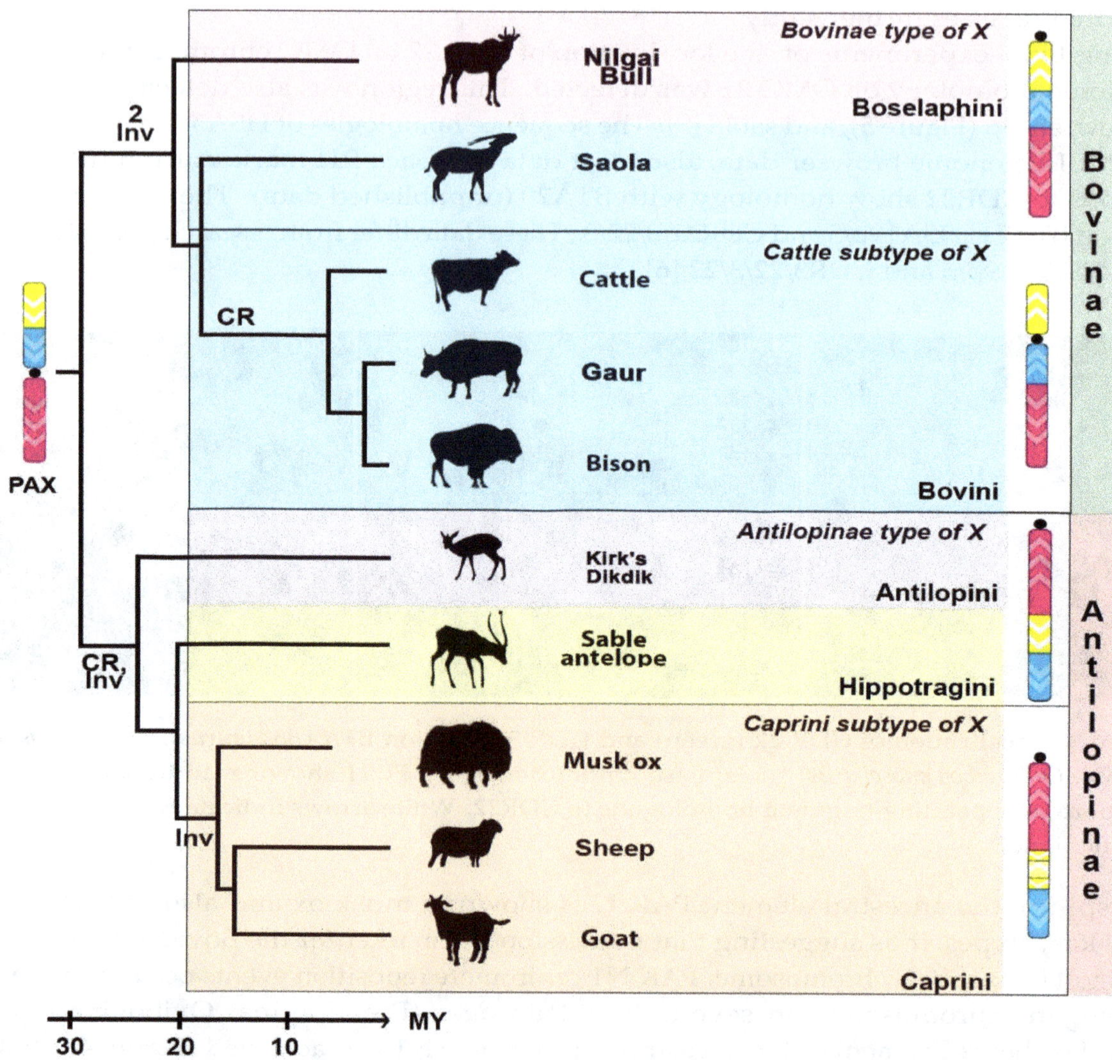

Figure 6. Changes in the structure of the Bovidae X chromosome are depicted on the phylogenetic tree of the family (the tree topology is from [40]). PAX is Pecoran ancestral X chromosome [15]. Major conservative segments are shown in yellow, blue, and pink. Centromere positions are designated by a black circle. White arrowheads show the orientation of the conservative segments. Chromosome changes are shown on phylogenetic tree near respective branches: CR—centromere reposition; and Inv—inversion. Frames show types and subtypes of the bovid X chromosome. The timescale is in million years (MY) of evolution. Nilgai bull X chromosome is shown without autosomal translocation.

The cattle subtype of the Xchromosome is formed by centromere reposition of the ancestral X chromosome in Bovinae. This type of the X chromosome is presented in cattle, American bison [15], and gaur (Figure 6). Thus, it appears to be characteristic not only of cattle, but of the whole subtribe Bovina.

The centromere reposition and one inversion resulted in the formation of an acrocentric caprine type of the X chromosome [15]. This type of the X chromosome is retained in Kirk's Dikdik and sable antelope [15]. We suggest calling this type of bovid the X – Antilopine type. Another inversion, which occurred within the XSB1 in the Caprini lineage, is an apomorphic phylogenetic marker for this tribe [15] and marks the formation of the Caprini subtype of X. The mapping of *Panthalops hodgsonii* X chromosome would assert this type for the whole Caprini tribe.

In general, the X chromosome is highly conserved in eutherians [19], but several different types of chromosome rearrangements on the cetartiodactyl X have been shown [15]. It was suggested that the evolutionary chromosome rearrangements may reduce gene flow by suppressing recombination and contributing to species isolation [47]. However, in ruminant species, several evolutionary breakpoint regions (EBR) on the X chromosome associated with enhancers were described that may change gene expression [16]. Therefore, these rearrangements may have an adaptive value and an evolutionary meaning.

Overall, we can distinguish four types of the X chromosome in Bovidae: Bovinae type with derived cattle subtype; and Antilopinae type with Caprini subtype. The Bovinae type was formed from the ancestral pecoran X by two inversions, whereas the Antilopinae type was formed by inversion and centromere reposition. The Cattle and Caprini subtypes were created by centromere repositions and inversion in XSB1, respectively.

5. Conclusions

Detailed comparative maps were obtained for musk ox karyotype and X chromosomes of four bovids: Kirk's Dikdik, gaur, saola, and nilgai bull. Large structural rearrangements leading to the formation of the karyotype of the musk ox were shown. In general, its karyotype is close to the putative ancestral karyotype of Pecora infraorder. The detailed analysis of the BAC-clones order across four species and published data allowed illustrating chromosomal rearrangements during the formation of four main types of X chromosomes in the Bovidae family. In summary, conservation in BACs order was shown in the Bovinae and Antilopinae subfamilies.

Author Contributions: Conceptualization, A.I.K. and A.S.G.; Data curation, A.A.P. and P.L.P.; Formal analysis, A.A.P.; Funding acquisition, D.M.L. and A.S.G.; Investigation, A.A.P., A.I.K. and N.A.L.; Methodology, A.A.P. and A.I.K.; Project administration, A.S.G.; Resources, P.L.P., D.V.Y., N.A.L., I.M.O., E.V.K., M.F., D.M.L., M.B., M.E.R.-P. and S.J.O.; Software, M.F. and D.M.L.; Supervision, A.I.K.; Visualization, A.A.P.; Writing—original draft, A.A.P.; Writing—review & editing, P.L.P., M.F., D.M.L. and S.J.O.

Acknowledgments: We would like to sincerely thank Malcolm A. Ferguson-Smith, Fengtang Yang and Patricia CM O'Brien for kindly providing the sets of human and dromedary camel chromosome-specific painting probes. We would like to sincerely thank Nicolai Mamaev (Institute for Biological Problems of Cryolithozone Siberian Branch of RAS, Russia) to helping in providing musk ox samples. We would like to sincerely thank Mitchell Bush (Conservation and Research Center, National Zoological Park, Virginia, USA), Doug Armstrong (Henry Doorly Zoo, OMAHA, NE, USA), Melody Roelke (NIH, Bethesda, USA), June Bellizzi and director Richard Hahn (Catoctin wildlife Zoo and Preserve, MD, USA) for kindly providing samples. We would like to thank Mary Thompson (NCI-Frederick, USA) for establishing cell lines.

References

1. Wilson, D.E.; Reeder, D.M. *Mammal Species of The World: A Taxonomic and Geographic Reference*; JHU Press: Baltimore, MD, USA, 2005; Volume 1.
2. Robinson, T.J.; Ropiquet, A. Examination of hemiplasy, homoplasy and phylogenetic discordance in chromosomal evolution of the Bovidae. *Syst. Biol.* **2011**, *60*, 439–450. [CrossRef]
3. Zurano, J.P.; Magalhães, F.M.; Asato, A.E.; Silva, G.; Bidau, C.J.; Mesquita, D.O.; Costa, G.C. Cetartiodactyla: Updating a time-calibrated molecular phylogeny. *Mol. Phylogenet. Evol.* **2019**, *133*, 256–262. [CrossRef]
4. Chi, J.X.; Huang, L.; Nie, W.; Wang, J.; Su, B.; Yang, F. Defining the orientation of the tandem fusions that occurred during the evolution of Indian muntjac chromosomes by BAC mapping. *Chromosoma* **2005**, *114*, 167–172. [CrossRef]

5. Kulemzina, A.I.; Perelman, P.L.; Grafodatskaya, D.A.; Nguyen, T.T.; Thompson, M.; Roelke-Parker, M.E.; Graphodatsky, A.S. Comparative chromosome painting of pronghorn (Antilocapra americana) and saola (Pseudoryx nghetinhensis) karyotypes with human and dromedary camel probes. *BMC Genet.* **2014**, *15*, 68. [CrossRef]
6. Balmus, G.; Trifonov, V.A.; Biltueva, L.S.; O'Brien, P.C.; Alkalaeva, E.S.; Fu, B.; Skidmore, J.A.; Allen, T.; Graphodatsky, A.S.; Yang, F.; et al. Cross-species chromosome painting among camel, cattle, pig and human: Further insights into the putative Cetartiodactyla ancestral karyotype. *Chromosome Res.* **2007**, *15*, 499–514. [CrossRef] [PubMed]
7. Iannuzzi, L.; Di Meo, G.P.; Perucatti, A.; Schibler, L.; Incarnato, D.; Cribiu, E.P. Comparative FISH mapping in river buffalo and sheep chromosomes: Assignment of forty autosomal type I loci from sixteen human chromosomes. *Cytogenet. Genome Res.* **2001**, *94*, 43–48. [CrossRef]
8. Rubes, J.; Kubickova, S.; Pagacova, E.; Cernohorska, H.; Di Berardino, D.; Antoninova, M.; Vahala, J.; Robinson, T.J. Phylogenomic study of spiral-horned antelope by cross-species chromosome painting. *Chromosome Res.* **2008**, *16*, 935–947. [CrossRef] [PubMed]
9. Cernohorska, H.; Kubickova, S.; Vahala, J.; Robinson, T.J.; Rubes, J. Cytotypes of Kirk's dik-dik (Madoqua kirkii, Bovidae) show multiple tandem fusions. *Cytogenet. Genome Res.* **2011**, *132*, 255–263. [CrossRef]
10. Iannuzzi, L.; Di Meo, G.P.; Perucatti, A.; Bardaro, T. ZOO-FISH and R-banding reveal extensive conservation of human chromosome regions in euchromatic regions of river buffalo chromosomes. *Cytogenet. Genome Res.* **1998**, *82*, 210–214. [CrossRef]
11. Chaves, R.; Frönicke, L.; Guedes-Pinto, H.; Wienberg, J. Multidirectional chromosome painting between the Hirola antelope (Damaliscus hunteri, Alcelaphini, Bovidae), sheep and human. *Chromosome Res.* **2004**, *12*, 495–503. [CrossRef] [PubMed]
12. Buckland, R.A.; Evans, H.J. Cytogenetic aspects of phylogeny in the Bovidae. *Cytogenet. Genome Res.* **1978**, *21*, 42–63. [CrossRef] [PubMed]
13. Pagacova, E.; Cernohorska, H.; Kubickova, S.; Vahala, J.; Rubes, J. Centric fusion polymorphism in captive animals of family Bovidae. *Conserv. Genet.* **2011**, *12*, 71–77. [CrossRef]
14. Rubes, J.; Musilova, P.; Kopecna, O.; Kubickova, S.; Cernohorska, H.; Kulemsina, A.I. Comparative molecular cytogenetics in Cetartiodactyla. *Cytogenet. Genome Res.* **2012**, *137*, 194–207. [CrossRef]
15. Proskuryakova, A.A.; Kulemzina, A.I.; Perelman, P.L.; Makunin, A.I.; Larkin, D.M.; Farré, M.; Kukekova, A.V.; Johnson, J.L.; Lemskaya, N.A.; Beklemisheva, V.R.; et al. X Chromosome Evolution in Cetartiodactyla. *Genes* **2017**, *8*, 216. [CrossRef]
16. Farré, M.; Kim, J.; Proskuryakova, A.A.; Zhang, Y.; Kulemzina, A.I.; Li, Q.; Zhou, Y.; Xiong, Y.; Johnson, J.L.; Perelman, P.L. Evolution of gene regulation in ruminants differs between evolutionary breakpoint regions and homologous synteny blocks. *Genome Res.* **2019**, *29*, 576–589. [CrossRef]
17. Slate, J.; Van Stijn, T.C.; Anderson, R.M.; McEwan, K.M.; Maqbool, N.J.; Mathias, H.C.; Bixley, M.J.; Stevens, D.R.; Molenaar, A.J.; Beever, J.E.; et al. A deer (subfamily Cervinae) genetic linkage map and the evolution of ruminant genomes. *Genetics* **2002**, *160*, 1587–1597. [PubMed]
18. Gallagher, D.S., Jr.; Womack, J.E. Chromosome conservation in the Bovidae. *J. Hered.* **1992**, *83*, 287–298. [CrossRef]
19. Ohno, S.; Beçak, W.; Beçak, M.L. X-autosome ratio and the behavior pattern of individual X-chromosomes in placental mammals. *Chromosoma* **1964**, *15*, 14–30. [CrossRef]
20. Robinson, T.J.; Harrison, W.R.; Ponce de Leon, F.A.; Davis, S.K.; Elder, F.F.B. A molecular cytogenetic analysis of X chromosome repatterning in the Bovidae: Transpositions, inversions, and phylogenetic inference. *Cytogenet. Genome Res.* **1998**, *80*, 179–184. [CrossRef] [PubMed]
21. Iannuzzi, L.; King, W.A.; Di Berardino, D. Chromosome evolution in domestic bovids as revealed by chromosome banding and FISH-mapping techniques. *Cytogenet. Genome Res.* **2009**, *126*, 49–62. [CrossRef]
22. Cernohorska, H.; Kubickova, S.; Vahala, J.; Rubes, J. Molecular insights into X; BTA5 chromosome rearrangements in the tribe Antilopini (Bovidae). *Cytogenet. Genome Res.* **2012**, *136*, 188–198. [CrossRef] [PubMed]
23. Gallagher, D.S.; Davis, S.K.; De Donato, M.; Burzlaff, J.D.; Womack, J.E.; Taylor, J.F.; Kumamoto, A.T. A karyotypic analysis of nilgai, Boselaphus tragocamelus (Artiodactyla: Bovidae). *Chromosome Res.* **1998**, *6*, 505–514. [CrossRef]
24. Lent, P.C. Ovibos moschatus. *Mamm. Species* **1988**, *302*, 1–9. [CrossRef]

25. Biltueva, L.S.; Sharshov, A.A.; Graphodatsky, A.S. G-Banding Homologies in Musk Ox, Ovibos moschatus, and other Bovids. *Hereditas* **1995**, *122*, 185–187. [CrossRef]
26. Pasitschniak-Arts, M.; Flood, P.F.; Schmutz, S.M.; Seidel, B. *A Comparison of G-Band Patterns of The Muskox and Takin and Their Evolutionary Relationship to Sheep*; Oxford University Press: Oxford, UK, 1994.
27. Desaulniers, D.M.; King, W.A.; Rowell, J.E.; Flood, P.F. The banded chromosomes of the muskox (Ovibos moschatus). *Can. J. Zool.* **1989**, *67*, 1155–1158. [CrossRef]
28. Huang, L.; Nie, W.; Wang, J.; Su, W.; Yang, F. Phylogenomic study of the subfamily Caprinae by cross-species chromosome painting with Chinese muntjac paints. *Chromosome Res.* **2005**, *13*, 389–399. [CrossRef]
29. Kulemzina, A.I.; Yang, F.; Trifonov, V.A.; Ryder, O.A.; Ferguson-Smith, M.A.; Graphodatsky, A.S. Chromosome painting in Tragulidae facilitates the reconstruction of Ruminantia ancestral karyotype. *Chromosome Res.* **2011**, *19*, 531. [CrossRef]
30. Kulemzina, A.I.; Trifonov, V.A.; Perelman, P.L.; Rubtsova, N.V.; Volobuev, V.; Ferguson-Smith, M.A.; Stanyon, R.; Yang, F.; Graphodatsky, A.S. Cross-species chromosome painting in Cetartiodactyla: Reconstructing the karyotype evolution in key phylogenetic lineages. *Chromosome Res.* **2009**, *17*, 419–436. [CrossRef]
31. Yang, F.; O'Brien, P.C.M.; Milne, B.S.; Graphodatsky, A.S.; Solanky, N.; Trifonov, V.; Rens, W.; Sargan, D.; Ferguson-Smith, M.A. A complete comparative chromosome map for the dog, red fox, and human and its integration with canine genetic maps. *Genomics* **1999**, *62*, 189–202. [CrossRef] [PubMed]
32. Yang, F.; Graphodatsky, A.S. Animal probes and ZOO-FISH. In *Fluorescence In Situ Hybridization (FISH)*; Springer: Berlin/Heidelberg, Germany, 2017; pp. 323–346.
33. Seabright, M. A rapid banding technique for human chromosomes. *The Lancet* **1971**, *2*, 971–972. [CrossRef]
34. Lemskaya, N.A.; Kulemzina, A.I.; Beklemisheva, V.R.; Biltueva, L.S.; Proskuryakova, A.A.; Perelman, P.L.; Graphodatsky, A.S. The combined method of heterogeneous heterochromatin detection (CDAG) in different mammalian species. *Chromosoma* **2018**, *26*, 307–315. [CrossRef] [PubMed]
35. Proskuryakova, A.; Kulemzina, A.; Perelman, P.; Serdukova, N.; Ryder, O.; Graphodatsky, A. The Case of X and Y Localization of Nucleolus Organizer Regions (NORs) in Tragulus javanicus (Cetartiodactyla, Mammalia). *Genes* **2018**, *9*, 312. [CrossRef] [PubMed]
36. Ijdo, J.W.; Wells, R.A.; Baldini, A.; Reeders, S.T. Improved telomere detection using a telomere repeat probe (TTAGGG) n generated by PCR. *Nucleic Acids Res.* **1991**, *19*, 4780. [CrossRef]
37. Telenius, H.; Ponder, B.A.; Tunnacliffe, A.; Pelmear, A.H.; Carter, N.P.; Ferguson-Smith, M.A.; Behmel, A.; Nordenskjöld, M.; Pfragner, R. Cytogenetic analysis by chromosome painting using DOP-PCR amplified flow-sorted chromosomes. *Genes Chromosomes Cancer* **1992**, *4*, 257–263. [CrossRef] [PubMed]
38. Berrios del Solar, M.S.; Manieu Seguel, C.; López Fenner, J.; Ayarza Ramírez, G.; Page, J.; González Vergara, M.; Manterola Zúñiga, M.; Fernández Donoso, R. Robertsonian chromosomes and the nuclear architecture of mouse meiotic prophase spermatocytes. *Biol. Res.* **2014**, *47*, 16. [CrossRef]
39. De Gortari, M.J.; Freking, B.A.; Cuthbertson, R.P.; Kappes, S.M.; Keele, J.W.; Stone, R.T.; Leymaster, K.A.; Dodds, K.G.; Crawford, A.M.; Beattie, C.W. A second-generation linkage map of the sheep genome. Mamm. *Genome* **1998**, *9*, 204–209.
40. Hassanin, A.; Delsuc, F.; Ropiquet, A.; Hammer, C.; van Vuuren, B.J.; Matthee, C.; Ruiz-Garcia, M.; Catzeflis, F.; Areskoug, V.; Nguyen, T.T.; et al. Pattern and timing of diversification of Cetartiodactyla (Mammalia, Laurasiatheria), as revealed by a comprehensive analysis of mitochondrial genomes. *C. R. Biol.* **2012**, *335*, 32–50. [CrossRef] [PubMed]
41. Yang, F.; Müller, S.; Just, R.; Ferguson-Smith, M.A.; Wienberg, J. Comparative chromosome painting in mammals: Human and the Indian muntjac (Muntiacus muntjak vaginalis). *Genomics* **1997**, *39*, 396–401. [CrossRef]
42. Frohlich, J.; Kubickova, S.; Musilova, P.; Cernohorska, H.; Muskova, H.; Vodicka, R.; Rubes, J. Karyotype relationships among selected deer species and cattle revealed by bovine FISH probes. *PLoS ONE* **2017**, *12*, e0187559. [CrossRef]
43. Kulemzina, A.I.; Proskuryakova, A.A.; Beklemisheva, V.R.; Lemskaya, N.A.; Perelman, P.L.; Graphodatsky, A.S. Comparative Chromosome Map and Heterochromatin Features of the Gray Whale Karyotype (Cetacea). *Cytogenet. Genome Res.* **2016**, *148*, 25–34. [CrossRef]

44. Iannuzzi, L.; Di Meo, G.P.; Perucatti, A.; Incarnato, D.; Schibler, L.; Cribiu, E.P. Comparative FISH mapping of bovid X chromosomes reveals homologies and divergences between the subfamilies Bovinae and Caprinae. *Cytogenet. Genome Res.* **2000**, *89*, 171–176. [CrossRef] [PubMed]
45. Vozdova, M.; Ruiz-Herrera, A.; Fernandez, J.; Cernohorska, H.; Frohlich, J.; Sebestova, H.; Kubickova, S.; Rubes, J. Meiotic behaviour of evolutionary sex-autosome translocations in Bovidae. *Chromosome Res.* **2016**, *24*, 325–338. [CrossRef] [PubMed]
46. O'Brien, S.J.; Menninger, J.C.; Nash, W.G. *Atlas of Mammalian Chromosomes*; John Wiley & Sons: Hoboken, NJ, USA, 2006.
47. Rieseberg, L.H. Chromosomal rearrangements and speciation. *Trends Ecol. Evol.* **2001**, *16*, 351–358. [CrossRef]

Structural Variation of the X Chromosome Heterochromatin in the *Anopheles gambiae* Complex

Atashi Sharma [1], Nicholas A. Kinney [2], Vladimir A. Timoshevskiy [1], Maria V. Sharakhova [1,3,4] and Igor V. Sharakhov [1,2,3,5,*]

1 Department of Entomology, Virginia Polytechnic and State University, Blacksburg, VA 24061, USA; atashi04@vt.edu (A.S.); v.a.timoshevsky@gmail.com (V.A.T.); msharakh@vt.edu (M.V.S.)
2 Genomics Bioinformatics and Computational Biology, Virginia Polytechnic and State University, Blacksburg, VA 24061, USA; nak3c@vt.edu
3 Laboratory of Evolutionary Genomics of Insects, the Federal Research Center Institute of Cytology and Genetics, Siberian Branch of the Russian Academy of Sciences, 630090 Novosibirsk, Russia
4 Laboratory of Ecology, Genetics and Environmental Protection, Tomsk State University, 634050 Tomsk, Russia
5 Department of Cytology and Genetics, Tomsk State University, 634050 Tomsk, Russia
* Correspondence: igor@vt.edu

Abstract: Heterochromatin is identified as a potential factor driving diversification of species. To understand the magnitude of heterochromatin variation within the *Anopheles gambiae* complex of malaria mosquitoes, we analyzed metaphase chromosomes in *An. arabiensis*, *An. coluzzii*, *An. gambiae*, *An. merus*, and *An. quadriannulatus*. Using fluorescence *in situ* hybridization (FISH) with ribosomal DNA (rDNA), a highly repetitive fraction of DNA, and heterochromatic Bacterial Artificial Chromosome (BAC) clones, we established the correspondence of pericentric heterochromatin between the metaphase and polytene X chromosomes of *An. gambiae*. We then developed chromosome idiograms and demonstrated that the X chromosomes exhibit qualitative differences in their pattern of heterochromatic bands and position of satellite DNA (satDNA) repeats among the sibling species with postzygotic isolation, *An. arabiensis*, *An. merus*, *An. quadriannulatus*, and *An. coluzzii* or *An. gambiae*. The identified differences in the size and structure of the X chromosome heterochromatin point to a possible role of repetitive DNA in speciation of mosquitoes. We found that *An. coluzzii* and *An. gambiae*, incipient species with prezygotic isolation, share variations in the relative positions of the satDNA repeats and the proximal heterochromatin band on the X chromosomes. This previously unknown genetic polymorphism in malaria mosquitoes may be caused by a differential amplification of DNA repeats or an inversion in the sex chromosome heterochromatin.

Keywords: *Anopheles*; heterochromatin; mosquito; mitotic chromosome; sex chromosome; satellite DNA; X chromosome

---

1. Introduction

The major malaria vector in Africa *Anopheles gambiae* has been a subject of extensive research over the past few decades. Initially described as a single species, *An. gambiae* was later subdivided into a complex of morphologically indistinguishable species by crossing experiments [1,2], fixed differences in polytene chromosome arrangement and distinct adaptations [3,4], differences in Intergenic Sequence (IGS) and Internal Transcribed Spacer (ITS) regions in the ribosomal DNA (rDNA) [5,6], and by whole-genome divergence [7–9]. Members of the *An. gambiae* complex differ in genetic diversity and they include major vectors, minor vector, and nonvectors [3,4,10]. The current list of the species includes *An. arabiensis*, *An. amharicus*, *An. bwambae*, *An. coluzzii*, *An. fontenillei*, *An. gambiae*, *An. merus*, *An. melas*, and *An. quadriannulatus* [4,7,9].

The majority of interspecies crosses in the *An. gambiae* complex produce fertile females and sterile males [2,11–14] in agreement with Haldane's rule [15]. However, the most closely related species *An. coluzzii* and *An. gambiae* do not have postzygotic reproductive barriers as in laboratory conditions they readily mate and lay viable eggs that produce fertile hybrid males and females [16,17]. If the *An. gambiae* complex originated as recently as 526 thousand years ago, the *An. coluzzii* and *An. gambiae* lineages split from the common ancestor only ~61 thousand years ago [18]. The recently diverged species *An. gambiae sensu stricto* and *An. coluzzii* have been initially considered as the S and M molecular forms of *An. gambiae* [5,6] based on specific single nucleotide polymorphism (SNP) differences of ITS2 sequences [19,20]. Later studies demonstrated that the two forms also differ in genome sequence [8], gene expression [21], larval ecology [22], larval behavior in the presence of predators [23,24], adult swarming behavior [25], and adult mate recognition [17,26]. Thus, these data show evidence for ecological, behavioral, and genome-wide differentiation between *An. coluzzii* and *An. gambiae*.

Cytogenetic analysis of the polytene chromosomes banding patterns, including fixed chromosomal inversions, is an established tool for distinguishing species of malaria mosquitoes [3,4,27–29]. In addition, cytogenetic analysis of mitotic chromosomes demonstrated interspecific differences in sex chromosome heterochromatin between *An. gambiae* and *An. arabiensis* (named *An. gambiae* species a and B at that time) [30]. Staining with the Hoechst fluorescent dye has shown that the presence and brightness of X-chromosome heterochromatic bands differ between the two species [31]. More recently, heterochromatin variation of mitotic Y chromosomes among species of the *An. gambiae* complex has been clearly demonstrated [32]. Intraspecific polymorphism in the X and Y chromosome heterochromatin has also been observed in both natural and laboratory populations of *An. gambiae* [30,31], but is unclear if the previously observed variations can be related to possible differences between the more recently described incipient species *An. coluzzii* and *An. gambiae s.s.* [7]. This information can be useful because vector control to be successful must consider full spectrum of genetic and phenotypic variation within vector species. Further, it has been suggested that polymorphism in sex chromosome heterochromatin may affect fertility and sexual behavior of mosquitoes [33], but the lack of understanding of heterochromatin structure and function prevent researchers from mechanistic understanding of the phenotypic effects.

Sequencing of the *An. gambiae* genome provided important information regarding its organization [34]. It has been demonstrated that repetitive DNA component represent a substantial portion in of the *An. gambiae* genome (33%) that is higher than in *Drosophila melanogaster* genome where repeats make up approximately 24% of the genome [35,36]. The majority of repetitive DNA in the *An. gambiae* genome is tightly packed in heterochromatin around the centromeres [37]. Difficulty with sequencing of heterochromatin led to underrepresentation of heterochromatic sequences in the *An. gambiae* genome assembly [34]. Moreover, the repetitive nature of these sequences poses an impediment in mapping them correctly to chromosomes. Subsequent attempts to map heterochromatic genomic scaffolds led to an addition of ~16 Mb of heterochromatin to the genome of the *An. gambiae* PEST strain [38]. Further progress was made by predicting functions of 232 heterochromatin genes [39] and by mapping genes on the heterochromatin-euchromatin boundary of the polytene chromosome map [40]. Bioinformatics analysis of so called "unknown chromosome" or ~42 Mb of unmapped sequences in *An. gambiae* PEST genome demonstrated that it has characteristics of heterochromatin [39]. Although both *An. gambiae* and *An. coluzzii* genome assemblies are now available [39,41], corresponding information of their entire heterochromatin on a chromosome map is still missing. Likewise, sequencing of five other species from the *An. gambiae* complex also excluded the majority of heterochromatic sequences from the genome assembly [9,42].

Comparison of *An. gambiae* (the former S form) and *An. coluzzii* (the former M form) genomes identified pericentromeric autosomal and X-chromosome regions of high differentiation, termed "speciation islands," or "islands of genomic divergence" [43,44]. These regions largely correspond to pericentromeric heterochromatin. Overlaps between the heterochromatin and the islands of genomic divergence are 91% in the X chromosome, 97% in the 2L arm, and 94% in the 3L arm [39]. Other studies emphasized that the highest genomic divergence between these nascent species occurred within the 4 Mb of mapped heterochromatin on the X chromosome [8,45,46]. A recent population genomic analysis of 765 field-collected mosquitoes across Africa has demonstrated that sequences in the pericentromeric X heterochromatin show the greatest separation between *An. coluzzii* and *An. gambiae* populations on a neighbor-joining tree [47]. The study also analyzed the *An. coluzzii* and *An. gambiae* genomes for CRISPR/Cas9 target sites; it found that 5474 genes had at least one viable target after excluding target sites with nucleotide variation. Interestingly, targetable genes are spread non-uniformly across the genome, falling predominantly in pericentromeric heterochromatic regions, where levels of variation are lower [47]. However, the full spectrum of heterochromatin variation and molecular composition of the X-chromosome heterochromatin in malaria mosquitoes remains unknown. Because repetitive DNA is underreplicated [48], polytene chromosomes have limited applications in studies of heterochromatin. In comparison, mitotic chromosomes are a promising system for evolutionary studies of heterochromatin in malaria mosquitoes.

In this study, we examined similarities and differences in heterochromatin patterns within X mitotic chromosomes among the major malaria vectors *An. gambiae*, *An. coluzzii*, *An. arabiensis*, minor vector *An. merus*, and zoophilic non-vector *An. quadriannulatus*. The correspondence of pericentric heterochromatin between the metaphase and polytene X chromosome was established for *An. gambiae*. We report quantitative differences in molecular organization of heterochromatin among members of the *An. gambiae* complex and highlight shared polymorphism of the heterochromatin structure among strains of the incipient species *An. gambiae* and *An. coluzzii*. The identified variation in heterochromatin is discussed with respect to the phylogenetic relationships among the species and a possible role of heterochromatin in speciation.

2. Materials and Methods

2.1. Mosquito Strains and Colony Maintenance

Laboratory colonies examined for this study were provided by the Biodefense and Emerging Infections Research Resources Repository and included colonies of *An. gambiae* PIMPERENA (MRA-861), *An. gambiae* KISUMU1 (MRA-762), *An. gambiae* ZANU (MRA-594), *An. coluzzii* SUA2La (MRA-765), *An. coluzzii* MOPTI (MRA-763), *An. coluzzii* MALI-NIH (MRA-860), *An. arabiensis* DONGOLA 2Ra2Rb3R (MRA-1235), *An. quadriannulatus* SANGWE (MRA-1155), and *An. merus* MAF (MRA-1156). The species and strains used in this study are presented in Table S1. All mosquitoes were reared at 27 °C, with 12:12 light:dark cycle and 70% relative humidity. Authentication of the species was done using PCR diagnostics [49,50]. Larvae were fed fish food and adult mosquitoes were fed 1% sugar water. To induce oviposition, females were fed defibrinated sheep blood (Colorado Serum Co., Denver, Colorado, USA) using artificial blood feeders. To perform interspecies crosses between *An. gambiae* and *An. coluzzii*, male and female pupae were separated to ensure virginity of adult mosquitoes. We differentiated males and females at the pupal stage using sex-specific differences in the shape of their terminalia [51]. After the emergence of adults, crossing experiments were performed by combining 30 females and 15 males in a single cage. Five days after random mating, the females were fed sheep blood. Two days later, an egg dish, covered with moist filter paper to keep the eggs from drying out, was put into the cage to obtain F1 hybrid larvae for cytogenetic analyses. Any possible colony contamination during the experiment was ruled out by verifying all the strains for their respective species using intentional mismatch primers (IMP) as described elsewhere [19]. All the strains showed the expected band size, with *An. gambiae* strains around 330 bp and *An. coluzzii* strains around 460 bp.

2.2. Chromosome Preparation

Preparations from early 4th instar larvae of lab colonies were made from leg and wing imaginal discs, as previously described [52]. Both male and female larvae were immobilized on ice for 10 min, and then dissected in a drop of cold, freshly prepared, hypotonic solution (0.075 M potassium chloride). After decapitating the head, the thorax was opened using dissecting scissors (Fine Science Tools, Foster City, CA, USA), followed by removal of the gut and fat from the body. a fresh drop of hypotonic solution was added to the preparation for 10 min, followed by fixation in a drop of modified Carnoy's solution (ethanol:glacial acetic acid, 3:1) for 1 min. Next, a drop of freshly prepared 50% propionic acid was added, and the imaginal discs were covered with a 22 × 22 mm coverslip. After 5 min, the preparations were squashed using the flat rubber end of a pencil and dipped in liquid nitrogen until the bubbling stopped. Coverslips were removed using a sharp blade and slides were transferred to cold 50% ethanol and stored at −20 °C. After 2 h, slides were dehydrated in a series of 70%, 80%, and 100% ethanol. Preparations with the highest number of metaphase plates were chosen for *in situ* hybridization.

2.3. C_0t DNA Preparation and DNA Probe Labeling

To identify the position of the centromere in chromosome X, we prepared the repetitive DNA fraction using a previously described method [52]. Genomic DNA was isolated from 500 g of non-bloodfed adult *An. coluzzii* MOPTI mosquitoes using a Qiagen Blood and Cell culture DNA Maxi kit (Qiagen, Hilden, Germany). Isolated C_0t1 repetitive fraction was precipitated with isopropanol and labelled by nick-translation in 50 µl, containing 1 µg DNA, 0.05 mM each of unlabeled dATP, dCTP, and dGTP, and 0.015 mM of dTTP, 1 µl of fluorescein-dUTP, 0.05 mg/ml of BSA, 5 µl of 10× nick-translation buffer, 20 U of DNA-polymerase I, and 0.0012 U of DNase, at 15 °C for 2.5 h. Primers for 18S rDNA and satDNA repeats from *An. gambiae* AgY53A, AgY477-AgY53B, AgY477, and AgY53C were obtained, as previously described [14,53] (Table S2). An Immomix PCR kit (Bioline, Inc., Taunton, MA, USA) was used to label satellites by incorporating Cy3 and Cy5 fluorescently labeled nucleotides (Enzo Life Sciences, Inc., Farmingdale, NY, USA) directly into the PCR reaction. Each 25 µl PCR mix consisted of 35–40 ng genomic DNA, 0.3 U Taq polymerase, 1× PCR buffer, 200 µM each of dATP, dCTP, and dGTP, and 65 µM dTTP, and 0.5 µl Cy3-dUTP or 0.5 µl Cy5-dUTP (Enzo Life Sciences, Inc., Farmingdale, NY, USA). Thermocycling was performed using ImmoMix TM (Bioline Inc., Taunton, MA, USA) beginning with a 95 °C incubation for 10 min followed by 35 cycles of 95 °C for 30 sec, 52 °C for 30 sec, 72 °C for 45 sec, 72 °C for 5 min, and a final hold at 4 °C. Bacterial Artificial Chromosome (BAC) clones 05F01 (GenBank accession: AL142298, AL142299), 179F22 (GenBank accession: BH373300, BH373306), and 01K23 (GenBank accession: AL607293, AL607294) [34] were labelled by nick-translation in 50 µl, containing 1 µg DNA, 0.05 mM each of unlabeled dATP, dCTP, and dGTP and 0.015 mM of dTTP, 1 µl of Cy3-dUTP (or another fluorochrome), 0.05 mg/ml of BSA, 5 µl of 10× nick-translation buffer, 20 U of DNA-polymerase I, and 0.0012 U of DNase at 15 °C for 2.5 h.

2.4. Fluorescence in situ Hybridization (FISH)

Suitable slides with >10 metaphase plates were selected for FISH, which was performed as previously described [52,54]. Briefly, slides with good preparations were treated with 0.1 mg/ml RNase at 37 °C for 30 min. After washing twice with 2× saline-sodium citrate (SSC) for 5 min, slides were digested with 0.01% pepsin and 0.037% HCl solution for 5 min at 37 °C. After washing slides twice in 1× phosphate-buffered saline (PBS) for 5 min at room temperature, preparations were fixed in 3.7% formaldehyde for 10 min at RT. Slides were then washed in 1× PBS and dehydrated in a series of 70%, 80%, and 100% ethanol for 5 min at RT. Then, 10 µl of probes were mixed, added to the preparations, and incubated overnight at 37 °C. After washing slides in 1× SSC at 60 °C for 5 min, 4× SSC/NP40 solution at 37 °C for 10 min, and 1× PBS for 5 min at room temperature (RT), preparations were counterstained with a DAPI-antifade solution (Life Technologies, Carlsbad, CA, USA) and kept in the dark for at least 2 h before visualization with a fluorescence microscope.

2.5. Image Acquisition and Chromosome Measurements

Chromosome slide preparations were viewed with an Olympus BX61 fluorescence microscope (Olympus, Tokyo, Japan) using BioView software (BioView Inc., Billerica, MA, USA) at 1000× magnification. For idiogram development, metaphase plates were measured and analyzed from 5 larvae per strain. a total of 40 metaphase plates were used to measure chromosome lengths in *An. coluzzii*. For sibling species *An. arabiensis*, *An. quadriannulatus*, and *An. merus*, 11–13 metaphase plates were used to measure chromosome lengths (Table S3). Chromosome images were inverted using Adobe Photoshop CS6 (Adobe Inc., San Jose, CA, USA) and measured using the ruler tool at 1000× magnification. a statistical Tukey's test or a nonparametric Kruskal–Wallis rank sum test followed by a Dunn's test were performed to compare chromosome length between the species using JMP13 software (SAS Institute Inc., Cary, NC, USA) or R software (RStudio, Boston, MA, USA). Fluorescence intensities of proximal and distal X chromosome heterochromatin bands were measured using Adobe Photoshop CS6 (Adobe Inc., San Jose, CA, USA) and compared between three strains of *An. gambiae* and three strains of *An. coluzzii*. About forty measurements for each band were taken from forty sister chromatids of twenty X chromosomes for each strain. Pairwise comparisons between strains were performed using Student's t-test.

2.6. Quantative Analysis of Fluorescent Signal Positions

Chromosomal positions of satDNA AgY477–AgY53B and AgY477, and the proximal DAPI band were identified and compared in three strains of *An. gambiae* and three strains of *An. coluzzii*. a custom MATLAB script [55] was written and used to measure the position of fluorescence peaks in individual channels [56]. The program was automated to measure the position of each satDNA and DAPI peak along the length of the X chromosome. The output provided the maximum likelihood of a combination of probe positions. The results were compared within and between species. The MATLAB program consisted of two user-guided steps followed by a third step of automated analysis. In the first step, sex chromosomes were identified by the user on mitotic slide preparations from *An. gambiae* and *An. coluzzii* strains. In the second step, the boundary of each identified sex chromosome was traced by a trained user. In an automated step, the pattern of heterochromatin and satDNA fluorescence was averaged longitudinally along the user-defined boundary. Program output displayed fluorescence intensity graphically from chromosome centromere to telomere. Peak fluorescence intensity was used to automatically infer the order of the heterochromatin and satDNA probes. User guidance was deliberately introduced to eliminate the pitfalls of full automation. In particular, user guidance was critical for accurate identification of sex chromosomes in the imaged mitotic slide preparations. Longitudinal averaging of fluorescence intensity during the program's automated analysis was robust to the boundary of each chromosome traced by the user.

3. Results

3.1. Correspondence Between Heterochromatin of the Polytene and Mitotic X Chromosome in Anopheles gambiae

Using FISH with a C_0t1 fraction of repetitive DNA, 18S rDNA, and heterochromatic BAC clones, we attempted to establish the correspondence of pericentric heterochromatin between the metaphase and polytene X chromosome in *An. gambiae*. The most repetitive C_0t1 DNA fraction should presumably represent the centromeric satellites of the mitotic chromosomes corresponding to the most proximal regions of polytene arms. C_0t analysis is the process of renaturation of single stranded DNA to its complementary sequence [57]. The analysis is based on the observation that the more repetitive DNA sequences require a shorter time to re-anneal following denaturation. DNA fractions with C_0t1 values equal to 1.0×10^{-4}–1.0×10^{-1} are considered as highly repetitive DNA fractions [57]. We expected C_0t1 to map to narrow regions corresponding to centromeres in mitotic chromosomes. While that was the case for autosomes, in X chromosomes the C_0t1 DNA probe localized to a wide region, including the entire proximal heterochromatin band adjacent to the rDNA locus (Figure 1A). Thus, hybridization of the C_0t1 DNA fraction could be indicative of the centromere positions in autosomes, but we could not precisely determine the location of the centromere on the X chromosome in *An. gambiae*. We mapped three heterochromatic BAC clones, 05F01, 179F22, and 01K23, to the distal heterochromatin band of mitotic chromosomes in *An. coluzzii* (Figure 1B). According to the *An. gambiae* PEST genome assembly [38], these BAC clones are located in the most proximal heterochromatic region 6 in the X polytene chromosome map (Figure 1C). BAC clone 05F01 is mapped to scaffold AAAB01008973 (X: 20,698,730–20,818,594), BAC clone 01K23 is mapped to scaffold AAAB01008967 (X: 23,512,846–23569864), and BAC clone 179F22 is mapped to scaffold AAAB01008976 (X: 23,930,050–24,043,835) [38]. These results demonstrate that the assembled euchromatic part of the X chromosome occupies no more than half of the total X chromosome length (Figure 1C). Although polytene chromosomes are significantly larger than mitotic chromosomes, heterochromatin is under-replicated [48] and represented by diffuse structures without clear banding patterns [39], while heterochromatin in mitotic chromosomes is much more prevalent and detailed (Figure 1D). Because the *An. gambiae* PEST X chromosome genome assembly ends at the coordinate 24,393,108, our data indicate that the assembled heterochromatin largely corresponds to the distal heterochromatic band of the mitotic X chromosome. Only few copies of 18S rDNA sequences are found in the assembled X chromosome. Thus, most of the rDNA repeats and more proximal regions of the X chromosome heterochromatin are absent from the mapped genome assembly.

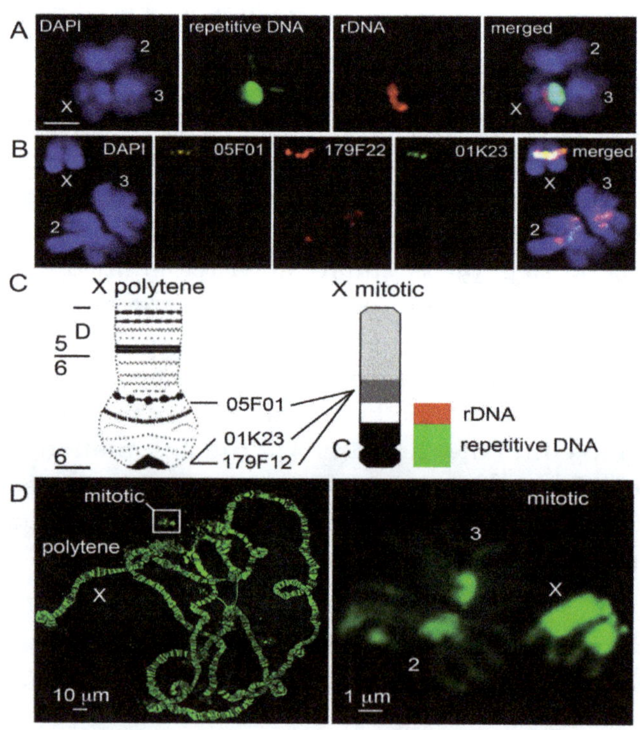

Figure 1. Comparison of the X heterochromatin structure between mitotic and polytene chromosomes. **(A)** Multicolor fluorescence *in situ* hybridization (FISH) of a C_0t1 fraction of repetitive DNA and 18S rDNA with mitotic chromosomes of *An. gambiae* PIMPERENA; scale bar is 2 µm. **(B)** FISH mapping of BAC clones 05F01 (yellow), 179F22 (red), and 01K23 (green) to the distal heterochromatin block on the X chromosome of the *An. coluzzii* MOPTI strain. Chromosomes are counter-stained with DAPI. **(C)** Relative positions of heterochromatic blocks and DNA probes of the polytene and mitotic X chromosome idiograms. rDNA = 18S rDNA probe. Repetitive DNA = C_0t1 fraction of repetitive DNA. C = putative centromere. **(D)** Relative sizes of mitotic and polytene chromosomes in the KISUMU strain of *An. gambiae*. Chromosomes are counter-stained with YOYO-1. Polytene chromosome physical map is from [38].

3.2. *Variation in Heterochromatin Morphology Among Sibling Species of the An. gambiae Complex*

We analyzed the heterochromatin morphology in species of the *An. gambiae* complex originating from different parts of Africa (Figure 2A) and diverging from each other from ~61,000 to ~526,000 years ago [18] (Figure 2B).

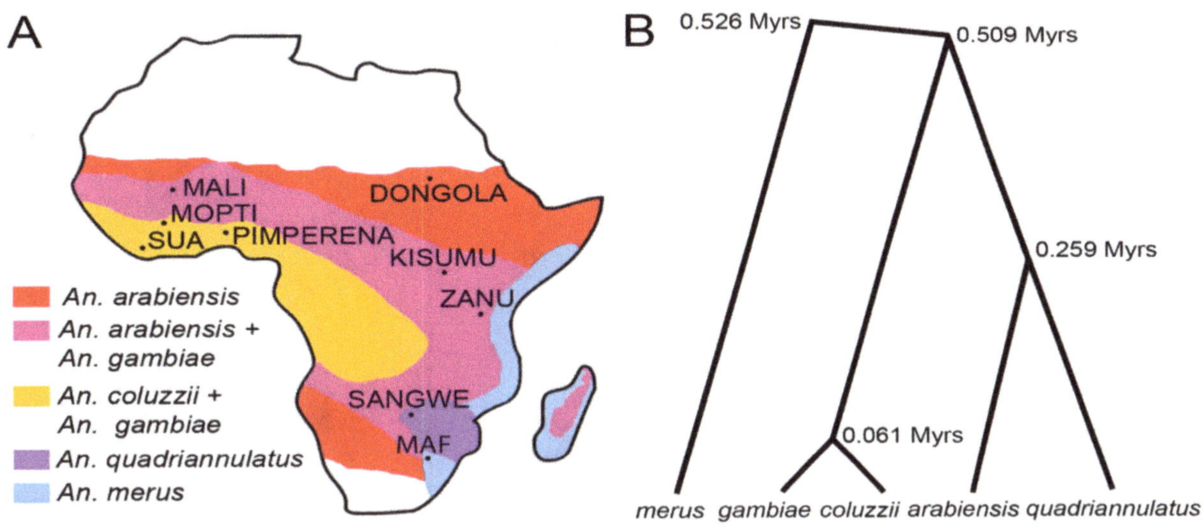

Figure 2. Geographical distribution of species and phylogeny of the *An. gambiae* complex. (**A**) a map of Africa with approximal distribution of species and places of origin of the laboratory strains. (**B**) Species phylogeny based on the X chromosome genomic sequences (redrawn from [18]). Times of species divergence in million years (Myrs) are shown at the tree nodes.

We studied mid-metaphase chromosomes obtained from imaginal discs of the 4th instar larvae of *An. gambiae*, *An. coluzzii*, *An. arabiensis*, *An. merus*, and *An. quadriannulatus*. Chromosomes at this stage provide reproducible heterochromatin patterns verified across multiple individuals. DAPI, a counter-stain that preferentially stains AT-rich DNA, was utilized to augment the natural banding patterns resulting from heterochromatin variation between and within species. To simplify the comparison, all fluorescence images were inverted, and converted to gray-scale images (Figure 3). The number and size of the X chromosome heterochromatin bands varied across the sibling species. Differences in the X chromosome heterochromatin pattern were clearly visible between all species with postzygotic reproductive isolation (Figure 3). Two relatively small heterochromatin bands can be seen in *An. gambiae* and *An. coluzzii* (Figure 3A,B). The KISUMU strain of *An. gambiae* often shows polymorphism in the size of the heterochromatin between different X chromosomes, even within one individual (Figures 1D and 3A) Extended heterochromatin is observed in X chromosomes of *An. arabiensis*, *An. quadriannulatus*, and *An. merus* (Figure 3C). The pattern of fluorescence intensity of the heterochromatic bands differs among species as well. *An. gambiae* and *An. coluzzii* possess a dense heterochromatin block on the end proximal to the putative centromere. This is followed by a dull fluorescence area to which the 18S rDNA locus is mapped. a light distal band is present immediately next to the rDNA locus, marking the assembled heterochromatin in the current *An. gambiae* PEST genome assembly [38]. a large dark heterochromatin band is visible in the *An. arabiensis* X chromosome. Our results for *An. gambiae* and *An. arabiensis* corroborated with those for the natural populations of these species in a previous study [38]. In contrast, *An. quadriannulatus*, a zoophilic non-vector species of the *An. gambiae* complex, has one light and two dark heterochromatin bands. The longest X chromosome was found in *An. merus*, the marshy saltwater resident of the *An. gambiae* complex. The *An. merus* X chromosome contains three large heterochromatin blocks and a large rDNA locus making it comparable in length to the autosomes.

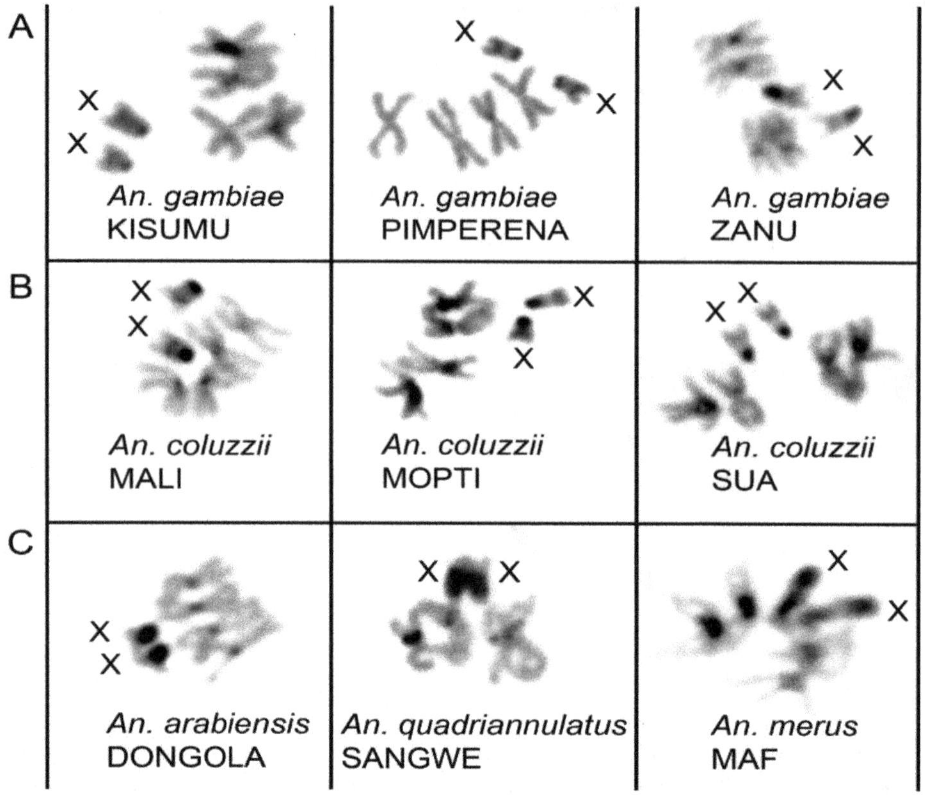

Figure 3. Metaphase karyotypes of females in species from the *An. gambiae* complex. (**A**) *An. gambiae* strains. (**B**) *An. coluzzii* strains. (**C**) *An. arabiensis*, *An. quadriannulatus*, and *An. merus* strains. Black and dark gray blocks correspond to compact and diffuse heterochromatin, respectively. X chromosomes are labeled. Species names are indicated in italics and strain names are capitalized.

3.3. Molecular Variation of Heterochromatin Among Species of the An. gambiae Complex

We analyzed similarities and differences in the molecular content of the heterochromatin among sibling species. We performed FISH with 18S rDNA and previously identified satDNA sequences AgY53A, Ag53C, AgY477, and the genomic fragment AgY477–AgY53B that contains partial sequences of satellites AgY477 and AgY53B as well as a junction region [53] (Table S2). In *An. gambiae* PIMPERENA, satellites AgY53A and AgY477–AgY53B hybridized to single locations in the tip and base of the X chromosome proximal heterochromatin band, respectively (Figure 4A,B). This is in sharp contrast to *An. merus*, where AgY477–AgY53B hybridized with four distinct regions of the X chromosome heterochromatin (Figure 4C). Interestingly, AgY477–AgY53B hybridized differently in *An. coluzzii* SUA compared with *An. gambiae* PIMPERENA; AgY477–AgY53B is found in the base of the X chromosome proximal heterochromatin band of *An. coluzzii* SUA (Figure 4D). Satellite AgY477 has also been mapped to the base of the X chromosome proximal heterochromatin band of *An. gambiae* ZANU [32]. As described in the following results section, localization of these satellites represents shared variation among strains of *An. gambiae* and *An. coluzzii.* Unlike AgY53A and AgY477–AgY53B, satellite Ag53C is not sex chromosome-specific [53]; it hybridized with pericentromeric regions of autosomes in *An. coluzzii* MOPTI (Figure 4E). The 18S rDNA probe was mapped to the space between two dense heterochromatic bands on the X chromosome in *An. quadriannulatus* (Figure 4F). Previously, it was demonstrated that the AgY477–AgY53B sequence is immediately adjacent to the rDNA locus on the X chromosome of *An. quadriannulatus* [32]. Both AgY477–AgY53B and AgY477 co-localize at the very tip of the X chromosome outside the major dense heterochromatin band in *An. arabiensis* (Figure 4G). Hybridization of each of these satellites, together with 18S rDNA, demonstrated that, unlike in *An. gambiae* or *An. coluzzii*, AgY477 (Figure 4H) and AgY477–AgY53B (Figure 4I) are separated from the rDNA locus by the large band of dense heterochromatin in *An. arabiensis*.

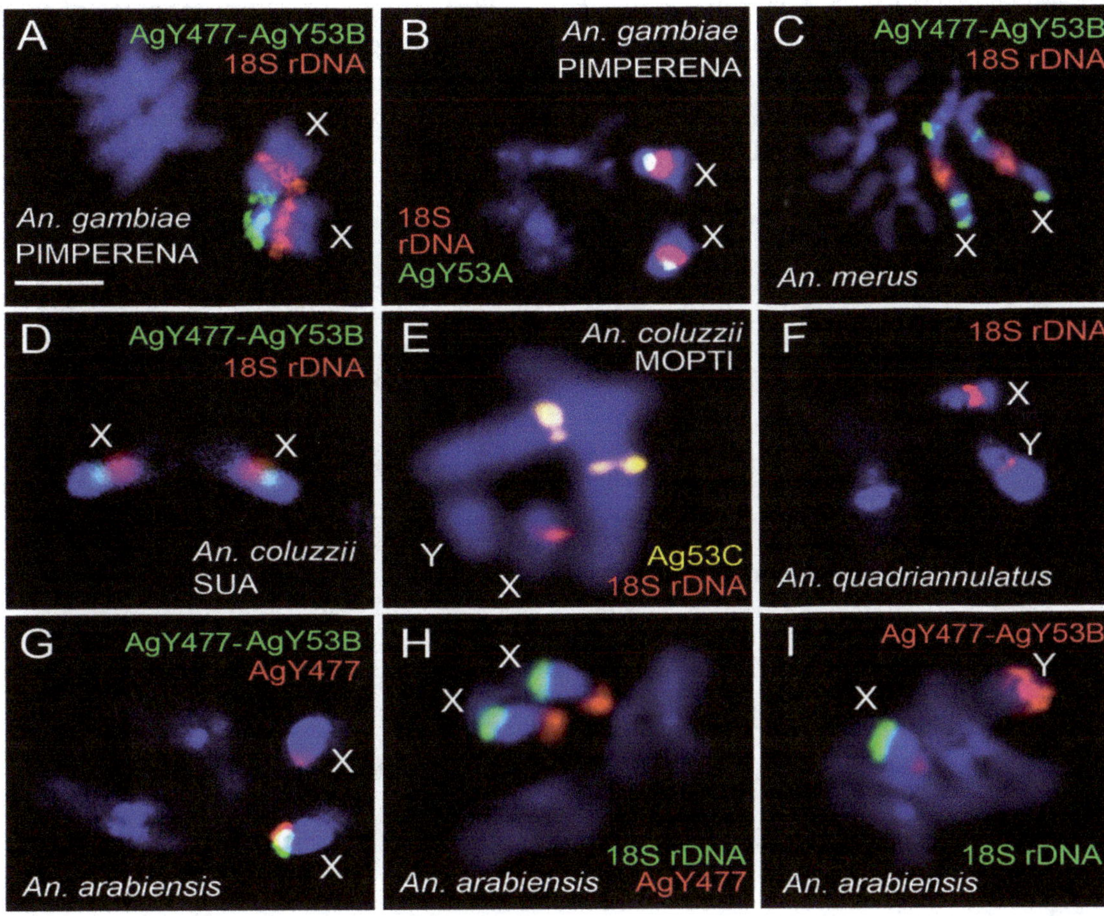

Figure 4. Mapping of repetitive DNA elements to mitotic chromosomes of species from the *An. gambiae* complex. (**A**) FISH of AgY477–AgY53B (green) and 18S rDNA (red) in *An. gambiae* PIMPERENA. (**B**) FISH of satellite AgY53A (green) and 18S rDNA (red) in *An. gambiae* PIMPERENA. (**C**) FISH of AgY477–AgY53B (green) and 18D rDNA (red) in *An. merus*. (**D**) FISH of AgY477–AgY53B (green) and 18S rDNA (red) in *An. coluzzii* SUA. (**E**) FISH of satellite Ag53C (yellow) and 18D rDNA (red) in *An. coluzzii* MOPTI. (**F**) FISH of 18S rDNA (red) in *An. quadriannulatus*. (**G**) FISH of satellite AgY477 (red) and AgY477–AgY53B (green) in *An. arabiensis*. (**H**) FISH of satellite AgY477 (red) and 18D rDNA (green) in *An. arabiensis*. (**I**) FISH of 18S rDNA (green) and AgY477–AgY53B (red) in *An. arabiensis*. Scale bar = 2 µm.

3.4. Shared Heterochromatin Variation Between An. coluzzii and An. gambiae

To visualize the rDNA locus, we performed FISH of the 18S rDNA probe with mitotic chromosomes. Within all strains of *An. gambiae* and *An. coluzzii*, the rDNA locus was mapped between the proximal and distal heterochromatin bands (Figure S1). We noticed variation in the size of the rDNA locus both among the strains of *An. gambiae* and *An. coluzzii* and between homologous chromosomes within individual mosquitoes. The intra-strain polymorphism in the size of the rDNA locus was especially obvious in KISUMU, ZANU, and MOPTI. This polymorphism may suggest variation in the rDNA copy number.

To better resolve the relative satDNA positions among multiple mosquito strains, we generated F1 hybrids between laboratory strains of *An. gambiae* and *An. coluzzii*. F1 hybrid larvae resulting from the *An. gambiae* × *An. coluzzii* crosses were analyzed for heterochromatin morphology and satDNA location. Qualitative and quantitative differences in the pattern of heterochromatin blocks, rDNA loci, and satDNA location were clearly observed between homoeologous X chromosomes within the F1 hybrids. Different positions and sizes of the AgY477–AgY53B and AgY477 FISH signals with respect to the proximal heterochromatin band can be seen between homoeologous X chromosomes in F1 ♀*An. gambiae* PIMPERENA × ♂*An. coluzzii* MOPTI (Figure 5A). Different sizes of the proximal heterochromatin band and the rDNA locus can be seen between homoeologous X chromosomes in F1 ♀*An. gambiae* ZANU × ♂*An. coluzzii* MALI (Figure 5B). Also, different sizes of the hybridization signals from AgY477 and 18S rDNA can be seen between homologous X chromosomes in F1 ♀*An. coluzzii* MOPTI × ♂*An. gambiae* KISUMU (Figure 5C).

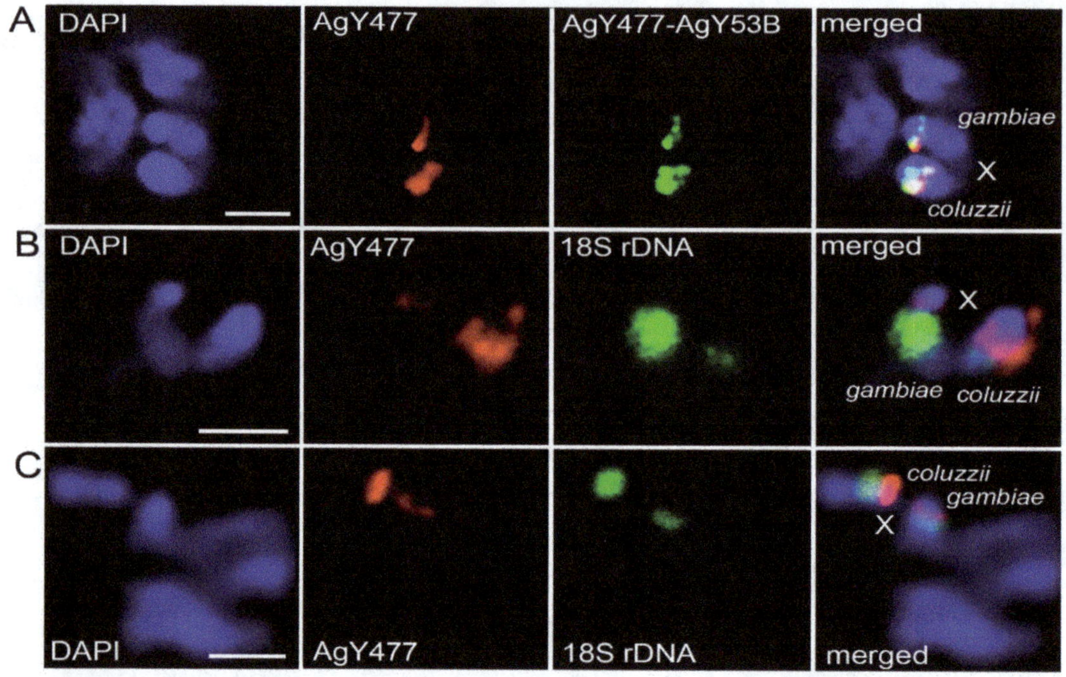

Figure 5. Variation in the pattern of heterochromatin blocks, rDNA loci, and satDNA location on the X chromosomes in F1 female hybrids between *An. gambiae* and *An. coluzzii.* (**A**) FISH of AgY477–AgY53B (green) and AgY477 (red) in F1 ♀*An. gambiae* PIMPERENA × ♂*An. coluzzii* MOPTI. (**B**) FISH of AgY477 (red) and 18S rDNA (green) in F1 ♀*An. gambiae* ZANU × ♂*An. coluzzii* MALI. (**C**) FISH of AgY477 (red) and 18S rDNA (green) in F1 ♀*An. coluzzii* MOPTI × ♂*An. gambiae* KISUMU. Scale bar = 2 μm.

Using a custom MATLAB script [56], the position of the satDNA repeats with respect to the proximal heterochromatin band was mapped on the X chromosomes in *An. gambiae* and *An. coluzzii* strains. Because sequences of AgY477–AgY53B and AgY477 overlap, it is technically unfeasible to distinguish between the signals when mapping their distance from the proximal heterochromatin band. Therefore, we combined the fluorescence signals of AgY477–AgY53B and AgY477 and mapped them with respect to the fluorescence DAPI peak of the heterochromatin. The advantages of our automated analysis are twofold. First, fluorescent signals in each image possess a degree of overlap, which masks the order of fluorescent peaks from centromere to telomere; however, the order of peaks is easily detected using our automated analysis. Second, observation alone is insufficient to determine the distance between fluorescent peaks, which are generally separated by fewer than 10 pixels. Determining the distance between nearby peaks is made tractable in our pipeline; however, even this computational approach remains limited by microscopic resolution. Images of the mitotic sex chromosomes were captured using 1000× magnification; neighboring pixels at this resolution represent a distance of 0.05 μm. Thus, peaks spaced closer than 0.05 μm were prone to falsely inverted order. Nonetheless, peak orders were generally resolvable by aggregating multiple images averaged for multiple individuals of each strain. We measured the distance from satellites to the fluorescence DAPI peak and the linear order of the peaks. For example, the order of the satDNA and DAPI peaks is switched between *An. gambiae* PIMPERENA and *An. coluzzii* MOPTI (Figure 6A). Distribution patterns of signal intensities for satellites and DAPI along the X chromosomes also differ between *An. gambiae* PIMPERENA and *An. coluzzii* MOPTI (Figure 6B). Our mapping results revealed that the strains differed in the pattern of satDNA and DAPI peak relative positions and distance on the X chromosome (Figure 6C). Clustering analysis showed that *An. gambiae* PIMPERENA and *An. coluzzii* MALI depicted the same pattern of satDNA, followed by the DAPI band, while *An. coluzzii* MOPTI, *An. coluzzii* SUA, and *An. gambiae* ZANU clustered together showing the DAPI peak followed by the satDNA locus (Figure 6D). Interestingly, *An. gambiae* KISUMU had an almost 1:1 ratio of the direct and the reverse satDNA-DAPI band order indicating that this strain is maintaining high polymorphism in their X chromosome heterochromatin. Thus, our data shows the dynamism of the heterochromatin at the molecular level in the important malaria vectors. We found no evidence for clustering the strains by species, indicating that this heterochromatin variation is shared between *An. coluzzii* and *An. gambiae* either as ancestral polymorphism or by ongoing hybridization between the incipient species.

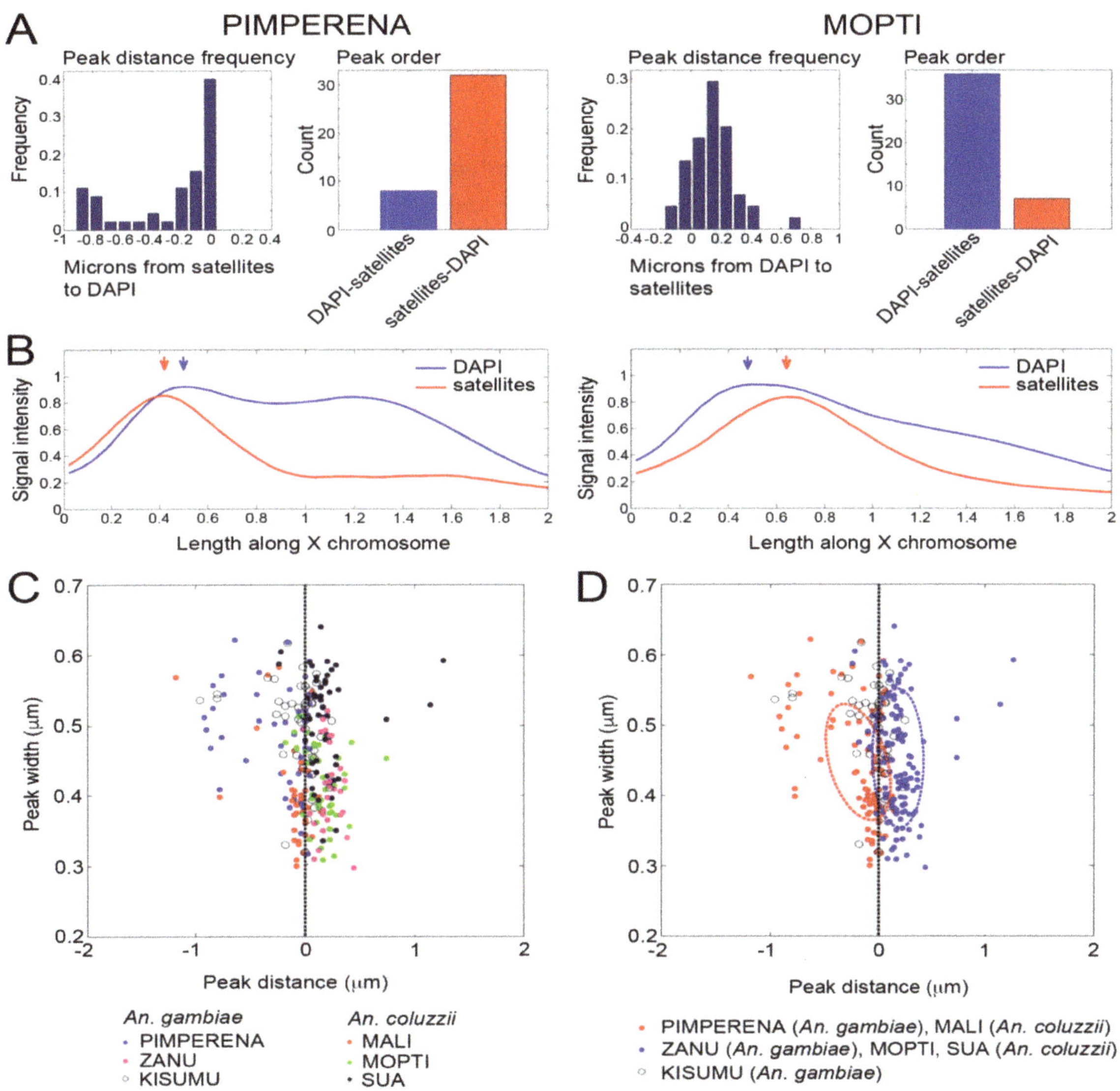

Figure 6. Polymorphism of the X heterochromatin structure among the *An. gambiae* and *An. coluzzii* strains. (**A**) Distance and order frequency of FISH signal and DAPI band intensity peaks in *An. gambiae* PIMPERENA and *An. coluzzii* MOPTI. (**B**) Distribution patterns of signal intensities for satellites and DAPI along the X chromosomes in *An. gambiae* PIMPERENA and *An. coluzzii* MOPTI. (**C**) Relative positions of satellite peak distance and width with respect to the proximal heterochromatin band for six strains of *An. gambiae* and *An. coluzzii*. (**D**) Clustering of the *An. gambiae* and *An. coluzzii* strains based on the relative positions of satellite peak distance and width with respect to the proximal heterochromatin band.

We also compared the average fluorescence intensities of the proximal and distal X chromosome heterochromatin between three strains of *An. gambiae* and three strains of *An. coluzzii*. a statistical analysis, using Student's t-test, revealed no specific pattern when strains were compared with respect to the species. Most strains were significantly different from each other in their fluorescence intensities of the proximal band ($P < 0.05$), except for SUA-PIMPERENA ($P = 0.06$), SUA-ZANU ($P = 0.16$), ZANU-PIMPERENA ($P = 0.61$), and MALI-KISUMU ($P = 0.74$). MOPTI was significantly different from every other strain (Figure S2A). Comparison of fluorescence intensities of the distal heterochromatin band revealed a similar pattern for pairwise comparison between strains of both species. Most strains were significantly different from each other ($P < 0.05$), except for PIMPERENA-KISUMU ($P = 0.07$), ZANU-MALI ($P = 0.11$), and KISUMU-SUA ($P = 0.89$) (Figure S2B). Overall, the fluorescence intensities of both X chromosome heterochromatin bands vary among the strains, regardless of their species identity.

3.5. X Chromosome Idiograms for Species of the An. gambiae Complex

To construct idiograms, we measured the lengths of the chromosomes (Table S3) and conducted a statistical pairwise Tukey's test between the sibling species. Our analysis revealed that the total X chromosome lengths were significantly different for all species pairs, except between *An. quadriannulatus* and *An. arabiensis* ($P = 0.64$). Due to the possibility that the X chromosome length data violated the assumption of equal variance, we repeated statistical analyses using a nonparametric Kruskal–Wallis rank sum test followed by a Dunn's test. These analyses resulted in qualitatively identical results (Figure 7, Table S4). The polytene X chromosomes in all these species are similar in length; hence, the difference in mitotic X chromosomes can be attributed to the variation in heterochromatin.

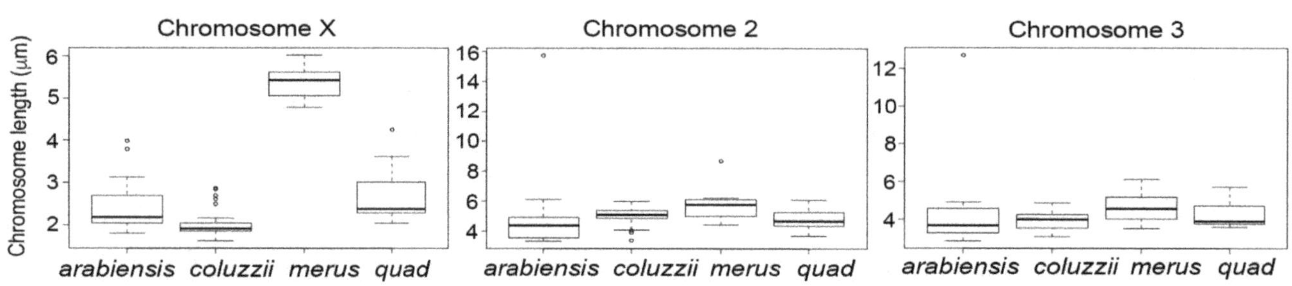

Figure 7. Lengths of the metaphase chromosomes in species from the *An. gambiae* complex.

We also conducted a statistical pairwise Tukey's test for autosomes between the sibling species. We found that, unlike the X chromosome, the autosomal lengths do not significantly differ between the species (Figure 7, Table S5). This finding demonstrates the contrasting patterns of the sex chromosome and autosome evolution and highlights the rapids evolutionary changes in the X chromosome heterochromatin.

Based on mitotic chromosome measurements, banding patterns, and repeat mapping, we constructed X chromosome idiograms for *An. gambiae*, *An. coluzzii*, *An. arabiensis*, *An. quadriannulatus*, and *An. merus*. FISH with the genomic fragment AgY477–AgY53B revealed its multiple locations in the X chromosome heterochromatin bands in *An. merus* but a single location in the pericentric heterochromatin in other sibling species. These features were placed on respective idiograms for easy comparison of molecular features within heterochromatin among species (Figure 8). The identified differences in the X chromosome size and heterochromatin structure are consistent with *An. gambiae*, *An. coluzzii*, *An. arabiensis*, *An. quadriannulatus* grouping together while *An. merus* being a separate lineage in the species phylogeny [18] (Figure 2B). The *An. merus* X chromosome has the largest amount of heterochromatin among the X chromosomes of sibling species. Moreover, *An. merus* is the only species in this study with the rDNA locus located in the L arm of the X chromosome suggesting a fixed pericentric inversion that differentiate *An. merus* from the other species. The developed idiograms will facilitate future studies of genetic variation in repeat-rich heterochromatic regions of malaria mosquitoes from the *An. gambiae* complex.

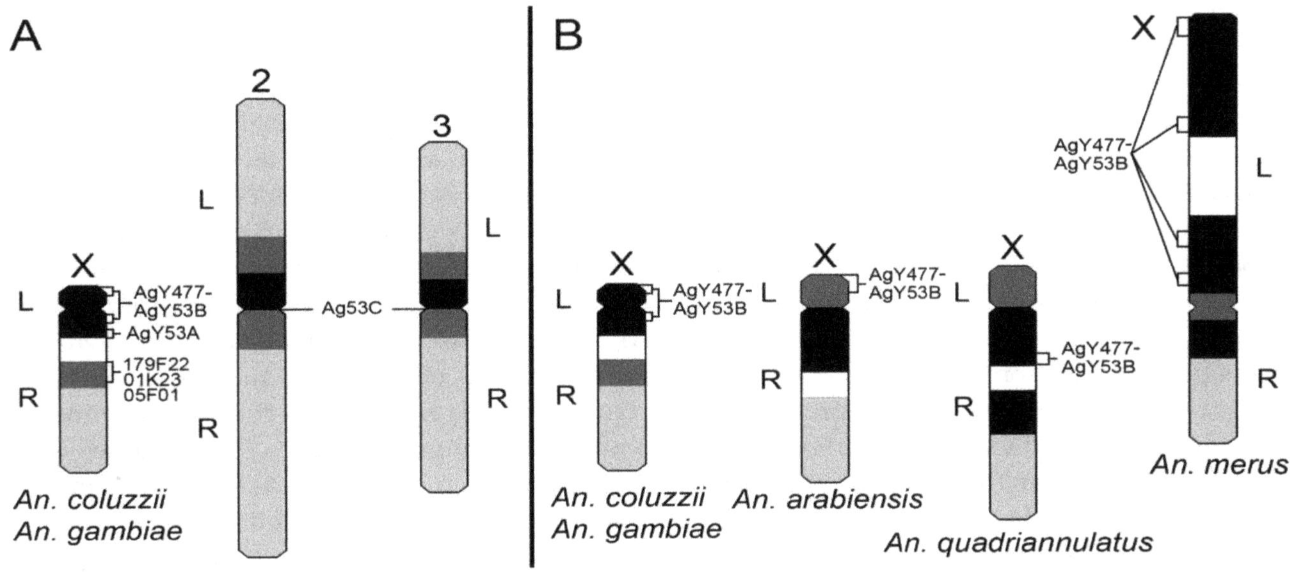

Figure 8. Idiograms of metaphase karyotypes of species from the *An. gambiae* complex. (**A**) An idiogram of metaphase karyotypes of *An. gambiae* and *An. coluzzii*. (**B**) Comparison of the X chromosome idiograms among species from the *An. gambiae* complex. Left and right arms are labeled with L and R, respectively. Black bands correspond to condensed heterochromatin. Dark gray bands correspond to diffuse heterochromatin. White areas represent the rDNA locus. Light gray areas show euchromatin. The area of constriction represents the putative centromere. Mapping of the BAC clones and satDNA repeats AgY53A and AgY53C were performed only with *An. coluzzii* or *An. gambiae*. Two different positions of AgY477–AgY53B on the X chromosome of *An. gambiae* and *An. coluzzii* represent polymorphism among strains.

4. Discussion

Mosquitoes, with their small number of diploid chromosomes, $2n = 6$, present a convenient model to investigate the molecular organization of their genomes and chromosome evolution [58]. Since polytene chromosomes are well-developed in *Anopheles* mosquitoes, they are often preferred over mitotic chromosomes for gene mapping [40,59]. Analysis of polytene chromosomes has been useful for identifying genomic variations in euchromatin. Comparative genome mapping within the *Anopheles* genus has demonstrated that the euchromatic portion of the X chromosome is the fastest evolving by genomic rearrangements [42,60]. However, under-replication of heterochromatin in polytene chromosomes makes them less suitable for mapping repetitive elements of the genome. In this study, we utilized metaphase mitotic chromosomes for repeat mapping and for studying heterochromatin variation among species in the *An. gambiae* complex. Since the genome of *An. gambiae* was first published [34], polytene chromosomes have been used to improve the heterochromatin assembly and to characterize its genetic content [38–40]. Here, we established the correspondence of pericentric heterochromatin between the metaphase and polytene X chromosome of *An. gambiae*, and demonstrated that at least half of the *An. gambiae* X chromosome length is heterochromatic. Future long-read sequencing and assembly of the heterochromatin in malaria mosquitoes may yield important insights into structural and functional organization of the sex chromosomes, as demonstrated for *Drosophila* [61].

Our comparative analysis of the X chromosome pericentric heterochromatin identified qualitative differences among sibling species, *An. arabiensis, An. merus, An. quadriannulatus,* and *An. coluzzii* or *An. gambiae*, with postzygotic isolation. The identified fixed variations included the number and pattern of heterochromatic bands and chromosomal location of the satDNA repeats. A common emerging theme from studies performed in *Drosophila* suggests that heterochromatin plays an important role in hybrid incompatibility [62–66]. Rapidly evolving heterochromatic repeats and heterochromatin-binding proteins are considered major players in the evolution of intrinsic reproductive isolation. Reproductively isolated *Drosophila* species have different satDNA content and localization within heterochromatin [64,67]. Evolutionary forces lead to a great variation in the copy number and types of heterochromatin repeats between closely related species. As repetitive DNA elements replicate, mechanisms such as unequal crossover, rolling circle replication, and segmental duplication become sources of genome evolution [68,69]. Thus, our study supports observations from other organisms that sex chromosomes have a propensity to accumulate species-specific differences in heterochromatin [70].

SatDNA repeats are seen as prime candidates to trigger genome instability in interspecies hybrids [71]. They have been implicated in disruption of chromosome pairing, abnormal heterochromatin packaging, and selfish post-meiotic drive substitutions in *Drosophila* hybrids [67]. Disruption of chromosome pairing and synapsis is observed frequently in interspecies hybrids of various organisms. Interestingly, sex chromosomes more often than autosomes drive speciation and hybrid incompatibilities [70]. For example, pairing of X and Y chromosomes is more adversely affected than that of autosomes during the prophase I in male hybrids between Campbell's dwarf hamster and the Djungarian hamster [72]. a recent work has clarified that the autosomes of male hybrids between these hamster species undergo pairing and recombination normally as their parental forms do, but the heterochromatic arms of the X and Y chromosomes show a high frequency of asynapsis and recombination failure [73]. It was proposed that asynapsis of heterospecific chromosomes in prophase I may provide a recurrently evolving trigger for the meiotic arrest of interspecific F1 hybrids of mice [74,75]. Our recent study showed that meiotic failures in sterile F1 ♀*A. merus* × ♂*A. coluzzii* hybrids are accompanied by the disruption of sex chromosome pairing and insufficient chromosome condensation. These cytogenetic abnormalities lead to a premature equational division and formation of diploid immotile sperm [14]. Demonstrated here, qualitative inter-species differences in heterochromatin suggest that they may be responsible for incompatibilities of sibling species from the *An. gambiae* complex.

Unlike species with postzygotic isolation, *An. coluzzii* and *An. gambiae* do not show fixed differences in the pattern of the X chromosome pericentric heterochromatin. An early study observed an intra-specific polymorphism of the X chromosome heterochromatin among lab strains of *An. gambiae* [31]. The described polymorphism was based on the Hoechst staining of the heterochromatin. Here, we studied the pattern of heterochromatic bands or location of satDNA in multiple strains of *An. coluzzii* and *An. gambiae*, incipient species with prezygotic isolation. We found that, although the strains may differ from each other, the species do share variation in the relative positions of the fluorescence intensity peaks for the satDNA repeats and the proximal DAPI-positive heterochromatin band of the X chromosomes. The mechanism behind this inversion-like genetic polymorphism in malaria mosquitoes could be a rapid differential amplification of satellite repeats or actual structural rearrangement in the sex chromosome heterochromatin. It is possible that further evolution of heterochromatin in *An. coluzzii* and *An. gambiae* will contribute to higher genetic differentiation, and will eventually lead to postzygotic reproductive isolation between some populations of these important disease vectors. Extensive studies of *Drosophila* species show that heterochromatin, in general, and satellite DNA, in particular, can be a source of tremendous genetic variation [76–80]. Heterochromatin can also modulate differential gene expression and cause variable phenotypes, including differences in immune response [81,82]. It is possible that changes in repetitive sequences can have fitness consequences upon which selection will act. a population genomics study in *D. melanogaster* found that satDNA repeats often show population differentiation, but the population structure inferred from overall satellite quantities does not recapitulate the expected population relationships based on the demographic history of this species [79]. Moreover, some satellites have likely been involved in antagonistic interactions, as inferred from negative correlations among them.

The rDNA locus is immediately adjacent to the heterochromatin bands of the X chromosome. Its variation in size, among and within species, suggests polymorphism of the rDNA copy number. Natural polymorphism in the number of rDNA copies has been recorded in vertebrate and invertebrate animals, both among and within species [83]. Associations between the heterochromatin or rDNA variation and organismal phenotypes have also been demonstrated. a study in *Drosophila* showed that heterochromatin formation prolongs lifespan and controls ribosomal RNA synthesis [84]. Another study demonstrated that rDNA copy number decreases during aging, and that this age-dependent decrease in rDNA copy number is transgenerationally heritable [85]. In humans, the number of rDNA repeats can vary from 250 to 670 copies per diploid genome [86], and the changes in rDNA copy number can affect genome-wide gene expression [87]. Natural variation in rDNA and heterochromatin may contribute to genome evolution, formation of reproductive barriers, and eventually to diversification of species [88–90]. Genomics studies of the heterochromatin in natural populations of *An. coluzzii* and *An. gambiae* can determine the actual levels of polymorphism and divergence in satDNA and rDNA copy number variation. These studies may also identify heterochromatin variations that correlate with specific environmental adaptations or behaviors of the malaria vectors.

5. Conclusions

We discovered a new type of shared cytogenetic polymorphism in the incipient species, *An. gambiae* and *An. coluzzii*—an inversion of the satDNA location with respect to the proximal heterochromatin band. This finding suggests mechanisms of rapid differentiation of the sex chromosome heterochromatin during evolution: genomic rearrangement or differential amplification of heterochromatic repeats. Future genome sequencing using long-read technologies will be crucial for determining the precise nature of the observed polymorphism. Our study also demonstrated X chromosome heterochromatin divergence among mosquito species with post-zygotic isolation, which manifests in species-specific localization of repetitive DNA as well as size and number of heterochromatin bands. The identified differences in heterochromatin support the basal phylogenetic position of *An. merus* and point to the role of heterochromatin in speciation. The differences in molecular organization of the sex chromosome heterochromatin may impair meiotic synapsis during meiosis in sterile inter-species male hybrids.

We suggest that inter-species divergence of heterochromatin structure represents a cytogenetic threshold that triggers the evolutionary shift from prezygotic to postzygotic isolation. The new idiograms developed here for the sibling species will aid in future studies of heterochromatin organization and evolution in these malaria mosquitoes.

Supplementary Materials
Figure S1: FISH of 18S rDNA with the X chromosomes in strains of *An. gambiae* and *An. coluzzii.* (**A**) *An. gambiae* strains: PIMPERENA, ZANU, KISUMU. (**B**) *An. coluzzii* strains: MALI, SUA, MOPTI. Scale bar = 1 μM. Red = 18S rDNA, Blue = DAPI., Figure S2: Fluorescence intensities of X chromosome heterochromatin in three strains of *An. gambiae* and three strains of *An. coluzzii.* (**A**) Fluorescence intensities of the proximal heterochromatin band. (**B**) Fluorescence intensities of the distal heterochromatin band., Table S1: Mosquito strains used in the study of the X chromosome heterochromatin., Table S2: Probes and primer sequences used for FISH., Table S3: Length of mitotic chromosomes in species of the *An. gambiae* complex. Table S4: Statistical analyses of the X chromosome lengths using a nonparametric Kruskal-Wallis rank sum test followed by a Dunn's test., Table S5: Statistical analyses of the chromosome lengths using Tukey's honestly significant difference test.

Author Contributions: Conceptualization, M.V.S., and I.V.S.; methodology, A.S., N.A.K., V.A.T., and I.V.S.; software, N.A.K.; validation, A.S., M.V.S., and I.V.S.; formal analysis, A.S., N.A.K. and I.V.S.; investigation, A.S., N.A.K., and V.A.T.; resources, I.V.S.; data curation, A.S., N.A.K, M.V.S., and I.V.S.; writing—original draft preparation, A.S., M.V.S., and I.V.S.; writing—review and editing, I.V.S.; visualization, A.S., N.A.K., V.A.T.; supervision, M.V.S., and I.V.S.; project administration, I.V.S.; funding acquisition, M.V.S. and I.V.S. All authors have read and agreed to the published version of the manuscript.

Acknowledgments: The following reagents were obtained through the Biodefense and Emerging Infections Research Resources Repository, NIAID, NIH: *An. arabiensis*, Strain DONGOLA 2Ra, 2Rb/b and 3R, MRA-1235, contributed by Ellen M. Dotson; *An. coluzzii*, Strain MALI-NIH, Eggs, MRA-860, contributed by Nora J. Besansky; *An. coluzzii*, Strain MOPTI, Eggs, MRA-763, contributed by Gregory C. Lanzaro; *An. coluzzii*, Strain SUA2La, Eggs, MRA-765, contributed by Alessandra della Torre; *An. gambiae*, Strain KISUMU1, Eggs, MRA-762, contributed by Vincent Corbel; *An. gambiae*, Strain PIMPERENA (S form), Eggs, MRA-861, contributed by Nora J. Besansky; *An. gambiae*, Strain ZANU, MRA-594, contributed by Hilary Ranson and Frank H. Collins; *An. merus*, Strain MAF, MRA-1156, contributed by Maureen Coetzee; and *An. quadriannulatus*, Strain SANGWE, Eggs, MRA-1155, contributed by Willem Takken. We thank Semen Bondarenko for his help with statistical analysis using R, Jiangtao Liang for useful discussions, as well as Jean Clarke, Kristin Rose, and Janet Webster for proofreading the text.

References

1. Davidson, G. The Five Mating Types in the Anopheles Gambiae Complex. *Riv. Malariol.* **1964**, *43*, 167–183. [PubMed]
2. Davidson, G.; Paterson, H.E.; Coluzzi, M.; Mason, G.F.; Micks, D.W. The Anopheles gambiae Complex. In *Genetics of Insect Vectors of Diesease*; Wright, J.W., Pal, R., Eds.; Elsevier Publishing Company: Amsterdam, The Netherlands; London, UK; New York, NY, USA, 1967; pp. 211–249.
3. Coluzzi, M.; Sabatini, A.; Petrarca, V.; Di Deco, M.A. Chromosomal differentiation and adaptation to human environments in the Anopheles gambiae complex. *Trans. R. Soc. Trop. Med. Hyg.* **1979**, *73*, 483–497. [CrossRef]
4. Coluzzi, M.; Sabatini, A.; della Torre, A.; Di Deco, M.A.; Petrarca, V. a polytene chromosome analysis of the Anopheles gambiae species complex. *Science* **2002**, *298*, 1415–1418. [CrossRef] [PubMed]
5. Della Torre, A.; Costantini, C.; Besansky, N.J.; Caccone, A.; Petrarca, V.; Powell, J.R.; Coluzzi, M. Speciation within Anopheles gambiae—The glass is half full. *Science* **2002**, *298*, 115–117. [CrossRef]
6. Della Torre, A.; Fanello, C.; Akogbeto, M.; Dossou-yovo, J.; Favia, G.; Petrarca, V.; Coluzzi, M. Molecular evidence of incipient speciation within Anopheles gambiae s.s. in West Africa. *Insect Mol. Biol.* **2001**, *10*, 9–18. [CrossRef]
7. Coetzee, M.; Hunt, R.H.; Wilkerson, R.; Della Torre, A.; Coulibaly, M.B.; Besansky, N.J. Anopheles coluzzii and Anopheles amharicus, new members of the Anopheles gambiae complex. *Zootaxa* **2013**, *3619*, 246–274. [CrossRef]
8. Lawniczak, M.K.; Emrich, S.J.; Holloway, A.K.; Regier, A.P.; Olson, M.; White, B.; Redmond, S.; Fulton, L.; Appelbaum, E.; Godfrey, J.; et al. Widespread divergence between incipient Anopheles gambiae species revealed by whole genome sequences. *Science* **2010**, *330*, 512–514. [CrossRef]
9. Barron, M.G.; Paupy, C.; Rahola, N.; Akone-Ella, O.; Ngangue, M.F.; Wilson-Bahun, T.A.; Pombi, M.; Kengne, P.; Costantini, C.; Simard, F.; et al. a new species in the major malaria vector complex sheds light on reticulated species evolution. *Sci. Rep.* **2019**, *9*, 14753. [CrossRef]
10. White, B.J.; Collins, F.H.; Besansky, N.J. Evolution of Anopheles gambiae in Relation to Humans and Malaria. *Annu. Rev. Ecol. Evol. Syst.* **2011**, *42*, 111–132. [CrossRef]
11. Slotman, M.; Della Torre, A.; Powell, J.R. The genetics of inviability and male sterility in hybrids between Anopheles gambiae and An. arabiensis. *Genetics* **2004**, *167*, 275–287. [CrossRef]
12. Bernardini, F.; Galizi, R.; Wunderlich, M.; Taxiarchi, C.; Kranjc, N.; Kyrou, K.; Hammond, A.; Nolan, T.; Lawniczak, M.N.K.; Papathanos, P.A.; et al. Cross-Species Y Chromosome Function Between Malaria Vectors of the Anopheles gambiae Species Complex. *Genetics* **2017**, *207*, 729–740. [CrossRef] [PubMed]
13. Presgraves, D.C.; Orr, H.A. Haldane's rule in taxa lacking a hemizygous X. *Science* **1998**, *282*, 952–954. [CrossRef] [PubMed]
14. Liang, J.; Sharakhov, I.V. Premeiotic and meiotic failures lead to hybrid male sterility in the Anopheles gambiae complex. *Proc. Biol. Sci.* **2019**, *286*, 20191080. [CrossRef] [PubMed]
15. Haldane, J.B.S. Sex ratio and unisexual sterility in hybrid animals. *J. Genet.* **1922**, *12*, 101–109. [CrossRef]
16. Diabate, A.; Dabire, R.K.; Millogo, N.; Lehmann, T. Evaluating the effect of postmating isolation between molecular forms of Anopheles gambiae (Diptera: Culicidae). *J. Med. Entomol.* **2007**, *44*, 60–64. [CrossRef]

17. Aboagye-Antwi, F.; Alhafez, N.; Weedall, G.D.; Brothwood, J.; Kandola, S.; Paton, D.; Fofana, A.; Olohan, L.; Betancourth, M.P.; Ekechukwu, N.E.; et al. Experimental swap of Anopheles gambiae's assortative mating preferences demonstrates key role of X-chromosome divergence island in incipient sympatric speciation. *PLoS Genet.* **2015**, *11*, e1005141. [CrossRef]
18. Thawornwattana, Y.; Dalquen, D.; Yang, Z. Coalescent Analysis of Phylogenomic Data Confidently Resolves the Species Relationships in the Anopheles gambiae Species Complex. *Mol. Biol. Evol.* **2018**, *35*, 2512–2527. [CrossRef]
19. Wilkins, E.E.; Howell, P.I.; Benedict, M.Q. IMP PCR primers detect single nucleotide polymorphisms for Anopheles gambiae species identification, Mopti and Savanna rDNA types, and resistance to dieldrin in Anopheles arabiensis. *Malar. J.* **2006**, *5*, 125. [CrossRef]
20. Santolamazza, F.; Mancini, E.; Simard, F.; Qi, Y.; Tu, Z.; della Torre, A. Insertion polymorphisms of SINE200 retrotransposons within speciation islands of Anopheles gambiae molecular forms. *Malar. J.* **2008**, *7*, 163. [CrossRef]
21. Caputo, B.; Nwakanma, D.; Caputo, F.P.; Jawara, M.; Oriero, E.C.; Hamid-Adiamoh, M.; Dia, I.; Konate, L.; Petrarca, V.; Pinto, J.; et al. Prominent intraspecific genetic divergence within Anopheles gambiae sibling species triggered by habitat discontinuities across a riverine landscape. *Mol. Ecol.* **2014**, *23*, 4574–4589. [CrossRef]
22. Gimonneau, G.; Pombi, M.; Choisy, M.; Morand, S.; Dabire, R.K.; Simard, F. Larval habitat segregation between the molecular forms of the mosquito Anopheles gambiae in a rice field area of Burkina Faso, West Africa. *Med. Vet. Entomol.* **2012**, *26*, 9–17. [CrossRef] [PubMed]
23. Dabire, K.R.; Diabate, A.; Agostinho, F.; Alves, F.; Manga, L.; Faye, O.; Baldet, T. Distribution of the members of Anopheles gambiae and pyrethroid knock-down resistance gene (kdr) in Guinea-Bissau, West Africa. *Bull. Soc. Pathol. Exot.* **2008**, *101*, 119–123. [PubMed]
24. Gimonneau, G.; Pombi, M.; Dabire, R.K.; Diabate, A.; Morand, S.; Simard, F. Behavioural responses of Anopheles gambiae sensu stricto M and S molecular form larvae to an aquatic predator in Burkina Faso. *Parasit. Vectors* **2012**, *5*, 65. [CrossRef] [PubMed]
25. Tripet, F.; Toure, Y.T.; Taylor, C.E.; Norris, D.E.; Dolo, G.; Lanzaro, G.C. DNA analysis of transferred sperm reveals significant levels of gene flow between molecular forms of Anopheles gambiae. *Mol. Ecol.* **2001**, *10*, 1725–1732. [CrossRef]
26. Cassone, B.J.; Kamdem, C.; Cheng, C.; Tan, J.C.; Hahn, M.W.; Costantini, C.; Besansky, N.J. Gene expression divergence between malaria vector sibling species Anopheles gambiae and An. coluzzii from rural and urban Yaounde Cameroon. *Mol. Ecol.* **2014**, *23*, 2242–2259. [CrossRef]
27. Sharakhova, M.V.; Peery, A.; Antonio-Nkondjio, C.; Xia, A.; Ndo, C.; Awono-Ambene, P.; Simard, F.; Sharakhov, I.V. Cytogenetic analysis of Anopheles ovengensis revealed high structural divergence of chromosomes in the Anopheles nili group. *Infect. Genet. Evol.* **2013**, *16*, 341–348. [CrossRef]
28. Stegnii, V.N.; Novikov, Y.M.; Kabanova, V.M. Cytogenetic analysis and distribution of the malaria mosquito Anopheles beklemishevi. *Zool. ZH* **1978**, *57*, 873–876.
29. Stegnii, V.N.; Kabanova, V.M. Chromosome analysis of the malarial mosquitoes Anopheles atroparvus and A. maculipennis (Diptera, Culicidae). *Zool. ZH* **1978**, *8*, 207–225.
30. Gatti, M.; Santini, G.; Pimpinelli, S.; Coluzzi, M. Fluorescence banding techniques in the identification of sibling species of the Anopheles gambiae complex. *Heredity (Edinb)* **1977**, *38*, 105–108. [CrossRef]
31. Bonaccorsi, S.; Santini, G.; Gatti, M.; Pimpinelli, S.; Colluzzi, M. Intraspecific polymorphism of sex chromosome heterochromatin in two species of the Anopheles gambiae complex. *Chromosoma* **1980**, *76*, 57–64. [CrossRef]
32. Hall, A.B.; Papathanos, P.A.; Sharma, A.; Cheng, C.; Akbari, O.S.; Assour, L.; Bergman, N.H.; Cagnetti, A.; Crisanti, A.; Dottorini, T.; et al. Radical remodeling of the Y chromosome in a recent radiation of malaria mosquitoes. *Proc. Natl. Acad. Sci. USA* **2016**, *113*, E2114–E2123. [CrossRef] [PubMed]
33. Fraccaro, M.; Tiepolo, L.; Laudani, U.; Marchi, A.; Jayakar, S.D. Y chromosome controls mating behaviour on Anopheles mosquitoes. *Nature* **1977**, *265*, 326–328. [CrossRef] [PubMed]
34. Holt, R.A.; Subramanian, G.M.; Halpern, A.; Sutton, G.G.; Charlab, R.; Nusskern, D.R.; Wincker, P.; Clark, A.G.; Ribeiro, J.M.; Wides, R.; et al. The genome sequence of the malaria mosquito Anopheles gambiae. *Science* **2002**, *298*, 129–149. [CrossRef] [PubMed]
35. Smith, C.D.; Shu, S.; Mungall, C.J.; Karpen, G.H. The Release 5.1 annotation of Drosophila melanogaster heterochromatin. *Science* **2007**, *316*, 1586–1591. [CrossRef] [PubMed]

36. Manning, J.E.; Schmid, C.W.; Davidson, N. Interspersion of repetitive and nonrepetitive DNA sequences in the Drosophila melanogaster genome. *Cell* **1975**, *4*, 141–155. [CrossRef]
37. Besansky, N.J.; Collins, F.H. The mosquito genome: Organization, evolution and manipulation. *Parasitol. Today* **1992**, *8*, 186–192. [CrossRef]
38. Sharakhova, M.V.; Hammond, M.P.; Lobo, N.F.; Krzywinski, J.; Unger, M.F.; Hillenmeyer, M.E.; Bruggner, R.V.; Birney, E.; Collins, F.H. Update of the Anopheles gambiae PEST genome assembly. *Genome Biol.* **2007**, *8*, R5. [CrossRef]
39. Sharakhova, M.V.; George, P.; Brusentsova, I.V.; Leman, S.C.; Bailey, J.A.; Smith, C.D.; Sharakhov, I.V. Genome mapping and characterization of the Anopheles gambiae heterochromatin. *BMC Genom.* **2010**, *11*, 459. [CrossRef]
40. George, P.; Sharakhova, M.V.; Sharakhov, I.V. High-resolution cytogenetic map for the African malaria vector Anopheles gambiae. *Insect Mol. Biol.* **2010**, *19*, 675–682. [CrossRef]
41. Kingan, S.B.; Heaton, H.; Cudini, J.; Lambert, C.C.; Baybayan, P.; Galvin, B.D.; Durbin, R.; Korlach, J.; Lawniczak, M.K.N. a High-Quality De novo Genome Assembly from a Single Mosquito Using PacBio Sequencing. *Genes* **2019**, *10*. [CrossRef]
42. Neafsey, D.E.; Waterhouse, R.M.; Abai, M.R.; Aganezov, S.S.; Alekseyev, M.A.; Allen, J.E.; Amon, J.; Arca, B.; Arensburger, P.; Artemov, G.; et al. Mosquito genomics. Highly evolvable malaria vectors: The genomes of 16 Anopheles mosquitoes. *Science* **2015**, *347*, 1258522. [CrossRef]
43. Turner, T.L.; Hahn, M.W.; Nuzhdin, S.V. Genomic islands of speciation in Anopheles gambiae. *PLoS Biol.* **2005**, *3*, e285. [CrossRef]
44. White, B.J.; Cheng, C.; Simard, F.; Costantini, C.; Besansky, N.J. Genetic association of physically unlinked islands of genomic divergence in incipient species of Anopheles gambiae. *Mol. Ecol.* **2010**, *19*, 925–939. [CrossRef]
45. Nwakanma, D.C.; Neafsey, D.E.; Jawara, M.; Adiamoh, M.; Lund, E.; Rodrigues, A.; Loua, K.M.; Konate, L.; Sy, N.; Dia, I.; et al. Breakdown in the process of incipient speciation in Anopheles gambiae. *Genetics* **2013**, *193*, 1221–1231. [CrossRef]
46. Neafsey, D.E.; Lawniczak, M.K.; Park, D.J.; Redmond, S.N.; Coulibaly, M.B.; Traore, S.F.; Sagnon, N.; Costantini, C.; Johnson, C.; Wiegand, R.C.; et al. SNP genotyping defines complex gene-flow boundaries among African malaria vector mosquitoes. *Science* **2010**, *330*, 514–517. [CrossRef]
47. Miles, A.; Harding, N.J.; Botta, G.; Clarkson, C.; Antao, T.; Kozak, K.; Schrider, D.; Kern, A.; Redmond, S.; Sharakhov, I.V.; et al. Genetic diversity of the African malaria vector Anopheles gambiae. *Nature* **2017**, *552*, 96–100.
48. Zhimulev, I.F. Polytene chromosomes, heterochromatin, and position effect variegation. *Adv. Genet.* **1998**, *37*, 1–555.
49. Scott, J.A.; Brogdon, W.G.; Collins, F.H. Identification of single specimens of the Anopheles gambiae complex by the polymerase chain reaction. *Am. J. Trop. Med. Hyg.* **1993**, *49*, 520–529. [CrossRef]
50. Fanello, C.; Santolamazza, F.; della Torre, A. Simultaneous identification of species and molecular forms of the Anopheles gambiae complex by PCR-RFLP. *Med. Vet. Entomol.* **2002**, *16*, 461–464. [CrossRef]
51. Benedict, M.; Dotson, E.M. Methods in Anopheles Research. 2015, p. 408. Available online: https://www.beiresources.org/Portals/2/VectorResources/2016%20Methods%20in%20Anopheles%20Research%20full%20manual.pdf (accessed on 12 December 2019).
52. Timoshevskiy, V.A.; Sharma, A.; Sharakhov, I.V.; Sharakhova, M.V. Fluorescent *in situ* Hybridization on Mitotic Chromosomes of Mosquitoes. *J. Vis. Exp.* **2012**, e4215. [CrossRef]
53. Krzywinski, J.; Sangare, D.; Besansky, N.J. Satellite DNA from the Y chromosome of the malaria vector Anopheles gambiae. *Genetics* **2005**, *169*, 185–196. [CrossRef]
54. Sharakhova, M.V.; George, P.; Timoshevskiy, V.; Sharma, A.; Peery, A.; Sharakhov, I.V. Mosquitoes (Diptera). In *Protocols for Cytogenetic Mapping of Arthropod Genomes*; Sharakhov, I.V., Ed.; CRC Press; Taylor & Francis Group: Boca Raton, FL, USA, 2015; pp. 93–170.
55. MATLAB. *MATLAB:2010 Version 7.10.0 (R2010a)*; The MathWorks Inc.: Natick, MA, USA, 2010.
56. Kinney, N.; Sharakhov, I.V. MATLAB Script to Measure the Position of Fluorescence Peaks on Chromosomes. Figshare. Software. Available online: https://doi.org/10.6084/m9.figshare.11782140.v1 (accessed on 31 January 2020).
57. Trifonov, V.A.; Vorobieva, N.V.; Serdyukova, N.A.; Rens, W. FISH with and Without COT1 DNA. In *Fluorescence In Situ Hybridization (FISH)—Application Guide*; Springer: Berlin/Heidelberg, Germany, 2009. [CrossRef]
58. Kumar, A.; Rai, K.S. Chromosomal localization and copy number of 18S + 28S ribosomal RNA genes in evolutionarily diverse mosquitoes (Diptera, Culicidae). *Hereditas* **1990**, *113*, 277–289. [CrossRef]

59. Grushko, O.G.; Rusakova, A.M.; Sharakhova, M.V.; Sharakhov, I.V.; Stegnii, V.N. Localization of repetitive DNA sequences in the pericentromeric heterochromatin of malarial mosquitoes of the "Anopheles maculipennis" complex. *Tsitologiia* **2006**, *48*, 240–245.
60. Jiang, X.; Peery, A.; Hall, A.; Sharma, A.; Chen, X.G.; Waterhouse, R.M.; Komissarov, A.; Riehl, M.M.; Shouche, Y.; Sharakhova, M.V.; et al. Genome analysis of a major urban malaria vector mosquito, Anopheles stephensi. *Genome Biol.* **2014**, *15*, 459. [CrossRef]
61. Chang, C.H.; Larracuente, A.M. Heterochromatin-Enriched Assemblies Reveal the Sequence and Organization of the Drosophila melanogaster Y Chromosome. *Genetics* **2019**, *211*, 333–348. [CrossRef]
62. Brideau, N.J.; Flores, H.A.; Wang, J.; Maheshwari, S.; Wang, X.; Barbash, D.A. Two Dobzhansky-Muller genes interact to cause hybrid lethality in Drosophila. *Science* **2006**, *314*, 1292–1295. [CrossRef]
63. Cattani, M.V.; Presgraves, D.C. Genetics and lineage-specific evolution of a lethal hybrid incompatibility between Drosophila mauritiana and its sibling species. *Genetics* **2009**, *181*, 1545–1555. [CrossRef]
64. Satyaki, P.R.; Cuykendall, T.N.; Wei, K.H.; Brideau, N.J.; Kwak, H.; Aruna, S.; Ferree, P.M.; Ji, S.; Barbash, D.A. The Hmr and Lhr hybrid incompatibility genes suppress a broad range of heterochromatic repeats. *PLoS Genet.* **2014**, *10*, e1004240. [CrossRef]
65. Cattani, M.V.; Presgraves, D.C. Incompatibility between X chromosome factor and pericentric heterochromatic region causes lethality in hybrids between Drosophila melanogaster and its sibling species. *Genetics* **2012**, *191*, 549–559. [CrossRef]
66. Blum, J.A.; Bonaccorsi, S.; Marzullo, M.; Palumbo, V.; Yamashita, Y.M.; Barbash, D.A.; Gatti, M. The Hybrid Incompatibility Genes Lhr and Hmr Are Required for Sister Chromatid Detachment During Anaphase but Not for Centromere Function. *Genetics* **2017**, *207*, 1457–1472. [CrossRef]
67. Ferree, P.M.; Prasad, S. How can satellite DNA divergence cause reproductive isolation? Let us count the chromosomal ways. *Genet. Res. Int.* **2012**, *2012*, 430136. [CrossRef]
68. Kejnovsky, E.; Hobza, R.; Cermak, T.; Kubat, Z.; Vyskot, B. The role of repetitive DNA in structure and evolution of sex chromosomes in plants. *Heredity* **2009**, *102*, 533–541. [CrossRef]
69. Matsunaga, S. Junk DNA promotes sex chromosome evolution. *Heredity* **2009**, *102*, 525–526. [CrossRef]
70. Deakin, J.E.; Potter, S.; O'Neill, R.; Ruiz-Herrera, A.; Cioffi, M.B.; Eldridge, M.D.B.; Fukui, K.; Marshall Graves, J.A.; Griffin, D.; Grutzner, F.; et al. Chromosomics: Bridging the Gap between Genomes and Chromosomes. *Genes* **2019**, *10*. [CrossRef]
71. Dion-Cote, A.M.; Barbash, D.A. Beyond speciation genes: An overview of genome stability in evolution and speciation. *Curr. Opin. Genet. Dev.* **2017**, *47*, 17–23. [CrossRef]
72. Ishishita, S.; Tsuboi, K.; Ohishi, N.; Tsuchiya, K.; Matsuda, Y. Abnormal pairing of X and Y sex chromosomes during meiosis I in interspecific hybrids of Phodopus campbelli and P. sungorus. *Sci. Rep.* **2015**, *5*, 9435. [CrossRef]
73. Bikchurina, T.I.; Tishakova, K.V.; Kizilova, E.A.; Romanenko, S.A.; Serdyukova, N.A.; Torgasheva, A.A.; Borodin, P.M. Chromosome Synapsis and Recombination in Male-Sterile and Female-Fertile Interspecies Hybrids of the Dwarf Hamsters (Phodopus, Cricetidae). *Genes* **2018**, *9*. [CrossRef]
74. Bhattacharyya, T.; Gregorova, S.; Mihola, O.; Anger, M.; Sebestova, J.; Denny, P.; Simecek, P.; Forejt, J. Mechanistic basis of infertility of mouse intersubspecific hybrids. *Proc. Natl. Acad. Sci. USA* **2013**, *110*, E468–E477. [CrossRef]
75. Bhattacharyya, T.; Reifova, R.; Gregorova, S.; Simecek, P.; Gergelits, V.; Mistrik, M.; Martincova, I.; Pialek, J.; Forejt, J. X chromosome control of meiotic chromosome synapsis in mouse inter-subspecific hybrids. *PLoS Genet.* **2014**, *10*, e1004088. [CrossRef]
76. Jagannathan, M.; Warsinger-Pepe, N.; Watase, G.J.; Yamashita, Y.M. Comparative Analysis of Satellite DNA in the Drosophila melanogaster Species Complex. *G3 (Bethesda)* **2017**, *7*, 693–704. [CrossRef]
77. Craddock, E.M.; Gall, J.G.; Jonas, M. Hawaiian Drosophila genomes: Size variation and evolutionary expansions. *Genetica* **2016**, *144*, 107–124. [CrossRef] [PubMed]
78. De Lima, L.G.; Svartman, M.; Kuhn, G.C.S. Dissecting the Satellite DNA Landscape in Three Cactophilic Drosophila Sequenced Genomes. *G3 (Bethesda)* **2017**, *7*, 2831–2843. [CrossRef] [PubMed]
79. Wei, K.H.; Grenier, J.K.; Barbash, D.A.; Clark, A.G. Correlated variation and population differentiation in satellite DNA abundance among lines of Drosophila melanogaster. *Proc. Natl. Acad. Sci. USA* **2014**, *111*, 18793–18798. [CrossRef] [PubMed]
80. Talbert, P.B.; Kasinathan, S.; Henikoff, S. Simple and Complex Centromeric Satellites in Drosophila Sibling Species. *Genetics* **2018**, *208*, 977–990. [CrossRef]

81. Lemos, B.; Branco, A.T.; Hartl, D.L. Epigenetic effects of polymorphic Y chromosomes modulate chromatin components, immune response, and sexual conflict. *Proc. Natl. Acad. Sci. USA* **2010**, *107*, 15826–15831. [CrossRef]
82. Francisco, F.O.; Lemos, B. How do y-chromosomes modulate genome-wide epigenetic states: Genome folding, chromatin sinks, and gene expression. *J. Genom.* **2014**, *2*, 94–103. [CrossRef]
83. Sochorova, J.; Garcia, S.; Galvez, F.; Symonova, R.; Kovarik, A. Evolutionary trends in animal ribosomal DNA loci: Introduction to a new online database. *Chromosoma* **2018**, *127*, 141–150. [CrossRef]
84. Larson, K.; Yan, S.J.; Tsurumi, A.; Liu, J.; Zhou, J.; Gaur, K.; Guo, D.; Eickbush, T.H.; Li, W.X. Heterochromatin formation promotes longevity and represses ribosomal RNA synthesis. *PLoS Genet.* **2012**, *8*, e1002473. [CrossRef]
85. Lu, K.L.; Nelson, J.O.; Watase, G.J.; Warsinger-Pepe, N.; Yamashita, Y.M. Transgenerational dynamics of rDNA copy number in Drosophila male germline stem cells. *Elife* **2018**, *7*. [CrossRef]
86. Chestkov, I.V.; Jestkova, E.M.; Ershova, E.S.; Golimbet, V.E.; Lezheiko, T.V.; Kolesina, N.Y.; Porokhovnik, L.N.; Lyapunova, N.A.; Izhevskaya, V.L.; Kutsev, S.I.; et al. Abundance of ribosomal RNA gene copies in the genomes of schizophrenia patients. *Schizophr. Res.* **2018**, *197*, 305–314. [CrossRef]
87. Gibbons, J.G.; Branco, A.T.; Yu, S.K.; Lemos, B. Ribosomal DNA copy number is coupled with gene expression variation and mitochondrial abundance in humans. *Nat. Commun.* **2014**, *5*, 1–12. [CrossRef] [PubMed]
88. Dion-Cote, A.M.; Symonova, R.; Rab, P.; Bernatchez, L. Reproductive isolation in a nascent species pair is associated with aneuploidy in hybrid offspring. *Proc. R. Soc. B Biol. Sci.* **2015**, *282*, 20142826. [CrossRef] [PubMed]
89. Dion-Cote, A.M.; Symonova, R.; Lamaze, F.C.; Pelikanova, S.; Rab, P.; Bernatchez, L. Standing chromosomal variation in Lake Whitefish species pairs: The role of historical contingency and relevance for speciation. *Mol. Ecol.* **2017**, *26*, 178–192. [CrossRef] [PubMed]
90. Symonova, R. Integrative rDNAomics-Importance of the Oldest Repetitive Fraction of the Eukaryote Genome. *Genes* **2019**, *10*. [CrossRef] [PubMed]

11

A Comprehensive Cytogenetic Analysis of Several Members of the Family Columbidae (Aves, Columbiformes)

Rafael Kretschmer [1,2,*], Ivanete de Oliveira Furo [3,4], Anderson José Baia Gomes [5], Lucas G. Kiazim [1], Ricardo José Gunski [6], Analía del Valle Garnero [6], Jorge C. Pereira [7], Malcolm A. Ferguson-Smith [8], Edivaldo Herculano Corrêa de Oliveira [4,9], Darren K. Griffin [1], Thales Renato Ochotorena de Freitas [2,†] and Rebecca E. O'Connor [1,†]

1 School of Biosciences, University of Kent, Canterbury CT2 7NJ, UK; lgk3@kent.ac.uk (L.G.K.); d.k.griffin@kent.ac.uk (D.K.G.); r.o'connor@kent.ac.uk (R.E.O.)
2 Departamento de Genética, Universidade Federal do Rio Grande do Sul, Porto Alegre 91509-900, Brazil; thales.freitas@ufrgs.br
3 Instituto de Ciências Biológicas, Universidade Federal do Pará, Belém 66075-110, Brazil; ivanetefuro100@gmail.com
4 Laboratório de Cultura de Tecidos e Citogenética, SAMAM, Instituto Evandro Chagas, Ananindeua 67030-000, Brazil; ehco@ufpa.br
5 Instituto Federal do Pará, Abaetetuba 68440-000, Brazil; anderson.gomes@ifpa.edu.br
6 Laboratório de Diversidade Genética Animal, Universidade Federal do Pampa, São Gabriel 97300-162, Brazil; ricardogunski@unipampa.edu.br (R.J.G.); analiagarnero@unipampa.edu.br (A.d.V.G.)
7 Animal and Veterinary Research Centre (CECAV), University of Trás-os-Montes and Alto Douro (UTAD), 5000-801 Vila Real, Portugal; jorgecpereira599@gmail.com
8 Cambridge Resource Centre for Comparative Genomics, University of Cambridge Department of Veterinary Medicine, Cambridge CB3 0ES, UK; maf12@cam.ac.uk
9 Instituto de Ciências Exatas e Naturais, Universidade Federal do Pará, Belém 66075-110, Brazil
* Correspondence: r.kretschmer@kent.ac.uk
† These authors are joint last authors and contributed equally to this work.

Abstract: The Columbidae species (Aves, Columbiformes) show considerable variation in their diploid numbers (2n = 68–86), but there is limited understanding of the events that shaped the extant karyotypes. Hence, we performed whole chromosome painting (wcp) for paints GGA1-10 and bacterial artificial chromosome (BAC) probes for chromosomes GGA11-28 for *Columbina passerina*, *Columbina talpacoti*, *Patagioenas cayennensis*, *Geotrygon violacea* and *Geotrygon montana*. *Streptopelia decaocto* was only investigated with paints because BACs for GGA10-28 had been previously analyzed. We also performed phylogenetic analyses in order to trace the evolutionary history of this family in light of chromosomal changes using our wcp data with chicken probes and from *Zenaida auriculata*, *Columbina picui*, *Columba livia* and *Leptotila verreauxi*, previously published. G-banding was performed on all these species. Comparative chromosome paint and G-banding results suggested that at least one interchromosomal and many intrachromosomal rearrangements had occurred in the diversification of Columbidae species. On the other hand, a high degree of conservation of microchromosome organization was observed in these species. Our cladistic analysis, considering all the chromosome rearrangements detected, provided strong support for *L. verreauxi* and *P. cayennensis*, *G. montana* and *G. violacea*, *C. passerina* and *C. talpacoti* having sister taxa relationships, as well as for all Columbidae species analyzed herein. Additionally, the chromosome characters were mapped in a consensus phylogenetic topology previously proposed, revealing a pericentric inversion in the chromosome homologous to GGA4 in a chromosomal signature unique to small New World ground doves.

Keywords: birds; doves and pigeons; evolution; genome organization; macrochromosomes; microchromosomes

1. Introduction

Birds have an enigmatic karyotype structured in two chromosomal groups distinguished by size-macrochromosomes (size from ~23 to 200 Mb) and microchromosomes (size from ~3 to 12 Mb), the latter representing the largest number of chromosomes in the karyotype [1–3]. This karyotypic structure is found in most avian species and is estimated to have been maintained since the diapsid common ancestor [4]. However, although a rare event, chromosomal rearrangements do occur and are often driven by breakpoint regions, usually associated with genomic features, including transposable elements and conserved noncoding elements [5].

The order Columbiformes (doves, pigeons and dodos) represents one of the oldest and most diverse extant lineages of birds, including approximately 300 living species [6–8], inhabiting a range of ecological environments in all continents except Antarctica [6]. The traditional taxonomic classification divides the Columbiformes into two families: Raphidae, which includes the dodo and the solitaire, and Columbidae, which includes doves and pigeons [6,9]. However, more recent phylogenetic studies support the inclusion of the dodo and the solitaire into the family Columbidae [7,8,10]. According to Pereira et al. [7], there are three major clades within the family Columbidae: clade A, containing genera from the Old and New World pigeons and doves, clade B includes the small Neotropical ground doves and clade C includes mostly genera found in the Old World (Afro-Eurasian and Australasian), the dodo and the solitaire. In this report, we analyze four genera belonging to clade A (*Geotrygon*, *Streptopelia* and *Patagioenas*) and clade B (*Columbina*).

Cytogenetic studies based on conventional staining has demonstrated that the diploid numbers in Columbiformes species range from 2n = 68 in *Uropelia campestris* to 2n = 86 in the genus *Geotrygon* [11–15]. Molecular cytogenetic characterization with *Gallus gallus* or *Zenaida auriculata* chromosome paints have been performed in five Columbidae species with a typical diploid number (2n = 76–80), and interchromosomal rearrangements were found in only two species [13,15,16]. *Streptopelia roseogrisea* has at least two chromosome fusions between chicken chromosomes 6, 7, 8 and 9, however, the particular chromosomes involved in these fusions could not be identified, due to the similarity in size and morphology of the derivative chromosomes and the use of only one fluorescent dye [13]. In *Leptotila verreauxi* a fusion was detected between chromosomes 6 and 7 [16]. Additionally, whole chromosome painting with *Leucopternis albicollis* probes in four Columbiformes species from different genera (*Leptotila*, *Zenaida*, *Columbina* and *Columba*) show a series of intrachromosomal rearrangements involving the ancestral chromosome 1 in all analyzed species [16]. The microchromosome organization has been analyzed only in *Streptopelia decaocto* and *Columba livia*, using bacterial artificial chromosome (BAC) probes for chicken chromosomes 10–28, and no rearrangements involving microchromosomes were detected [17].

Although substantial progress has been made in terms of understanding karyotype evolution within Columbiformes, it still remains poorly understood, especially with regard to the microchromosomes [13,14,16,17]. Therefore, the purpose of this study was to investigate the chromosome organization in six Columbidae species, representing four different genera, three of them with the typical diploid number (*Columbina*, *Patagioenas* and *Streptopelia*, 2n = 76) and one genus (*Geotrygon*, 2n = 86) with the highest diploid number for the family Columbidae. Furthermore, using chromosomal rearrangements as characters, we constructed a phylogenetic tree in order to compare with previous phylogenies based on molecular approaches. The results provide a comprehensive cytogenetic analysis of Columbiformes species based on molecular cytogenetics for macrochromosomes and microchromosomes and represent novel insights into chromosome evolution of Columbiformes.

2. Material and Methods

2.1. Cell Culture and Chromosome Preparation

The experiments were approved by the Ethics Committee on Animal Experimentation of Universidade Federal do Rio Grande do Sul (CEUA number 30750), and the samplings were authorized by the System of Authorization and Information in Biodiversity (SISBIO, number 33860-1 and 44173-1). Chromosome preparations were established from fibroblast cultures generated from skin biopsies according to Sasaki et al. [18]. The cells lines were established at 37 °C in Dulbecco's Modified Eagle's Medium (DMEM), supplemented with 15% fetal bovine serum, 2% Penicillin Streptomycin, and 1% L-glutamine. Chromosomes suspension were obtained after treatment with colcemid (1 h), hypotonic solution (0.075 M KCl, 15 min) and fixation with 3:1 methanol/acetic acid. The diploid number and chromosome morphology of each individual was determined in at least 20 metaphase chromosomes stained with Giemsa 10% in 0.07 M phosphate buffer, at pH 6.8. The species analyzed and methods performed are listed in Table 1.

Table 1. List of Columbidae species and cytogenetics methods performed in this study. WCP = whole chromosome probes.

Species	Number of Individuals/Sex	2n	Wcp	Micro BACs	G-Banding
Columbina talpacoti	2 M	76	Present study	Present study	Present study
Columbina passerina	1 M	76	Present study	Present study	Present study
Columbina picui	1 M and 1 F	76	[16]	-	Present study
Columba livia	1 M	80	[16]	[17]	Present study
Geotrygon montana	1 M	86	Present study	Present study	Present study
Geotrygon violacea	1 F	86	Present study	Present study	Present study
Leptotila verreauxi	2 M	78	[16]	-	Present study
Patagioenas cayennensis	2 M	76	Present study	Present study	Present study
Streptopelia decaocto	1 F	76	Present study	[17]	Present study
Zenaida auriculata	2 M	76	[16]	-	Present study

M = male, F = female.

2.2. Comparative Chromosome Painting

Chromosome specific paints of *Z. auriculata* (ZAU1-5 and Z) and *G. gallus* (GGA6-10) were generated from flow-sorted chromosomes and amplified by degenerate oligonucleotide-primed polymerase chain reaction (DOP-PCR). The paints were labeled with biotin-16-dUTP or digoxygenin dNTPs during secondary DOP-PCR amplification. Standard techniques were used for denaturation, hybridization, stringency washes and detection using Cy3-streptavidin for biotin-labeled probes or anti-digoxigenin FITC for digoxygenin labeled probes. Briefly, slides were pepsinized for 3 min, washed three times in 2 × SSC (5 min each), dehydrated in an ethanol series (2 min in 70% and 90%, and 4 min in 100% ethanol at room temperature) and incubated for 1 h at 65 °C. Probes were diluted in a hybridization buffer after incubation at 75 °C for 10 min and then pre-annealed at 37 °C for 30 min. Slides were denatured in 70% formamide/2 × SSC solution at 68 °C for 1 min and 20 sec, dehydrated through ethanol series and air-dried (2 min each in ice-cold ethanol 70%, 70%, 85% and 100% ethanol at room temperature). The probe mix was pipetted onto slides and covered with coverslips, sealed with rubber cement and incubated in a humidified chamber at 37 °C for 72 h. After that, they were washed for two times for 5 min in 50% formamide/2 × SSC followed by two times for 5 min in 2 × SSC at 40 °C, incubated three times in 4 × SSC Tween (0.05% Tween) for 5 min at room temperature. Chromosomes were counterstained with DAPI and analyzed using a Zeiss Axioplan2 fluorescence microscope and ISIS software (Metasystems).

Although we used some ZAU probes (ZAU1-5 and Z), the chromosomal comparisons were performed with the GGA homologous chromosomes [16], because most studies with comparative chromosome mapping in birds have concentrated on GGA probes, including the homology to the putative avian ancestral karyotype.

2.3. FISH with BAC Probes

Two BACs, selected from chicken or Zebra finch, were chosen for each of the microchromosomes GGA11-28 (except GGA16) according to O'Connor et al. [17] and applied to the selected species. The BAC clone isolation, amplification and labeling were performed following O'Connor et al. [17].

Chromosome preparations were fixed to slides and dehydrated through an ethanol series (2 min each in 2 × SSC, 70%, 85%, and 100% ethanol at room temperature). Probes were dissolved in a hybridization buffer (Cytocell) with chicken hybloc (Insight Biotech) and applied onto slides before sealing with rubber cement. The probe mix was simultaneously denatured on a 75 °C hotplate (2 min) prior to hybridization in a humidified chamber for 72 h at 37 °C. Slides were washed post-hybridization for 30 s in 2 × SSC w/0.05% Tween 20 at room temperature and counterstained with DAPI. The BACs FISH images were captured using an Olympus BX61 epifluorescence microscope with a cooled CCD camera and SmartCapture (Digital Scientific UK) system.

2.4. G-Banding

The G-banding was performed for ten Columbidae species with a combination of DAPI and propidium iodide [19] in order to detect intrachromosomal rearrangements not observed by chromosome painting. Images were captured by an Olympus BX61 epifluorescence microscope with a cooled CCD camera and SmartCapture (Digital Scientific UK) system. Afterward, the images were converted to grayscale using Corel Photo-Paint 2019.

2.5. Phylogenetic Analysis

A binary matrix of 28 discrete chromosomal rearrangements was conducted following Dobigny et al. [20]. A Maximum Parsimony (MP) tree was performed using PAUP 4.0b10 program [21]. The chromosomal rearrangements were established using the chicken chromosome painting (GGA 1–10) results obtained herein and from literature data [13,16] and the comparative chromosome analysis using G-banding. A heuristic search to find the most parsimonious tree(s) was performed using Tree Bisection Reconnection (TBR) branch-swapping. The bootstrap probability was performed with one thousand replicates, using chicken as an outgroup.

Additionally, the synapomorphic characters obtained in the MP analysis were also mapped onto a well-supported molecular phylogeny tree proposed by Pereira et al. [7], in order to verify which chromosomal characters give support to each branch in that topology. However, we only considered the chromosomal rearrangements detected by chromosome painting using chicken probes and G-banding since they were used in all Columbidae species analyzed so far.

3. Results

3.1. Macrochromosome Organization (GGA1-10)

Chromosomal homologies were examined among six Columbidae species by whole chromosome painting with *G. gallus* and *Z. auriculata* chromosome-specific DNA paints. Representative results of FISH experiments are shown in Figure 1 and detailed below.

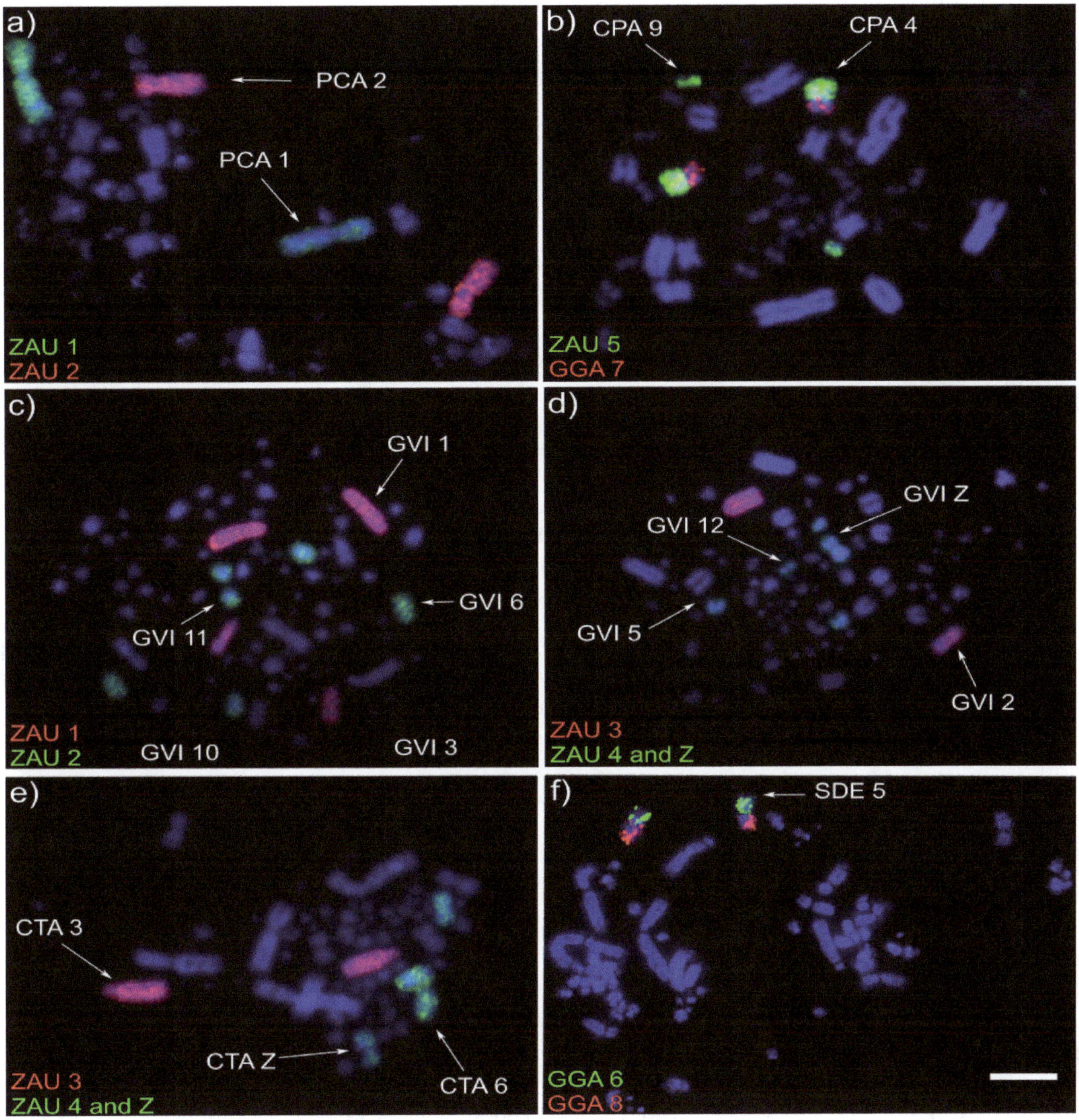

Figure 1. Representative FISH results using different sets of chromosome probes from *Gallus gallus* (GGA) and *Zenaida auriculata* (ZAU) probes on chromosomes of different Columbidae species: (**a**) *Patagioenas cayennensis* (PCA), (**b**) *Columbina passerina* (CPA), (**c**) *Geotrygon violacea* (GVI), (**d**) *Geotrygon violacea* (GVI), (**e**) *Columbina talpacoti* (CTA), (**f**) *Streptopelia decaocto* (SDE). The chromosome probes used are indicated on the left bottom, in green (fluoroscein labeled) or red (biotin-cy3 labeled). Scale bar 10 µm.

All GGA paints produced identical hybridization patterns in *Columbina talpacoti* (CTA) and *Columbina passerina* (CPA). GGA1–3, 9 and 10 paints each hybridized to a single chromosome pair, while paints GGA4 and 5 hybridized to two chromosomes pairs. One segment of GGA5 produced signals in the same chromosome pair as GGA7 (CPA4 and CTA4), while paints GGA6 and GGA8 hybridized to CPA5 and CTA5, revealing the occurrence of centric fusions. Figure 2 shows the homology between the *G. gallus* and *C. talpacoti* and *C. passerina*.

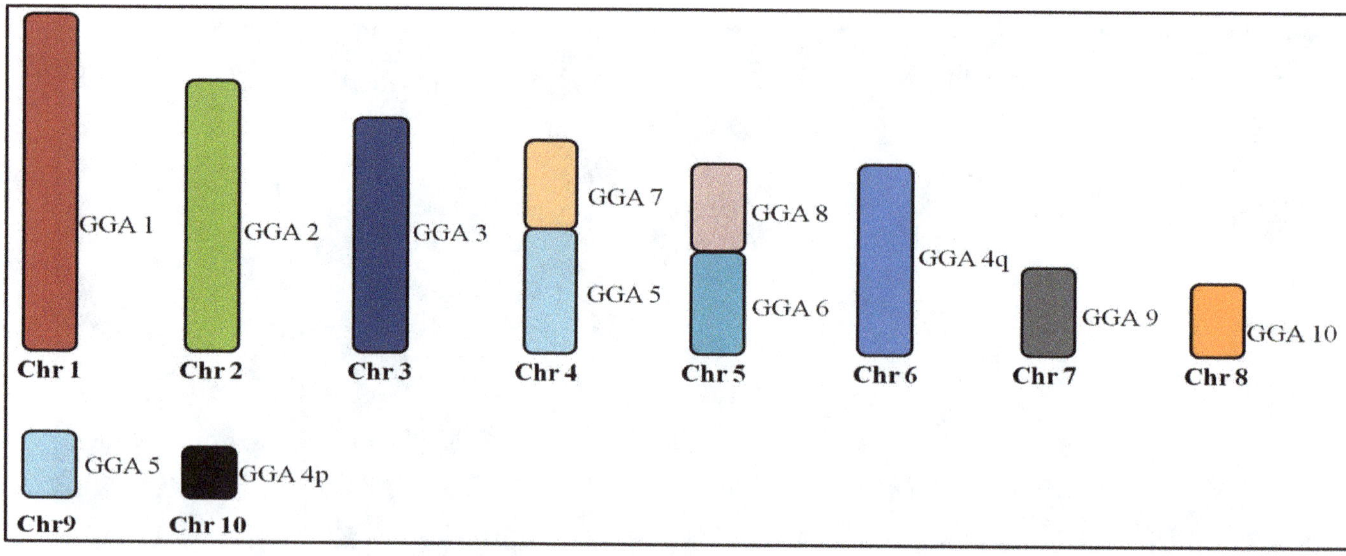

Figure 2. Homologous chromosomal segments of *Gallus gallus* (GGA) in *Columbina talpacoti* and *Columbina passerina* macrochromosomes as detected by fluorescence in situ hybridization (FISH) using GGA and *Zenaida auriculata* whole chromosome paints. Chr = Chromosome.

Hybridization of chicken paints GGA1–3 and 5, 8–10, each hybridized to a single chromosome pair in Patagioenas cayennensis (PCA). On the other hand, paint 4 hybridized to two chromosome pairs, while paints GGA6 and 7 were associated in one pair (PCA4), indicating the occurrence of centric fusion. Figure 3 shows the homology between the *G. gallus* and *P. cayennensis*.

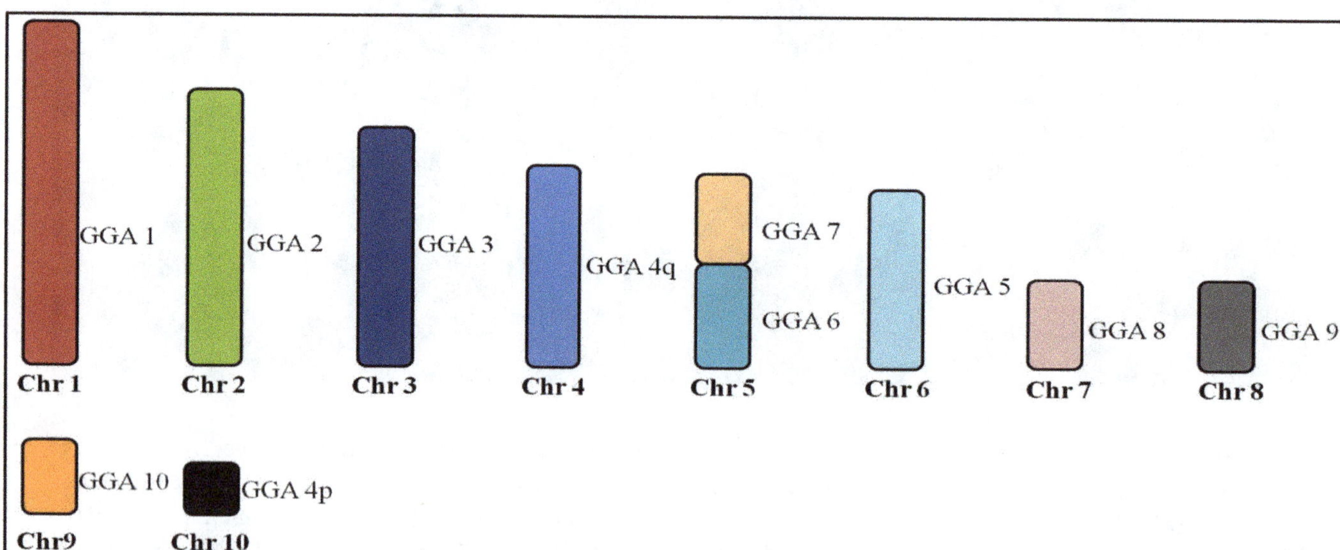

Figure 3. Homologous chromosomal segments of *Gallus gallus* (GGA) in *Patagioenas cayennensis* macrochromosomes as detected by fluorescence in situ hybridization (FISH) using GGA and *Zenaida auriculata* whole chromosome paints.

The same hybridization pattern of the chicken probes was observed in *Geotrygon violacea* (GVI) and *Geotrygon montana* (GMO). GGA3, 5–10 probes each hybridized to a single chromosome pair, while paint GGA1 hybridized to two chromosome pairs, and paints GGA2 and 4 hybridized to three chromosomes pairs. Figure 4 shows the homology between the *G. gallus* and *G. violacea* and *G. montana*.

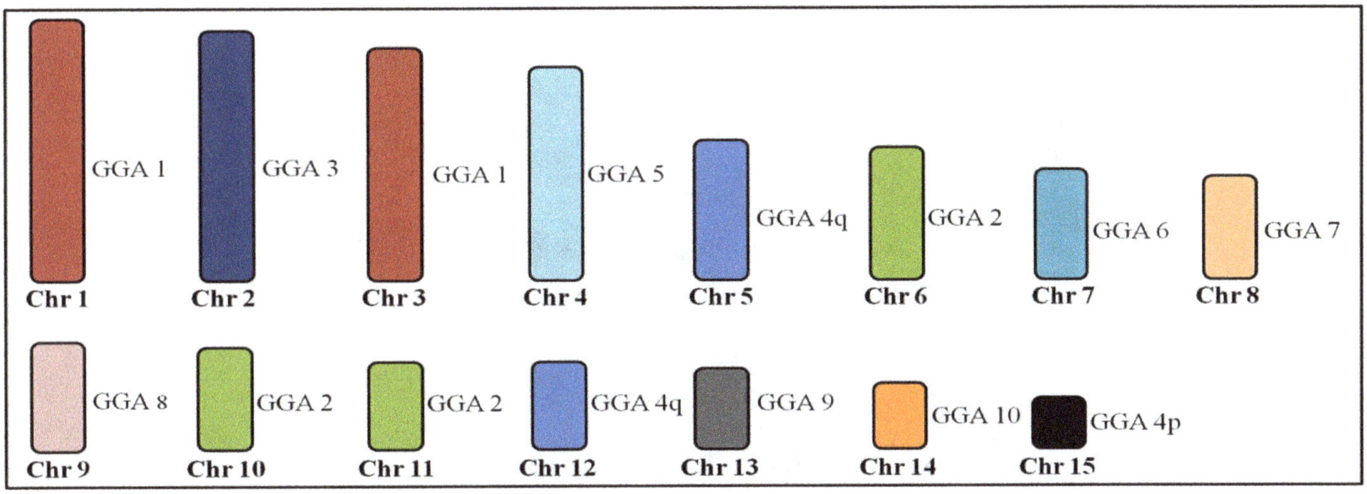

Figure 4. Homologous chromosomal segments of *Gallus gallus* (GGA) in *Geotrygon violacea* and *Geotrygon montana* macrochromosomes as detected by fluorescence in situ hybridization (FISH) using GGA and *Zenaida auriculata* whole chromosome paints.

In *S. decaocto* (SDE), chicken chromosome paints 1–3 and 5, each hybridized to a single chromosome pair, while paints GGA4 hybridized to two chromosome pairs. Chicken paints GGA6–9 confirmed that these chromosomes are involved in two fusions, as observed in *Streptopelia roseogrisea* [12]. However, here the fusions involving GGA6/8 and GGA7/9 were identified. It is likely that these fusions are also present in *S. roseogrisea*, since both have similar karyotypes and are considered sister species. Figure 5 shows the homology between the *G. gallus* and *S. decaocto*.

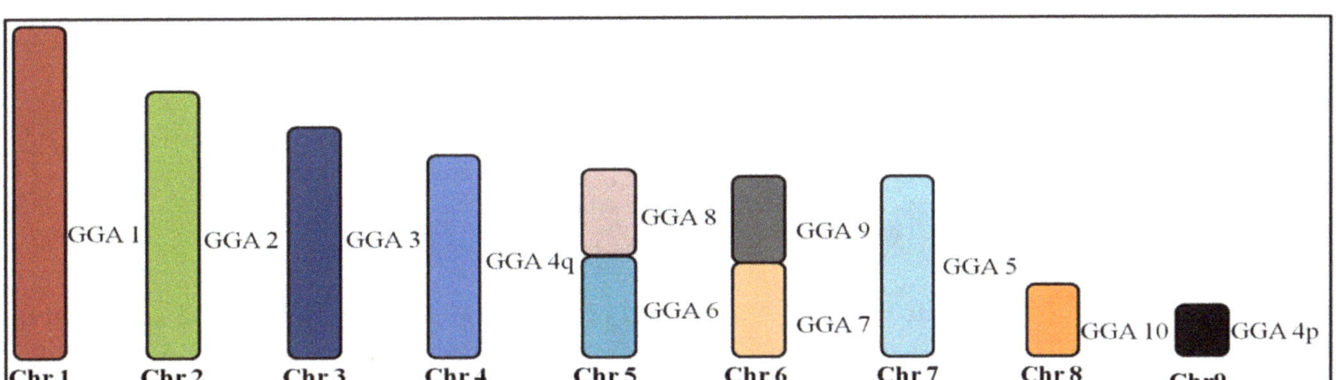

Figure 5. Homologous chromosomal segments of *Gallus gallus* (GGA) in *Streptopelia decaocto* macrochromosomes as detected by fluorescence in situ hybridization (FISH) using GGA and *Zenaida auriculata* whole chromosome paints.

3.2. Microchromosome Organization (GGA11-28, Except GGA16)

Results for *C. talpacoti*, *C. passerina*, *G. violacea*, *G. montana* and *P. cayennensis* provide no evidence of interchromosomal rearrangements in the microchromosomes in any of the five species here analyzed. All chicken microchromosome BACs GGA11-28 (except GGA16) were efficiently cross-hybridized to chromosomes of all five Columbidae species, revealing the ancestral microchromosomal pattern, similar to results observed in *S. decaocto* and *C. livia*, previously performed by O'Connor et al. [17]. Examples of the BACs FISH results are demonstrated in Figure 6 for chromosome 26 for all species.

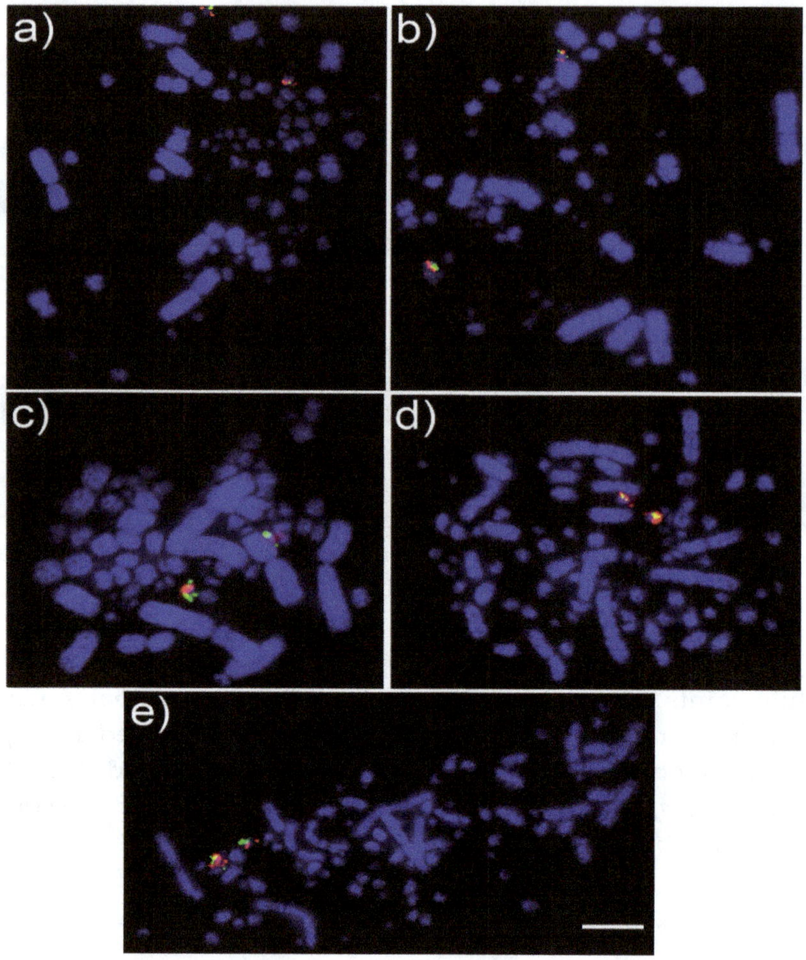

Figure 6. FISH using BACs for chicken chromosome 26 (CH261-186M13 FITC and CH261-170L23 Texas Red) on chromosomes of different Columbidae species revealing no evidence of interchromosomal rearrangements: (**a**) *Columbina talpacoti,* (**b**) *Patagioenas cayennensi,* (**c**) *Columbina passerina,* (**d**) *Geotrygon violacea,* (**e**) *Geotrygon montana.* Scale bar 10 μm.

3.3. G-Banding

G-banding patterns generated with a combination of DAPI and propidium iodide (Supplementary Figure S1) in ten Columbidae species and chicken were used to detect chromosomal homology between these species after visual inspection (Supplementary Table S1).

3.4. Phylogenetic Analysis

The data on chromosome painting and G-banding obtained from ten species of Columbiformes and the outgroup (*G. gallus*) were used to generate a matrix with 28 discrete chromosome characters (Supplementary Table S1 and Supplementary Figure S2). Only macrochromosomes were used in this analysis since no interchromosomal rearrangement involving microchromosomes was observed in the Columbidae species. The MP analysis resulted in 5 most parsimonious trees (tree length = 35, consistence index = 0.800, retention index = 0.6957 and homoplasy index = 0.200) (Supplementary Figure S3). Overall, 16 characters were parsimoniously informative. Despite the fact that Columbiformes correspond to a monophyletic group supported by six synapomorphies and a bootstrap value of 100, the species form an unresolved polytomy (Figure 7). However, three monophyletic clades, supported by high bootstrap values, are observed among this politomy: *L. verreauxi* and *P. cayennensis*, *G. montana* and *G. violacea*, *C. passerina* and *C. talpacoti*.

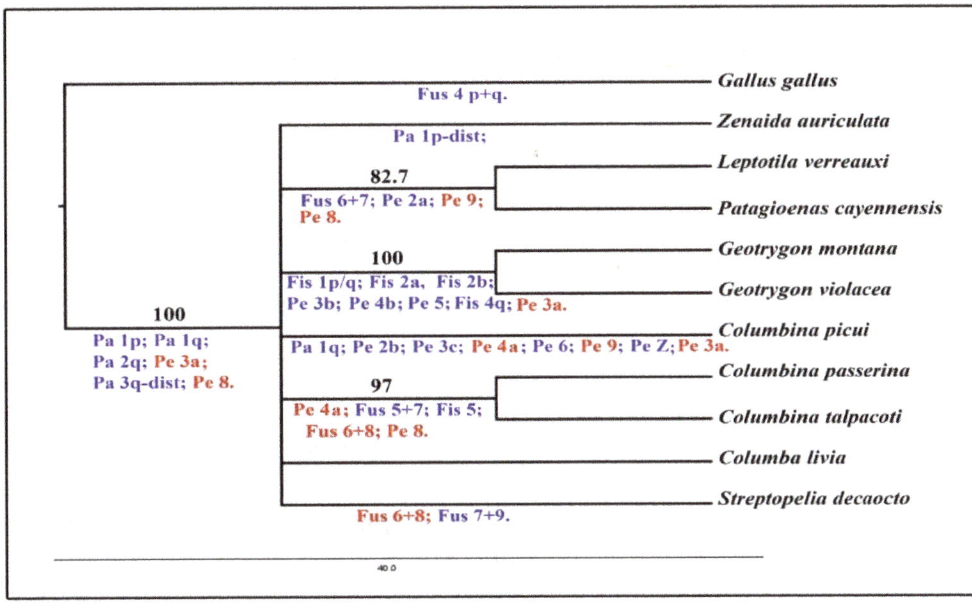

Figure 7. Phylogenetic analysis of maximum parsimony using PAUP based on chromosome rearrangements present in Columbidae species according to results obtained by whole chromosome painting with *Gallus gallus* and *Zenaida auriculata* probes and G-banding. All rearrangement characters are mapped a posteriori and are shown below the branches. Characters in red indicate chromosome rearrangements shared for more than one branch (homoplasic characters). Black numbers above the branch indicate the bootstrap value with one thousand replicates.

4. Discussion

In this study, we have described the karyotypes of six Columbidae species (*C. talpacoti*, *C. passerina*, *P. cayennensis*, *G. violacea*, *G. montana* and *S. decaocto*) using *G. gallus* and *Z. auriculata* chromosome painting and have performed an integrative analysis with previously published maps of another four species (*Z. auriculata*, *L. verreauxi*, *C. livia* and *C. picui*) [16]. The whole chromosome painting performed here demonstrated that each Columbidae species showed at least one interchromosomal rearrangement involving macrochromosomes when compared with *G. gallus*, as observed previously in *L. verreauxi*, but in contrast to *Z. auriculata*, *C. livia* and *C. picui*, in which no interchromosomal rearrangements were found. However, the G-banding results (10 species) indicated that intrachromosomal rearrangement is the main driver of chromosome evolution in Columbidae species, being evident in all Columbidae species, even in species without interchromosomal rearrangements (i.e., *Z. auriculata*, *C. livia* and *C. picui*). On the other hand, we observed a high degree of genome stability in the microchromosomes of all these species.

4.1. Macrochromosome Organization

Typical karyotypes are found in *C. talpacoti* (2n = 76), *C. passerina* (2n = 76), *P. cayennensis* (2n = 76) and *S. decaocto* (2n = 76), since 61.3% of birds have 2n = 76–82. However, atypical karyotypes are found in *Geotrygon* species because only 1.2% of birds have 2n = 86 [22]. There are clear differences in chromosomal morphologies among the *Geotrygon* species in relation to the other Columbidae species [15]. While Columbidae species generally have between five and eight biarmed chromosomes, all the autosomal chromosomes of *Geotrygon* species are telocentric [15]. The comparative chromosome painting performed in this study has brought to light the extent of evolutionary karyotype organization among the different species of the Columbidae family.

Whole chromosome paints derived from GGA1-10 produced identical results in *C. talpacoti* and *C. passerina*, including a fusion between GGA6/GGA8. Interchromosomal fusions are exceptionally rare in birds, and evidence of fusion between GGA6/GGA8 has only been previously described in three not-closely related species, *Tetrao urogallus* (Galliformes), *Falco columbarius* (Falconiformes) and *Opisthocomus hoazin* (Opisthocomiformes) [23–25], hence, this fusion in *C. talpacoti* and *C. passerina* likely occurred in their common ancestor. This hypothesis is reinforced by the absence of this fusion in other species of the same genus, *C. picui* [16]. Apart from the GGA6/GGA8 fusion, we have found the fission of ancestral chromosome 5 (GGA5) and a fusion between a segment of GGA5 with GGA7 in *C. talpacoti* and *C. passerina*. These two rearrangements were also not found in *C. picui*, reinforcing the phylogenetic proximity between *C. talpacoti* and *C. passerina* [26]. To our knowledge, the association between GGA5 and GGA7 is a rare event in birds, having been reported only in *Melopsittacus undulatus* (Psittaciformes) in this species, however, these chromosomes are also fused with segments from other chromosomes [27].

The chromosome complement of *P. cayennensis* is derived from the ancestral karyotype (2n = 80) by a centric fusion between GGA6/7, which was recently found in *L. verreauxi* [16]. Although this rearrangement may have a common origin in both species, we cannot discard the possibility of convergent evolution, since it seems to be one of the most common associations in birds, being found also in Galliformes [23], Gruiformes [28], Strigiformes [29], Trogoniformes [30] and Psittaciformes [27,31].

The karyotypes of *G. violacea* and *G. montana* are derived from the putative ancestral avian karyotype by fissions of the chromosomes 1, 2 and 4. This fact explains the higher diploid number observed in *G. violacea* and *G. montana*, with 2n = 86, compared to other Columbidae species (around 2n = 76).

Results of chromosome painting shown here demonstrate that the fusions between GGA6/8 and GGA7/9 are present in *S. decaocto*, and are probably also present in *S. roseogrisea*, because both species are considered to be sister species [26] and present similar karyotypes [13]. Besides that, the fusion between GGA7/GGA9 in *S. decaocto* was detected for the first time in any avian species and may be a synapomorphic trait for the genus *Streptopelia*. In contrast to *S. roseogrisea*, we did not find the fusion between GGA4q and GGA4p in *S. decaocto*.

4.2. Conservation of Microchromosome Organization

No interchromosomal rearrangements involving microchromosome pairs GGA11-28 (GGA16 not tested) were found in the species analyzed, as previously observed in two other Columbidae species, *C. livia* and *S. decaocto* [17]. The lack of interchromosomal rearrangement observed in the Columbidae species corroborates the possible evolutionary advantage of retaining this pattern of microchromosome organization, as previously proposed [17].

4.3. Phylogenetic Relationships in the Family Columbidae

Our chromosomal phylogeny gives strong support for the monophyly of doves and pigeons, with high bootstrap value (Figure 7). This ancient branch received 100% of bootstrap support, having six chromosome characters supporting the basal position for the group (Figure 7). This data corroborates previous molecular phylogenies that also demonstrate the monophyly of the group with strong support [7,8]. Despite the chromosome similarities of Columbiformes and the outgroup *G. gallus*, the main rearrangements in Columbidae species are intrachromosomal rearrangements, as revealed by G-banding.

However, we could not resolve the topology of the Columbidae family using chromosomal rearrangements. After the divergence of the basal branch of the Columbiformes, we found a polytomy, in which *Z. auriculata*, *C. picui*, *C. livia* and *S. decaocto* do not group with any other species. This is a commonly observed scenario in chromosomal phylogenies, especially in birds in which the karyotype changes occur at a low rate when compared with other groups [32,33]. Furthermore, the diversification of pigeon and doves occurred in the late Oligocene and continued to diversify into Miocene around 24.7 Mya [8], and hence the common ancestor of the main lineages may not have been present long enough to accumulate chromosome differences detectable by G-banding or chromosome painting.

On the other hand, three clades, formed by *L. verreauxi* and *P. cayennensis*, *C. talpacoti* and *C. passerina*, and *G. montana* and *G. violacea* were observed in our consensus tree (Figure 7). Within the Holarctic doves, *Geotrygon* species (*G. violacea* and *G. montana*) presented the most derived karyotypes, resulting in seven synapomorphies supporting this clade (fission on GGA1, two fission on GGA2, pericentric inversion on GGA 3, pericentric inversion on GGA 4, pericentric inversion on GGA 5 and GGA4q fission). Another clade was formed by *L. verreauxi* and *P. cayennensis*, supported by three derived characters (two synapomorphies, pericentric inversion on GGA2 and GGA9 and one homoplasy-fusion between GGA6/7). This clade is unusual when we compare it with consensual topologies based on DNA sequence, which usually support the close affinities of the genus *Geotrygon*, *Zenaida* and *Leptotila* allied with another branch containing genus *Patagioenas*, *Streptopelia* and *Columba* [7,8,10].

In the three species analyzed of New World ground doves, only two species group together, with five characters supporting *C. talpacoti* and *C. passerina* as sister groups (pericentric inversion on GGA 4 and 8, fusion of GGA 6/8 and GGA5/7, and fission of GGA5). The species *C. picui* does not group with other species of the genus, probably due to the loss of some chromosome characters or its basal position in previous phylogenies [34]. Hence, the fusion between GGA 5/7 and the fission in GGA5 may have arisen only in the common ancestor of *C. talpacoti* and *C. passerina*. The pericentric inversion on GGA4 is shared among *Columbina* species analyzed herein and could be a synapomorphy for the genus. However, despite the highly reshuffled karyotype of *C. picui* and due to the accumulated homoplasic characters, the affinity of the genus *Columbina* was not retrieved through parsimony analyses. Hence, future studies are necessary to reconstruct the chromosomal history of Columbiformes using more taxa (including closely related outgroups) and other methodologies in order to test conflicting homologies and to increase the number of informative characters.

An interesting approach in the investigation of the karyotype evolution of the Columbidae family is plotting the chromosomal rearrangements in a well-resolved phylogeny, such as proposed by Pereira et al. [7], using nuclear and mitochondrial DNA genes (Figure 8). For instance, these authors recovered three major clades within the Columbiformes: clade A, with genera from Old and New World pigeons and doves, clade B with small New World ground doves and clade C with genera found mostly in the Old World. However, chromosomal characters did not give support to clade A, although data from other Columbidae species are still required for a more thorough appreciation of the role of chromosomal rearrangements in this clade. On the other hand, a pericentric inversion in the chromosome homologous to GGA4 (character 15) supported clade B, which includes a group of small ground dove species.

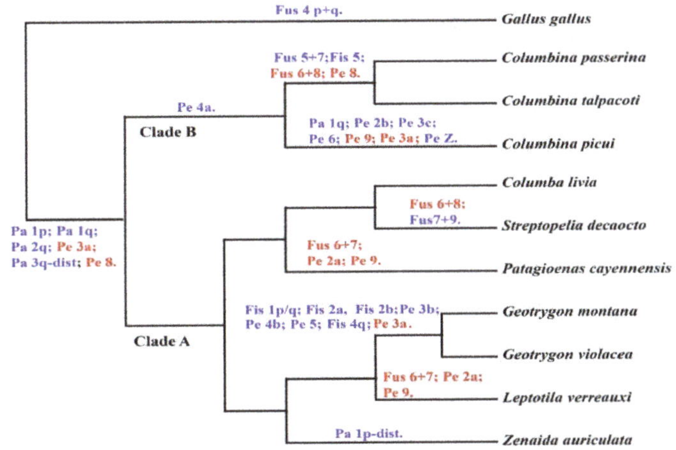

Figure 8. Chromosomal rearrangements in Columbidae species detected by chromosome painting using *Gallus gallus* and *Zenaida auriculata* probes and G-banding plotted in a molecular phylogeny [2]. Characters in red indicate chromosome rearrangements shared for more than one branch (homoplasic characters). Published chromosome painting data from Kretschmer et al. [10] and the data obtained in this study were used for this figure.

The fusion between GGA6/GGA8 may have a common origin in *C. talpacoti* and *C. passerina*, since this fusion is not common in birds [2,25]. However, if the phylogeny proposed by Pereira et al. [7] is correct, this fusion is likely to be a convergent chromosome rearrangement in *S. decaocto* and in *C. talpacoti* and *C. passerina*. In the same way, the pericentric inversion on GGA2 and GGA9 and the fusion between GGA6/7 also would represent a convergent chromosome rearrangement in *L. verreauxi* and *P. cayennensis*.

5. Conclusions

In summary, out of 10 Columbidae species analyzed by chromosome painting so far (including the six species analyzed here), we present the first report of chromosome fission in Columbidae species, i.e., *C. talpacoti*, *C. passerina*, *G. violacea* and *G. montana*. Chromosomal fusions were observed in *C. talpacoti*, *C. passerina*, *P. cayennensis* and *S. decaocto*, and previously in *L. verreauxi* and *S. roseogrisea*. In view of the conservation of microchromosome organization in the Columbidae species, we suggest that the main forces in the chromosome evolution in Columbidae species are interchromosomal rearrangements involving macrochromosomes and also intrachromosomal rearrangements, as observed in all Columbidae species analyzed here by G-banding.

The comparisons of the chromosome organizations presented here allow us to speculate on the process of karyotype evolution and the overall picture of the phylogeny in several members of the family Columbidae. Although *C. talpacoti*, *C. passerina*, *P. cayennensis* and *S. decaocto* have undergone interchromosomal rearrangements in the macrochromosomes, the typical 'avian-like' karyotype (~80 chromosomes) has been maintained in these species. However, due to interchromosomal fissions in the macrochromosomes, atypical karyotypes (2n = 86) were found in *G. violacea* and *G. montana*. Nevertheless, taking into account the possibility of a high percentage of convergent chromosome rearrangement observed in Columbidae species, we could not resolve the phylogeny of Columbidae members using chromosome rearrangements. Considering the high frequency of intrachromosomal rearrangements in avian genome evolution [35] and in Columbidae [16,36], we believe that the key to reconstructing the evolutionary history of the Columbiformes must be in the intrachromosomal rearrangements. Hence, future strategies that expand the ability to detect smaller rearrangements in macrochromosomes, for example, using BACs probes to each macrochromosomes, are needed to increase our knowledge about karyotype evolution and to resolve the phylogeny using chromosome rearrangements. Furthermore, molecular cytogenetic studies in species from clade C are necessary to improve our knowledge about the direction of chromosomal changes in all three clades in the Columbidae family.

Supplementary Materials

Figure S1: G-banded macrochromosomes of *Gallus gallus* (GGA), ZAU, *Zenaida auriculata*; CPA, *Columbina passerina*; CTA, *Columbina talpacoti*; CPI, *Columbina picui*; LVE, *Leptotila verreauxi*; CLI, *Columba livia*; SDE *Streptopelia decaocto*; PCA, *Patagioenas cayennensis*; GMO, *Geotrygon montana*, and; GVI, *Geotrygon violacea*. Figure S2: G-banded macrochromosomes of *Gallus gallus* (GGA), ZAU, *Zenaida auriculata*; CPA, *Columbina passerina*; CTA, *Columbina talpacoti*; CPI, *Columbina picui*; LVE, *Leptotila verreauxi*; CLI, *Columba livia*; SDE *Streptopelia decaocto*; PCA, *Patagioenas cayennensis*; GMO, *Geotrygon montana*, and; GVI, *Geotrygon violacea* indicating the chromosome characters used in the phylogenetic analysis. An asterisk indicates a heteromorphism in GMO1 and 3. Figure S3: The five most parsimonious tree (A–E) resulted in the maximum parsimony analysis (MP) using PAUP based on chromosome rearrangements present in Columbidae species according to results obtained by whole chromosome painting with *Gallus gallus* and *Zenaida auriculata* probes and G-banding. All rearrangement characters were mapped a posteriori and are shown below the branches. Characters in red indicate chromosome rearrangements shared for more than one branch (Homoplasic characters). Table S1: Binary Matrix of chromosome characters obtained by chromosome painting and G-band. *Gallus gallus* (GGA), ZAU, *Zenaida auriculata*; CPA, *Columbina passerina*; CTA, *Columbina talpacoti*; CPI, *Columbina picui*; LVE, *Leptotila verreauxi*; CLI, *Columba livia*; SDE *Streptopelia decaocto*; PCA, *Patagioenas cayennensis*; GMO, *Geotrygon montana*, and; GVI, *Geotrygon violacea*.

Author Contributions: R.K. conceived and designed the experiments; R.K. analyzed the data and wrote the manuscript with contributions of I.d.O.F. and A.J.B.G.; R.K., I.d.O.F., A.d.V.G. and R.J.G. collected the specimens; R.K. was responsible for the classical cytogenetics; R.K. and L.G.K. were responsible for the FISH experiments; A.J.B.G. performed the phylogenetic analyses; J.C.P. and M.A.F.-S. were responsible for *Gallus gallus* whole chromosome probes generation; J.C.P., R.K. and M.A.F.-S. were responsible for *Zenaida auriculata* whole chromosome probes generation; R.E.O., T.R.O.d.F. and D.K.G. coordinated the research; Funding acquisition E.H.C.d.O., A.d.V.G., R.J.G., M.A.F.-S., R.E.O., T.R.O.d.F. and D.K.G. All authors corrected, revised and discussed the data. All authors have read and agreed to the published version of the manuscript.

Acknowledgments: We are grateful to Conselho Nacional de Desenvolvimento Científico e Tecnológico (CNPq) and the Coordenação de Aperfeiçoamento de Pessoal de Nível Superior (CAPES) by the scholarships to R. Kretschmer and Sistema de Autorização e Informação em Biodiversidade (SISBIO) for the authorization to sample the specimens studied in this manuscript.

References

1. Tegelstrom, H.; Ryttman, H. Chromosomes in birds (Aves): Evolutionary implications of macro- and microchromosome numbers and lengths. *Hereditas* **1981**, *94*, 225–233. [CrossRef]
2. Kretschmer, R.; Ferguson-Smith, M.A.; de Oliveira, E.H.C. Karyotype evolution in birds: From conventional staining to chromosome painting. *Genes* **2018**, *9*, 181. [CrossRef]
3. Hillier, L.W.; Miller, W.; Birney, E.; Warren, W.; Hardison, R.C.; Ponting, C.P.; Bork, P.; Burt, D.W.; Groenen, M.A.; Delany, M.E.; et al. Sequence and comparative analysis of the chicken genome provide unique perspectives on vertebrate evolution. *Nature* **2004**, *432*, 695–716.
4. O'Connor, R.E.; Romanov, M.N.; Kiazim, L.G.; Barrett, P.M.; Farré, M.; Damas, J.; Ferguson-Smith, M.; Valenzuela, N.; Larkin, D.M.; Griffin, D.K. Reconstruction of the diapsid ancestral genome permits chromosome evolution tracing in avian and non-avian dinosaurs. *Nat. Commun.* **2018**, *9*, 1883. [CrossRef] [PubMed]
5. Farré, M.; Narayan, J.; Slavov, G.T.; Damas, J.; Auvil, L.; Li, C.; Jarvis, E.D.; Burt, D.W.; Griffin, D.K.; Larkin, D.M. Novel Insights Into Chromosome Evolution in Birds, Archosaurs, and Reptiles. *Genome Biol. Evol.* **2016**, *8*, 2442–2451. [CrossRef] [PubMed]
6. Gibbs, D.; Barnes, E.; Cox, J.D. *Pigeons and Doves: A Guide to the Pigeons and Doves of the World*; Pica Press: Mountfield, UK, 2001.
7. Pereira, S.L.; Johnson, K.P.; Clayton, D.H.; Baker, A.J. Mitochondrial and nuclear DNA sequences support a cretaceous origin of Columbiformes and a dispersal driven radiation in the paleogene. *Syst. Biol.* **2007**, *56*, 656–672. [CrossRef]
8. Soares, A.E.R.; Novak, B.J.; Haile, J.; Heupink, T.H.; Fjeldså, J.; Gilbert, M.T.P.; Poinar, H.; Church, G.M.; Shapiro, B. Complete mitochondrial genomes of living and extinct pigeons revise the timing of the columbiform radiation. *BMC Evol. Biol.* **2016**, *16*, 230. [CrossRef] [PubMed]
9. Sibley, C.G.; Ahlquist, J.E. *Phylogeny and Classification of Birds: A Study in Molecular Evolution*; Yale University Press: New Haven, CT, USA, 1990.
10. Shapiro, B.; Sibthorpe, D.; Rambaut, A.; Austin, J.; Wragg, G.M.; Bininda-Emonds, O.R.; Lee, P.L.; Cooper, A. Flight of the dodo. *Science* **2002**, *295*, 1683. [CrossRef] [PubMed]
11. De Lucca, E.J.; de Aguiar, M.L.R. Chromosomal evolution in Columbiformes (Aves). *Caryologia* **1976**, *29*, 59–68. [CrossRef]
12. De Lucca, E.J. Chromosomal evolution of South American Columbiformes (Aves). *Genetica* **1984**, *62*, 177–185. [CrossRef]
13. Guttenbach, M.; Nanda, I.; Feichtinger, W.; Masabanda, J.S.; Griffin, D.K.; Schmid, M. Comparative chromosome painting of chicken autosomal paints 1–9 in nine different bird species. *Cytogenet. Genome Res.* **2003**, *103*, 173–184. [CrossRef] [PubMed]
14. Derjusheva, S.; Kurganova, A.; Haberman, F.; Gaginskaia, E. High chromosome conservation detected by comparative chromosome painting in chicken, pigeon and passerine birds. *Chromosome Res.* **2004**, *12*, 715–723. [CrossRef] [PubMed]
15. Kretschmer, R.; de Oliveira, T.D.; Furo, I.O.; Silva, F.A.O.; Gunski, R.J.; Garnero, A.D.V.; Cioffi, M.B.; de Oliveira, E.H.C.; de Freitas, T.R.O. Repetitive DNAs and shrink genomes: A chromosomal analysis in nine Columbidae species (Aves, Columbiformes). *Genet. Mol. Biol.* **2018**, *41*, 98–106. [CrossRef]

16. Kretschmer, R.; Furo, I.O.; Gunski, R.J.; Garnero, A.D.V.; Pereira, J.C.; O'Brien, P.C.M.; Ferguson-Smith, M.A.; de Oliveira, E.H.C.; de Freitas, T.R.O. Comparative chromosome painting in Columbidae (Columbiformes) reinforces divergence in Passerea and Columbea. *Chromosome Res.* **2018**, *26*, 211–223. [CrossRef] [PubMed]
17. O'Connor, R.E.; Kiazim, L.; Skinner, B.; Fonseka, G.; Joseph, S.; Jennings, R.; Larkin, D.M.; Griffin, D.K. Patterns of microchromosome organization remain highly conserved throughout avian evolution. *Chromosoma* **2019**, *128*, 21–29. [CrossRef] [PubMed]
18. Sasaki, M.; Ikeuchi, T.; Maino, S. A feather pulp culture for avian chromosomes with notes on the chromosomes of the peafowl and the ostrich. *Experientia* **1968**, *24*, 1923–1929. [CrossRef]
19. Joseph, S.; O'Connor, R.E.; Al Mutery, A.F.; Watson, M.; Larkin, D.M.; Griffin, D.K. Chromosome Level Genome Assembly and Comparative Genomics between Three *Falcon* Species Reveals an Unusual Pattern of Genome Organisation. *Diversity* **2018**, *10*, 113. [CrossRef]
20. Dobigny, G.; Ozouf-Costaz, C.; Bonillo, C.; Volouev, V. Viability of X-autosome translocations in mammals: An epigenomic hypothesis from a rodent case-study. *Chromosoma* **2004**, *113*, 34–41. [CrossRef]
21. Swoford, D.L. *PAUP*. Phylogenetic Analysis Using Parsimony (*and Other Methods)*; Version 4; Sinauer Associates: Sunderland, MA, USA, 2002.
22. Degrandi, T.M.; Barcelos, S.A.; Costa, A.L.; Garnero, A.D.V.; Hass, I.; Gunski, R.J. Introducing the Bird Chromosome Database: An overview of cytogenetic studies on birds. *Cytogenet. Genome Res.* **2020**, in press. [CrossRef]
23. Shibusawa, M.; Nishibori, M.; Nishida-Umehara, C.; Tsudzuk, M.; Masaband, J.; Griffin, D.K.; Matsuda, Y. Karyotypic evolution in the Galliformes: An examination of the process of karyotypic evolution by comparison of the molecular cytogenetic findings with the molecular phylogeny. *Cytogenet. Genome Res.* **2004**, *106*, 111–119. [CrossRef]
24. Nishida, C.; Ishijima, J.; Kosaka, A.; Tanabe, H.; Habermann, F.A.; Griffin, D.K.; Matsuda, Y. Characterization of chromosome structures of Falconinae (Falconidae, Falconiformes, Aves) by chromosome painting and delineation of chromosome rearrangements during their differentiation. *Chromosome Res.* **2008**, *16*, 171–181. [CrossRef] [PubMed]
25. Dos Santos, M.S.; Furo, I.O.; Tagliarini, M.M.; Kretschmer, R.; O'Brien, P.C.M.; Ferguson-Smith, M.A.; de Oliveira, E.H.C. The Karyotype of the Hoatzin (*Opisthocomus hoazin*)—A Phylogenetic Enigma of the Neornithes. *Cytogenet. Genome Res.* **2018**, *156*, 158–164. [CrossRef] [PubMed]
26. Lapiedra, O.; Sol, D.; Carranza, S.; Beaulieu, J.M. Behavioural changes and the adaptive diversification of pigeons and doves. *Proc. R. Soc. B* **2013**, *280*, 20122893. [CrossRef] [PubMed]
27. O'Connor, R.E.; Farré, M.; Joseph, S.; Damas, J.; Kiazim, L.; Jennings, R.; Bennett, S.; Slack, E.A.; Allanson, E.; Larkin, D.M.; et al. Chromosome-level assembly reveals extensive rearrangement in saker falcon and budgerigar, but not ostrich, genomes. *Genome Biol.* **2018**, *19*, 171. [CrossRef]
28. Nanda, I.; Benisch, P.; Fetting, D.; Haaf, T.; Schmid, M. Synteny conservation of chicken macrochromosomes 1–10 in different Avian lineages revealed by cross-species chromosome painting. *Cytogenet. Genome Res.* **2011**, *132*, 165–181. [CrossRef]
29. De Oliveira, E.H.; de Moura, S.P.; dos Anjos, L.J.; Nagamachi, C.Y.; Pieczarka, J.C.; O'Brien, P.C.; Ferguson-Smith, M.A. Comparative chromosome painting between chicken and spectacled owl (*Pulsatrix perspicillata*): Implications for chromosomal evolution in the Strigidae (Aves, Strigiformes). *Cytogenet. Genome Res.* **2008**, *122*, 157–162. [CrossRef]
30. Degrandi, T.M.; Garnero, A.D.V.; O'Brien, P.C.M.; Ferguson-Smith, M.A.; Kretschmer, R.; de Oliveira, E.H.C.; Gunski, R.J. Chromosome Painting in *Trogon s. surrucura* (Aves, Trogoniformes) Reveals a Karyotype Derived by Chromosomal Fissions, Fusions, and Inversions. *Cytogenet. Genome Res.* **2017**, *151*, 208–215.
31. Nanda, I.; Karl, E.; Griffin, D.K.; Schartl, M.; Schmid, M. Chromosome repatterning in three representative parrots (Psittaciformes) inferred from comparative chromosome painting. *Cytogenet. Genome Res.* **2007**, *117*, 43–53. [CrossRef]
32. Ellegren, H. Evolutionary stasis: The stable chromosomes of birds. *Trends Ecol. Evol.* **2010**, *25*, 283–291.
33. Nie, W.; O'Brien, P.C.M.; Fu, B.; Wang, J.; Su, W.; He, K.; Bed'Hom, B.; Volobouev, V.; Ferguson-Smith, M.A.; Dobigny, G.; et al. Multidirectional chromosome painting substantiates the occurrence of extensive genomic reshuffling within Accipitriformes. *BMC Evol. Biol.* **2015**, *15*, 205. [CrossRef]
34. Sweet, A.D.; Johnson, K.P. Patterns of diversification in small New World ground doves are consistent with major geologic events. *Auk* **2014**, *132*, 300–312. [CrossRef]

35. Warren, W.C.; Clayton, D.F.; Ellegren, H.; Arnold, A.P.; Hillier, L.W.; Künstner, A.; Searle, S.; White, S.; Vilella, A.J.; Fairley, S.; et al. The genome of a songbird. *Nature* **2010**, *464*, 757–762. [CrossRef] [PubMed]
36. Damas, J.; O'Connor, R.; Farré, M.; Lenis, V.P.E.; Martell, H.J.; Mandawala, A.; Fowler, K.E.; Jospeh, S.; Swain, M.; Griffin, D.K.; et al. Upgrading short-read animal genome assemblies to chromosome level using comparative genomics and a universal probe set. *Genome Res.* **2017**, *27*, 875–884. [CrossRef] [PubMed]

12

Decoding the Role of Satellite DNA in Genome Architecture and Plasticity—An Evolutionary and Clinical Affair

Sandra Louzada [1,2], Mariana Lopes [1,2], Daniela Ferreira [1,2], Filomena Adega [1,2], Ana Escudeiro [1,2], Margarida Gama-Carvalho [2] and Raquel Chaves [1,2,*]

1 Laboratory of Cytogenomics and Animal Genomics (CAG), Department of Genetics and Biotechnology (DGB), University of Trás-os-Montes and Alto Douro (UTAD), 5000-801 Vila Real, Portugal; slouzada@utad.pt (S.L.); lopesfmariana@gmail.com (M.L.); daniela_p_ferreira@hotmail.com (D.F.); filadega@utad.pt (F.A.); anac.escudeiro@gmail.com (A.E.)

2 Biosystems and Integrative Sciences Institute (BioISI), Faculty of Sciences, University of Lisboa, 1749-016 Lisbon, Portugal; mhcarvalho@fc.ul.pt

* Correspondence: rchaves@utad.pt

Abstract: Repetitive DNA is a major organizational component of eukaryotic genomes, being intrinsically related with their architecture and evolution. Tandemly repeated satellite DNAs (satDNAs) can be found clustered in specific heterochromatin-rich chromosomal regions, building vital structures like functional centromeres and also dispersed within euchromatin. Interestingly, despite their association to critical chromosomal structures, satDNAs are widely variable among species due to their high turnover rates. This dynamic behavior has been associated with genome plasticity and chromosome rearrangements, leading to the reshaping of genomes. Here we present the current knowledge regarding satDNAs in the light of new genomic technologies, and the challenges in the study of these sequences. Furthermore, we discuss how these sequences, together with other repeats, influence genome architecture, impacting its evolution and association with disease.

Keywords: satellite DNA; genome architecture; chromosome restructuring; Robertsonian translocations; satellite DNA transcription

1. Introduction

The linear organization of DNA sequences in the genome and how these sequences are packed into chromosomes define their architecture and influence its evolution. Repetitive DNA represents a major organizational component of eukaryotic genomes and includes sequences dispersed throughout the genome like transposable elements (TEs) and tandemly repeated sequences, such as satellite DNA (satDNA) [1,2]. Together with TEs, satDNAs contribute significantly to the differences in genome size between species, accounting for more than 50% of some species total DNA [3]. SatDNAs can be found in varied locations in the chromosomes, such as pericentromeric, subtelomeric and interstitial regions, forming blocks of constitutive heterochromatin (CH) [2–7] that are part of vital structures like centromeres and telomeres [2]. However, satDNA location is not restricted to CH with some satDNAs being found also dispersed throughout euchromatic regions in different species [5,8]. Multiple lines of evidence show that satDNAs have key roles in centromere function, heterochromatin formation and maintenance and chromosome pairing [9–12]. Interestingly, despite their association to critical chromosomal structures, satDNA families can display an astounding sequence variation even among closely related species. This results from their highly dynamic behavior, leading to rapid changes in sequence composition and array size within short evolutionary periods, which can lead to speciation (reviewed in [13]). Moreover, these sequences have been consistently correlated with fragile sites and evolutionary breakpoint regions in diverse species [14–19] and are intrinsically involved in frequent chromosomal rearrangements like Robertsonian translocations [20,21]. SatDNA dynamics has been shown to promote genome plasticity and to have an active involvement in the modulation of genomic architecture by promoting rearrangements.

Nevertheless, some satDNAs seem to have been preserved or "frozen" across different taxa during long evolutionary periods [22–24] with some of them being transcribed into satellite non-coding RNAs (satncRNAs). Indeed, transcripts of satDNAs have been reported in different species, highlighting a possible role for satncRNAs in the regulation of gene expression, cancer outcomes and aging [25–27]. This suggests that functional constraints may be causing the preservation of these sequences over the time [24,28]. Accordingly, some species centromeric satDNAs have been found to share a 17 bp motif known as the centromere protein B (CENP-B) box, representing the binding site for centromere protein B (CENP-B) [29,30]. It has been demonstrated that the CENP-B box is required for de novo centromere chromatin assembly and CENP-B protein is involved in centromere functions [31]. In this case, the conservation of a sequence motif across diverse mammalian species satDNAs [32] seems to be related to a specific function.

Over the years, different techniques have been used to address satDNA sequences. The advances in sequencing technology and computational approaches have revolutionized the study of these regions, known as the "black holes" of the genome. The increasing number of studies assessing the genomic abundance and sequence variation of satDNAs in different species has led to the coining of new terms to describe the whole collection of repeats (repeatome) and satDNAs (satellitome) in a species genome [33,34], and contributed to improve our knowledge regarding the evolution and function of these sequences [35].

In this review, we contextualize satDNA sequences in the genomes/chromosomes of different species in the light of recent data provided by new technologies and bioinformatic tools and the challenges of studying these DNA sequences and their associated non-coding RNAs. We also discuss the contribution of repetitive sequences to the organization of genomes and their participation in the restructuring of species karyotypes during evolution, focusing on their involvement in rearrangements with evolutionary and clinical significance: Robertsonian translocations. Finally, we address the structural role of satDNA transcripts in the genome.

2. SatDNA Features and Organization in the Genome and Chromosomes: Emerging Technologies and Changing Concepts

The concept of satDNA suffered considerable changes through time. Early experiments historically coined the term "satellite DNA" referring to tandemly arranged sequences that formed satellite bands separate from the rest of the genomic DNA during density gradient centrifugation [36]. Given that no function was initially attributed to these sequences, they were considered as genomic "junk", representing parasites proliferating independently in the genomes [37]. Today, satDNAs are viewed as important genomic functional components. In order to understand participation of these sequences in genome architecture and evolution, we need to briefly address their organizational features, localization and mode of evolution.

SatDNA is typically organized as long arrays of head-to-tail linked repeats and usually present in the genomes in several million copies [1]. The length of the repeating unit (monomer) can range from a few base pairs up to more than 1 kb, forming arrays that may reach 100 Mb in length (reviewed by [38]), and that can form higher-order repeat (HOR) units (e.g., [39–41]). Human chromosome centromeres are populated by α satDNA (αSAT) organized in HORs that are structurally distinct and confer chromosome specificity [39,42]. Complex HORs have been found in non-human mammals such as insects, mouse, swine, bovids, horse, dog and elephant (reviewed in [43]), and more recently in Callitrichini monkeys [44] and Teleostei fish [45]. SatDNA arrays are mainly found clustered in heterochromatin, although studies also report the presence of short satDNA arrays dispersed along euchromatic regions [2–7]. These sequences can be found in varied locations in the chromosomes, such as pericentromeric, subtelomeric and interstitial regions [2,46–48], as well as being part of vital structures like centromeres and telomeres [2].

Usually more than one family of satDNAs can be found in the same genome, thus forming a library, which can be shared among closely related species. The satDNAs within the library may differ in monomer sequence, size, abundance, distribution and location (reviewed in [12]). Expansions and contractions of satDNA arrays can dramatically change the landscape of repetitive sequences, leading to significant differences of satDNA copy number among related species [49,50]. That is the case of the *Drosophila* genus, which contains very dissimilar satDNAs, varying from 0.5% in some species genomes to as high as 50% in others [51,52]. Such striking differences in satDNA abundance in *Drosophila* sp. were proposed to result predominantly from lineage-specific gains accumulated over the past 40 MY of evolution [53], ultimately causing species reproductive barriers [54,55].

The mechanisms proposed to be responsible for the amplification/deletion of repetitive DNA, consequently leading to their rapid evolutionary turnover, are unequal crossing over, replication slippage and rolling circle amplification [56]. SatDNA sequence divergence among species is quite variable, as some repeats are species-specific, while others are widely conserved, being shared across distantly related species [22,24,57]. SatDNAs have a unique mode of evolution, known as concerted evolution, a two-level process in which mutations are homogenized throughout monomers of a repetitive family and concomitantly fixed within a group of reproductively linked organisms [58,59].

The study and characterization of satDNA has lagged behind when compared with other genomic sequences. Throughout time, different methodological approaches have generated insights into the structure, organization, function and evolution of these sequence elements, although this characterization has been significantly hampered by their highly repetitive nature. The advent of high-throughput sequencing technologies and associated bioinformatics tools opened the door to whole genome sequencing projects, and as the technology became more robust and cheaper, the number of sequenced species increased exponentially. In 2018, the Earth BioGenome project was launched, aiming to increase the number of sequenced eukaryotic genomes from 2534 species (of which only 25 comply with the standard for contig and scaffold N50 established by the Genome 10K organization) to characterize the genomes of the 1.5 million known species within a 10 year time frame [60]. Of note, satDNA, as well as other repetitive sequences, have been systematically omitted from the genome projects, due to difficulties in sequence alignment and assembly, given that the read length of current sequencing technologies is unable to span the longer repeats and tandem arrays [61,62]. Nevertheless, high-throughput sequencing contributed significantly to increase our knowledge regarding satDNA sequences [63]. Next generation sequencing (NGS; e.g., Illumina), allied to newly developed bioinformatics tools capable of identifying satDNA sequences in unassembled data (e.g., RepeatExplorer) [64–66], helped uncover the extent of satDNAs present in the genome of different species, revealing unpredicted levels of satDNA diversity (e.g., [34,67–71]). For instances, 62 satDNA families were identified in the genome of the migratory locust, leading to the coining of the term 'satellitome' to refer to the whole collection of satDNA families found in a single genome [34], a part of the 'repeatome', a term proposed previously [33] to refer to the collection of all repetitive sequences in a genome (TEs, satDNAs, etc.). This number has been surpassed by a recent study where 164 satDNA families have been identified in Teleostei fish, being this the biggest satellitome characterized for a given species so far [70]. The availability of a methodology capable of assessing satDNA array abundance and diversity led to an explosion of comparative studies across a wide range of clades, including mammals, insects and plants (e.g., [44,45,69,71–73]) providing insights into these sequences.

The development of sequencing technologies that generate long-range data has allowed the community to overcome some of the limitations imposed by NGS and is fueling the study of repeats. Single-molecule real-time sequencing and nanopore sequencing technologies (commercialized by PacBio and Oxford Nanopore Technologies (ONT), respectively) can generate longer reads capable of spanning repetitive regions, thus enabling their assembly into contigs (reviewed in [62]). For instances, ONT nanopore sequencers have been shown to generate unprecedented ultra-long reads that can reach mega-base lengths, leading to significant improvements in the human genome assembly [74–77], with some of the repetitive-containing gaps being closed [78,79]. By using long-read methods we are gaining access to important repeated-rich structures, like centromeres, revealing further insights into their sequence content and structure [80]. For instances, *Drosophila* centromeric satDNAs were recently shown to be intermingled with TEs [81]. Other recent studies report the improvement of human Y chromosome centromere assembly [78] and the reconstruction of a 2.8 megabase centromeric satDNA array, with the potential to achieve for the first-time telomere-to-telomere sequencing of the X chromosome [79].

Several studies demonstrate that the combination of different high-throughput sequencing methods (e.g., Illumina, ONT and PacBio) with other techniques, such as optical mapping, cytogenetics and molecular techniques, is beneficial and sometimes essential to determine satDNA features. The use of PacBio long-read sequencing together with optical mapping proved to be helpful in the assembly of satDNA arrays with large monomers and provided insights regarding recombination rates in the Eurasian crow [82]. Positional data derived from fluorescent in situ hybridization (FISH) remains vital to determine the physical location of satDNAs, since such information cannot be achieved for genomes that have not yet been properly assembled (e.g., [34,44,71,81,83]), and sequences mapping by FISH on extended DNA fibers can provide significant assistance to the process of genome assembly, aiding in contig ordering (e.g., [84,85]). Improved techniques based on FISH, helped shedding light into repetitive-rich chromosome regions with centromeric function (e.g., [86]). Other methods have also shown to provide a valid and expedite analysis of repetitive sequences profile, such as PCR-based approaches, that have been used to determine satDNA copy number differences between healthy and cancer cells/tissues [87]. In particular, the use of droplet digital PCR (ddPCR) combined with other methodologies has contributed to the validation and quantification of rare retrotransposon insertion events in different tissues including tumors [88] and the detection and accurate quantification of human *SATII* ncRNA in cancer patients [89]. The integration of genomic, cytogenetic and cell biology data helps to establish a connection between sequence information, its localization in the chromosomes and their interaction with other components of the genome, defining the field of chromosomics [90]. We believe that this approach is essential to fully understand the organization of repetitive sequences.

Other aspects of satDNA biology are also becoming accessible through the use of recent methodologies, such as the characterization of their expression and chromatin state, namely by using RNA sequencing (RNA-seq) and chromatin immunoprecipitation approaches followed by DNA sequencing (CHIP-seq) [91,92]. In particular, for CHIP-seq experiments several studies report the use of a specific antibody against DNA binding centromere-specific histone H3 (CENH3), which is an ortholog for human CENP-A. This methodology has proven to be useful for clarifying the satDNA content in the centromere, improving some organisms reference sequence and uncovering satDNA variability (e.g., [93,94]).

The data generated is now being used to determine satDNA sequences organization in the genome [95], explore predicted evolutionary patterns and hypothesis (e.g., [35,68,96,97]), as well as to shed light into the function of these sequences [81,98]. We are now closer than ever to fully access the sequence information hidden within repetitive-rich chromosome structures like centromeres and telomeres. However, we still need to further develop and adapt currently available approaches to achieve a combination of genomic, cytogenetic and molecular techniques to optimally address these regions, which we propose could be referred to as centrOMICs and telOMICs (Figure 1). SatDNAs represent one of the most intriguing and also interesting components of the genome and their full characterization will help us to better understand genome organization, architecture and evolution.

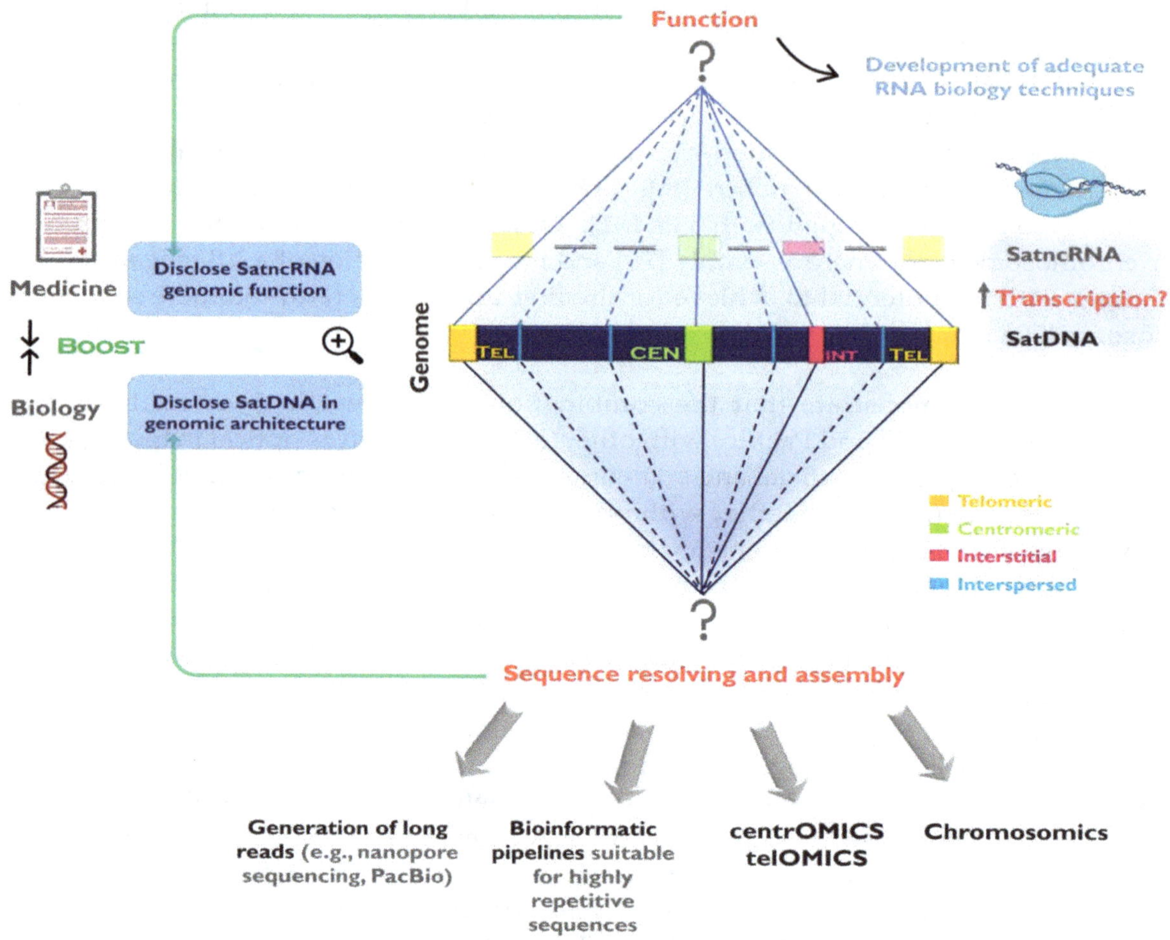

Figure 1. Challenges in the study of satellite DNA (satDNA) sequences and the importance to fully understand the repetitive genomic fraction. SatDNAs can be found clustered at the centromeres, telomeres and forming interstitial heterochromatin (CH) blocks, as well as scattered (interspersed) throughout the chromosomes. The full characterization of satDNAs needs to be addressed in two levels: 1-Disclose satDNAs linear sequence and improve their representation in genome assemblies. Despite currently used sequencing strategies (e.g., next generation sequencing (NGS)) contributed for satDNA studies, the full characterization of these sequences will only be achieved by using sequencing technologies capable of long reads, bioinformatics pipelines suitable for highly repetitive sequences, together with other techniques (e.g., FISH, optical mapping). These strategies need to be directed to specific chromosome structures such as centromeres (centrOMICs) and telomeres (telOMICs), which harbor large amounts of satDNA. Important also is the integration of genomic data with sequence localization in the chromosomes, and their interaction with other components of the genome (chromosomics); 2- Clarify satDNAs function(s) in the genome by studying the satellite non-coding RNAs (satncRNA) and their interaction with other components and structures in the genome. In this field there is the need to develop adequate biology techniques to address repetitive sequences transcription study. The disclosure of satDNA sequences will help to better understand its genomic architecture ant its role in genome restructuring in evolution and disease.

3. Modulating Genome Architecture with SatDNAs

The architecture of genomes confers identity to species. From a generalist point of view, the genomic architectural configuration is the product of a series of sequential molecular events that occurred during the evolutionary process. The impact of these events on genome organization is reflected by chromosome size, number and morphology. Eukaryotic genomes, and particularly, karyotypes, can be viewed as a set of homologous chromosomes, each harboring a combination of syntenic blocks—conserved blocks that can be differently assembled between species [99]. The events with capability for shaping genomes are based on structural and quantitative chromosomal alterations (e.g., [100]) of variable dimensions, from small to large regions that may completely change the morphology and number of species chromosomes and karyotypes. Amongst these, chromosome fusions (i.e., Robertsonian translocations), fissions (reviewed in [99]) and inversions [101], are perhaps the ones with a stronger impact on the architectural appearance of genomes during species evolution.

Chromosome structural variation may originate from illegitimate non-homologous recombination between different chromosome fields, such as centromeres, chromosome arms and telomeres during meiosis, requiring double strand breaks in at least two chromosomes or chromosome regions [102–105]. The resulting rearranged chromosomes are transmitted either as potentially harmful alterations, or as new variants associated with a selective advantage that will eventually conduct to speciation [99,106,107].

Even before the routine use of advanced molecular technologies, cytogeneticists could realize that the regions where chromosomes break and rearrange (the so-called chromosomal breakpoints) were enriched in constitutive heterochromatin, evidenced by C-bands [57]. Molecular technologies demonstrated that evolutionary breakpoint regions are composed of repeats [107–111]. The involvement of repetitive sequences, including TEs (e.g., [112–115]), segmental duplications (e.g., [108,110,116]) and tandem repeats (e.g., [14–17]), in genome restructuring and evolution is now widely recognized.

The evolutionary rate of tandemly repeated satDNA was shown to be higher than in other genomic sequences, presenting significant changes in short evolutionary times. It is thought that the mechanisms leading to the rapid turnover of these sequences promote chromosome rearrangements and consequently contribute to re-shaping of the genomes. Unequal crossing-over events seem to be responsible for the rapid evolution and divergence found among satDNA families, specifically at the levels of monomer length, nucleotide sequence, complexity and copy number [1,14,49,117,118]. DNA polymerase slippage during DNA replication and recombination in meiosis caused by faulty alignment of repetitive elements further contributes to the instability of these repeat rich regions and to chromosome rearrangements (e.g., [107,119,120]).

SatDNAs can display complex structural organization resulting from the formation of secondary DNA structures, including hairpins, triplexes [121] and even tetraplexes (G-quadruplexes) [122,123]. The formation of such structures can cause problems during genome duplication in the S-phase by slowing down or even stalling the replication fork, resulting in double-strand breaks [124,125]. This damage is then targeted for repair by means of homologous recombination-based mechanisms, which may lead to chromosome and genome architecture alterations due to the selection of identical sequences in non-homologous regions as the template for repair [1,19,126].

Several studies document the presence of TEs intermingled with centromeric satDNA [127,128], in some cases forming complex structures [24,129–131]. TEs are highly represented in some vertebrate species, making up to 60% or more of their genomes. They are characterized by their mobility within genomes using either a direct cut-and-paste mechanism to alter their position (transposons) or requiring an RNA intermediate (retrotransposons) [127,132]. This intrinsic feature makes them active elements of the genome and has been associated with genomic instability. TEs may cause double strand breaks, not only during the transposition process itself, but also by TE–TE ectopic recombination, which may lead to chromosomal rearrangements and consequently to alterations in the genome architecture [133–136]. The integration of TEs in the genome may also result in the disruption of a functional DNA sequence (reviewed in [128]), which can have adverse consequences. Together with segmental duplications, these elements share a high degree of similarity between different intra- and inter-chromosomal regions, making them the perfect templates for non-allelic homologous recombination [137–139]. TEs dynamics has shown to be linked with satDNA origin and evolution. Evidences suggest that some mobile elements may lead to the generation of new repetitive sequences that can be amplified into long arrays of satDNAs [140]. Moreover, it has been suggested that the autonomous LINE-1 retrotransposons could enable amplification and intragenomic movements of satDNA sequences throughout the genome [141]. It thus seems plausible to think that

TEs, especially retrotransposons may, in fact be an adjuvant for satDNA evolution and consequently lead to the creation of genomic innovations.

The dynamic nature of repetitive elements is clearly a basilar reason for genomic plasticity (e.g., [14,102,142,143]) and it is in fact a way of having a low impact on the euchromatic genome [14,97]. Today, an increasing body of evidence strongly validates the involvement of satDNA in the modulation of genomic architectures of a large number of taxa, as in the case of bovids [21,104,144,145], rodents [17,111,143,146], suiformes [57] or genets [147]. This largely extends beyond mammalian evolution, as it can also be observed in insects (e.g., [54,148]), reptiles (e.g., [149]), plants [150] and many other lineages. SatDNAs and TEs can thus be considered the 'engine' triggering genome evolution [14,107], with the regions harboring these sequences functioning as 'hotspots' or fragile sites for structural chromosome rearrangements, leading to species-specific genome architectures [14,17,57,107,139,143,151] and contributing to the generation of key variations responsible for the success of vertebrates [152].

3.1. Repetitive Sequences, Chromosome Instability and Disease

Alterations of genomic architecture can also be pathogenic and have a detrimental effect in organisms, either if occurring at the germinal lineage or somatically. This is the case of many diseases caused or boosted by genomic instability that impacts on nuclear architecture, such as cancer, neurodegenerative disorders and other genetic diseases [153–155]. In fact, alterations in genome architecture can interfere both with chromosomal territories and with topological positioning of chromosomes and genes in the nucleus. Due to the constraints in the regulation of genes and gene networks and to differences in somatic mutation frequencies between genome regions located at the nuclear periphery or core (higher in the periphery) [156], structural variations of critical genome regions can in fact threaten normal cell function. Again, in these situations, repeats seem to be the main actors at play [125,128,153,157]. The repetitive fraction of the eukaryotic genome and in particular, of the mammalian genome, is usually methylated and repressed by a highly condensed chromatin state, which seems to be essential to maintain genome integrity (reviewed in [158,159]). When perturbation of the epigenetic landscape of specific genomic regions occurs (e.g., [160]), repeats that are usually silenced can become active and unconstrained, which may lead to mobilization (in the case of TEs) and an open chromatin state that allows the occurrence of double strand breaks at fragile or hotspot regions. This results in chromosome rearrangements with impact on the three-dimensional genome architecture and gene expression regulation [100], which may lead to disease onset and progression.

3.1.1. Remodeling Genome Architecture Through Robertsonian Translocations from a SatDNA Perspective

The most frequent rearrangements occurring in genomes are Robertsonian translocations (rob). These rearrangements are commonly found in two different genomic scenarios: as an evolutionary rearrangement involved in mammalian karyotypic evolution; and as a chromosomal abnormality with clinical/polymorphic meaning [161,162]. The occurrence of Robertsonian translocations involves a break near or at the centromeric region, followed by the fusion of the entire long (q) arms of two acrocentric chromosomes, forming a dicentric or monocentric chromosome. The associated breakpoints, as well as the subsequent mechanistic steps, have been shown to involve reorganization of satDNA sequences at the centromere level [14,20]. The illegitimate recombination between homologous sequences, such as satDNA on non-homologous chromosomes, has been suggested as a possible path for the occurrence of Robertsonian translocations in mice and humans [163,164]. In fact, the high frequency of rob chromosomes linked to genome remodeling events can be caused not only by the homology of the satDNA sequences shared by the acrocentric chromosomes involved in each translocation, but also by the nicking activity of the centromere protein B (CENP-B), originating the double-strand breaks that precede the fusion events [165]. Robertsonian translocations are complex rearrangements that require, in addition to the double-strand breaks, mechanisms of repair, the silencing of possible additional centromeric sequences and the adjustment of the amount of CH/satDNA over time, in order to maintain chromosome viability [20,162]. This assigns a primordial task to satDNA in the control, success and viability of Robertsonian translocation events [14,162].

One of the well-known examples of the dual character of these rearrangements is the rob (1;29), which assumes a special relevance as it is the most widespread chromosome rearrangement occurring in domestic cattle with clinical significance [166–169]. In parallel, the rob (1;29) is also a constitutional chromosome rearrangement fixed in several wild bovid species, such as most of the Tragelaphini [170].

The analyses of the sequences at the breakpoint regions preceding a translocation are of great importance in understanding the translocation event [21,131]. These sequences are essentially centromeric satDNAs, whose detailed physical and organizational analysis contributed much to better comprehend the chromosomal mechanism behind the rob (1;29) translocation [20,145]. In 2000, Chaves and colleagues suggested that this chromosomal abnormality might not be a single event [144] and in 2003, using centromeric satDNA sequences, the same group proposed, for the first time, a two-step mechanism for this rearrangement [20]. This translocation mechanism involved, besides the centric fusion of the two acrocentric chromosomes, the loss and reorganization of specific satDNA families that were retained in the translocated chromosome [20]. Later, Di Meo and colleagues [145], using both satDNA and BAC probes, validated the pericentric inversion previously proposed [20]. This event would probably be necessary for satDNA reorganization at the centromeric level, highlighting the active role of satDNA sequences in the translocation mechanism and reinforcing their functional relevance in genome reorganization [14,20].

In humans, the Robertsonian translocations are also the most common structural chromosome abnormality [171,172], with rob (13;14) and rob(14;21) being the most frequent examples [163]. During several decades, aspects such as the high frequency of de novo robs in the human population, their origin during oogenesis, and the non-random participation of the acrocentric chromosomes, have supported the hypothesis that there must be a specific mechanism leading to the formation of these robs [163,164,173]. However, and despite the high frequency of these rearrangements and their clinical implications, there is still insufficient information on the molecular mechanism and exact genomic location of the breakpoints [174]. The rob translocation event has been deeply connected with satDNA sequence homology and consequent recombination [174] giving rise to two alternative explanations: (i) the presence of a homologous inversely-oriented segment on chromosome 14 shared with chromosomes 13 and 21 [163,175]; (ii) the human satellite DNA *SATIII* ability to form uncommon DNA structures that could facilitate the illegitimate recombination [176,177]. However, these hypotheses need further research to be validated. Indeed, in the study of Robertsonian translocations, finding the breakpoint location is a problematic task due to the low resolution of the physical maps at the centromere and short arms of the acrocentric chromosomes [174]. Highly repetitive satDNA undoubtedly represents a major gap in the current human genome assemblies, significantly contributing to the lack of high-resolution sequencing studies in the field of centromere genomics [74,178].

4. Transcribing SatDNAs: Targeting Genomic Functions

The previous sections highlight the role of satDNA sequences in genome architecture and in specific chromosomal rearrangements. However, the participation of these sequences in shaping genome architecture goes beyond their DNA molecule. Currently the transcription of satDNA is a widely accepted feature across species. Different functions have been assigned to satellite non-coding RNAs (satncRNAs) in several cellular contexts, such as cell proliferation, stress response, development or cancer [27] (Figure 2). In fact, satDNA transcripts seem to participate in the most primordial concept of genomic function, being related to centromere structure, chromosome pairing/segregation and kinetochore assembly [2,27,179]. Moreover, recently satDNA transcripts have shown to be associated with male fertility in *Drosophila* sp. [180]. Unfortunately, the function of most satncRNAs remains unknown or unclear due to the inefficient methodologies currently available to analyze molecules of such repetitive nature.

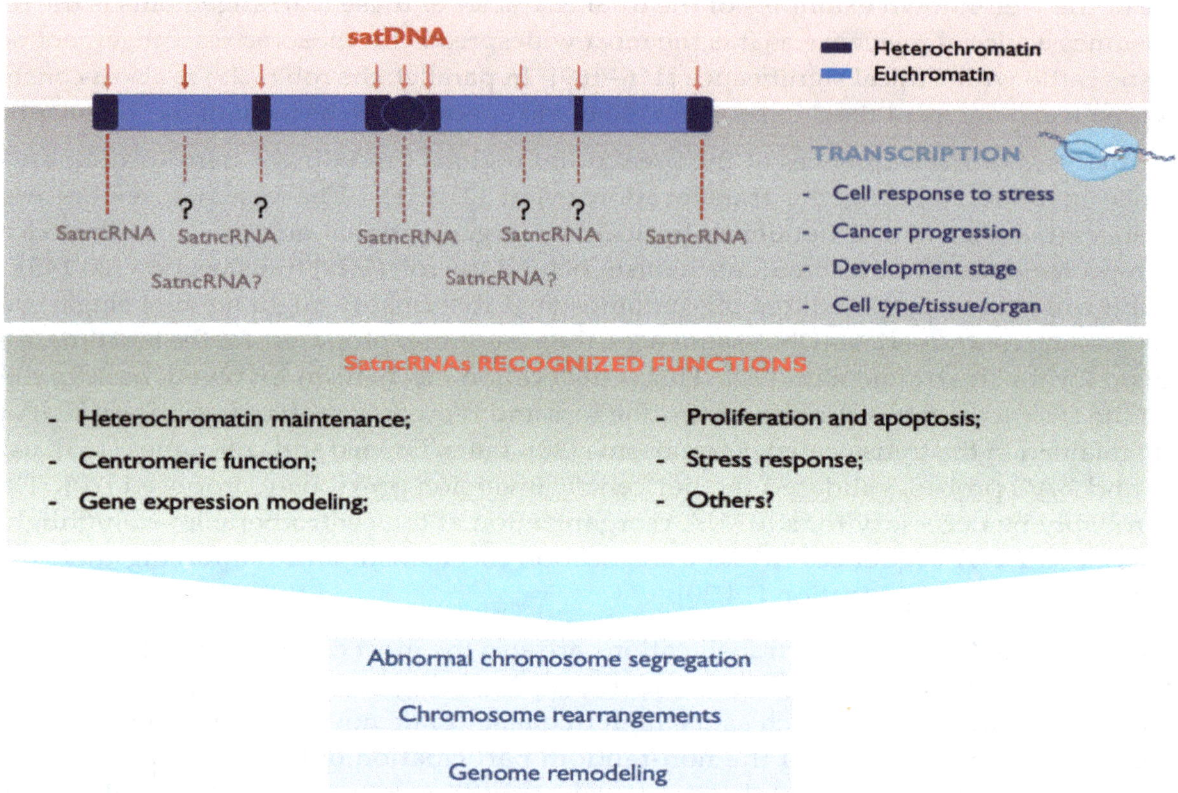

Figure 2. Summary of current knowledge regarding satellite non-coding RNAs (satncRNAs) and how they can contribute to genome remodeling. Even though satDNAs present in the heterochromatin and euchromatin can be transcribed, the most studied satncRNAs are the ones originated from pericentromeric and centromeric satDNAs families. For some satncRNAs reported, chromosome location of the origin satDNA cannot be determined. SatDNA transcription has been shown to be associated to cells response to stress, cancer progression, particular developmental stage and some are differentially expressed in specific cell types, tissues and organs. General recognized functions attributed to satncRNAs are listed. The aberrant expression of satncRNAs may result in abnormal chromosome segregation, and chromosome rearrangements that re-shape the genome and can lead to cancer progression or be fixed during species evolution. Further effort is needed to identify and better characterize satncRNA and their involvement in cellular functions and disease.

Concerning centromeric satDNAs, the human α satellite transcripts (αSAT) have been shown to be crucial for cell cycle progression, as depletion of αSAT resulted in defective centromeric protein A (CENP-A) loading and cell cycle arrest [181]. αSAT ncRNAs also seem to regulate spindle microtubule attachment and sister chromatid disjunction through association with AURORA B proteins [182]. These molecules were further shown to be associated with the SUV39H1 histone methyltransferase, thereby suggesting a regulatory function in heterochromatin maintenance [183–185].

Contrary to what is believed for most condensed genomic regions, centromeric sequences remain transcriptionally active during mitosis [186,187], essentially promoting kinetochore stabilization and centromere cohesion [188,189]. These functions have been similarly attributed to transcriptionally active centromeric satDNAs from other species [190–192], in spite of the observed sequence differences. This suggests that satncRNAs are involved in critical functions, which appear to be associated with their intrinsic molecular characteristics and most probably also with the genomic location of their satDNA sequence.

Pericentromeric satDNA transcripts have been related with pericentric chromatin formation [193–195], acting as molecular scaffolds for the accumulation of HP1 [194]. The presence of human *SATIII* ncRNA can be closely associated with cell response to stress. Particularly, heat shock can trigger *SATIII* transcription by the action of HSF1 (Heat Shock Factor 1) [196,197], giving rise to nuclear stress bodies (nSBs) close to *SATIII* DNA regions [198]. The splicing of relevant genes for stress response may be influenced by *SATIII* ncRNAs, which have been proposed to sequester RNA processing factors and downregulate global transcription [199], providing protection against stress-induced cell death [200]. However, *SATIII* transcription is not thermal stress-exclusive, as a basal level of expression is detectable even in the absence of cellular stress [201]. This same satDNA family can exhibit different genomic locations and its transcripts can be involved in multiple functions, making their study even more difficult. *SatDNA III* from *Drosophila melanogaster* is located at the centromere and pericentromere of the X chromosome and at the pericentromere of chromosomes 2 and 3 [202]. Its transcripts have been shown to play different roles in chromatin silencing/heterochromatinization, centromeric function and upregulation of X-linked genes [191,202–204].

Another interesting case is the *FA-SAT*, the major satDNA sequence of *Felis catus* (cat) genome, located at the (sub)telomeres and (peri)centromeres of chromosomes [153] and also in an interspersed fashion [24]. This satDNA is highly conserved in its primary sequence among Bilateria species (e.g., human, *Drosophila*, oyster, cattle, among others), a rare event observed in satDNA sequences [24]. In these species (non-*Felis* species), an interspersed distribution of this satDNA was proposed, with the exception of the other carnivore analyzed, *Genetta genetta*, in which it also presents a centromeric location. Of note, *FA-SAT* is transcribed in all these species [24], and an important conserved function (in cat and human) was ascribed to this ncRNA as a PKM2 interactor involved in the cross-talk between proliferation and apoptosis [205]. In fact, the absence of this satncRNA in both species results in cell death [205]. These transcripts possibly originate from the transcription of *FA-SAT* interspersed DNA (at current knowledge, the common location among these species). A putative connection between *FA-SAT* ncRNAs with cancer was also recently hypothesized [205].

With the progressive acceptance of satDNA transcriptional activity, the aberrant expression of satncRNAs has been increasingly associated with cancer progression (reviewed in [27]). In fact, the observed overexpression of satncRNAs in stress conditions may be comparable to cancer, since loss of sister chromatids cohesion, incorrect chromosome segregation or aneuploidy are common features of both states [206]. Overexpression of satncRNAs in cancer has been reported alongside decondensation and hypomethylation of pericentromeric DNA [154,207]. The transcription of satDNAs, the change in nuclear architecture and the altered sequestration of transcription factors may all be related to gene expression deregulation induced by the hypomethylation of *SATII* and *SATIII* satDNA sequences [208,209]. SatncRNAs may thus be involved in more general disease contexts associated to chromatin decondensation, DNA breaks and subsequent genomic rearrangements [209] (Figure 2). However, the value of satncRNAs as cancer biomarkers is still an unexplored field [27].

Due to their repetitive nature, high copy number and multiple genomic locations (different chromosomes and/or genomic regions), the study of satncRNAs remains a difficult challenge, namely regarding the original genomic location of their DNA sequence and the determination of their primary sequence(s) (Figure 2). Indeed, the most common next-generation sequencing platforms presents significant limitations in the analysis of satncRNA sequences in RNA-seq libraries, namely regarding their inability to assemble large repetitive transcripts from very short reads. This could be overcome in the future with the application of ultra-long read sequencing technology. The fact that most of the methods currently available that are mainly directed towards the analysis of gene coding sequences set a requirement for essential improvements or adjustments in order to support the efficient study of satncRNAs (Figure 1).

Although the importance of satncRNA in normal cell function and disease states is becoming increasingly accepted by the scientific community, in the wake of recent studies on these transcripts, much remains to be understood about their functions in different contexts. This will only be overcome through the development of improved methods for the study of repetitive sequences, as well as the commitment of the scientific community to this field of research.

5. Concluding Remarks

In this review we outlined the critical importance of satDNA sequences in driving karyotype evolution and genomic architecture, as well as their involvement in various basic genomic functions. However, to this day, significant technological limitations hinder the progress of this important field in biology and medicine, in particular in the study of diseases involving this genomic fraction. It is imperative to boost the study of satDNA sequences and their transcripts by adapting and developing sequencing technologies and bioinformatics pipelines capable of assembling chromosomes from telomere-to-telomere as well as focused approaches that follow the concept of chromosomics. Significant effort is needed from the entire scientific community to value these important genomic elements, which have been so neglected over time. Only then can we begin to fully understand the largest fraction of our genome.

Author Contributions: The concept and idea and the drafting of this review was done by R.C. and S.L. S.L., M.L., D.F., F.A., A.E. contributed to writing and M.G.-C. and R.C. revised the drafts of the manuscript. R.C. and M.L. prepared figures. All authors have read and agreed to the published version of the manuscript.

References

1. Charlesworth, B.; Sniegowski, P.; Stephan, L.W. The evolutionary dynamics of repetitive DNA in eukaryotes. *Nature* **1994**, *371*, 215–220. [CrossRef] [PubMed]
2. Plohl, M.; Mestrovic, N.; Mravinac, B. Satellite DNA evolution. *Genome Dyn.* **2012**, *7*, 126–152. [CrossRef]
3. López-Flores, I.; Garrido-Ramos, M.A. The Repetitive DNA Content of Eukaryotic Genomes. *Genome Dyn.* **2012**, *7*, 1–28. [CrossRef] [PubMed]
4. Brajković, J.; Feliciello, I.; Bruvo-Mađarić, B.; Ugarković, D. Satellite DNA-like elements associated with genes within euchromatin of the beetle Tribolium castaneum. *G3 Genes Genomes Genet.* **2012**, *2*, 931–941. [CrossRef]
5. Kuhn, G.C.; Küttler, H.; Moreira-Filho, O.; Heslop-Harrison, J.S. The 1.688 repetitive DNA of *Drosophila*: Concerted evolution at different genomic scales and association with genes. *Mol. Biol. Evol.* **2012**, *29*, 7–11. [CrossRef] [PubMed]
6. Larracuente, A.M. The organization and evolution of the Responder satellite in species of the *Drosophila* melanogaster group: Dynamic evolution of a target of meiotic drive. *BMC Evol. Biol.* **2014**, *14*, 233. [CrossRef] [PubMed]
7. Pavlek, M.; Gelfand, Y.; Plohl, M.; Meštrović, N. Genome-wide analysis of tandem repeats in Tribolium castaneum genome reveals abundant and highly dynamic tandem repeat families with satellite DNA features in euchromatic chromosomal arms. *DNA Res.* **2015**, *22*, 387–401. [CrossRef]
8. Feliciello, I.; Akrap, I.; Ugarkovic, D. Satellite DNA modulates gene expression in the beetle *Tribolium castaneum* after heat stress. *PLoS Genet.* **2015**, *11*, e1005466. [CrossRef]
9. Volpe, T.A.; Kidner, C.; Hall, I.M.; Teng, G.; Grewal, S.I.; Martienssen, R.A. Regulation of heterochromatic silencing and histone H_3 lysine-9 methylation by RNAi. *Science* **2002**, *297*, 1833–1837. [CrossRef]
10. Martienssen, R.A. Maintenance of heterochromatin by RNA interference of tandem repeats. *Nat. Genet.* **2003**, *35*, 213–214. [CrossRef]
11. Allshire, R.C.; Madhani, H.D. Ten Principles of Heterochromatin Formation and Function. *Nat. Rev. Mol. Cell Biol.* **2017**, *19*, 229–244. [CrossRef] [PubMed]
12. Garrido-Ramos, M.A. Satellite DNA: An Evolving Topic. *Genes* **2017**, *8*, 230. [CrossRef] [PubMed]
13. Ugarković, Đ. Evolution of α-Satellite DNA. In *Encyclopedia of Life Sciences*; John Wiley & Sons, Ltd.: Chichester, UK, 2013.

14. Adega, F.; Guedes-Pinto, H.; Chaves, R. Satellite DNA in the karyotype evolution of domestic animals—Clinical considerations. *Cytogenet. Genome Res.* **2009**, *126*, 12–20. [CrossRef] [PubMed]
15. Ruiz-Herrera, A.; Castresana, J.; Robinson, T.J. Is mammalian chromosomal evolution driven by regions of genome fragility? *Genome Biol.* **2006**, *7*. [CrossRef] [PubMed]
16. Farré, M.; Bosch, M.; López-Giráldez, F.; Ponsà, M.; Ruiz-Herrera, A. Assessing the role of tandem repeats in shaping the genomic architecture of great apes. *PLoS ONE* **2011**, *6*, e27239. [CrossRef] [PubMed]
17. Vieira-da-Silva, A.; Louzada, S.; Adega, F.; Chaves, R. A high-resolution comparative chromosome map of Cricetus cricetus and Peromyscus eremicus reveal the involvement of constitutive heterochromatin in breakpoint regions. *Cytogenet. Genome Res.* **2015**, *145*, 59–67. [CrossRef] [PubMed]
18. De La Fuente, R.; Baumann, C.; Viveiros, M.M. ATRX contributes to epigenetic asymmetry and silencing of major satellite transcripts in the maternal genome of the mouse embryo. *Development* **2015**, *142*, 1806–1817. [CrossRef]
19. Giunta, S.; Funabiki, H. Integrity of the human centromere DNA repeats is protected by CENP-A, CENP-C, and CENP-T. *Proc. Natl. Acad. Sci. USA* **2017**, *11*, 1928–1933. [CrossRef]
20. Chaves, R.; Adega, F.; Heslop-Harrison, J.S.; Guedes-Pinto, H.; Wienberg, J. Complex satellite DNA reshuffling in the polymorphic t (1; 29) Robertsonian translocation and evolutionarily derived chromosomes in cattle. *Chromosome Res.* **2003**, *11*, 641–648. [CrossRef]
21. Escudeiro, A.; Ferreira, D.; Mendes-da-Silva, A.; Heslop-Harrison, J.S.; Adega, F.; Chaves, R. Bovine satellite DNAs—A history of the evolution of complexity and impact in the *Bovidae* Family. *Eur. Zool. J.* **2019**, *86*, 20–37. [CrossRef]
22. Mravinac, B.; Plohl, M.; Mestrović, N.; Ugarković, D. Sequence of PRAT Satellite DNA 'Frozen' in Some Coleopteran Species. *J. Mol. Evol.* **2002**, *54*, 774–783. [CrossRef] [PubMed]
23. Plohl, M.; Petrović, V.; Luchetti, A.; Ricci, A.; Satović, E.; Passamonti, M.; Mantovani, B. Long-term conservation vs high sequence divergence: The case of an extraordinarily old satellite DNA in bivalve mollusks. *Heredity* **2010**, *104*, 543–551. [CrossRef] [PubMed]
24. Chaves, R.; Ferreira, D.; Mendes-da-Silva, A.; Meles, S.; Adega, F. FA-SAT Is an Old Satellite DNA Frozen in Several Bilateria Genomes. *Genome Biol. Evol.* **2017**, *9*, 3073–3087. [CrossRef] [PubMed]
25. Pezer, Z.; Brajković, J.; Feliciello, I.; Ugarkovć, D. Satellite DNA-Mediated Effects on Genome Regulation. *Genome Dyn.* **2012**, *7*, 153–169. [CrossRef] [PubMed]
26. Grenfell, A.W.; Heald, R.; Strzelecka, M. Mitotic noncoding RNA processing promotes kinetochore and spindle assembly in Xenopus. *J. Cell Biol.* **2016**, *214*, 133–141. [CrossRef] [PubMed]
27. Ferreira, D.; Meles, S.; Escudeiro, A.; Mendes-da-Silva, A.; Adega, F.; Chaves, R. Satellite non-coding RNAs: The emerging players in cells, cellular pathways and cancer. *Chromosome Res.* **2015**, *23*, 479–493. [CrossRef] [PubMed]
28. Ferreira, D.; Escudeiro, A.; Adega, F.; Anjo, S.I.; Manadas, B.; Chaves, R. FA-SAT ncRNA interacts with PKM2 protein: Depletion of this complex induces a switch from cell proliferation to apoptosis. *Cell Mol. Life Sci.* **2019**. [CrossRef]
29. Masumoto, H.; Masukata, H.; Muro, Y.; Nozaki, N.; Okazaki, T. Ahumancentromereantigen(CENP-B) interacts with a short specific sequence in alphoid DNA, a human centromeric satellite. *J. Cell Biol.* **1989**, *109*, 1963–1973. [CrossRef]
30. Muro, Y.; Masumoto, H.; Yoda, K.; Nozaki, N.; Ohashi, M.; Okazaki, T. Centromereprotein Bassembles human centromeric α-satellite DNA at the 17-bp sequence, CENP-B box. *J. Cell Biol.* **1992**, *116*, 585–596. [CrossRef]
31. Fachinetti, D.; Han, J.S.; McMahon, M.A.; Ly, P.; Abdullah, A.; Wong, A.J.; Cleveland, D.W. DNA sequence-specific binding of CENP-B enhances the fidelity of human centromere function. *Dev. Cell* **2015**, *33*, 314–327. [CrossRef]
32. Kipling, D.; Warburton, P.E. Centromeres, CENP-B and Tigger too. *Trends Genet.* **1997**, *13*, 141–145. [CrossRef]
33. Kim, Y.B.; Oh, J.H.; McIver, L.J.; Rashkovetsky, E.; Michalak, K.; Garner, H.R.; Kang, L.; Nevo, E.; Korol, A.B.; Michalak, P. Divergence of *Drosophila* melanogaster repeatomes in response to a sharp microclimate contrast in Evolution Canyon, Israel. *Proc. Natl. Acad. Sci. USA* **2014**, *111*, 10630–10635. [CrossRef] [PubMed]
34. Ruiz-Ruano, F.J.; López-León, M.D.; Cabrero, J.; Camacho, J.P.M. High-Throughput Analysis of the Satellitome Illuminates Satellite DNA Evolution. *Sci. Rep.* **2016**, *6*, 28333. [CrossRef] [PubMed]

35. Belyayev, A.; Josefiová, J.; Jandová, M.; Kalendar, R.; Krak, K.; Mandák, B. Natural History of a Satellite DNA Family: From the Ancestral Genome Component to Species-Specific Sequences, Concerted and Non-Concerted Evolution. *Int. J. Mol. Sci.* **2019**, *20*, 1201. [CrossRef]
36. Kit, S.J. Equilibrium Sedimentation in Density Gradients of DNA Preparations from Animal Tissues. *J. Mol. Biol.* **1961**, *3*, 711–716. [CrossRef]
37. Orgel, L.E.; Crick, F.H.C. Selfish DNA: The Ultimate Parasite. *Nature* **1980**, *284*, 604–607. [CrossRef]
38. Plohl, M.; Luchetti, A.; Mestrovic, N.; Mantovani, B. Satellite DNAs between selfishness and functionality: Structure, genomics and evolution of tandem repeats in centromeric (hetero) chromatin. *Gene* **2008**, *409*, 72–82. [CrossRef]
39. Willard, H.F. Chromosome-specific organization of human α satellite DNA. *Am. J. Hum. Genet.* **1985**, *37*, 524–532.
40. Alexandrov, I.A.; Medvedev, L.; Mashkova, T.D.; Kisselev, L.L.; Romanova, L.Y.; Yurov, Y.B. Definition of a new α satellite suprachromosomal family characterized by monomeric organization. *Nucleic Acids Res.* **1993**, *21*, 2209–2215. [CrossRef]
41. McNulty, S.M.; Sullivan, B.A. α satellite DNA biology: Finding function in the recesses of the genome. *Chromosome Res.* **2018**, *26*, 115–138. [CrossRef]
42. Waye, J.S.; Willard, H.F. Chromosome-specific α satellite DNA: Nucleotide sequence analysis of the 2.0 kilobasepair repeat from the human X chromosome. *Nucleic Acids Res.* **1985**, *13*, 2731–2743. [CrossRef] [PubMed]
43. Vlahovic, I.; Gluncic, M.; Rosandic, M.; Ugarkovic, Đ.; Paar, V. Regular Higher Order Repeat Structures in Beetle Tribolium castaneum Genome. *Genome Biol Evol.* **2017**, *9*, 2668–2680. [CrossRef] [PubMed]
44. Araújo, N.P.; de Lima, L.G.; Dias, G.B.; Kuhn, G.C.S.; de Melo, A.L.; Yonenaga-Yassuda, Y.; Stanyon, R.; Svartman, M. Identification and characterization of a subtelomeric satellite DNA in *Callitrichini monkeys*. *DNA Res.* **2017**, *24*, 377–385. [CrossRef] [PubMed]
45. Utsunomia, R.; Ruiz-Ruano, F.J.; Silva, D.M.Z.A.; Serrano, É.A.; Rosa, I.F.; Scudeler, P.E.S.; Hashimoto, D.T.; Oliveira, C.; Camacho, J.P.M.; Foresti, F. A Glimpse into the Satellite DNA Library in *Characidae* Fish (*Teleostei, Characiformes*). *Front. Genet.* **2017**, *8*, 103. [CrossRef]
46. Adega, F.; Chaves, R.; Guedes-Pintos, H. Chromosomal evolution and phylogenetic analysis in *Tayassu pecari* and *Pecari tajacu* (*Tayassuidae*): Tales from constitutive heterochromatin. *J. Genet.* **2007**, *86*, 19–26. [CrossRef]
47. Henikoff, S.; Dalal, Y. Centromeric chromatin: What makes it unique? *Curr. Opin. Genet. Dev.* **2005**, *15*, 177–184. [CrossRef]
48. Paço, A.; Adega, F.; Guedes-Pinto, H.; Chaves, R. Hidden heterochromatin Characterization in the *Rodentia* species *Cricetus cricetus*, *Peromyscus eremicus* (*Cricetidae*) and *Praomys tullbergi* (*Muridae*). *Genet. Mol. Biol.* **2009**, *32*, 58–68. [CrossRef]
49. Ugarković, Đ.; Plohl, M. Variation in satellite DNA profiles—Causes and effects. *EMBO J.* **2002**, *21*, 5955–5959. [CrossRef]
50. Louzada, S.; Vieira-da-Silva, A.; Mendes-da-Silva, A.; Kubickova, S.; Rubes, J.; Adega, F.; Chaves, R. A Novel Satellite DNA Sequence in the Peromyscus Genome (PMSat): Evolution via Copy Number Fluctuation. *Mol. Phylogenet. Evol.* **2015**, *92*, 193–203. [CrossRef]
51. Gall, J.G.; Cohen, E.G.; Polan, M.L. Repetitive DNA sequences in *Drosophila*. *Chromosoma* **1971**, *33*, 319–344. [CrossRef]
52. Lohe, A.R.; Brutlag, D.L. Identical satellite DNA sequences in sibling species of *Drosophila*. *J. Mol. Biol.* **1987**, *194*, 161–170. [CrossRef]
53. Wei, K.H.-C.; Lower, S.E.; Caldas, I.V.; Sless, T.J.S.; Barbash, D.A.; Clark, A.G. Variable rates of simple satellite gains across the *Drosophila* phylogeny. *Mol. Biol. Evol.* **2018**, *35*, 925–941. [CrossRef] [PubMed]
54. Ferree, P.M.; Barbash, D.A. Species-specific heterochromatin prevents mitotic chromosome segregation to cause hybrid lethality in *Drosophila*. *PLoS Biol.* **2009**, *7*, e1000234. [CrossRef] [PubMed]
55. Ferree, P.M.; Prasad, S. How can satellite DNA divergence cause reproductive isolation? Let us count the chromosomal ways. *Genet Res Int* **2012**, 430136. [CrossRef]
56. Walsh, J.B. Persistence of tandem arrays: Implications for satellite and simple-sequence DNAs. *Genetics* **1987**, *115*, 553–567.
57. Adega, F.; Chaves, R.; Guedes-Pinto, H. Suiformes orthologous satellite DNAs as a hallmark of collared and white-lipped peccaries (*Tayassuidae*) evolutionary rearrangements. *Micron* **2008**, *39*, 1281–1287. [CrossRef]

58. Dover, G.A. Molecular drive in multigene families: How biological novelties arise, spread and are assimilated. *Trends Genet.* **1986**, *2*, 159–165. [CrossRef]
59. Elder, J.F., Jr.; Turner, B.J. Concerted evolution of repetitive DNA sequences in eukaryotes. *Q. Rev. Biol.* **1995**, *70*, 297–320. [CrossRef]
60. Lewin, H.A.; Robinson, G.E.; Kress, W.J.; Baker, W.J.; Coddington, J.; Crandall, K.A.; Durbin, R.; Edwards, S.V.; Forest, F.; Gilbert, M.T.P.; et al. Earth BioGenome Project: Sequencing life for the future of life. *Proc. Natl. Acad. Sci. USA* **2018**, *115*, 4325–4333. [CrossRef]
61. Treangen, T.J.; Salzberg, S.L. Repetitive DNA and next-generation sequencing: Computational challenges and solutions. *Nat. Rev. Genet.* **2011**, *13*, 36–46. [CrossRef]
62. Van Dijk, E.L.; Jaszczyszyn, Y.; Naquin, D.; Thermes, C. The Third Revolution in Sequencing Technology. *Trends Genet.* **2018**, *34*, 666–681. [CrossRef] [PubMed]
63. Sedlazeck, F.J.; Lee, H.; Darby, C.A.; Schartz, M.C. Piercing the dark matter: Bioinformatics of long-range sequencing and mapping. *Nat. Rev. Genet.* **2018**, *19*, 329–346. [CrossRef] [PubMed]
64. Novák, P.; Neumann, P.; Pech, J.; Steinhaisl, J.; Macas, J. RepeatExplorer: A Galaxy-based web server for genome-wide characterization of eukaryotic repetitive elements from next-generation sequence reads. *Bioinformatics* **2013**, *29*, 792–793. [CrossRef]
65. Lower, S.S.; McGurk, M.P.; Clark, A.G.; Barbash, D.A. Satellite DNA Evolution: Old Ideas, New Approaches. *Curr. Opin. Genet. Dev.* **2018**, *49*, 70–78. [CrossRef] [PubMed]
66. Salmela, L.; Rivals, E. LoRDEC: Accurate and efficient long read error correction. *Bioinformatics* **2014**, *30*, 3506–3514. [CrossRef]
67. Cook, D.E.; Zdraljevic, S.; Tanny, R.E.; Seo, B.; Riccardi, D.D.; Noble, L.M.; Rockman, M.V.; Alkema, M.J.; Braendle, C.; Kammenga, J.E.; et al. The Genetic Basis of Natural Variation in Caenorhabditis elegans Telomere Length. *Genetics* **2016**, *204*, 371–383. [CrossRef]
68. Wei, K.H.; Grenier, J.K.; Barbash, D.A.; Clark, A.G. Correlated Variation and Population Differentiation in Satellite DNA Abundance among Lines of *Drosophila melanogaster*. *Proc. Natl. Acad. Sci. USA* **2014**, *111*, 18793–18798. [CrossRef]
69. Silva, D.M.Z.A.; Utsunomia, R.; Ruiz-Ruano, F.J.; Daniel, S.N.; Porto-Foresti, F.; Hashimoto, D.T.; Oliveira, C.; Camacho, J.P.M.; Foresti, F. High-throughput analysis unveils a highly shared satellite DNA library among three species of fish genus Astyanax. *Sci. Rep.* **2017**, *7*, 12726. [CrossRef]
70. Utsunomia, R.; Silva, D.M.Z.A.; Ruiz-Ruano, F.J.; Goes, C.A.G.; Melo, S.; Ramos, L.P.; Oliveira, C.; Porto-Foresti, F.; Foresti, F.; Hashimoto, D.T. Satellitome landscape analysis of *Megaleporinus macrocephalus* (*Teleostei, Anostomidae*) reveals intense accumulation of satellite sequences on the heteromorphic sex chromosome. *Sci. Rep.* **2019**, *9*, 5856. [CrossRef]
71. Palacios-Gimenez, O.M.; Dias, G.B.; De Lima, L.G.; Ramos, É.; Martins, C.; Cabral-de-Mello, D.C. High-throughput analysis of the satellitome revealed enormous diversity of satellite DNAs in the neo-Y chromosome of the cricket *Eneoptera surinamensis*. *Sci. Rep.* **2017**, *7*, 6422. [CrossRef]
72. Usai, G.; Mascagni, F.; Natali, L.; Giordani, T.; Cavallini, A. Comparative Genome-Wide Analysis of Repetitive DNA in the *Genus Populus* L. *Tree Genet. Genomes* **2017**, *13*, 96. [CrossRef]
73. Vozdova, M.; Kubickova, S.; Cernohorska, H.; Fröhlich, J.; Rubes, J. Satellite DNA Sequences in Canidae and Their Chromosome Distribution in Dog and Red Fox. *Cytogenet. Genome Res.* **2017**, *150*, 118–127. [CrossRef] [PubMed]
74. Miga, K.H. Completing the human genome: The progress and challenge of satellite DNA assembly. *Chromosome Res.* **2015**, *23*, 421–426. [CrossRef] [PubMed]
75. Jain, M.; Olsen, H.E.; Paten, B.; Akeson, M. The Oxford Nanopore MinION: Delivery of nanopore sequencing to the genomics community. *Genome Biol.* **2016**, *17*, 239. [CrossRef]
76. Kuderna, L.F.K.; Lizano, E.; Julià, E.; Gomez-Garrido, J.; Serres-Armero, A.; Kuhlwilm, M.; Alandes, R.A.; Alvarez-Estape, M.; Juan, D.; Simon, H.; et al. Selective Single Molecule Sequencing and Assembly of a Human Y Chromosome of African Origin. *Nat. Commun.* **2019**, *10*, 4. [CrossRef]
77. Miga, K.H. Centromeric Satellite DNAs: Hidden Sequence Variation in the Human Population. *Genes* **2019**, *10*, 352. [CrossRef]
78. Jain, M.; Koren, S.; Miga, K.H.; Quick, J.; Rand, A.C.; Sasani, T.A.; Tyson, J.R.; Beggs, A.D.; Dilthey, A.T.; Fiddes, I.T.; et al. Nanopore Sequencing and Assembly of a Human Genome with Ultra-Long Reads. *Nat. Biotechnol.* **2018**, *36*, 338–345. [CrossRef]

79. Miga, K.H.; Koren, S.; Rhie, A.; Vollger, M.R.; Gershman, A.; Bzikadze, A.; Brooks, S.; Howe, E.; Porubsky, D.; Logsdon, G.A.; et al. Telomere-to-Telomere Assembly of a Complete Human X Chromosome. *bioRxiv* **2019**, *735928*. [CrossRef]
80. Muller, H.; Gil, J., Jr.; Drinnenberg, I.A. The impact of centromeres on spatial genome architecture. *Trends Genet.* **2019**, *35*, 565–578. [CrossRef]
81. Chang, C.H.; Chavan, A.; Palladino, J.; Wei, X.; Martins, N.M.C.; Santinello, B.; Chen, C.C.; Erceg, J.; Beliveau, B.J.; Wu, C.T.; et al. Islands of retroelements are major components of *Drosophila* centromeres. *PLoS Biol.* **2019**, *17*, e3000241. [CrossRef]
82. Weissensteiner, M.H.; Pang, A.W.C.; Bunikis, I.; Höijer, I.; Vinnere-Petterson, O.; Suh, A.; Wolf, J.B.W. Combination of Short-Read, Long-Read, and Optical Mapping Assemblies Reveals Large-Scale Tandem Repeat Arrays with Population Genetic Implications. *Genome Res.* **2017**, *27*, 697–708. [CrossRef] [PubMed]
83. Ruiz-Ruano, F.J.; Cuadrado, Á.; Montiel, E.E.; Camacho, J.P.; López-León, M.D. Next Generation Sequencing and FISH Reveal Uneven and Nonrandom Microsatellite Distribution in Two Grasshopper Genomes. *Chromosoma* **2015**, *124*, 221–234. [CrossRef] [PubMed]
84. Skinner, B.M.; Sargent, C.A.; Churcher, C.; Hunt, T.; Herrero, J.; Loveland, J.E.; Dunn, M.; Louzada, S.; Fu, B.; Chow, W.; et al. The Pig X and Y Chromosomes: Structure, Sequence, and Evolution. *Genome Res.* **2016**, *26*, 130–139. [CrossRef] [PubMed]
85. Louzada, S.; Komatsu, J.; Yang, F. Fluorescence in situ hybridization onto DNA fibres generated using molecular combing. In *Fluorescence In Situ Hybridization (FISH)*; Liehr, T., Ed.; Springer: Berlin/Heidelberg, Germany, 2017; pp. 275–293.
86. Giunta, S. Centromere Chromosome Orientation Fluorescent In Situ Hybridization (Cen-CO-FISH) Detects Sister Chromatid Exchange at the Centromere in Human Cells. *Bio-protocol* **2018**, *8*. [CrossRef]
87. Saha, A.; Mourad, M.; Kaplan, M.H.; Chefetz, I.; Malek, S.N.; Buckanovich, R.; Markovitz, D.M.; Contreras-Galindo, R. The Genomic Landscape of Centromeres in Cancers. *bioRxiv* **2019**, 505800. [CrossRef]
88. White, T.B.; McCoy, A.M.; Streva, V.A.; Fenrich, J.; Deininger, P.L. A droplet digital PCR detection method for rare L1 insertions in tumors. *Mobile DNA* **2014**, *5*, 30. [CrossRef]
89. Kishikawa, T.; Otsuka, M.; Yoshikawa, T.; Ohno, M.; Yamamoto, K.; Yamamoto, N.; Yamamoto, N.; Kotani, A.; Koike, K. Quantitation of circulating satellite RNAs in pancreatic cancer patients. *JCI Insight* **2016**, *1*, e86646. [CrossRef]
90. Deakin, J.E.; Potter, S.; O'Neill, R.; Ruiz-Herrera, A.; Cioffi, M.B.; Eldridge, M.D.B.; Fukui, K.; Marshall Graves, J.A.; Griffin, D.; Grutzner, F.; et al. Chromosomics: Bridging the Gap between Genomes and Chromosomes. *Genes* **2019**, *10*, 627. [CrossRef]
91. Barski, A.; Cuddapah, S.; Cui, K.; Roh, T.Y.; Schones, D.E.; Wang, Z.; Wei, G.; Chepelev, I.; Zhao, K. High-resolution profiling of histone methylations in the human genome. *Cell* **2007**, *129*, 823–837. [CrossRef]
92. Palacios-Gimenez, O.M.; Bardella, V.B.; Lemos, B.; Cabral-de-Mello, D.C. Satellite DNAs are conserved and differentially transcribed among *Gryllus* cricket species. *DNA Res.* **2018**, *25*, 137–147. [CrossRef]
93. Kowar, T.; Zakrzewski, F.; Macas, J.; Kobližková, A.; Viehoever, P.; Weisshaar, B.; Schmidt, T. Repeat Composition of $CenH_3$-chromatin and $H_3K_9me_2$-marked heterochromatin in Sugar Beet (β vulgaris). *BMC Plant Biol.* **2016**, *16*, 120. [CrossRef] [PubMed]
94. Su, H.; Liu, Y.; Liu, C.; Shi, Q.; Huang, Y.; Han, F. Centromere satellite repeats have undergone rapid changes in polyploid wheat subgenomes. *Plant Cell* **2019**, *31*, 2035–2051. [CrossRef] [PubMed]
95. Khost, D.E.; Eickbush, D.G.; Larracuente, A.M. Single-Molecule Sequencing Resolves the Detailed Structure of Complex Satellite DNA Loci in *Drosophila*. *Genome Res.* **2017**, *27*, 709–721. [CrossRef] [PubMed]
96. De Lima, L.G.; Svartman, M.; Kuhn, C.S. Dissecting the Satellite DNA Landscape in Three Cactophilic *Drosophila* Sequenced Genomes. *G3 Genes Genomes Genet.* **2017**, *7*, 2831–2843. [CrossRef]
97. Flynn, J.M.; Long, M.; Wing, R.A.; Clark, A.G. Evolutionary dynamics of abundant 7 bp satellites in the genome of *Drosophila virilis*. *bioRxiv* **2019**, 693077. [CrossRef]
98. Hartley, G.; O'neill, R.J. Centromere Repeats: Hidden Gems of the Genome. *Genes* **2019**, *10*, 223. [CrossRef]
99. Ferguson-Smith, M.A.; Trifonov, V. Mammalian karyotype evolution. *Nat. Rev. Genet.* **2007**, *8*, 950–962. [CrossRef]
100. Spielmann, M.; Lupiáñez, D.G.; Mundlos, S. Structural variation in the 3D genome. *Nat. Rev. Genet.* **2018**, *19*, 453–467. [CrossRef]
101. Dobzhansky, T. *Genetics of the Evolutionary Process*; Columbia University Press: New York, NY, USA, 1970.

102. Wichman, H.; Payne, C.; Ryder, O.; Hamilton, M.; Maltbie, M.; Baker, R. Genomic distribution of heterochromatic sequences in equids: Implications to rapid chromosomal evolution. *J. Hered.* **1991**, *82*, 369–377. [CrossRef]
103. Rossi, M.; Redi, C.; Viale, G.; Massarini, A.; Capanna, E. Chromosomal distribution of the major satellite DNA of South American rodents of the genus Ctenomys. *Cytogenet. Cell Genet.* **1995**, *69*, 179–184. [CrossRef]
104. Chaves, R.; Guedes-Pinto, H.; Heslop-Harrison, J.S. Phylogenetic relationships and the primitive X chromosome inferred from chromosomal and satellite DNA analysis in Bovidae. *Proc. Biol. Sci.* **2005**, *272*, 2009–2016. [CrossRef] [PubMed]
105. Adega, F.; Chaves, R.; Guedes-Pinto, H.; Heslop-Harrison, J. Physical organization of the 1.709 satellite IV DNA family in Bovini and Tragelaphini tribes of the Bovidae: Sequence and chromosomal evolution. *Cytogenet. Genome Res.* **2006**, *114*, 140–146. [CrossRef] [PubMed]
106. Maynard Smith, J.; Burian, R.; Kaufman, S.; Alberch, P.; Campbell, J.; Goodwin, B. Developmental constraints and evolution. *Q. Rev. Biol.* **1985**, *60*, 265–287. [CrossRef]
107. Farré, M.; Robinson, T.J.; Ruiz-Herrera, A. An Integrative Breakage Model of genome architecture, reshuffling and evolution: The Integrative Breakage Model of genome evolution, a novel multidisciplinary hypothesis for the study of genome plasticity. *Bioessays* **2015**, *37*, 479–488. [CrossRef] [PubMed]
108. Kehrer-Sawatzki, H.; Cooper, D.N. Molecular mechanisms of chromosomal rearrangement during primate evolution. *Chromosome Res.* **2008**, *16*, 41–56. [CrossRef]
109. Carbone, L.; Harris, R.A.; Vessere, G.M.; Mootnick, A.R.; Humphray, S.; Rogers, J.; Kim, S.K.; Wall, J.D.; Martin, D.; Jurka, J.; et al. Evolutionary breakpoints in the gibbon suggest association between cytosine methylation and karyotype evolution. *PLoS Genet.* **2009**, *5*, e1000538. [CrossRef]
110. Zhao, H.; Bourque, G. Recovering genome rearrangements in the mammalian phylogeny. *Genome Res.* **2009**, *19*, 934–942. [CrossRef]
111. Paço, A.; Chaves, R.; Vieira-da-Silva, A.; Adega, F. The involvement of repetitive sequences in the remodelling of karyotypes: The *Phodopus* genomes (*Rodentia, Cricetidae*). *Micron* **2013**, *46*, 27–34. [CrossRef]
112. Bourque, G. Transposable elements in gene regulation and in the evolution of vertebrate genomes. *Curr. Opin. Genet. Dev.* **2009**, *19*, 607–612. [CrossRef]
113. Longo, M.S.; Carone, D.M.; Green, E.D.; O'Neill, M.J.; O'Neill, R.J.; NISC Comparative Sequencing Program. Distinct retroelement classes define evolutionary breakpoints demarcating sites of evolutionary novelty. *BMC Genomics* **2009**, *10*, 334. [CrossRef]
114. Garcia-Perez, J.L.; Widmann, T.J.; Adams, I.R. The impact of transposable elements on mammalian development. *Development* **2016**, *143*, 4101–4114. [CrossRef] [PubMed]
115. Platt, R.N.; Vandewege, M.W.; Ray, D.A. Mammalian transposable elements and their impacts on genome evolution. *Chromosome Res.* **2018**, *26*, 25–43. [CrossRef] [PubMed]
116. Bailey, J.A.; Eichler, E.E. Primate segmental duplications: Crucibles of evolution, diversity and disease. *Nat. Rev. Genet.* **2006**, *7*, 552–564. [CrossRef] [PubMed]
117. Louzada, S.; Paço, A.; Kubickova, S.; Adega, F.; Guedes-Pinto, H.; Rubes, J.; Chaves, R. Different evolutionary trails in the related genomes *Cricetus cricetus* and *Peromyscus eremicus* (*Rodentia, Cricetidae*) uncovered by orthologous satellite DNA repositioning. *Micron* **2008**, *39*, 1149–1155. [CrossRef] [PubMed]
118. Mashkova, T.; Oparina, N.; Alexandrov, I.; Zinovieva, O.; Marusina, A.; Yurov, Y.; Lacroix, M.H.; Kisselev, L. Unequal cross-over is involved in human α satellite DNA rearrangements on a border of the satellite domain. *FEBS Lett.* **1998**, *441*, 451–457. [CrossRef]
119. Kelkar, Y.D.; Tyekucheva, S.; Chiaromonte, F.; Makova, K.D. The genome-wide determinants of human and chimpanzee microsatellite evolution. *Genome Res.* **2008**, *18*, 30–38. [CrossRef]
120. Sankoff, D. The where and wherefore of evolutionary breakpoints. *J. Biol.* **2009**, *8*, 66. [CrossRef]
121. Catasti, P.; Chen, X.; Mariappan, S.V.; Bradbury, E.M.; Gupta, G. DNA repeats in the human genome. *Genetica* **1999**, *106*, 15–36. [CrossRef]
122. Yang, D.; Okamoto, K. Structural insights into G-quadruplexes: Towards new anticancer drugs. *Future Med. Chem.* **2010**, *2*, 619–646. [CrossRef]
123. Zhang, Z.; Dai, J.; Veliath, E.; Jones, R.A.; Yang, D.Z. Structure of a two-G-tetrad intramolecular G-quadruplex formed by a variant human telomeric sequence in K+ solution: Insights into the interconversion of human telomeric G-quadruplex structures. *Nucleic Acids Res.* **2010**, *38*, 1009–1021. [CrossRef]

124. Zhao, J.; Bacolla, A.; Wang, G.; Vasquez, K.M. Non-B DNA structure-induced genetic instability and evolution. *Cell. Mol. Life Sci.* **2010**, *67*, 43–62. [CrossRef] [PubMed]
125. Black, E.M.; Giunta, S. Repetitive Fragile Sites: Centromere Satellite DNA as a Source of Genome Instability in Human Diseases. *Genes* **2018**, *9*, 615. [CrossRef] [PubMed]
126. Khodaverdian, V.Y.; Hanscom, T.; Yu, A.M.; Yu, T.L.; Mak, V.; Brown, A.J.; Roberts, S.A.; McVey, M. Secondary structure forming sequences drive SD-MMEJ repair of DNA double-strand breaks. *Nucleic Acids Res.* **2017**, *45*, 12848–12861. [CrossRef] [PubMed]
127. De Sotero-Caio, C.G.; Cabral-de-Mello, D.C.; Calixto, M.D.S.; Valente, G.T.; Martins, C.; Loreto, V.; de Souza, M.J.; Santos, N. Centromeric enrichment of LINE-1 retrotransposons and its significance for the chromosome evolution of *Phyllostomid bats*. *Chromosome Res.* **2017**, *25*, 313–325. [CrossRef]
128. Klein, S.J.; O'Neill, R.J. Transposable elements: Genome innovation, chromosome diversity, and centromere conflict. *Chromosome Res.* **2018**, *26*, 5–23. [CrossRef]
129. Mayorov, V.I.; Adkison, L.R.; Vorobyeva, N.V.; Khrapov, E.A.; Kholodhov, N.G.; Rogozin, I.B.; Nesterova, T.B.; Protopopov, A.I.; Sablina, O.V.; Graphodatsky, A.S.; et al. Organization and chromosomal localization of a B1-like containing repeat of Microtus subarvalis. *Mamm. Genome* **1996**, *7*, 593. [CrossRef]
130. Marchal, J.A.; Acosta, M.J.; Bullejos, M.; Puerma, E.; Díaz de la Guardia, R.; Sánchez, A. Distribution of L1-retroposons on the giant sex chromosomes of *Microtus cabrerae* (*Arvicolidae, Rodentia*): Functional and evolutionary implications. *Chromosome Res.* **2006**, *14*, 177–186. [CrossRef]
131. Escudeiro, A.; Adega, F.; Robinson, T.J.; Heslop-Harrison, J.S.; Chaves, R. Conservation, divergence and functions of centromeric satellite DNA families in the *Bovidae*. *Genome Biol. Evol.* **2019**, *11*, 1152–1165. [CrossRef]
132. Wessler, S.R. Transposable elements and the evolution of eukaryotic genomes. *Proc. Natl. Acad. Sci. USA* **2006**, *103*, 17600–17601. [CrossRef]
133. Lim, J.K.; Simmons, J.M. Gross chromosome rearrangements mediated by transposable elements in *Drosophila melanogaster*. *Bioessays* **1994**, *16*, 269–275. [CrossRef]
134. Gray, Y.H.M. It takes two transposons to tango: Transposable-element-mediated chromosomal rearrangements. *Trends Genet.* **2000**, *16*, 461–468. [CrossRef]
135. Hedges, D.J.; Deininger, P.L. Inviting instability: Transposable elements, double-strand breaks, and the maintenance of genome integrity. *Mutat. Res.* **2007**, *616*, 46–59. [CrossRef] [PubMed]
136. Carbone, L.; Harris, R.A.; Gnerre, S.; Veeramah, K.R.; Lorente-Galdos, B.; Huddleston, J.; Meyer, T.J.; Herrero, J.; Roos, C.; Aken, B.; et al. Gibbon genome and the fast karyotype evolution of small apes. *Nature* **2014**, *513*, 195. [CrossRef] [PubMed]
137. Bailey, J.A.; Baertsch, R.; Kent, W.J.; Haussler, D.; Eichler, E.E. Hotspots of mammalian chromosomal evolution. *Genome Biol.* **2004**, *5*, R23. [CrossRef] [PubMed]
138. Cordaux, R.; Batzer, M.A. The impact of retrotransposons on human genome evolution. *Nat. Rev. Genet.* **2009**, *10*, 691–703. [CrossRef]
139. Warren, I.; Naville, M.; Chalopin, D.; Levin, P.; Berger, C.; Galiana, D.; Volff, J.N. Evolutionary impact of transposable elements on genomic diversity and lineage-specific innovation in vertebrates. *Chromosome Res.* **2015**, *23*, 505–531. [CrossRef]
140. Meštrović, N.; Mravinac, B.; Pavlek, M.; Vojvoda-Zeljko, T.; Šatović, E.; Plohl, M. Structural and functional liaisons between transposable elements and satellite DNAs. *Chrom. Res.* **2015**, *23*, 583–596. [CrossRef]
141. Paço, A.; Adega, F.; Chaves, R. Line-1 Retrotransposons: From "parasite" sequences to Functional Elements. *J. Appl. Genet.* **2015**, *56*, 133–145. [CrossRef]
142. Slamovits, C.H.; Rossi, M.S. Satellite DNA: Agent of chromosomal evolution in mammals. *A review. J. Neotrop. Mammal.* **2002**, *9*, 297–308.
143. Chaves, R.; Louzada, S.; Meles, S.; Wienberg, J.; Adega, F. *Praomys tullbergi* (*Muridae, Rodentia*) genome architecture decoded by comparative chromosome painting with *Mus* and *Rattus*. *Chromosome Res.* **2012**, *20*, 673–683, Epub 31 July 2012. [CrossRef]
144. Chaves, R.; Heslop-Harrison, J.S.; Guedes-Pinto, H. Centromeric heterochromatin in the cattle rob (1;29) translocation: α-satellite I sequences, In-Situ MspI digestion patterns, chromomycin staining and C-bands. *Chromosome Res.* **2000**, *8*, 621–626. [CrossRef] [PubMed]

145. Di Meo, G.P.; Perucatti, A.; Chaves, R.; Adega, F.; De Lorenzi, L.; Molteni, L.; De Giovanni, A.; Incarnato, D.; Guedes-Pinto, H.; Eggen, A.; et al. Cattle rob(1;29) originating from complex chromosome rearrangements as revealed by both banding and FISH-mapping techniques. *Chromosome Res.* **2006**, *14*, 649–655. [CrossRef] [PubMed]
146. Smalec, B.M.; Heider, T.N.; Flynn, B.L.; O'Neil, R.J. A centromere satellite concomitant with extensive karyotypic diversity across the *Peromyscus* genus defies predictions of molecular drive. *Chromosome Res.* **2019**, *27*, 237–252. [CrossRef] [PubMed]
147. Adega, F.; Matoso, S.R.; Kjöllerström, H.; Vercammen, P.; Raudsepp, T.; Collares-Pereira, M.J.; Fernandes, C.; Oom, M.; Chaves, R. Comparative chromosome painting in genets (*Carnivora, Viverridae, Genetta*), the only known feliforms with a highly rearranged karyotype. *Cytogenet. Genome Res.* **2018**, *156*, 35–44. [CrossRef] [PubMed]
148. Caceres, M.; Ranz, J.M.; Barbadilla, A.; Long, M.; Ruiz, A. Generation of a widespread *Drosophila* inversion by a transposable element. *Science* **1999**, *285*, 415–418. [CrossRef] [PubMed]
149. Rojo, V.; Martínez-Lage, A.; Giovannotti, M.; González-Tizón, A.M.; Cerioni, P.N.; Barucchi, V.C.; Galán, P.; Olmo, E.; Naveira, H. Evolutionary dynamics of two satellite DNA families in rock lizards of the genus *Iberolacerta* (*Squamata, Lacertidae*): Different histories but common traits. *Chromosome Res.* **2015**, *23*, 441–461. [CrossRef] [PubMed]
150. Dela Paz, J.S.; Stronghill, P.E.; Douglas, S.J.; Saravia, S.; Hasenkampf, C.A.; Riggs, C.D. Chromosome fragile sites in *Arabidopsis* harbor matrix attachment regions that may be associated with ancestral chromosome rearrangement events. *PLoS Genet.* **2012**, *8*, e1003136. [CrossRef]
151. Chaves, R.; Santos, S.; Guedes-Pinto, H. Comparative analysis (*Hippotragini* versus *Caprini, Bovidae*) of X-chromosome's constitutive heterochromatin by in situ restriction endonuclease digestion: X-chromosome constitutive heterochromatin evolution. *Genetica* **2004**, *121*, 315–325. [CrossRef]
152. Biscotti, M.A.; Olmo, E.; Heslop-Harrison, J.S. Repetitive DNA in Eukaryotic Genomes. *Chromosome Res.* **2015**, *23*, 415–420. [CrossRef]
153. Santos, S.; Chaves, R.; Guedes-Pinto, H. Chromosomal localization of the major satellite DNA family (FA-SAT) in the domestic cat. *Cytogenet. Genome Res.* **2004**, *107*, 119–122. [CrossRef]
154. Ting, D.T.; Lipson, D.; Paul, S.; Brannigan, B.W.; Akhavanfard, S.; Coffman, E.J.; Contino, G.; Deshpande, V.; Iafrate, A.J.; Letovsky, S.; et al. Aberrant Overexpression of Satellite Repeats in Pancreatic and Other Epithelial Cancers. *Science* **2011**, *331*, 593–596. [CrossRef] [PubMed]
155. Zhu, Q.; Pao, G.; Huynh, A.; Suh, H.; Tonnu, N.; Nederlof, P.M.; Gage, F.H.; Verma, I.M. BRCA1 tumour suppression occurs via heterochromatin-mediated silencing. *Nature* **2011**, *477*, 179–184. [CrossRef] [PubMed]
156. Smith, K.S.; Liu, L.L.; Ganesan, S.; Michor, F.; De, S. Nuclear topology modulates the mutational landscapes of cancer genomes. *Nat. Struct. Mol. Biol.* **2017**, *24*, 1000–1006. [CrossRef] [PubMed]
157. Guo, C.; Jeong, H.H.; Hsieh, Y.C.; Klein, H.U.; Bennett, D.A.; De Jager, P.L.; Liu, Z.; Shulman, J.M. Tau Activates Transposable Elements in Alzheimer's Disease. *Cell Rep.* **2018**, *23*, 2874–2880. [CrossRef]
158. Francastel, C.; Magdinier, F. DNA methylation in satellite repeats disorders. *Essays Biochem.* **2019**. [CrossRef]
159. Payer, L.M.; Burns, K.H. Transposable elements in human genetic disease. *Nat. Rev. Genet.* **2019**, *20*, 760–772. [CrossRef]
160. Fournier, A.; Mcleer-florin, A.; Lefebvre, C.; Duley, S.; Debernardi, A.; Rousseaux, S.; De Fraipont, F.; Figeac, M.; Kerckaert, J.; De Vos, J.; et al. 1q12 chromosome translocations form aberrant heterochromatic foci associated with changes in nuclear architecture and gene expression in B cell lymphoma. *EMBO Mol. Med.* **2010**, *2*, 159–171. [CrossRef]
161. Ducos, A.; Revay, T.; Kovacs, A.; Hidas, A.; Pinton, A.; Bonnet-Garnier, A.; Molteni, L.; Slota, E.; Switonski, M.; Arruga, M.V.; et al. Cytogenetic screening of livestock populations in Europe: An overview. *Cytogenet. Genome Res.* **2008**, *120*, 26–41. [CrossRef]
162. Iannuzzi, L.; King, W.; Di Berardino, D. Chromosome evolution in domestic bovids as revealed by chromosome banding and FISH-mapping techniques. *Cytogenet. Genome Res.* **2009**, *126*, 49–62. [CrossRef]
163. Therman, E.; Susman, B.; Denniston, C. The nonrandom participation of human acrocentric chromosomes in Robertsonian translocations. *Ann. Hum. Genet.* **1989**, *53*, 49–65. [CrossRef]
164. Page, S.L.; Shin, J.C.; Han, J.Y.; Choo, K.A.; Shaffer, L.G. Breakpoint diversity illustrates distinct mechanisms for Robertsonian translocation formation. *Hum. Mol. Genet.* **1996**, *5*, 1279–1288. [CrossRef] [PubMed]

165. Garagna, S.; Marziliano, N.; Zuccotti, M.; Searle, J.B.; Capanna, E.; Redi, C.A. Pericentromeric organization at the fusion point of mouse Robertsonian translocation chromosomes. *Proc. Natl. Acad. Sci. USA* **2001**, *98*, 171–175. [CrossRef] [PubMed]
166. Gustavsson, I. Cytogenetics, distribution and phenotypic effects of a translocation in *Swedish cattle*. *Hereditas* **1969**, *63*, 68–169. [CrossRef] [PubMed]
167. Dyrendahl, I.; Guatavsson, I. Sexual functions, semen characteristics and fertility of bulls carrying the 1/29 chromosome translocation. *Hereditas* **1979**, *90*, 281–289. [CrossRef]
168. Rangel-Figueiredo, T.; Iannuzzi, L. A cattle breed close to 58 diploid number due to high frequency of rob (1; 29). *Hereditas* **1991**, *115*, 73–78. [CrossRef]
169. Rangel-Figueiredo, T.; Iannuzzi, L. Frequency and distribution of rob (1; 29) in three *Portuguese cattle* breeds. *Hereditas* **1993**, *119*, 233–237. [CrossRef]
170. Rubes, J.; Kubickova, S.; Pagacova, E.; Cernohorska, H.; Di Berardino, D.; Antoninova, M.; Vahala, J.; Robinson, T.J. Phylogenomic study of spiral-horned antelope by cross-species chromosome painting. *Chromosome Res.* **2008**, *16*, 935–947. [CrossRef]
171. Gravholt, C.H.; Friedrich, U.; Caprani, M.; Jørgensen, A.L. Breakpoints in Robertsonian translocations are localized to satellite III DNA by fluorescence in situ hybridization. *Genomics* **1992**, *14*, 924–930. [CrossRef]
172. Wang, B.; Nie, B.; Tang, D.; Li, R.; Liu, X.; Song, J.; Wang, W.; Liu, Z. Analysis of meiotic segregation patterns and interchromosomal effects in sperm from 13 robertsonian translocations. *Balk. J. Med. Genet.* **2017**, *20*, 43–50. [CrossRef]
173. Kim, S.R.; Shaffer, L.G. Robertsonian translocations: Mechanisms of formation, aneuploidy, and uniparental disomy and diagnostic considerations. *Genet. Test.* **2002**, *6*, 163–168. [CrossRef]
174. Jarmuz-Szymczak, M.; Janiszewska, J.; Szyfter, K.; Shaffer, L.G. Narrowing the localization of the region breakpoint in most frequent Robertsonian translocations. *Chromosome Res.* **2014**, *22*, 517–532. [CrossRef] [PubMed]
175. Choo, K.H.; Vissel, B.; Brown, R.; Filby, R.G.; Earle, E. Homologous α satellite sequences on human acrocentric chromosomes with selectivity for chromosomes 13, 14 and 21: Implications for recombination between nonhomologues and Robertsonian translocations. *Nucleic Acids Res.* **1988**, *16*, 1273–1284. [CrossRef] [PubMed]
176. Akgun, E.; Zahn, J.; Baumes, S.; Brown, G.; Liang, F.; Romanienko, P.J.; Lewis, S.; Jasin, M. Palindrome resolution and recombination in the mammalian germ line. *Mol. Cell. Biol.* **1997**, *17*, 5559–5570. [CrossRef] [PubMed]
177. Bandyopadhyay, R.; Heller, A.; Knox-DuBois, C.; McCaskill, C.; Berend, S.A.; Page, S.L.; Shaffer, L.G. Parental origin and timing of de novo Robertsonian translocation formation. *Am. J. Hum. Genet.* **2002**, *71*, 1456–1462. [CrossRef] [PubMed]
178. Li, H. Toward better understanding of artifacts in variant calling from high-coverage samples. *Bioinformatics* **2014**, *30*, 2843–2851. [CrossRef] [PubMed]
179. Biscotti, M.A.; Canapa, A.; Forconi, M.; Olmo, E.; Barucca, M. Transcription of tandemly repetitive DNA: Functional roles. *Chromosome Res.* **2015**, *23*, 463–477. [CrossRef] [PubMed]
180. Mills, W.K.; Lee, Y.C.G.; Kochendoerfer, A.M.; Dunleavy, E.M.; Karpen, G.H. RNA from a simple-tandem repeat is required for sperm maturation and male fertility in *Drosophila melanogaster*. *Elife* **2019**, *8*, 48940. [CrossRef]
181. McNulty, S.M.; Sullivan, L.L.; Sullivan, B.A. Human Centromeres Produce Chromosome-Specific and Array-Specific α Satellite Transcripts that Are Complexed with CENP-A and CENP-C. *Dev. Cell* **2017**, *42*, 226–240. [CrossRef]
182. Liu, H.; Qu, Q.; Warrington, R.; Rice, A.; Cheng, N.; Yu, H. Mitotic transcription installs Sgo1 at centromeres to coordinate chromosome segregation. *Mol. Cell* **2015**, *59*, 426–436. [CrossRef]
183. Camacho, O.V.; Galan, C.; Swist-Rosowska, K.; Ching, R.; Gamalinda, M.; Karabiber, F.; De La Rosa-Velazquez, I.; Engist, B.; Koschorz, B.; Shukeir, N. Major satellite repeat RNA stabilize heterochromatin retention of Suv39h enzymes by RNA-nucleosome association and RNA: DNA hybrid formation. *Elife* **2017**, *6*, e25293. [CrossRef]
184. Johnson, W.L.; Yewdell, W.T.; Bell, J.C.S.; McNulty, M.; Duda, Z.; O'Neill, R.J.; Sullivan, B.A.; Straight, A.F. RNA-dependent stabilization of SUV39H1 at constitutive heterochromatin. *Elife* **2017**, *6*, e25299. [CrossRef] [PubMed]

185. Shirai, A.; Kawaguchi, T.; Shimojo, H.; Muramatsu, D.; Ishida-Yonetani, M.; Nishimura, Y.; Kimura, H.; Nakayama, J.I.; Shinkai, Y. Impact of nucleic acid and methylated H3K9 binding activities of Suv39h1 on its heterochromatin assembly. *Elife* **2017**, *6*, e25317. [CrossRef] [PubMed]
186. Chan, F.L.; Marshall, O.J.; Saffery, R.; Kim, B.W.; Earle, E.; Choo, K.H.; Wong, L.H. Active transcription and essential role of RNA polymerase II at the centromere during mitosis. *Proc. Natl. Acad. Sci. USA* **2012**, *109*, 1979–1984. [CrossRef] [PubMed]
187. Chan, F.L.; Wong, L.H. Transcription in the maintenance of centromere chromatin identity. *Nucleic Acids Res.* **2012**, *40*, 11178–11188. [CrossRef] [PubMed]
188. Grenfell, A.W.; Strzelecka, M.; Heald, R. Transcription brings the complex(ity) to the centromere. *CellCycle* **2017**, *16*, 235–236. [CrossRef] [PubMed]
189. Smurova, K.; De Wulf, P. Centromere and pericentromere transcription: Roles and regulation ... in sickness and in health. *Front. Genet.* **2018**, *9*, 674. [CrossRef] [PubMed]
190. Bouzinba-Segard, H.; Guais, A.; Francastel, C. Accumulation of small murine minor satellite transcripts leads to impaired centromeric architecture and function. *Proc. Natl. Acad. Sci. USA* **2006**, *103*, 8709–8714. [CrossRef]
191. Rošič, S.; Köhler, F.; Erhardt, S. Repetitive centromeric satellite RNA is essential for kinetochore formation and cell division. *J. Cell Biol.* **2014**, *207*, 335–349. [CrossRef]
192. Chan, D.Y.L.; Moralli, D.; Khoja, S.; Monaco, Z.L. Noncoding Centromeric RNA Expression Impairs Chromosome Stability in Human and Murine Stem Cells. *Dis. Markers* **2017**, 7506976. [CrossRef]
193. Lu, J.; Gilbert, D.M. Proliferation-dependent and cell cycle regulated transcription of mouse pericentric heterochromatin. *J. Cell Biol.* **2007**, *179*, 411–421. [CrossRef]
194. Maison, C.; Bailly, D.; Roche, D.; Montes de Oca, R.; Probst, A.V.; Vassias, I.; Dingli, F.; Lombard, B.; Loew, D.; Quivy, J.P.; et al. Sumoylation promotes de novo targeting of HP1alpha to pericentric heterochromatin. *Nat. Genet.* **2011**, *43*, 220–227. [CrossRef] [PubMed]
195. Vourc'h, C.; Biamonti, G. Transcription of Satellite DNAs in Mammals. *Prog. Mol. Subcell. Biol.* **2011**, *51*, 95–118. [CrossRef] [PubMed]
196. Jolly, C.; Metz, A.; Govin, J.; Vigneron, M.; Turner, B.M.; Khochbin, S.; Vourc'h, C. Stress-induced transcription of satellite III repeats. *J. Cell Biol.* **2004**, *164*, 25–33. [CrossRef] [PubMed]
197. Rizzi, N.; Denegri, M.; Chiodi, I.; Corioni, M.; Valgardsdottir, R.; Cobianchi, F.; Riva, S.; Biamonti, G. Transcriptional activation of a constitutive heterochromatic domain of the human genome in response to heat shock. *Mol. Biol. Cell* **2004**, *15*, 543–551. [CrossRef] [PubMed]
198. Biamonti, G. Nuclear stress bodies: A heterochromatin affair? *Nat. Rev. Mol. Cell Biol.* **2004**, *5*, 493–498. [CrossRef]
199. Biamonti, G.; Vourc'h, C. Nuclear stress bodies. *Cold Spring Harb Perspect. Biol.* **2010**, *2*. [CrossRef]
200. Goenka, A.; Sengupta, S.; Pandey, R.; Parihar, R.; Mohanta, G.C.; Mukerji, M.; Ganesh, S. Human satellite-III non-coding RNAs modulate heat-shock-induced transcriptional repression. *J. Cell Sci.* **2016**, *129*, 3541–3552. [CrossRef]
201. Valgardsdottir, R.; Chiodi, I.; Giordano, M.; Rossi, A.; Bazzini, S.; Ghigna, C.; Riva, S.; Biamonti, G. Transcription of Satellite III non-coding RNAs is a general stress response in human cells. *Nucleic Acids Res.* **2008**, *36*, 423–434. [CrossRef]
202. Kuhn, G.C. Satellite DNA transcripts have diverse biological roles in *Drosophila*'. *Heredity* **2015**, *115*, 1–2. [CrossRef]
203. Usakin, L.; Abad, J.; Vagin, V.V.; de Pablos, B.; Villasante, A.; Gvozdev, V.A. Transcription of the 1.688 satellite DNA family is under the control of RNA interference machinery in Drosophila melanogaster ovaries. *Genetics* **2007**, *176*, 1343–1349. [CrossRef]
204. Menon, D.U.; Coarfa, C.; Xiao, W.; Gunaratne, P.H.; Meller, V.H. siRNAs from an X-linked satellite repeat promote X-chromosome recognition in *Drosophila melanogaster*. *Proc. Natl. Acad. Sci. USA* **2014**, *111*, 16460–16465. [CrossRef] [PubMed]
205. Ferreira, D.; Escudeiro, A.; Adega, F.; Chaves, R. DNA Methylation Patterns of a Satellite Non-coding Sequence—FA-SAT in Cancer Cells: Its Expression Cannot Be Explained Solely by DNA Methylation. *Front. Genet.* **2019**, *10*, 101. [CrossRef] [PubMed]
206. Sana, J.; Faltejskova, P.; Svoboda, M.; Slaby, O. Novel classes of non-coding RNAs and cancer. *J. Transl. Med.* **2012**, *10*, 103. [CrossRef] [PubMed]

207. Eymery, A.; Horard, B.; El Atifi-Borel, M.; Fourel, G.; Berger, F.; Vitte, A.L.; Van den Broeck, A.; Brambilla, E.; Fournier, A.; Callanan, M.; et al. A transcriptomic analysis of human centromeric and pericentric sequences in normal and tumor cells. *Nucleic Acids Res.* **2009**, *37*, 6340–6354. [CrossRef] [PubMed]
208. Ehrlich, M.; Sanchez, C.; Shao, C.; Nishiyama, R.; Kehrl, J.; Kuick, R.; Kubota, T.; Hanash, S.M. ICF, an immunodeficiency syndrome: DNA methyltransferase 3B involvement, chromosome anomalies, and gene dysregulation. *Autoimmunity* **2008**, *41*, 253–271. [CrossRef]
209. Saksouk, N.; Simboeck, E.; Dejardin, J. Constitutive heterochromatin formation and transcription in mammals. *Epigenet. Chromatin* **2015**, *8*, 3. [CrossRef]

13

Chromosomal Differentiation in Genetically Isolated Populations of the Marsh-Specialist *Crocidura suaveolens* (Mammalia: Soricidae)

Francisca Garcia [1], Luis Biedma [2], Javier Calzada [2], Jacinto Román [3], Alberto Lozano [4,5], Francisco Cortés [1], José A. Godoy [6,*] and Aurora Ruiz-Herrera [4,5,*]

1 Unitat de Cultius Cel.lulars (UCC), Universitat Autònoma de Barcelona, Campus UAB, 08193 Cerdanyola del Vallès, Spain; francisca.garcia@uab.cat (F.G.); francisco.cortes@uab.cat (F.C.)

2 Department of Integrated Sciences, Faculty of Experimental Sciences, University of Huelva, Avenida de las Fuerzas Armadas, S/N, 21007 Huelva, Spain; setebiedma@hotmail.com (L.B.); javier.calzada@dbasp.uhu.es (J.C.)

3 Department of Conservation Biology, Doñana Biological Station, CSIC, C. Americo Vespucio 26, 41092 Sevilla, Spain; jroman@ebd.csic.es

4 Departament de Biologia Cel·lular, Fisiologia i Immunologia, Universitat Autònoma de Barcelona, Campus UAB, 08193 Cerdanyola del Vallès, Spain; alberto.lozanom@e-campus.uab.cat

5 Genome Integrity and Instability Group, Institut de Biotecnologia i Biomedicina, Universitat Autònoma de Barcelona, 08193 Cerdanyola del Vallès, Spain

6 Department of Integrative Ecology, Doñana Biological Station, CSIC, C. Americo Vespucio 26, 41092 Sevilla, Spain

* Correspondence: godoy@ebd.csic.es (J.A.G.); aurora.ruizherrera@uab.cat (A.R.-H.)

Abstract: The genus *Crocidura* represents a remarkable model for the study of chromosome evolution. This is the case of the lesser white-toothed shrew (*Crocidura suaveolens*), a representative of the Palearctic group. Although continuously distributed from Siberia to Central Europe, *C. suaveolens* is a rare, habitat-specialist species in the southwesternmost limit of its distributional range, in the Gulf of Cádiz (Iberian Peninsula). In this area, *C. suaveolens* is restricted to genetically isolated populations associated to the tidal marches of five rivers (Guadiana, Piedras, Odiel, Tinto and Guadalquivir). This particular distributional range provides a unique opportunity to investigate whether genetic differentiation and habitat specialization was accompanied by chromosomal variation. In this context, the main objective of this study was to determinate the chromosomal characteristics of the habitat-specialist *C. suaveolens* in Southwestern Iberia, as a way to understand the evolutionary history of this species in the Iberian Peninsula. A total of 41 individuals from six different populations across the Gulf of Cádiz were collected and cytogenetically characterized. We detected four different karyotypes, with diploid numbers (2n) ranging from 2n = 40 to 2n = 43. Two of them (2n = 41 and 2n = 43) were characterized by the presence of B-chromosomes. The analysis of karyotype distribution across lineages and populations revealed an association between mtDNA population divergence and chromosomal differentiation. *C. suaveolens* populations in the Gulf of Cádiz provide a rare example of true karyotypic polymorphism potentially associated to genetic isolation and habitat specialization in which to investigate the evolutionary significance of chromosomal variation in mammals and their contribution to phenotypic and ecological divergence.

Keywords: *Crocidura suaveolens*; shrews; habitat specialist; chromosomes; chromosomal evolution; B-chromosomes; chromosomal polymorphism; mtDNA

1. Introduction

Large-scale chromosomal changes, such as inversions, translocations, fusions and fissions, contribute to the reshuffling of genomes, thus providing new chromosomal forms on which natural selection can work. In this context, genome reshuffling has important evolutionary and ecological implications, since gene flow can be reduced within the reorganized regions in the heterokaryotype, thus affecting co-adapted genes locked within the rearrangement that, if advantageous, increase in frequency in natural populations (reviewed in [1]). In fact, evidence on the role of large-scale chromosomal changes in adaptation and diversification has been reported, especially in the case of inversions [2–5]. As for chromosomal fusions, however, empirical studies are limited to the house mouse (*Mus musculus domesticus*) and shrews (*Sorex araneus*), two mammalian systems where the presence of chromosomal fusions (either fixed or in polymorphic state within populations) are widespread [6–13]. Understanding the genetic and mechanistic basis of these processes will provide insights into how biodiversity originates and is maintained.

Shrews (family Soricidae) represent a clear example of chromosomal diversification within mammals, with diploid numbers ranging from 2n = 19 (*Blarina hylophaga*) to 2n = 68 (*Crocidura yankariensis*), showing both inter- and intra-specific chromosomal diversity [14]. Within Soricidae, the genus *Crocidura* represents a remarkable model where karyotypic variation correlates with phylogenetic relationships within the group. Initial studies based on nuclear and mitochondrial sequences (mtDNA) suggested a common ancestry of all *Crocidura* species [15,16], with a clear dichotomy between Afrotropical and Palearctic taxa, which display contrasting patterns of chromosomal differentiation [17]. With the exception of *C. luna* (2n = 28 or 36 [18]) Afrotropical species are characterized by high diploid numbers (from 2n = 42 to 2n = 68), 2n = 50 being the most common chromosomal form. Palearctic species, on the other hand, present a tendency for low diploid numbers (from 2n = 22 to 2n = 42, [17], with 2n = 40 as the predominant karyotype. It has been suggested that this bimodal distribution of diploid numbers within the genus *Crocidura* is probably the result of two contrasting tendencies in chromosomal evolution: (i) chromosomal fissions in Afrotropical species, and (ii) chromosomal fusions (tandem fusions centric fusions, and/or whole arm translocations) in Palearctic species [19].

Among Palearctic species, the lesser white-toothed shrew (*C. suaveolens*) is one of the most widely distributed shrews in Eurasia that presents the characteristic 2n = 40 chromosomal form [20,21]. Chromosomal polymorphisms, mainly involving pericentric inversions and heterochromatin distribution, have been reported as rare for white-toothed shrews [21]. Although continuously distributed from Siberia to Central Europe [22], the lesser white-toothed shrew is less common and has a more fragmented distribution in Western Europe, including the Iberian Peninsula [23]. In the southwesternmost limit of its distributional range, in the Gulf of Cádiz, *C. suaveolens* is rare, and only occurs in restricted, isolated populations associated to the tidal marches of five rivers (Guadiana, Piedras, Odiel, Tinto and Guadalquivir) [24] (see Figure 1). It has been proposed that its restricted distribution, the strict tidal marsh association and partly the genetic isolation of the populations, could be the consequence of competitive exclusion by the greater white-toothed shrew (*C. russula*, 2n = 42) [25,26]. In fact, recent studies based on mtDNA revealed the presence of two differentiated sub-lineages in Southwestern Iberia that diverged from other Iberian lineages around 140 Ka (50–240 Ka), and between themselves around 110 Ka (30–190 Ka). One of these sub-lineages occupies four river mouths closely located to one another (from 1 to 12 km), in the province of Huelva (Guadiana, Piedras, Odiel and Tinto; sub-lineage C3 [25]), whereas a second sub-lineage was present in the Guadalquivir river mouth (sub-lineage C4 [25]). Subsequent studies using microsatellites revealed that *C. suaveolens* populations from Guadiana, Piedras, Odiel and Tinto rivers showed low differentiation among them, but high differentiation with both the distant Guadalquivir population and the closely located Estero Domingo Rubio (EDR) population [24]. Overall, genetic data suggest that the observed genetic patterns are the result of both historical isolation and current restrictions to gene flow imposed by an adverse landscape matrix [24]. How chromosomal reorganizations have contributed (if so) to this genetic diverge is currently unknown.

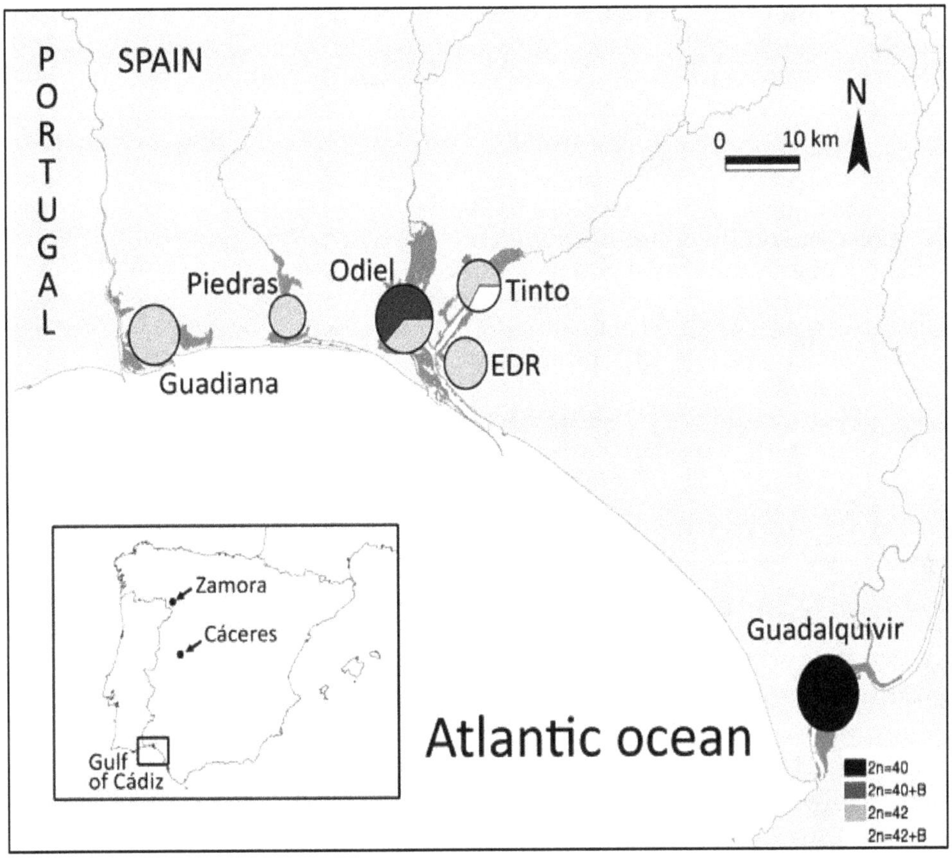

Figure 1. Study area and populations of *C. suaveolens* sampled in the tidal marsh (dark gray) of the Gulf of Cádiz, Southwestern Iberia. The individuals of Cáceres and Zamora are shown in the inset map. The sampling locations and diploid numbers are shown as pie charts. The size of each chart indicates the number of individuals sampled, which ranged between 5 and 8 individuals, in each location.

Despite their relevance, chromosomal data has often been neglected in phylogeographical studies (reviewed in [27]). The particular distributional range of the habitat-specialist *C. suaveolens* in Southwestern Iberia provides a unique opportunity to investigate whether the formation of genetically isolated populations and habitat specialization was accompanied by chromosomal differentiation. Given this context, in the present study, we aimed to do the following: (i) identify the extent of chromosomal variation *C. suaveolens* populations in Southwestern Iberia, and (ii) determine how chromosomal variation is distributed within and among genetic and phylogeographic groups. In this paper, we discuss the relative contribution of genetic drift and natural selection and the potential evolutionary significance of observed chromosomal differences.

2. Materials and Methods

2.1. Sampling

A total of 41 individuals from six different populations across the Gulf of Cádiz were collected in 2015 and 2016. This included different localities along the banks of the Guadalquivir, Guadiana, Odiel, Piedras and Tinto rivers (see Figure 1 and Table 1). The sample size ranged from two to six individuals per population. When possible, samples were obtained from both river banks, in order to evaluate whether rivers were acting as barriers to gene flow. Two additional individuals from the Northwest Iberian lineage (lineage B [24]) were also included in the study, for comparison (individuals from the Cáceres and Zamora populations, Table 1).

Table 1. Karyotype variants of *Crocidura suaveolens* found in the study area. Information includes population, region, locality, number of specimens (N), diploid chromosomal number (2n), autosomal fundamental number (FNa) and the presence of B-chromosomes (B-chr).

Population	Region	Locality	N	2n	FNa	B-chr
1	Guadalquivir	Eastern bank	4	2n = 40	46	0
1	Guadalquivir	Western bank	3	2n = 40	46	0
1	Guadalquivir	Faginao—Western bank	2	2n = 40	46	0
2	Guadiana	Isla de San Bruno	3	2n = 42	50	0
2	Guadiana	Salinas del Duque	4	2n = 42	50	0
3	Odiel	Cascajera	4	2n = 41/42	48/50	1
3	Odiel	Puntales	4	2n = 41/42	48/50	1
4	Piedras	Salinas	2	2n = 42	50	0
4	Piedras	El Terrón	3	2n = 42	50	0
5	Tinto	Eastern bank (Fosfoyesos)	6	2n = 42/43	48/52	1
6	EDR	Carabelas dock	6	2n = 42	50	0
7	Zamora	Flechas	1	2n = 40	46	0
8	Cáceres	Garganta de la Olla	1	2n = 40	46	0

Tissue samples (ear and tail) were obtained from each specimen on site and transferred to the cell culture laboratory in transport medium (Dulbecco's Modified Eagle Medium—DMEM— supplemented with 2 mm L-glutamine, 10% fetal bovine serum, 1% 100× Penicillin/Streptomycin/Amphotericin B solution and 0.07 mg/mL Gentamicin). Animals were immediately released after sample collection. Captures were performed with official permits issued by the corresponding nature conservation institutions, and research was conducted with approval of the bioethics committee of the University of Huelva and Universitat Autònoma de Barcelona.

2.2. *Cell Culture and Chromosomal Harvest*

Tissue samples were mechanically and enzymatically disaggregated. Briefly, tissue was washed with 5 mL of DPBS (Dulbecco's Phosphate-Buffered Saline) solution (DPBS with 1% 100x Penicillin/Streptomycin/Amphotericin B solution and 1mg/mL of Gentamicin) for 10 min, at 37 °C, in an orbital shaker, at 200 rpm. Biopsies were then shredded into small pieces, using a scalpel, in a Petri dish with 1 mL of DMEM without supplements, and incubated for 45 min in DMEM with 0.25% Collagenase Type II at 37 °C and at 200 rpm. Cell suspension was then centrifuged for 10 min at 300 G. The reaming cells were resuspended in 5 mL of completed growth medium (DMEM, supplemented with 20% of fetal bovine serum, 2 mm L-Glutamine) seeded on 25 cm^2 T-flasks and cultured at standard conditions (37 °C, 10% CO_2) for four weeks. Cultivated cells proliferated as an adherent monolayer. Subcultures of the adherent cells at early passages (3rd and 4th) were used to obtain chromosomes.

Chromosomal harvest was conducted as previously described [28]. In order to enhance the dispersion of chromosomes, cells were incubated in a hypotonic solution (KCl 0.075 M) for 20 min at 37 °C, inverting every 5 min. Subsequently, the cells were centrifuged (5 min at 300 g) and transferred into 15 mL tubes. The cell pellet was washed twice by adding 5 mL of fixative solution (methanol, acetic acid at 3:1 concentration, freshly prepared) and centrifuged (5 min at 300 g). Cells were centrifuged again and diluted in 1 mL of fixative solution and stored at −20 °C until use.

2.3. *Chromosomal Characterization*

Chromosomal spreads were obtained by dropping 15 μL of cell suspension onto a clean dry slide. Slides were baked at 65 °C during one hour and kept at −20 °C until use. Metaphases were stained homogenously with Giemsa solution for the analysis of the modal karyotype and then G-banded for karyotyping, as previously described [28].

An optical microscope (model Zeiss Axioskop) equipped with a charged coupled device camera (ProgResR CS10Plus, Jenoptik Optical Systems, Jena, Germany) was used for the microscope analysis.

A minimum of 25 good-quality metaphases were captured per specimen with the program Progress Capture 2.7.7 and analyzed in order to obtain the modal karyotype. In order to construct representative karyotype of each specimen analyzed, chromosomes were ordered by morphology and decreasing size, resulting in a representative karyotype.

2.4. Karyotype Distribution across Lineages and Populations

Several mitochondrial lineages in Iberia were defined previously by Biedma and collaborators [25], two of them (sub-lineages C3 and C4) are present in the study area. Populations were defined as genetic clusters, with each corresponding to one of the disjunct marshes associated to each of the five main rivers in the region (Guadalquivir, Tinto, Odiel, Guadiana and Piedras), and a sixth genetically differentiated population in EDR. Samples were grouped by mitochondrial lineage or geographical population for population cytogenetic analyses.

For the analysis of cytogenetic diversity and differentiation, the A diploid number (autosomes and sex chromosomes) and the presence/absence of B (supernumerary) chromosomes were treated as separate diploid and haploid traits, respectively. GENEPOP on the web (https://genepop.curtin.edu.au; [29]) was used to estimate allele (karyotype) frequencies, observed and expected heterozygosities, and to test for departure from Hardy–Weinberg (HW) expectations. We also tested for genotypic linkage disequilibrium between the two traits (A-chromosomal number and presence/absence of B-chromosomes), using the log likelihood ratio statistics. Differentiation between populations was assessed by the exact G or Fisher's tests and by estimating Wright's FST index.

3. Results

3.1. Chromosomal Diversity in C. suaveolens in the Gulf of Cádiz

Four karyotype variants were detected in the populations sampled: 2n = 40, 2n = 41 (i.e., 2n = 40 + B), 2n = 42 and 2n = 43 (i.e., 2n = 40 + B; Table 1). All nine individuals from the banks of the Guadalquivir River (sub-lineage C4) were characterized by presenting 2n = 40 (autosomal fundamental number, FNa = 46) (see Figure 2A), the same pattern found in the two individuals from the Northwest Iberian lineage (Cáceres and Zamora). The autosomes consisted of 15 pairs of acrocentric chromosomes and four pairs of bi-armed chromosomes, of which one pair was metacentric and three pairs sub-metacentric. The X-chromosome was a large sub-metacentric (see Figure 2A).

Specimens from three populations (Guadiana, Piedras and EDR) from sub-lineage C3 presented the same karyotype, consisting of 2n = 42 (FNa = 50) (see Figure 2B), with no polymorphic karyomorphs among the 18 individuals analyzed (see Table 1). The autosomes were 15 pairs of acrocentric chromosomes and five pairs of bi-armed chromosomes, of which one pair was metacentric and four pairs were sub-metacentric. The X-chromosome was a large sub-metacentric (see Figure 2B).

Interestingly, chromosomal polymorphisms were detected in both Odiel and Tinto river banks (Table 1), both populations also belonging to sub-lineage C3. In the case of the Odiel population, two distinct karyotypes were found in different proportion: 2n = 42 and 2n = 41. Three out of eight (37.5%) specimens presented the same 2n = 42 karyotype found in Guadiana, Piedras and EDR, whereas five individuals (62.5%) presented a karyotype consisting of 2n = 41 (FNa = 48) chromosomes (see Figure 2C). The 2n = 42 karyotype (FNa = 50) corresponded with the same one found in Guadiana, Piedras and EDR, and it was characterized by the presence of 15 pairs of acrocentric chromosomes and four pairs of bi-armed chromosomes, of which one pair was metacentric and three pairs were sub-metacentric. The 2n = 41 karyotype, however, corresponded to 15 pairs of acrocentric chromosomes and four pairs of bi-armed chromosomes, of which one pair was metacentric and three pairs were sub-metacentric, and the presence of one single B-chromosome was in all the individuals. Since the main difference with the 2n = 40 karyotype found in the Guadalquivir River was the presence of a single B-chromosome, we refer to the 2n = 41 karyotype from Odiel as 2n = 40 + B.

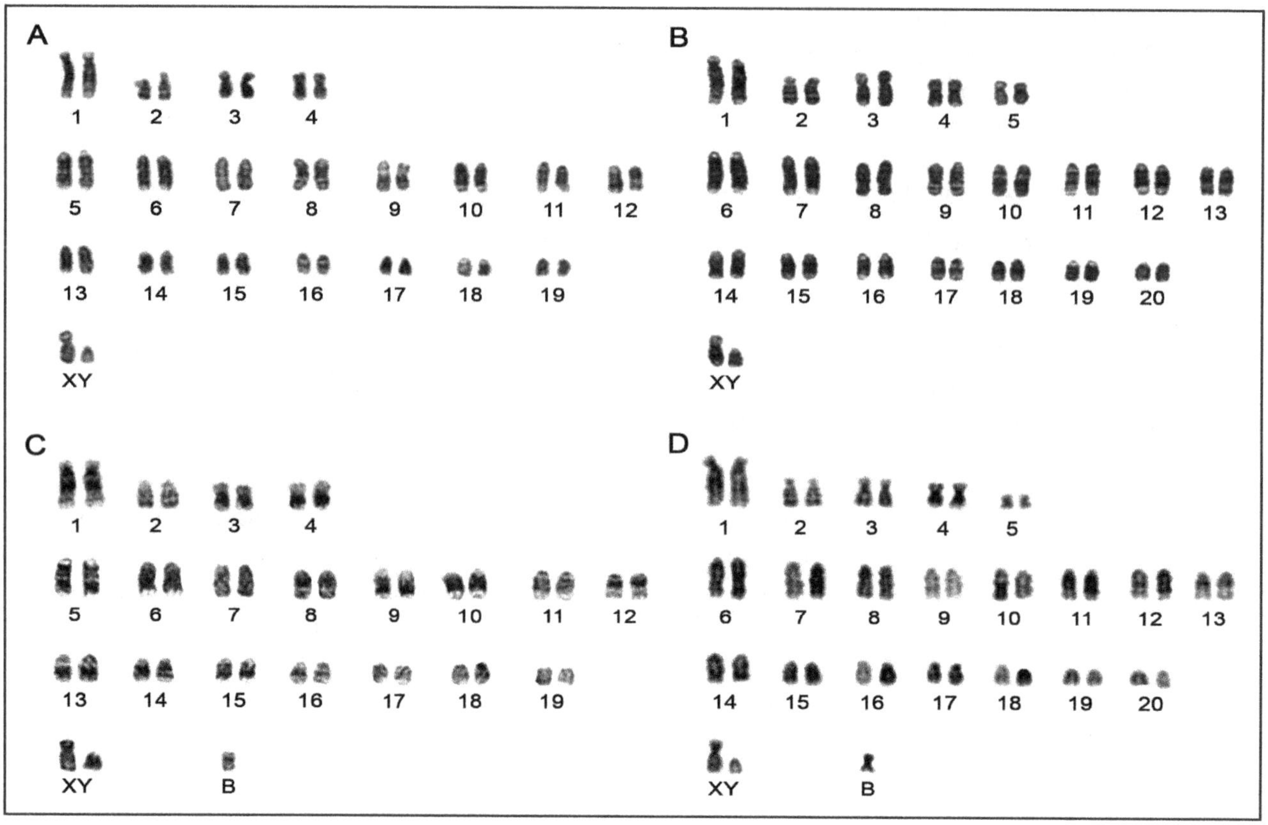

Figure 2. Examples of G-banded karyotypes of *C. suaveolens* found in the studied area. (**A**) Individual from the Guadalquivir River population (2n = 40). (**B**) Individual from the EDR population (2n = 42). (**C**) Individual from the Odiel River population (2n = 41). (**D**) Individual from the Tinto River population (2n = 43). Note the presence of a single B-chromosome in (**C**) and (**D**) karyotypes.

We also found two distinct karyotypes in the Tinto River population. Two out of six (33.3%) individuals surveyed in Tinto presented 2n = 43 (FNa = 52), whereas the rest (66.6%) were characterized by 2n = 42 (the same karyotype formula found in Guadiana, Piedras and EDR). The 2n = 43 karyotype was characterized by 15 pairs of acrocentric chromosomes and five pairs of bi-armed chromosomes, of which one pair was metacentric and four pairs were sub-metacentric, and the presence of a B-chromosome. The X-chromosome was a large sub-metacentric (see Figure 2D). Due to chromosomal G-banding homologies and the presence of a single B-chromosome, we refer to the 2n = 43 karyotype from Tinto as 2n = 42 + B.

G-banding comparison between 2n = 40 and 2n = 42 karyotypes suggests the presence of chromosomal fusion/fission events between bi-armed chromosomes.

3.2. *Karyotype Distribution and Diversity*

The four karyotypes observed were unevenly distributed across lineages and populations in the study area. The 2n = 40 A karyotype was the only one observed in the Guadalquivir population, where sub-lineage C4 occurs, and in Zamora and Cáceres samples, which are representative of the B lineage. In contrast, the 2n = 42 A karyotype was the most frequent in C3 populations, and the only one detected in EDR, Piedras and Guadiana (see Figure 3). The presence of B-chromosomes was restricted to the C3 lineage populations, where it occurred in the context of both 2n = 40 (five out of five occurrences) and 2n = 42 A karyotypes (two out of six occurrences). As a result, chromosomal differentiation between mtDNA subclades across both traits was highly significant (Fisher's exact test, Chi-Squared > 26.46, d.f. = 4, $p < 0.00002$), but it was high and significant for A-chromosome number

($F_{ST} = 0.763$, N = 82, $p < 0.00001$) and low and not significant for B-chromosome frequencies ($F_{ST} = 0.090$, N = 41, $p = 0.179$).

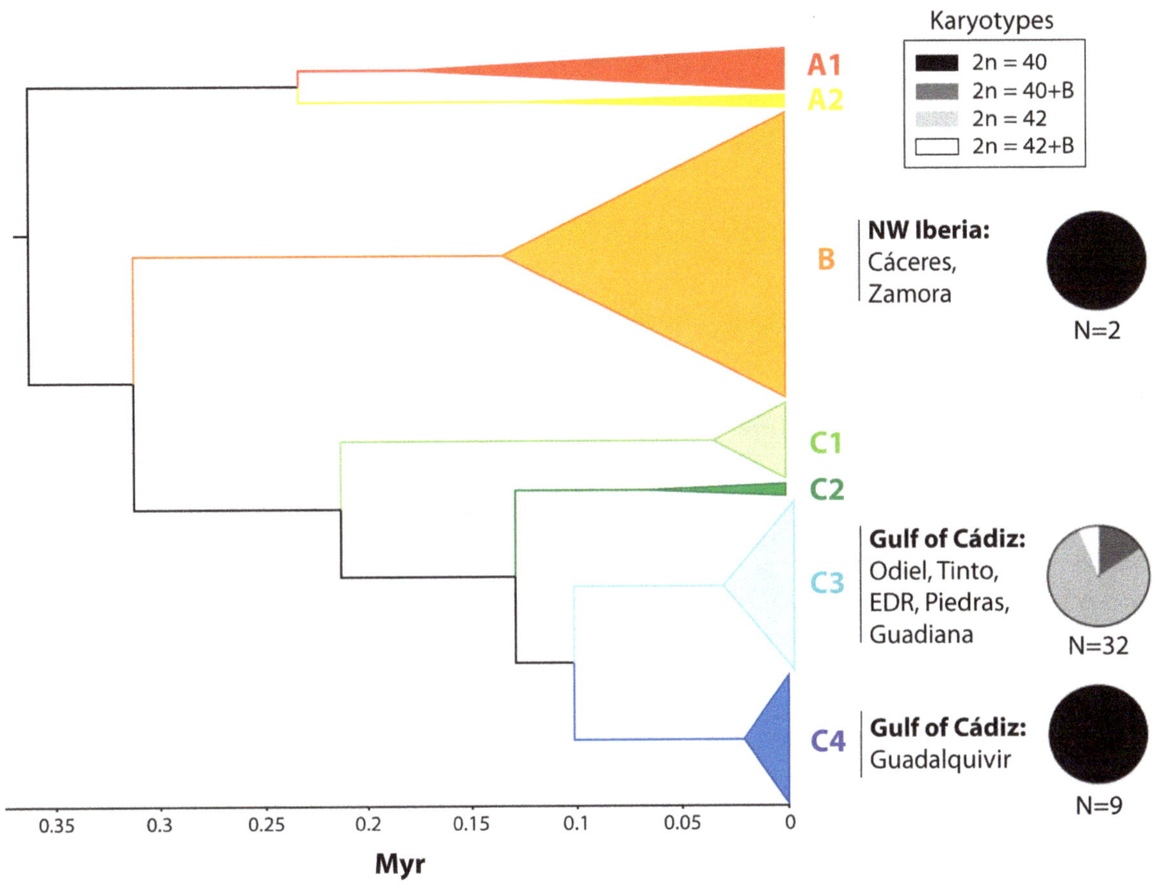

Figure 3. Distribution of karyotypes across the mitochondrial lineages described in Biedma et al. (2019). The geographical distribution of the sub-lineages sampled in this study is indicated. Pie charts represent the frequencies of karyotypes in each mitochondrial lineage, with N referring to the number of individuals sampled.

Most of the chromosomal diversity in the region is distributed among populations, as most populations showed no karyotypic diversity. Notable exceptions are the neighboring populations of Odiel and Tinto, each showing two composite karyotypes (2n = 40 + B and 2n = 42 in Odiel; 2n = 42 and 2n = 42 + B in Tinto). When only diploid A-chromosomal number is considered, the only population showing both 2n = 40 and 2n = 42 A-complements is Odiel ($K_R = 2$; $H_E = 0.47$; Table 2); although our limited sample sizes do not allow us to exclude the occurrence of both chromosomal types in other populations, results suggest highly unbalanced frequencies if this is the case. Departure from Hardy–Weinberg expectations was significant in Odiel ($F_{IS} = 1$, N = 8, p = 0.0074) due to the absence of heterokaryotypes despite rather equalized frequencies of the two karyotypes, although we cannot discard a spurious result due to small sample size. Odiel and Tinto were also variable with respect to the presence/absence of B-chromosomes (H = 0.47 and H = 0.44, respectively). The presence of B-chromosome appeared not completely independent of A-chromosome number in Odiel, as B-chromosome was observed there only on a 2n = 40 background (Chi-Squared = 8.02, d.f. = 2, $p = 0.018$).

Overall, karyotypic differentiation among populations was extremely high (Exact G test, N = 41, $p < 3.98 \times 10^{-9}$), both for A-chromosome number ($F_{ST} = 0.796$, N = 41, $p < 0.0001$) and B-chromosome ($F_{ST} = 0.408$, N = 41, $p = 0.0014$; Figure 4 chromosome ($F_{ST} = 0.408$, N = 41, $p = 0.0014$; Figure 4).

Table 2. Distribution of karyotypes across populations and karyotypic diversity. The frequency distribution of karyotypes for the composite karyotype and the frequency and diversity statistics for the A-chromosome number and B-chromosome presence are shown for each population, mtDNA lineage and for the pool of samples from the Gulf of Cádiz. Diversity is measured as expected heterozygosity and haplotype diversity for A-chromosome number and B-chromosome, respectively.

			Karyotype Frequency				**A-Chromosome Number**		**B- Chromosome**	
Population	N	mtDNA Clade	40	40+B	42	42+B	42 freq	HE	B freq	H
Guadalquivir	9	C4	1.00	0.00	0.00	0.00	0.00	0.00	0.00	0.00
Odiel	8	C3	0.00	0.63	0.38	0.00	0.38	0.47	0.63	0.47
Tinto	6	C3	0.00	0.00	0.67	0.33	1.00	0.00	0.33	0.44
EDR	6	C3	0.00	0.00	1.00	0.00	1.00	0.00	0.00	0.00
Piedras	5	C3	0.00	0.00	1.00	0.00	1.00	0.00	0.00	0.00
Guadiana	7	C3	0.00	0.00	1.00	0.00	1.00	0.00	0.00	0.00
Clade C4	9		1.00	0.00	0.00	0.00	0.00	0.00	0.00	0.00
Clade C3	32		0.00	0.16	0.78	0.06	0.84	0.00	0.22	0.00
Overall	41		0.11	0.06	0.30	0.02	0.66	0.45	0.17	0.28

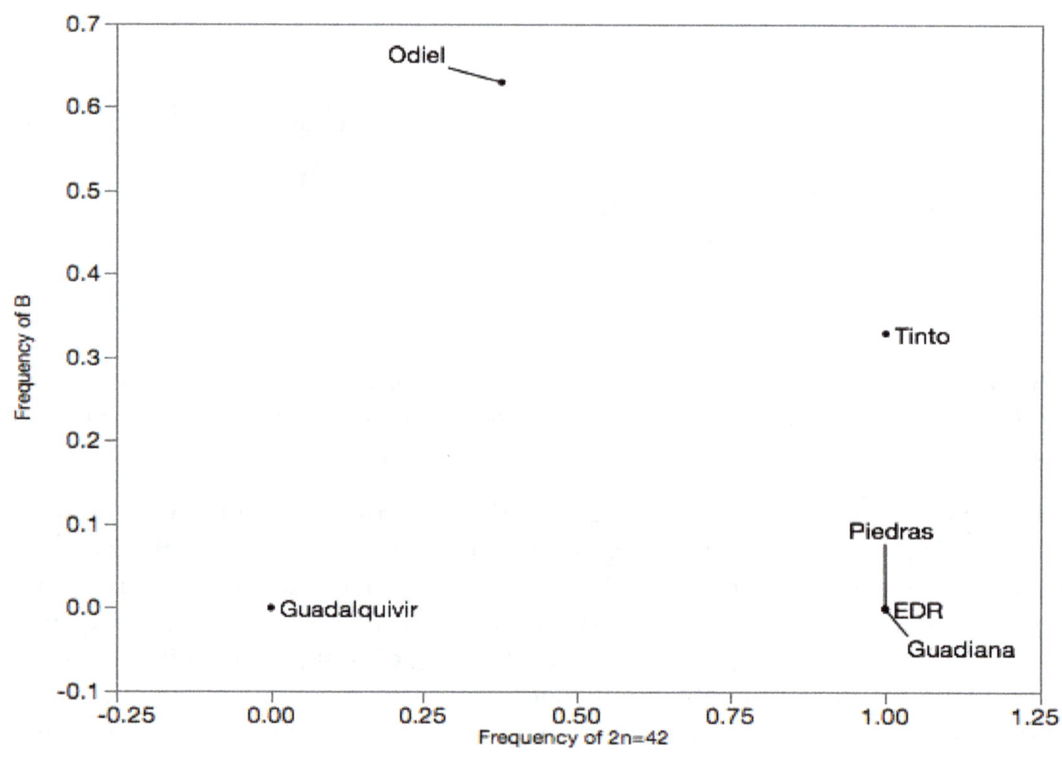

Figure 4. Separation of populations according to frequency of 2n = 42 and B-chromosome.

Karyotypic differentiation for A-chromosome number was maximal ($F_{ST} = 1$, N = 14–17, $p < 0.0001$) between Guadalquivir and all other populations, except Odiel ($F_{ST} = 0.308$, N = 17, $p = 0.0062$), and high between Odiel and Tinto, EDR, Piedras or Guadiana ($F_{ST} = 0.496$–0.550, N = 13–15, $p = 0.0014$–0.0004; Table 3). Slightly different patterns were observed for B-chromosome, with highest differentiation between Guadalquivir and Odiel ($F_{ST} = 0.591$, N= 17, $p = 0.0093$) and lowest between Odiel and Tinto ($F_{ST} = 0.013$, N = 14, $p = 0.591$; Table 3).

Table 3. Karyotypic differentiation among pairs of populations. F_{ST} values are shown for A-chromosomal number (above the diagonal) and B-chromosome (below the diagonal). Asterisks indicate the significance of exact G tests (* $p < 0.01$; ** $p < 0.001$). Pairwise comparisons indicated by "-" could not be estimated due to lack of variation in the pair.

	mtDNA Clade					
	C4	C3				
	Guadalquivir	*Odiel*	*Tinto*	*EDR*	*Piedras*	*Guadiana*
Guadalquivir		0.308 *	1000 **	1000 **	1000 **	1000 **
Odiel	0.591 *		0.525 **	0.525 **	0.496 *	0.550 **
Tinto	0.287	0.013		-	-	-
EDR	-	0.525	0.200		-	-
Piedras	-	0.496	0.161	-		-
Guadiana	-	0.550	0.233	-	-	

4. Discussion

4.1. Overview of Chromosomal Evolution in Crocidura

Understanding how genomes are organized and which types of chromosomal rearrangements are implicated in macroevolutionary events are fundamental to understanding the dynamics and emergence of new species [30]. Because Crocidura is the largest and one of the most karyotypically diverse genera of the family Soricidae [19,31], it offers a unique opportunity to test the role of chromosomal reorganization in species' diversification and habitat-specialization.

It has been long assumed that the widespread and monophyletic Palearctic *C. suaveolens* group is characterized by a karyotype of 2n = 40 [15,16]. Exceptions to this rule were initially reported in isolated populations from the Czech Republic (2n = 41) and Switzerland (2n = 42) [21]. Here we extend these initial observations and describe the presence of previously unreported chromosomal variation in the southwesternmost limit of its distributional range, in the Gulf of Cádiz, emphasizing the uniqueness of these genetically isolated and marsh-specialist populations. Karyotypic variability was reflected by the presence of four different karyotypes (2n = 40, 2n = 40 + B, 2n = 42 and 2n = 42 + B), the latter (2n = 42 + B) being reported here for the first time for C. suaveolens. As 2n = 40 is considered the chromosomal form ancestral for *C. suaveolens*, the G-banding comparisons between karyotypes suggest that chromosomal fissions and/or inversions changes in centromeric position together with the emergence of supernumerary (B) chromosomes have originated the chromosomal variability detected in our study. Most notably, our results provide a rare example of a true intrapopulation chromosomal polymorphism in *C. suaveolens*.

Remarkably, two of the karyotypes detected (2n = 40+B and 2n = 42+B) were characterized by the presence of B-chromosomes. Although previous studies recorded the occurrence of B-chromosomes for *C. crossei* [17], C. cf. *malayana* [32], *C. poensis* [33] and *C. suaveolens* itself [21], the present study is the first report focusing on the southwesternmost limit of the distributional range of *C. suaveolens*. Despite their widespread distribution in wild populations of several animal, plant and fungi species, the evolutionary origin and function of B-chromosomes are largely unknown. These dispensable chromosomes present a particular behavior, thus not following Mendelian segregation laws [34]. Most B-chromosomes are mainly or entirely heterochromatic (i.e., largely non-coding), although in some cases, B-chromosomes can provide some positive adaptive advantage, as suggested by associations with particular habitats [35] or with increases of crossing over and recombination frequencies [36–38]. Interestingly, the presence of B-chromosomes in our study was associated to the C3 mt DNA clade, more particularly in Tinto and Odiel populations. This, together with its generalized rarity in the rest of the distribution, suggests a recent and derivative origin of supernumerary chromosomes. Pending further functional and genomic studies, we can only speculate on the evolutionary implications of B-chromosomes in the *C. suaveolens* populations of Southwestern Iberia.

4.2. Karyotypic Diversity and Differentiation of C. suaveolens Populations in the Gulf of Cádiz

The karyotypic diversity detected in the *C. suaveolens* populations in the Gulf of Cadiz reveals the potential of chromosomal differentiation in the isolation of genetically distinct populations of the marsh-specialist *C. suaveolens* (Mammalia: Soricidae). In fact, we found an association between the two differentiated mtDNA sub-lineages (C3 and C4) and diploid numbers. All specimens from the Guadalquivir River's mouth (sub-lineage C4) were characterized by the ancestral chromosomal form for *C. suaveolens* (2n = 40), which was also present in the individuals sampled from the Northwest Iberian lineage (lineage B [24]). Remarkably, this ancestral karyotype was not detected in sub-lineage C3 populations (i.e., Guadiana, Odiel, Piedras, Tinto and EDR), which presented diploid numbers ranging from 2n = 41 to 2n = 43. Since it has been suggested that both mtDNA sub-lineages' ages diverged around 110 Ka (30–190 Ka) [25], the relationship between mtDNA divergence and chromosomal reorganizations adds support for the long-term isolation among C3 and C4 lineages.

Within populations of the C3 sub-lineage, the chromosomal form 2n = 42 was the most widespread, suggesting a common (and recent) origin in this lineage, most probably derived by a chromosomal fission from the ancestral form 2n = 40. The presence of 2n = 40 in these populations (always observed in combination with B-chromosome) is thus more parsimoniously explained as the retention of ancestral variation, especially since secondary contact with 2n = 40 populations in the Guadalquivir River is considered unlikely [24,25]. Chromosomal variation in Odiel and Tinto populations may thus constitute a transient "floating" polymorphism, as defined by [27], whose persistence may have been favored by different factors, such as its relatively recent origin (divergence of C3 sub-linage dated ca. 110 ka, [24]), relatively high population sizes and by the spatial structure within the Odiel–Tinto march complex. On the other hand, lack of detected polymorphisms in westernmost populations may be the consequence of a gradual diversity loss during westward colonization of more recent marshes, due to serial founder events [25].

Although chromosomal rearrangements have traditionally been considered to have a strong underdominance effect (e.g., [39]), there is increasing evidence to suggest that the small effect of certain types of rearrangements would favor their persistence within populations as polymorphisms [12]. This is the case, for example, of centric fusions/fissions (e.g., the physical joining of two acrocentric chromosomes by their centromeric regions and vice versa). Chromosomal fusions are particularly extended in nature (reviewed in [27]), occurring in as diverse taxa as mammals, reptiles, insects or mollusks [40,41]. Weak or null selection against centric fusions/fissions heterokaryotes due to mild meiotic pairing dysgenesis or reduction in meiotic recombination also might favor their presence in many lineages, as it has been previously suggested for Primates [42–44], Cetartiodactyla [45], the house mouse [12,46] and shrews [47]. Although the limited sample size included in our study calls for caution, the lack of observations of heterokaryotes, despite balanced frequencies of the two karyotypes detected in Odiel, could indicate underdominance in the case of *C. suaveolens* populations, a possibility that deserves further research.

5. Conclusions

Our observations for *C. suaveolens* provide an example of the persistence of chromosomal polymorphisms in mammals. The concurrence of chromosomal variation, recently diverged mitochondrial sub-lineages and habitat specialization in *C. suaveolens* populations in the Gulf of Cádiz provides a promising scenario to test the evolutionary significance of chromosomal variation in mammals and assess its contribution to phenotypic and ecological divergence. Future work should address the currently unresolved questions on the possible fitness differences among karyotypes, their role on postzygotic reproductive isolation and their association with morphological or ecological variation.

Author Contributions: Data curation, L.B., J.C., J.R., J.A.G. and A.R.-H.; formal analysis, F.G., L.B., J.C., J.R., J.A.G. and A.R.-H.; funding acquisition, J.A.G. and A.R.-H.; methodology, F.G., L.B., J.C., J.R., A.L., F.C., J.A.G and A.R.-H.;

project administration, J.A.G and A.R.-H.; resources, J.A.G. and A.R.-H.; supervision, A.R.-H; writing—original draft, J.A.G. and A.R.-H.; writing—review and editing, F.G., L.B., J.C. and J.R. All authors have read and agreed to the published version of the manuscript.

Acknowledgments: We acknowledge the valuable help in karyotyping of Ariadna Cortina, Marta Marín and Cinta Macià. We also thank the regional administrations of Andalucía, Castilla y León and Extremadura for the sampling permits, as well the managers of the protected natural areas sampled and environmental agents who assisted us in the field work.

References

1. Faria, R.; Johannesson, K.; Butlin, R.K.; Westram, A.M. Evolving Inversions. *Trends Ecol. Evol.* **2019**, *34*, 239–248. [CrossRef] [PubMed]
2. Hoffmann, A.A.; Rieseberg, L.H. Revisiting the impact of inversions in evolution: From population genetic markers to drivers of adaptive shifts and speciation? *Annu. Rev. Ecol. Evol. Syst.* **2008**, *39*, 21–24. [CrossRef] [PubMed]
3. Kapun, M.; Flatt, T. The adaptive significance of chromosomal inversion polymorphisms in *Drosophila melanogaster*. *Mol. Ecol.* **2019**, *28*, 1263–1282. [CrossRef] [PubMed]
4. Ayala, D.; Zhang, S.; Chateau, M.; Fouet, C.; Morlais, I.; Costantini, C.; Hahn, M.W.; Besansky, N.J. Association mapping desiccation resistance within chromosomal inversions in the African malaria vector *Anopheles gambiae*. *Mol. Ecol.* **2019**, *28*, 1333–1342. [CrossRef] [PubMed]
5. Christmas, M.J.; Wallberg, A.; Bunikis, I.; Olsson, A.; Wallerman, O.; Webster, M.T. Chromosomal inversions associated with environmental adaptation in honeybees. *Mol. Ecol.* **2019**, *28*, 1358–1374. [CrossRef] [PubMed]
6. Gündüz, I.; Auffray, J.C.; Britton-Davidian, J.; Catalan, J.; Ganem, G.; Ramalhinho, M.G.; Mathias, M.L.; Searle, J.B. Molecular studies on the colonization of the Madeiran archipelago by house mice. *Mol. Ecol.* **2001**, *10*, 2023–2029. [CrossRef]
7. Dumas, D.; Britton-Davidian, J. Chromosomal rearrangements and evolution of recombination: Comparison of chiasma distribution patterns in standard and Robertsonian populations of the house mouse. *Genetics.* **2002**, *162*, 1355–1366.
8. Giménez, M.D.; Mirol, P.M.; Bidau, C.J.; Searle, J.B. Molecular analysis of populations of *Ctenomys.* (Caviomorpha, Rodentia) with high karyotypic variability. *Cytogenet. Genome Res.* **2002**, *96*, 130–136. [CrossRef]
9. Borodin, P.M.; Karamysheva, T.V.; Belonogova, N.M.; Torgasheva, A.A.; Rubtsov, N.B.; Searle, J.B. Recombination map of the common shrew, *Sorex araneus.* (Eulipotyphla, Mammalia). *Genetics* **2008**, *178*, 621–632. [CrossRef]
10. Franchini, P.; Colangelo, P.; Solano, E.; Capanna, E.; Verheyen, E.; Castiglia, R. Reduced gene flow at pericentromeric loci in a hybrid zone involving chromosomal races of the house mouse *Mus musculus domesticus*. *Evolution* **2010**, *64*, 2020–2032.
11. Förster, D.W.; Mathias, M.L.; Britton-Davidian, J.; Searle, J.B. Origin of the chromosomal radiation of Madeiran house mice: A microsatellite analysis of metacentric chromosomes. *Heredity* **2013**, *110*, 380–388. [CrossRef] [PubMed]
12. Capilla, L.; Medarde, N.; Alemany-Schmidt, A.; Oliver-Bonet, M.; Ventura, J.; Ruiz-Herrera, A. Genetic recombination variation in wild Robertsonian mice: On the role of chromosomal fusions and Prdm9 allelic background. *Proc. R. Soc. B Biol. Sci.* **2014**, *281*, 20140297. [CrossRef] [PubMed]
13. Vara, C.; Capilla, L.; Ferretti, L.; Ledda, A.; Sánchez-Guillén, R.A.; Gabriel, S.I.; Albert-Lizandra, G.; Florit-Sabater, B.; Bello-Rodríguez, J.; Ventura, J.; et al. PRDM9 Diversity at fine geographical scale reveals contrasting evolutionary patterns and functional constraints in natural populations of house mice. *Mol. Biol. Evol.* **2019**, *36*, 1686–1700. [CrossRef] [PubMed]
14. O'Brien, S.J.; Menninger, J.C.; Nash, W.G. *An Atlas of Mammalian Chromosomes*; John Wiley & Sons Inc.: Hoboken, NJ, USA, 2006.
15. Dubey, S.; Salamin, N.; Ohdachi, S.D.; Barrière, P.; Vogel, P. Molecular phylogenetics of shrews (Mammalia: Soricidae) reveal timing of transcontinental colonizations. *Mol. Phylogenet. Evol.* **2007**, *44*, 126–137. [CrossRef] [PubMed]

16. Dubey, S.; Salamin, N.; Ruedi, M.; Barrière, P.; Colyn, M.; Vogel, P. Biogeographic origin and radiation of the Old World crocidurine shrews (Mammalia: Soricidae) inferred from mitochondrial and nuclear genes. *Mol. Phylogenet. Evol.* **2008**, *48*, 953–963. [CrossRef]
17. Maddalena, T.; Ruedi, M. Chromosomal Evolution in the Genus Crocidura. (Soricidae, Insectivora). In *Advances in the Biology of Shrews*; Merrit, J.F., Kirkland, G.L., Jr., Rose, R.K., Eds.; Carnegie Museum of Natural History: Pittsburgh, PA, USA, 1994; Volume 18, pp. 335–344.
18. Castiglia, R.; Annesi, F.; Sichilima, A.M.; Hutterer, R. A molecular and chromosomal study of the moonshine shrew, *Crocidura luna Dollman*, 1910 from Zambia with a description of a new remarkable karyotype. *Mammalia* **2009**, *73*, 56–59. [CrossRef]
19. Biltueva, L.; Vorobieva, N.; Perelman, P.; Trifonov, V.; Volobouev, V.; Panov, V.; Ilyashenko, V.; Onischenko, S.; O'Brien, P.; Yang, F.; et al. Karyotype evolution of eulipotyphla (insectivora): The genome homology of seven sorex species revealed by comparative chromosome painting and banding data. *Cytogenet. Genome Res.* **2001**, *135*, 51–64. [CrossRef]
20. Meylan, A.; Hausser, J. Cytotaxonomic position of several shrews of the genus Crocidura in Tessin (Mammalia, Insectivora). *Rev. Suisse Zool.* **1974**, *8*, 701–710. [CrossRef]
21. Zima, J.; Lukacova, L.; Machola, M. Chromosomal evolution in shrews. In *Evolution of Shrews*; Wojcik, J.M., Wolsan, M., Eds.; Mammal Research Institute, Polish Academy of Sciences: Bialowieza, Poland, 1998.
22. Palomo, L.; Kryštufek, B.; Amori, G.; Hutterer, R. *Crocidura suaveolens*. The IUCN Red List of Threatened Species 2016: E.T29656A22296429. Available online: http://www.iucnredlist.org/details/29656/ (accessed on 16 November 2016).
23. Libois, R.; Ramalhinho, M.G.; Fons, R. Crocidura. suaveolens. (Pallas, 1811), the lesser-white toothed shrew. In *The Atlas of European Mammals*; Mitchell-Jones, A.J., Amori, G., Bogdanowicz, W., Kryštufek, B., Reijnders, P.J.H., Spitzenberger, F., Stubbe, M., Thissen, J.B.M., Vohralik, V., Zima, J., Eds.; Poyser Natural History: London, UK, 1999; pp. 72–73.
24. Biedma, L.; Calzada, J.; Román, J.; Godoy, J.A. Rare and rear: Population genetics of marsh-specialist *Crocidura. suaveolens.* populations in the Gulf of Cádiz. *J. Mammal.* **2019**, *100*, 92–102. [CrossRef]
25. Biedma, L.; Román, J.; Calzada, J.; Friis, G.; Godoy, J.A. Phylogeography of *Crocidura suaveolens* (Mammalia: Soricidae) in Iberia has been shaped by competitive exclusion by *C. russula*. *Biol. J. Linn. Soc.* **2018**, mph123, 81–95. [CrossRef]
26. Biedma, L.; Calzada, J.; Godoy, J.A.; Román, J. Local habitat specialization as an evolutionary response to interspecific competition between two sympatric shrews. *J. Mammal.* **2020**, *101*, 80–91. [CrossRef]
27. Dobigny, G.; Britton-Davidian, J.; Robinson, T.J. Chromosomal polymorphism in mammals: An evolutionary perspective. *Biol. Rev.* **2004**, *92*, 1–21. [CrossRef] [PubMed]
28. Ruiz-Herrera, A.; Ponsà, M.; García, F.; Egozcue, J.; García, M. Fragile sites in human and *Macaca fascicularis* chromosomes are breakpoints in chromosome evolution. *Chromosome Res.* **2002**, *10*, 33–44. [CrossRef] [PubMed]
29. Raymond, M.; Rousset, F. GENEPOP (Version 1.2): Population genetics software for exact tests and ecumenicism. *Heredity* **1995**, *86*, 248–249. [CrossRef]
30. Ruiz-Herrera, A.; Farré, M.; Robinson, T.J. Molecular cytogenetic and genomic insights into chromosomal evolution. *Heredity* **2012**, *108*, 28–36. [CrossRef]
31. Hutterer, R. Order Soricomorpha. In *Mammal Species of the World: A Taxonomic and Geo-Graphic Reference*; Wilson, D.E., Reeder, D.M., Eds.; Johns Hopkins University Press: Baltimore, MD, USA, 2005; pp. 220–311.
32. Ruedi, M.; Vogel, P. Chromosomal evolution and zoogeographical origin of southeast Asian shrews (genus *Crocidura*). *Experientia* **1995**, *51*, 174–178. [CrossRef]
33. Maddalena, T. Systematics and biogeography of Afrotropical and Paleartic hrews of the genus *Crocidura* (Insectivora: Soricidae): An electrophoretic approach. In *Vertebrates in the Tropics*; Peters, G., Hutterer, R., Eds.; Museum Alexander Koenig Publication: Bonn, Germany, 1990; pp. 297–308.
34. Vujosevic, M.; Rajičić, M.; Blagojević, J.B. Chromosomes in Populations of Mammals Revisited. *Genes* **2018**, *9*, 487. [CrossRef]
35. Goodwin, S.; M'barek, S.B.; Dhillon, B.; Wittenberg, A.H.; Crane, C.F.; Hane, J.K.; Foster, A.J.; Van der Lee, T.A.; Grimwood, J.; Aerts, A.; et al. Finished genome of the fungal wheat pathogen *Mycosphaerella graminicola* reveals dispensome structure, chromosome plasticity, and stealth pathogenesis. *PLoS Genet.* **2011**, *7*, e1002070. [CrossRef]

36. Patton, J.L. A complex system of chromosomal variation in the pocket mouse, *Perognathus baileyi Merriam*. *Chromosoma* **1972**, *36*, 241–255. [CrossRef]
37. Dvorak, J.; Deal, K.R.; Luo, M.C. Discovery and mapping of wheat Ph1 suppressors. *Genetics* **2006**, *174*, 17–21. [CrossRef]
38. Thomson, R.L. B chromosomes in *Rattus fuscipes* II. The transmission of B chromosomes to offspring and population studies: Support for the "parasitic" model. *Heredity* **1984**, *52*, 363–372. [CrossRef]
39. Baker, R.J.; Bickham, J.W. Speciation by monobrachial centric fusions. *Proc. Natl. Acad. Sci. USA* **1979**, *83*, 8245–8248. [CrossRef] [PubMed]
40. White, M.J.D. *Animal Cytology and Evolution*; Cambridge University Press: New York, NY, USA, 1973.
41. King, M. *Species Evolution: The Role of Chromosome Change*; Cambridge University Press: New York, NY, USA, 1993.
42. Hamilton, A.E.; Beuttner-Janusch, J.; Chu, E.H. Chromosomes of lemuriformes. II. Chromosome polymorphism in Lemur fulvus collaris (E. Geoffroy 1812). *Am. J. Phys. Anthropol.* **1977**, *46*, 395–406. [CrossRef] [PubMed]
43. Farré, M.; Micheletti, D.; Ruiz-Herrera, A. Recombination rates and genomic shuffling in human and chimpanzee: A new twist in the chromosomal speciation theory. *Mol. Biol. Evol.* **2013**, *30*, 853–864. [CrossRef]
44. Ullastres, A.; Farré, M.; Capilla, L.; Ruiz-Herrera, A. Unraveling the effect of genomic structural changes in the rhesus macaque-implications for the adaptive role of inversions. *BMC Genom.* **2014**, *15*, 530. [CrossRef]
45. Rubes, J.; Pagacova, E.; Kopecna, O.; Kubickova, S.; Cernohorska, H.; Vahala, J.; Di Berardino, D. Karyotype, centric fusion polymorphism and chromosomal aberrations in captive-born mountain reedbuck (*Redunca fulvorufula*). *Cytogenet. Genome Res.* **2007**, *116*, 263–268. [CrossRef]
46. Medarde, N.; López-Fuster, M.J.; Muñoz-Muñoz, F.; Ventura, J. Spatio-temporal variation in the structure of a chromosomal polymorphism zone in the house mouse. *Heredity* **2012**, *109*, 78–89. [CrossRef]
47. Qumsiyeh, M.B.; Coate, J.L.; Peppers, J.A.; Kennedy, P.K.; Kennedy, M.L. Roberstonian chromosomal rearrangements in the short-tailed shrew *Blarina carolinensis*, in Western Tennessee. *Cytogenet. Genome Res.* **1997**, *76*, 153–158. [CrossRef]

14

Centric Fusions behind the Karyotype Evolution of Neotropical *Nannostomus* Pencilfishes (Characiforme, Lebiasinidae): First Insights from a Molecular Cytogenetic Perspective

Alexandr Sember [1], Ezequiel Aguiar de Oliveira [2,3], Petr Ráb [1], Luiz Antonio Carlos Bertollo [2], Natália Lourenço de Freitas [2], Patrik Ferreira Viana [4], Cassia Fernanda Yano [2], Terumi Hatanaka [2], Manoela Maria Ferreira Marinho [5], Renata Luiza Rosa de Moraes [2], Eliana Feldberg [4] and Marcelo de Bello Cioffi [2,*]

1 Laboratory of Fish Genetics, Institute of Animal Physiology and Genetics, Czech Academy of Sciences, Rumburská 89, 277 21 Liběchov, Czech Republic; sember@iapg.cas.cz (A.S.); rab@iapg.cas.cz (P.R.)

2 Departamento de Genética e Evolução, Universidade Federal de São Carlos, São Carlos, São Paulo 13565-905, Brazil; ezekbio@gmail.com (E.A.d.O.); bertollo@ufscar.br (L.A.C.B.); lfreitasnatalia@gmail.com (N.L.d.F.); yanocassia@gmail.com (C.F.Y.); hterumi@yahoo.com.br (T.H.); rlrdm@hotmail.com (R.L.R.d.M.)

3 Secretaria de Estado de Educação de Mato Grosso–SEDUC-MT, Cuiabá 78049-909, Brazil

4 Instituto Nacional de Pesquisas da Amazônia, Coordenação de Biodiversidade, Av. André Araújo 2936, Petrópolis, Manaus 69067-375, Brazil; patrik.biologia@gmail.com (P.F.V.); feldberg@inpa.gov.br (E.F.)

5 Universidade Federal da Paraíba (UFPB), Departamento de Sistemática e Ecologia (DSE), Laboratório de Sistemática e Morfologia de Peixes, João Pessoa 58051-090, Brazil; manoela.marinho@gmail.com

* Correspondence: mbcioffi@ufscar.br

Abstract: Lebiasinidae is a Neotropical freshwater family widely distributed throughout South and Central America. Due to their often very small body size, Lebiasinidae species are cytogenetically challenging and hence largely underexplored. However, the available but limited karyotype data already suggested a high interspecific variability in the diploid chromosome number ($2n$), which is pronounced in the speciose genus *Nannostomus*, a popular taxon in ornamental fish trade due to its remarkable body coloration. Aiming to more deeply examine the karyotype diversification in *Nannostomus*, we combined conventional cytogenetics (Giemsa-staining and C-banding) with the chromosomal mapping of tandemly repeated 5S and 18S rDNA clusters and with interspecific comparative genomic hybridization (CGH) to investigate genomes of four representative *Nannostomus* species: *N. beckfordi*, *N. eques*, *N. marginatus*, and *N. unifasciatus*. Our data showed a remarkable variability in $2n$, ranging from $2n = 22$ in *N. unifasciatus* (karyotype composed exclusively of metacentrics/submetacentrics) to $2n = 44$ in *N. beckfordi* (karyotype composed entirely of acrocentrics). On the other hand, patterns of 18S and 5S rDNA distribution in the analyzed karyotypes remained rather conservative, with only two 18S and two to four 5S rDNA sites. In view of the mostly unchanged number of chromosome arms (FN = 44) in all but one species (*N. eques*; FN = 36), and with respect to the current phylogenetic hypothesis, we propose Robertsonian translocations to be a significant contributor to the karyotype differentiation in (at least herein studied) *Nannostomus* species. Interspecific comparative genome hybridization (CGH) using whole genomic DNAs mapped against the chromosome background of *N. beckfordi* found a moderate divergence in the repetitive DNA content among the species' genomes. Collectively, our data suggest that the karyotype differentiation in *Nannostomus* has been largely driven by major structural rearrangements, accompanied by only low to moderate dynamics of repetitive DNA at the sub-chromosomal level. Possible mechanisms and factors behind the elevated tolerance to such a rate of karyotype change in *Nannostomus* are discussed.

Keywords: comparative genomic hybridization; karyotype variability; repetitive DNAs; Robertsonian translocation

1. Introduction

The Neotropical region harbors the richest freshwater ichthyofauna in the world, with approximately 5200 species belonging to 17 orders, thus representing about 40% of the freshwater biodiversity worldwide [1–3]. Moreover, the amount of cryptic and until now morphologically undistinguishable species suggests much higher species diversity (e.g., [4–9]). Fueled by these discoveries, the knowledge about the karyotype differentiation in Neotropical fishes has been rapidly growing (especially during the last few decades) and several important models for studying both sympatric and allopatric speciation, species complexes, and sex chromosome evolution have emerged [9–11]. As a prominent example, a remarkable cytogenetic variability has been found in the Erythrinidae family (Characiformes) and especially in *Erythrinus erythrinus* and *Hoplias malabaricus*, where several cases of multiple karyotype forms per species, high dynamics of repetitive DNA distribution, and intriguing diversity of chromosomal sex determination have been reported [10,12].

The family Lebiasinidae, which contains at least 72 valid species widely distributed throughout Central and South America, is divided into two subfamilies: The Lebiasininae (genera *Lebiasina*, *Piabucina*, and *Derhamia*) and the Pyrrhulininae (*Pyrrhulina*, *Nannostomus*, *Copeina*, and *Copella*) [13]. The latter represents the most diverse clade and it is also characterized by an extreme reduction of body size in some of its representatives. The most speciose genera in the subfamily are *Nannostomus* and *Pyrrhulina*, as each of them involves 19 species. Members of the genus *Nannostomus*, commonly referred to as pencilfishes, inhabit typically the flooded forests of the Amazon basin and they are valuable for the aquarist pet trade due to their colorful pigmentation. However, from the taxonomic viewpoint, it is one of the most challenging Lebiasinidae genera; therefore, a suite of complementary methodologies, such as cytogenetic comparisons and molecular analyses, including recently applied DNA barcoding, are highly valuable for clarifying this issue ([14,15] and references therein).

The small size of most Lebiasinidae fishes (i.e., ranging from 16 to 70 mm in length) makes the cytogenetic investigations of this group challenging and labor intensive, which may explain the large gaps in their cytogenetic data [16–18]. Nevertheless, a steadily growing body of information on karyotype characteristics in Lebiasinidae has been generated within recent years, using both conventional and molecular cytogenetic techniques, bringing new important pieces into the puzzle of lebiasinid karyotype differentiation and its underlying evolutionary mechanisms. More specifically, high rate of repetitive DNA dynamics and the occasional emergence of neo-sex chromosomes were found among four *Pyrrhulina* taxa; one of them may represent a new, yet undescribed species [19,20]. Furthermore, contrasting patterns of repetitive DNA content and distribution as well as a putative nascent sex chromosome system were also reported for *Lebiasina* species, supporting at the same time relationships between the Lebiasinidae and Ctenoluciidae families [21]. In addition, the first molecular cytogenetic report on *Copeina* species is filling another gap in this research [22]. Hence, for comparative purposes, similar data are necessary to be gathered in the remaining four lebiasinid genera (i.e., in *Copella*, *Derhamia*, *Piabucina*, and *Nannostomus*). A proper comparative cytogenetic survey might further contribute to cytotaxonomic comparisons between Lebiasinidae and evolutionarily related lineages.

In contrast to relative stability of the 2*n* in *Copeina*, *Lebiasina* and *Pyrrhulina* species karyotyped to date [16–22], representatives of *Copella* and *Nannostomus* display remarkable karyotype variability [16,18]. Indeed, even from limited karyotype data, it can be inferred that *Nannostomus* exhibits a wide range of 2*n*, from 22 (in *N. unifasciatus*) to 46 (in *N. trifasciatus*) [16,18,23], suggesting an important role of Robertsonian rearrangements in its karyotype differentiation.

The aim of the present study was to provide the first finer-scale cytogenetic investigation in the genus *Nannostomus*, performed both by conventional (Giemsa staining and C-banding) and molecular (fluorescence in situ hybridization (FISH) with 5S and 18S rDNA probes and comparative genomic hybridization (CGH)) methods in four species, namely *N. beckfordi'* *N. eques'* *N. marginatus*, and *N. unifasciatus*˙

2. Materials and Methods

2.1. Sampling

The collection sites, numbers, and sex of the individuals investigated are presented in Figure 1 and Table 1. All the specimens were collected under the appropriate authorization of the Brazilian environmental agency ICMBIO/SISBIO (License number 48628-2) and SISGEN (A96FF09). The specimens were taxonomically identified and sexed based on morphological characters and they were deposited in the fish collection site of the Museu de Zoologia da Universidade de São Paulo (MZUSP) under the voucher numbers 123071, 123079, 123083, and 123084. The experiments followed ethical and anesthesia conducts, in accordance with the Ethics Committee on Animal Experimentation of the Universidade Federal de São Carlos (Process number CEUA 1853260315) (Figure 1).

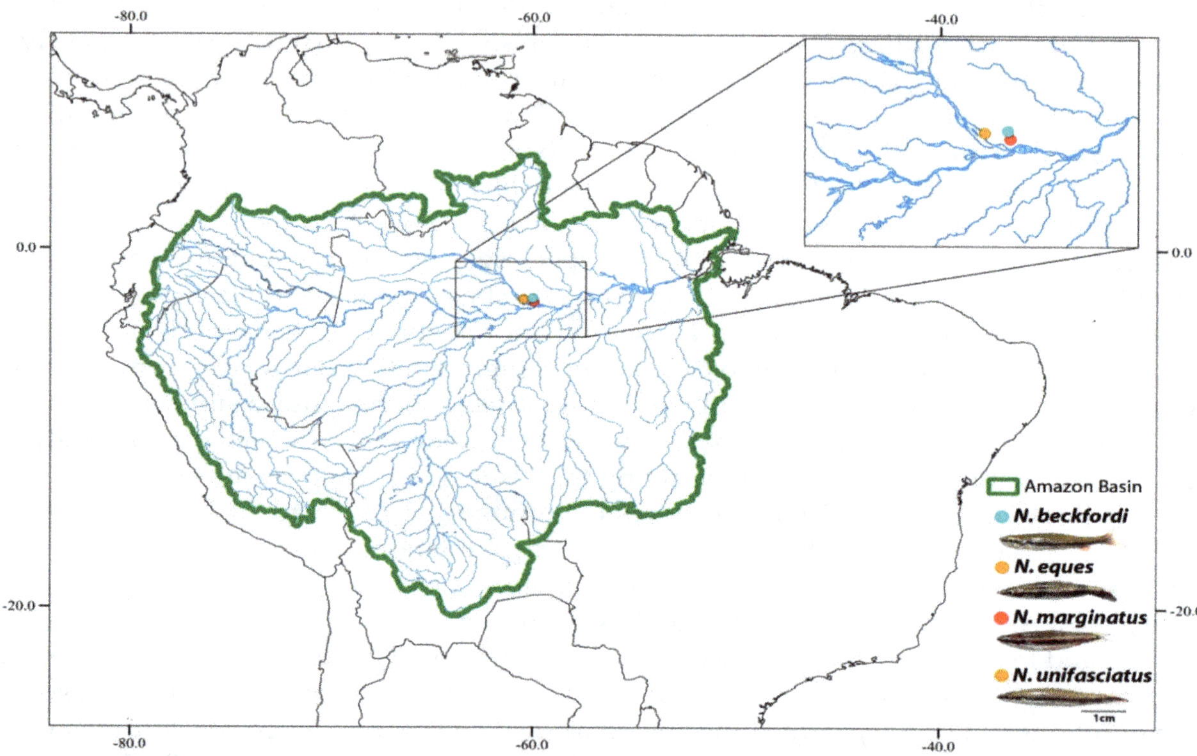

Figure 1. The map of Brazil with highlighted collection sites of *Nannostomus beckfordi* (blue circle), *N. eques*, *N. unifasciatus* (orange circle), and *N. marginatus* (red circle). The map was created using the following softwares: QGis 3.4.3, Inkscape 0.92, and Photoshop 7.0.

Table 1. Collection sites, 2*n* and the sample sizes (N) of the investigated *Nannostomus* species.

Species	2*n*	Sampling Site	N
Nannostomus beckfordi	44	Agenor Stream (Amazon River), AM	(09♀ 17♂)
Nannostomus eques	36	Cuieiras River, AM	(15♀ 12♂)
Nannostomus marginatus	42	Adolpho Ducke reserve (Negro River), AM	(08♀ 12♂)
Nannostomus unifasciatus	22	Cuieiras River, AM	(09♀ 13♂)

2.2. Chromosome Preparation and C-Banding

Mitotic chromosomes were obtained from kidney tissue using the air-drying technique according to Bertollo et al. [24]. Constitutive heterochromatin was visualized by C-banding following Sumner [25].

2.3. Repetitive DNA Mapping with Fluorescence In Situ Hybridization (FISH)

We mapped 5S and 18S rDNA tandem repeats generated from the genomic DNA of wolf fish *Hoplias malabaricus* by PCR amplification [26,27]. In the case of 5S rDNA, the resulting amplification product contained 120 base pairs (bp) of the 5S rRNA encoding region and 200 bp of the non-transcribed spacer (NTS). The second amplified fragment encompassed a 1400-bp-long segment of the 18S rRNA gene. 5S rDNA was labeled with digoxigenin-dUTP and 18S rDNA by biotin-dUTP, respectively, both by a nick translation kit, according to the manufacturer's recommendations (Roche, Mannheim, Germany). Fluorescence in situ hybridization (FISH) was performed under high stringency conditions, essentially following Yano et al. [28]. The hybridization mixture for each slide contained 100 ng of each probe, 50% deionized formamide, and 10% dextran sulphate (pH = 7.0), and it was denatured at 86 °C for 6 min prior to application. Chromosome preparations were denatured in 70% formamide in 2× SSC (pH = 7.0) for 3 min at 70 °C. Following overnight incubation at 37 °C in a moist chamber, post-hybridization washes were performed once in 2× SSC (5 min at 42 °C) and once in 1× SSC (5 min, Room Temperature). Prior to the probe detection, 3% non-fat dried milk (NFDM) in 2× SSC was applied on each slide (5 min, RT) to avoid the non-specific binding of antibodies. Probes were then detected using Avidin-FITC (Sigma, St. Louis, MO, USA) and Anti-Digoxigenin-Rhodamin (Roche, Basel, Switzerland). Finally, chromosomes were counterstained with 4.6-diamidino-2-phenylindole (DAPI) (1.2 μg/mL) and mounted in an antifade solution (Vector, Burlingame, CA, USA).

2.4. Comparative Genomic Hybridization (CGH)

We designed a set of experiments aimed at inter-specific genomic DNA comparison among all studied *Nannostomus* species. For this purpose, genomic DNAs (gDNA) from males and females of all species were isolated from liver tissue using a standard phenol/chloroform/isoamyl alcohol extraction [29]. We performed a set of separate experiments, where the *N. beckfordi* genomic probe was co-hybridized with the gDNA of one of the remaining species under study, against the chromosome background of *N. beckfordi*. The probes were generated again by nick translation reaction (Roche) as described above, with a differential labeling system employing biotin-dUTP (for *N. beckfordi*) and digoxigenin-dUTP (for *N. eques*, *N. marginatus*, and *N. unifasciatus*). Besides 500 ng of each labeled probe, the final hybridization mixture also contained 10 μg of unlabeled C_0t-1 DNA generated from a *N. beckfordi* female and 10 μg of unlabeled C_0t-1 DNA from the female of the compared species, in order to outcompete the excess of shared repetitive sequences. C_0t-1 DNA was prepared according to Zwick et al. [30]. The probes were precipitated with 100% ethanol and the air-dried pellets were mixed with a hybridization buffer containing 50% formamide, 10% SDS, 10% dextran sulfate, 2× SSC, and Denhardt's buffer (pH 7.0). The hybridization process took place in a moist chamber at 37 °C for 72 h. The hybridization procedure was performed according to Sember et al. [31]. After post-hybridization washes, done twice in 50% formamide in 2× SSC, pH 7.0 (44 °C, 10 min each) and three times in 1× SSC (44 °C, 7 min each), the probes were detected using Anti-Digoxigenin-Rhodamin (Roche, Basel, Switzerland) and Avidin-FITC (Sigma, St. Louis, MO, USA). Chromosomes were then counterstained with DAPI in antifade solution, as described above.

2.5. Microscopy and Image Processing

In total, 10 to 20 metaphases per individual were analyzed to confirm the 2*n*, chromosome morphology and FISH results. Images were captured using an Olympus BX50 microscope (Olympus Corporation, Ishikawa, Japan) with CoolSNAP and the images were processed using Image-Pro Plus 4.1 software (Media Cybernetics, Silver Spring, MD, USA). Chromosomes were classified as metacentric (m), submetacentric (sm), subtelocentric (st), or acrocentric (a) according to their centromere positions [32]. Karyotypes were arranged according to the chromosome size within each chromosome category.

3. Results

3.1. Conventional Cytogenetic Characteristics

The examined species differed markedly both in the 2*n* and karyotype composition (Figure 2). The karyotypes of *N. beckfordi* (2*n* = 44, FN = 44, where FN stands for the number of chromosome arms, i.e., fundamental number) and *N. eques* (2*n* = 36, FN = 36) were formed exclusively by acrocentric chromosomes. *N. marginatus* displayed, however, 2*n* = 42 and FN = 44, with only one (the largest) metacentric pair in an otherwise fully acrocentric set of chromosomes. In striking contrast, the karyotype of *N. unifasciatus* exhibited 2*n* = 22 and FN = 44, where all chromosomes were bi-armed only (i.e., metacentric and submetacentric ones) (Figure 2).

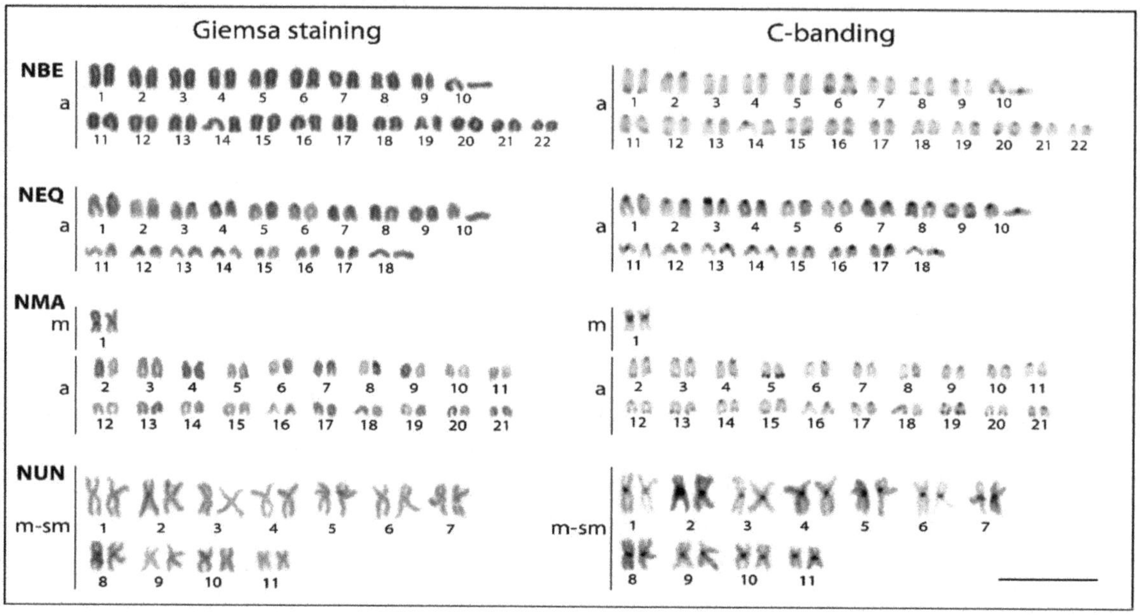

Figure 2. Karyotypes of *Nannostomus* species arranged after conventional cytogenetic protocols. Giemsa staining (left panel), C-banding (right panel). Abbreviations: NBE = *Nannostomus beckfordi*, NEQ = *N. eques*, NMA = *N. marginatus*, NUN = *N. unifasciatus*. Note a remarkable difference in the number, size, and morphology of chromosomes in *N. unifasciatus* in comparison to other studied species. Bar = 5 μm.

C-banding revealed that the constitutive heterochromatin is mainly confined to centromeric regions in all species. Terminal bands could be occasionally found in several (*N. beckfordi*) to few (*N. marginatus, N. eques*) chromosomes. Conspicuous heterochromatic blocks were found flanking the centromeres of all metacentric chromosomes in *N. unifasciatus* and the same counts also for a single metacentric pair in *N. marginatus* (Figure 2).

3.2. Patterns of 5S and 18S rDNA Distribution as Revealed by FISH

All karyotypes resulting from the rDNA FISH experiments are shown in Figure 3. The 5S rDNA probe revealed only one pair of signals in *N. beckfordi* and *N. unifasciatus* while the karyotypes of other species displayed two pairs with this repeat. The second pair of 5S rDNA signals was placed on the short (*p*) arms of the acrocentric chromosome pair No. 10 (in *N. eques*) and in the pericentromeric region of the acrocentric chromosome pair No. 19 (*N. marginatus*), respectively. Moreover, the karyotype of *N. unifasciatus* differed from those of the other species in that it had two syntenic sites in the large metacentric pair No. 2, one located in the proximal region and the second placed interstitially.

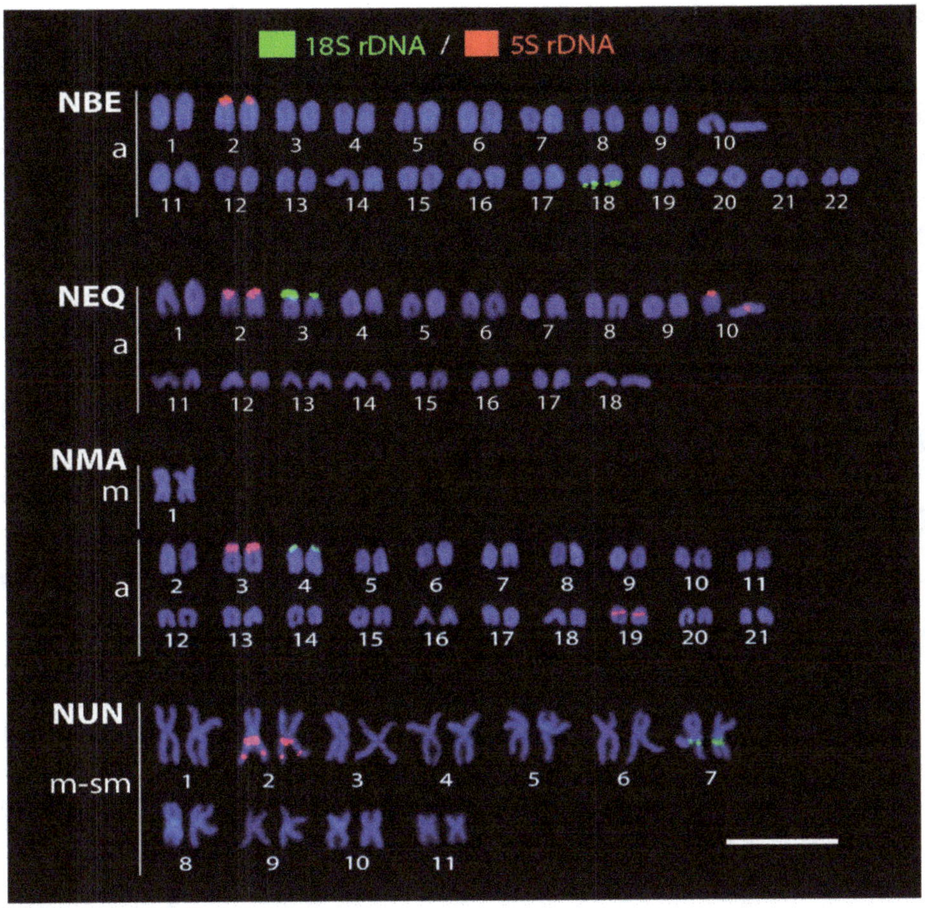

Figure 3. Karyotypes of *Nannostomus* species arranged after dual-color FISH with 5S and 18S rDNA probes. The FISH scheme includes 5S rDNA (red signals) and 18S rDNA (green signals) probes), and chromosomes were counterstained with DAPI. Abbreviations: NBE = *Nannostomus beckfordi*, NEQ = *N. eques*, NMA = *N. marginatus*, NUN = *N. unifasciatus*. Note the exceptional hybridization patterns in *N. unifasciatus* (specifically, the doubled 5S rDNA sites and the position of both rDNA classes near the centromeres of large metacentric chromosomes). Bar = 5 μm.

The 18S rDNA probe marked a single chromosomal pair in all species; however, the location of the signals differed slightly among species. While they were situated on the *p*-arms of the acrocentric pair No. 3 in *N. eques* and No. 4 in *N. marginatus*, respectively, *N. beckfordi* bore 18S rDNA sites on the terminal part of the long (*q*) arms of the acrocentric pair No. 18. Finally, *N. unifasciatus* displayed these cistrons in the proximal region of the metacentric pair No. 7.

3.3. *Patterns of Interspecific Genome Divergence as Revealed by CGH*

Cross-species CGH analysis revealed in each separate experiment rather equal binding of both co-hybridized genomic probes to all *N. beckfordi* chromosomes, thus yielding composite yellow signals (i.e., a combination of green and red). This hybridization pattern indicates the shared repetitive DNA content in the respective regions. Both probes hybridized preferentially to many centromeric and telomeric regions. In some cases, the intensity of the signals was biased towards either *N. beckfordi* genomic probe or to the probe of the compared species, probably reflecting the differential amount of specific repetitive DNA classes in the compared genomes. In addition, the chromosomes of *N. beckfordi* also showed many repetitive DNA accumulations which were stained exclusively by the *N. beckfordi* probe (Figure 4).

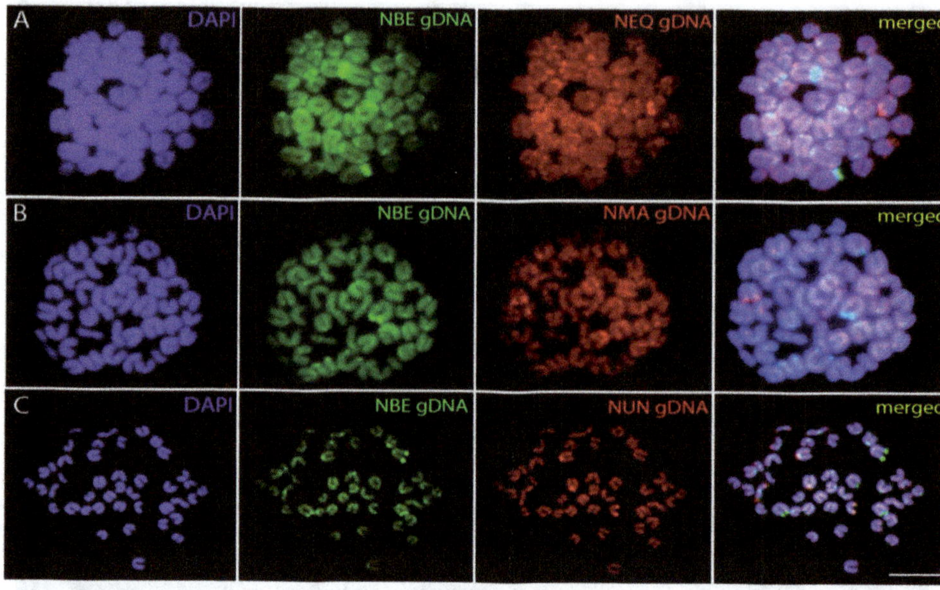

Figure 4. Mitotic chromosome spreads of *Nannostomus beckfordi* after interspecific CGH. Male-derived genomic DNA probe from **(A)** *N. eques*, **(B)** *N. marginatus*, and **(C)** *N. unifasciatus* mapped against male chromosomes of *N. beckfordi*. First column: DAPI images (blue); Second column: hybridization pattern produced by the genomic probe from one of the compared species; Third column: hybridization patterns produced by the genomic probe of *N. beckfordi*. Fourth column: merged images of both genomic probes and DAPI counterstaining. The common genomic regions are highlighted in yellow (i.e., a combination of the green and red hybridization probe). Bar = 5 μm.

4. Discussion

The herein studied *Nannostomus* species displayed a significant variability in the 2*n* values but with a stable FN equal to 44 in all but one species (*N. eques*; FN = 36). These patterns strongly indicated a series of Robertsonian rearrangements, for which the classification of chromosome arm numbers, i.e., NF value, was originally developed [33]. Nonetheless, because of a lack of clear landmarks to identify the individual chromosome pairs, the comparison across species is arbitrary and based on the chromosomal size and morphology only.

It is necessary to determine whether the evolutionary trajectory of karyotype change in *Nannostomus* is directed mainly towards centric fusions or fissions [34]. For this, the modal 2*n* of characiform fishes and the phylogenetic relationship of *Nannostomus* with the nearest lebiasinid lineages may provide a first useful indication (the principle reviewed in Dobigny et al. [35]). Thus, taking into account that (1) the modal 2*n* for characiforms very likely may be 2*n* = 54 [36], (2) *Lebiasina*, the most basal genus of Lebiasinidae [14] is characterized by 2*n* = 36 [21], and (3) the same 2*n* is also present among Ctenoluciidae species [37], a probable sister family of Lebiasinidae [38], we can infer that the reduction of 2*n* among the *Nannostomus* species was most likely achieved by a series of chromosome fusions. Specifically, according to Benzaquem et al. [15], *N. unifasciatus*, with 2*n* = 22 and with a karyotype formed exclusively by bi-armed chromosomes, is phylogenetically closely related to *N. beckfordi*, which possesses 2*n* = 44 and a karyotype formed by acrocentric chromosomes only. Therefore, considering their relationship, the chromosomal divergence between these species is clearly evidenced by their same NF and different karyotype compositions. Altogether, this suggests that centric fusions were the most probable mechanism behind the emergence of 22 m-sm chromosomes present in *N. unifasciatus*.

From the cytogenetic standpoint only, certain repetitive DNA markers, including 5S and 18S rDNA, have been formerly found to be involved in the formation of centric fusions (e.g., [39–41]). In the case of rDNAs, this may be possibly linked with the susceptibility of these tandemly repeated clusters to double-stranded DNA breaks, perhaps resulting from (1) a frequent rRNA transcription and thus break-prone R-loop emergence, (2) intermingling of NOR (Nuclear Organizer Region)-bearing chromosomes in the interphase nucleus, or (3) possible association of rDNA-bearing sites during the meiotic prophase I [42–48]. With a few exceptions, the terminal position of the 18S rDNA loci on chromosomes appears to be a common feature for all Lebiasinidae genera analyzed up to now (i.e., *Nannostomus, Pyrrhulina, Lebiasina,* and *Copeina*) ([19–22], this study). Altogether with Ctenoluciidae [37], this pattern can be considered as symplesiomorphy for both families. Although 5S rDNA displays a more dynamic evolution, with both terminal and interstitial signals among lebiasinids ([19–22], this study), it is noteworthy that *N. unifasciatus* underwent structural chromosome rearrangements involving both 18S and 5S rDNA loci, which have led to a derived pattern of rDNA distribution in this species. It is a rather expected scenario for *N. unifasciatus,* since this species exhibits the lowest $2n$ among Lebiasinidae fishes ($2n = 22$) and hence it may be speculated that the proximal 18S and 5S rDNA sites found in *N. unifasciatus* might rather represent hallmarks of fusion, suggesting the probable direction of chromosome change in this genus. However, despite this observation, it is obvious that, especially in *N. unifasciatus,* there might not be a preferential involvement of rDNA-bearing chromosomes in the formation of centric fusion, as many other uni-armed elements have been engaged in this process, leading to entirely bi-armed karyotype. Therefore, inversely, no major role of rDNA sites in the formation of fusions can so far be hypothesized in *Nannostomus*. Finer-scale analysis expanded in both taxonomic breadth and the number of cytogenetic markers is needed in order to better characterize the karyotype dynamics and to track whether there were also centric fissions or other types of rearrangements occurring in parallel in *Nannostomus* karyotype differentiation.

As another layer of evidence supporting the significant contribution of fusions in the karyotype dynamics of *Nannostomus,* the large blocks of constitutive heterochromatin flanking the centromeres of rather large-sized metacentric chromosomes, as found in the karyotypes of *N. marginatus* and *N. unifasciatus,* may be potentially considered as relics of two previously independent centromeres linked together by the process of fusion. In fact, such a situation has been repeatedly observed in many teleost species (sometimes, again, accompanied by the presence of rDNA sites in the fusion points) [49–54] and it was reported also in other animal taxa, e.g., amphibians [55] or mammals [56,57]. Nonetheless, other studies show that the large pericentromeric heterochromatic blocks can also be found evenly distributed throughout the karyotype regardless of the fusion events (see, e.g., Houck et al. [57] and Sousa et al. [58]).

Despite centric fusions not being a dominant type of chromosome rearrangement in teleosts, it seems that such a mechanism might indeed predominate in some lineages [59]. Within Teleostei, similar patterns of karyotype differentiation as those unraveled in *Nannostomus* have also been reported for African annual killifish genera *Nothobranchius* [60] and *Chromaphyosemion* [61,62], Gobiidae [63], Nothothenoidei [64], ophichthid eels (Ophichthidae) [54], Umbridae [65,66], and, in a broader context, also in the paleopolyploid Salmonidae family, where this process is apparently linked to the re-diploidization processes [67].

Gradual fixation of chromosome fusions may be linked to various selective pressures or to genetic drift [34,68,69]. It is also conceivable that the degree to which centric fusions are tolerated by the species' genome might be determined by specific properties linked to chromatin functional arrangement within the interphase nucleus, such as, e.g., elevated plasticity in the organization of chromosome territories, compartments, and topologically-associating domains [70–76]. It has been shown, for instance, that properly separated chromosome territories prevent the formation of inter-chromosomal fusions [77] and that changes in the architecture of the interphase genome may lead to severe consequences in gene expression [76]. Nonetheless, a recent study shows a high tolerance to disruption of the genome topology by rearrangements in the fruit fly *Drosophila melanogaster* [78]. Therefore, we may theorize that some organisms may better tolerate such alterations while others may be very sensitive to them, with selection acting strongly against the formation of inter-chromosomal rearrangements. Examples of both scenarios can be found among Teleostei, where some clades maintain constant $2n$ equal to 48 or 50 chromosomes while other lineages, including the genus *Nannostomus,* undergo frequent Robertsonian rearrangements [79]. It will therefore be an important aim for the future research to determine the main drivers behind such contrasting karyotype dynamics.

The distribution of repetitive DNAs may provide important clues about the pace of genome dynamics and it may also answer several taxonomic issues [9,10,80,81]. Chromosomal mapping of rDNA clusters has repeatedly helped to unveil diverse evolutionary issues (e.g., [82,83]). Particularly in fishes, it provided valuable clues about the incidence of cryptic, morphologically indistinguishable sibling species [5,6,8,10,84], polyploidization and interspecific hybridization events [85,86], a geographical gradient of genomic and morphological change [87], patterns of sex chromosome differentiation [80,88–90], and the correlation of genome dynamics in response to environmental cues [91,92].

Among the *Nannostomus* species investigated here, chromosomal mapping revealed somewhat uniform patterns of distribution for both rDNA classes, with one to few sites of accumulation, as found in most fishes [93,94], as well as in some other lebiasinids [21,22] investigated to date. While some of these sites may appear to be orthologous among the species under study, the frequently high dynamics of these repetitive DNA classes do not allow us to make certain conclusions without additional data (for an exemplary study, see Milhomem et al. [95]).

Nonetheless, in addition to the fact that some rDNA sites were clearly involved in Robertsonian fusions (as mentioned above), it may be inferred that like some other related lebiasinids [19], *Nannostomus* species do not show a substantial level of intrachromosomal dynamics that could be detected by the markers selected by us. This inference is further supported by the low to moderate amount of constitutive heterochromatin revealed by C-banding. Another supporting evidence for this assumption came from the CGH experiments. Despite CGH and related methods represent rather "rough" molecular tools, they may show patterns of the genomic divergence between species, as they rely on the presence of genome-specific repetitive DNA classes.

As repetitive DNA usually evolves rapidly in diverging genomes, such an approach may yield specific patterns of hybridization depending on the compared species, which (within a certain evolutionary timeframe) correlate with the degree of their divergence [31,96–98]. In the present study, rather minor interspecific differences in the composition of repetitive DNA among the compared *Nannostomus* species were shown. In summary, we propose that the karyotype differentiation in *Nannostomus*, at least in the species under study, was driven mainly by major structural rearrangements and the repetitive DNA content has not yet diverged significantly among the investigated genomes.

Chromosome rearrangements may not always be directly linked to speciation [68], but they may often provide an effective mechanism for post-zygotic reproductive isolation (in the case of fusions, e.g., [57,99,100]). By altering gene expression or by joining previously unlinked genetic material together, for instance, they might facilitate the emergence of evolutionarily advanced (e.g., locally adapted) sub-populations of a given species, thus contributing to diversification [69,101]. In addition, they might also be linked to the emergence of novel sex chromosome systems, such as that recently found in the lebiasinid genus *Pyrrhulina* [20]. Lastly, although additional detailed cytogenetic studies are still needed on a wider taxonomic scale, the present data reinforced the assumption that chromosomal fusions were important drivers of the karyotype evolution in the Neotropical family Lebiasinidae and, especially, in the pencil fishes of the genus *Nannostomus*.

Author Contributions: Conceptualization, A.S., P.R.; Formal analysis, A.S., E.A.d.O., P.F.V., C.F.Y., M.d.B.C., M.M.F.M., R.L.R.d.M. and E.F.; Funding acquisition, A.S., L.A.C.B. and M.d.B.C.; Investigation, A.S., E.A.d.O., L.A.C.B., N.L.d.F., P.F.V., C.F.Y., T.H., M.M.F.M. and E.F.; Methodology, A.S., E.A.d.O., N.L.d.F., P.F.V., C.F.Y., M.d.B.C., T.H., M.M.F.M. and R.L.R.d.M.; Project administration, M.d.B.C.; Supervision, A.S., L.A.C.B. and M.d.B.C.; Validation, A.S., P.R., E.A.d.O., N.L.d.F., C.F.Y., T.H., M.M.F.M., R.L.R.d.M., P.R. and E.F.; Visualization, L.A.C.B., P.F.V., C.F.Y., M.d.B.C. and T.H.; Writing—original draft, A.S., P.R. and N.L.d.F.; Writing—review and editing, E.A.d.O., L.A.C.B., P.F.V., C.F.Y., M.d.B.C., T.H., M.M.F.M., R.L.R.d.M., P.R. and E.F. All authors have read and agreed to the published version of the manuscript.

References

1. Albert, J.S.; Reis, R.E. *Historical Biogeography of Neotropical Freshwater Fishes*, 1st ed.; University of California Press: Berkeley, CA, USA, 2011.
2. Nelson, J.S.; Grande, T.C.; Wilson, M.V.H. *Fishes of the World*, 5th ed.; John Wiley & Sons: Hoboken, NJ, USA, 2016.
3. Reis, R.E.; Albert, J.S.; Di Dario, F.; Mincarone, M.M.; Petry, P.; Rocha, L.A. Fish biodiversity and conservation in South America. *J. Fish Biol.* **2016**, *89*, 12–47. [CrossRef] [PubMed]
4. Pereira, L.H.G.; Hanner, R.; Foresti, F.; Oliveira, C. Can DNA barcoding accurately discriminate megadiverse Neotropical freshwater fish fauna? *BMC Genet.* **2013**, *14*, 20. [CrossRef] [PubMed]
5. Ferreira, M.; Kavalco, K.F.; de Almeida-Toledo, L.F.; Garcia, C. Cryptic diversity between two *Imparfinis* species (Siluriformes, Heptapteridae) by cytogenetic analysis and DNA barcoding. *Zebrafish* **2014**, *11*, 306–317. [CrossRef] [PubMed]
6. Ferreira, M.; Garcia, C.; Matoso, D.A.; de Jesus, I.S.; Cioffi, M.B.; Bertollo, L.A.C.; Zuanon, J.; Feldberg, E. The *Bunocephalus coracoideus* species complex (Siluriformes, Aspredinidae). Signs of a speciation process through chromosomal, genetic and ecological diversity. *Front. Genet.* **2017**, *8*, 1–12. [CrossRef]
7. Ramirez, J.L.; Birindelli, J.L.; Carvalho, D.C.; Affonso, P.R.A.M.; Venere, P.C.; Ortega, H.; Carrillo-Avila, M.; Rodríguez-Pulido, J.A.; Galetti, P.M., Jr. Revealing hidden diversity of the underestimated neotropical ichthyofauna: DNA barcoding in the recently described genus *Megaleporinus* (Characiformes: Anostomidae). *Front. Genet.* **2017**, *8*, 1–11. [CrossRef]
8. Prizon, A.C.; Bruschi, D.P.; Borin-Carvalho, L.A.; Cius, A.; Barbosa, L.M.; Ruiz, H.B.; Zawadzki, C.H.; Fenocchio, A.S.; Portela-Castro, A.L.B. Hidden diversity in the populations of the armored catfish *Ancistrus* Kner, 1854 (Loricariidae, Hypostominae) from the Paraná River Basin revealed by molecular and cytogenetic data. *Front. Genet.* **2017**, *8*, 185. [CrossRef]
9. Cioffi, M.B.; Moreira-Filho, O.; Ráb, P.; Sember, A.; Molina, W.F.; Bertollo, L.A.C. Conventional cytogenetic approaches—Useful and indispensable tools in discovering fish biodiversity. *Curr. Genet. Med. Rep.* **2018**, *6*, 176–186. [CrossRef]
10. Cioffi, M.B.; Molina, W.F.; Artoni, R.F.; Bertollo, L.A.C. Chromosomes as tools for discovering biodiversity—The case of Erythrinidae fish family. In *Recent Trends in Cytogenetic Studies—Methodologies Applications*, 1st ed.; Tirunilai, P., Ed.; INTECH: London, UK, 2012; Volume 1, pp. 125–146.
11. Cioffi, M.B.; Yano, C.F.; Sember, A.; Bertollo, L.A.C. Chromosomal evolution in lower vertebrates: Sex chromosomes in Neotropical fishes. *Genes* **2017**, *8*, 258. [CrossRef]
12. Bertollo, L.A.C. Chromosome evolution in the neotropical Erythrinidae fish family: An overview. In *Fish Cytogenetics*, 1st ed.; Pisano, E., Ed.; CRC Press: Boca Raton, FL, USA, 2007; pp. 195–211.
13. Catalog of Fishes: Genera, Species, References. Available online: http://researcharchive.calacademy.org/research/ichthyology/catalog/fishcatmain.asp (accessed on 1 October 2019).
14. Netto-Ferreira, A.L. Revisão taxonômica e relações interespecíficas de Lebiasininae (Ostariophysi: Characiformes: Lebiasinidae). Ph.D. Thesis, Universidade de São Paulo, São Paulo, Brazil, 2010.
15. Benzaquem, D.C.; Oliveira, C.; da Silva Batista, J.; Zuanon, J.; Porto, J.I.R. DNA barcoding in pencilfishes (Lebiasinidae: *Nannostomus*) reveals cryptic diversity across the brazilian Amazon. *PLoS ONE* **2015**, *10*, e0112217. [CrossRef]
16. Scheel, J.J. *Fish Chromosomes and Their Evolution*; Danmarks Akvarium: Charlottenlund, Denmark, 1973.
17. Oliveira, C.; Andreata, A.; Almeida-Toledo, L.F.; Toledo Filho, S.A. Karyotype and nucleolus organizer regions of *Pyrrhulina* cf. *australis* (Pisces, Characiformes, Lebiasinidae). *Rev. Bras. Genet.* **1991**, *14*, 685–690.
18. Arai, R. *Fish Karyotypes: A Check List*, 1st ed.; Springer: Tokyo, Japan, 2011.
19. Moraes, R.L.R.; Bertollo, L.A.C.; Marinho, M.M.F.; Yano, C.F.; Hatanaka, T.; Barby, F.F.; Troy, W.P.; Cioffi, M.B. Evolutionary relationships and cytotaxonomy considerations in the genus *Pyrrhulina* (Characiformes, Lebiasinidae). *Zebrafish* **2017**, *14*, 536–546. [CrossRef] [PubMed]
20. Moraes, R.L.; Sember, A.; Bertollo, L.A.C.; de Oliveira, E.A.; Ráb, P.; Hatanaka, T.; Marinho, M.M.F.; Liehr, T.; Al-Rikabi, A.B.H.; Feldberg, E.; et al. Comparative cytogenetics and neo-Y formation in small-sized fish species of the genus *Pyrrhulina* (Characiformes, Lebiasinidae). *Front. Genet.* **2019**, *10*, 1–13. [CrossRef] [PubMed]
21. Sassi, F.M.C.; de Oliveira, E.A.; Bertollo, L.A.C.; Nirchio, M.; Hatanaka, T.; Marinho, M.M.F.; Moreira-Filho, O.; Aroutiounian, R.; Liehr, T.; Al-Rikabi, A.B.H.; et al. Chromosomal evolution and evolutionary relationships of *Lebiasina* species (Characiformes, Lebiasinidae). *Int. J. Mol. Sci.* **2019**, *20*, 2944. [CrossRef] [PubMed]

22. Toma, G.A.; Moraes, R.L.R.; Sassi, F.M.C.; Bertollo, L.A.C.; de Oliveira, E.A.; Ráb, P.; Sember, A.; Liehr, T.; Hatanaka, T.; Viana, P.F.; et al. Cytogenetics of the small-sized fish, *Copeina guttata* (Characiformes, Lebiasinidae): Novel insights into the karyotype differentiation of the family. *PLoS ONE* **2019**, *14*, e0226746. [CrossRef] [PubMed]
23. Arefjev, V.A. Karyotypic diversity of characid families (Pisces, Characidae). *Caryologia* **1990**, *43*, 291–304. [CrossRef]
24. Bertollo, L.A.C.; Cioffi, M.B.; Moreira-Filho, O. Direct chromosome preparation from Freshwater teleost fishes. In *Fish Cytogenetic Techniques (Ray-Fin Fishes and Chondrichthyans)*, 1st ed.; Ozouf-Costaz, C., Pisano, E., Foresti, F., Toledo, L.F.A., Eds.; CRC Press: Boca Raton, FL, USA, 2015; Volume 1, pp. 21–26.
25. Sumner, A.T. A simple technique for demonstrating centromeric heterochromatin. *Exp. Cell Res.* **1972**, *75*, 304–306. [CrossRef]
26. Martins, C.; Ferreira, I.A.; Oliveira, C.; Foresti, F.; Galetti, P.M., Jr. A tandemly repetitive centromeric DNA sequence of the fish *Hoplias malabaricus* (Characiformes: Erythrinidae) is derived from 5S rDNA. *Genetica* **2006**, *127*, 133–141. [CrossRef]
27. Cioffi, M.B.; Martins, C.; Centofante, L.; Jacobina, U.; Bertollo, L.A.C. Chromosomal variability among allopatric populations of Erythrinidae fish *Hoplias malabaricus*: Mapping of three classes of repetitive DNAs. *Cytogenet. Genome Res.* **2009**, *125*, 132–141. [CrossRef]
28. Yano, C.F.; Bertollo, L.A.C.; Cioffi, M.B. Fish-FISH: Molecular cytogenetics in fish species. In *Fluorescence in Situ Hybridization (FISH)—Application Guide*, 2nd ed.; Liehr, T., Ed.; Springer: Berlin, Germany, 2017; pp. 429–444.
29. Sambrook, J.; Russell, D.W. *Molecular Cloning: A Laboratory Manual*, 3rd ed.; Cold Spring Harbor Laboratory Press: New York, NY, USA, 2001.
30. Zwick, M.S.; Hanson, R.E.; Islam-Faridi, M.N.; Stelly, D.M.; Wing, R.A.; Price, H.J.; McKnight, T.D. A rapid procedure for the isolation of C0t-1 DNA from plants. *Genome* **1997**, *40*, 138–142. [CrossRef]
31. Sember, A.; Bertollo, L.A.C.; Ráb, P.; Yano, C.F.; Hatanaka, T.; de Oliveira, E.A.; Cioffi, M.B. Sex chromosome evolution and genomic divergence in the fish *Hoplias malabaricus* (Characiformes, Erythrinidae). *Front. Genet.* **2018**, *9*, 1–12. [CrossRef]
32. Levan, A.; Fredga, K.; Sandberg, A.A. Nomenclature for centromeric position on chromosomes. *Hereditas* **1964**, *52*, 201–220. [CrossRef]
33. Matthey, R. L' evolution de la formule chromosomiale chez les vertebrees. *Experientia* **1945**, *1*, 78–86. [CrossRef]
34. King, M. *Species Evolution: The Role of Chromosome Change*, 1st ed.; Cambridge University Press: Cambridge, UK, 1993.
35. Dobigny, G.; Ducroz, J.-F.; Robinson, T.J.; Volobouev, V. Cytogenetics and cladistics. *Syst. Biol.* **2004**, *53*, 470–484. [CrossRef] [PubMed]
36. Oliveira, C.; Almeida-Toledo, L.F.; Foresti, F. Karyotypic evolution in Neotropical fishes. In *Fish Cytogenetics*, 1st ed.; Pisano, E., Ozouf-Costaz, C., Foresti, F., Kapoor, B.G., Eds.; Science Publishers: Enfield, CT, USA, 2007; pp. 111–164.
37. Souza e Sousa, J.F.; Viana, P.F.; Bertollo, L.A.C.; Cioffi, M.B.; Feldberg, E. Evolutionary relationships among *Boulengerella* species (Ctenoluciidae, Characiformes): Genomic organization of repetitive DNAs and highly conserved karyotypes. *Cytogenet. Genome Res.* **2017**, *152*, 194–203. [CrossRef]
38. Betancur-R, R.; Arcila, D.; Vari, R.P.; Hughes, L.C.; Oliveira, C.; Sabaj, M.H.; Ortí, G. Phylogenomic incongruence, hypothesis testing, and taxonomic sampling: The monophyly of characiform fishes. *Evolution* **2019**, *73*, 329–345. [CrossRef]
39. De Barros, A.V.; Wolski, M.A.V.; Nogaroto, V.; Almeida, M.C.; Moreira-Filho, O.; Vicari, M.R. Fragile sites, dysfunctional telomere and chromosome fusions: What is 5S rDNA role? *Gene* **2017**, *608*, 20–27. [CrossRef]
40. Cavalcante, M.G.; Eduardo, C.; Carvalho, M.; Nagamachi, Y.; Pieczarka, J.C.; Vicari, M.R.; Noronha, R.C.R. Physical mapping of repetitive DNA suggests 2n reduction in Amazon turtles *Podocnemis* (Testudines: Podocnemididae). *PLoS ONE* **2018**, *13*, e0197536. [CrossRef]
41. Glugoski, L.; Giuliano-Caetano, L.; Moreira-Filho, O.; Vicari, M.R.; Nogaroto, V. Co-located *hAT* transposable element and 5S rDNA in an interstitial telomeric sequence suggest the formation of Robertsonian fusion in armored catfish. *Gene* **2018**, *650*, 49–54. [CrossRef]
42. Schweizer, D.; Loidl, J. A model for heterochromatin dispersion and the evolution of C-band patterns. In *Chromosomes Today*, 1st ed.; Stahl, A., Luciani, J.M., Vagner-Capodano, A.M., Eds.; Springer: Paris, France, 1987; Volume 9, pp. 61–74. [CrossRef]

43. Ráb, P.; Crossman, E.J.; Reed, K.M.; Rábová, M. Chromosomal characteristics of ribosomal DNA in two extant species of North American mudminows *Umbra pygmaea* and *U. limi* (Euteleostei: Umbridae). *Cytogenet. Genome Res.* **2002**, *98*, 194–198. [CrossRef]
44. Cazaux, B.; Catalan, J.; Veyrunes, F.; Douzery, E.J.; Britton-Davidian, J. Are ribosomal DNA clusters rearrangement hotspots? A case study in the genus *Mus* (Rodentia, Muridae). *BMC Evol. Biol.* **2011**, *11*, 124. [CrossRef]
45. Santos-Pereira, J.M.; Aguilera, A. R loops: New modulators of genome dynamics and function. *Nat. Rev. Genet.* **2015**, *16*, 583–597. [CrossRef] [PubMed]
46. Sawyer, I.A.; Dundr, M. Chromatin loops and causality loops: The influence of RNA upon spatial nuclear architecture. *Chromosoma* **2017**, *126*, 541–557. [CrossRef]
47. Blokhina, Y.P.; Nguyen, A.D.; Draper, B.W.; Burgess, S.M. The telomere bouquet is a hub where meiotic double-strand breaks, synapsis, and stable homolog juxtaposition are coordinated in the zebrafish, *Danio rerio*. *PLoS Genet.* **2019**, *15*, e1007730. [CrossRef] [PubMed]
48. Potapova, T.A.; Gerton, J.L. Ribosomal DNA and the nucleolus in the context of genome organization. *Chromosome Res.* **2019**, *27*, 109–127. [CrossRef] [PubMed]
49. Giles, V.; Thode, G.; Alvarez, M.C. A new Robertsonian fusion in the multiple chromosome polymorphism of a mediterranean population of *Gobius paganellus* (Gobiidae, Perciformes). *Heredity* **1985**, *55*, 255–260. [CrossRef]
50. Molina, W.F.; Galetti, P.M., Jr. Robertsonian rearrangements in the reef fish *Chromis* (Perciformes, Pomacentridae) involving chromosomes bearing 5S rRNA genes. *Genet. Mol. Biol.* **2002**, *25*, 373–377. [CrossRef]
51. Rosa, K.O.; Ziemniczak, K.; de Barros, A.V.; Nogaroto, V.; Almeida, M.C.; Cestari, M.M.; Artoni, R.F.; Vicari, M.R. Numeric and structural chromosome polymorphism in *Rineloricaria lima* (Siluriformes: Loricariidae): Fusion points carrying 5S rDNA or telomere sequence vestiges. *Rev. Fish Biol. Fish.* **2012**, *22*, 739–749. [CrossRef]
52. Sember, A.; Bohlen, J.; Šlechtová, V.; Altmanová, M.; Symonová, R.; Ráb, P. Karyotype differentiation in 19 species of river loach fishes (Nemacheilidae, Teleostei): Extensive variability associated with rDNA and heterochromatin distribution and its phylogenetic and ecological interpretation. *BMC Evol. Biol.* **2015**, *15*, 251. [CrossRef]
53. Getlekha, N.; Molina, W.F.; Cioffi, M.B.; Yano, C.F.; Maneechot, N.; Bertollo, L.A.C.; Supiwong, W.; Tanomtong, A. Repetitive DNAs highlight the role of chromosomal fusions in the karyotype evolution of *Dascyllus* species (Pomacentridae, Perciformes). *Genetica* **2016**, *144*, 203–211. [CrossRef]
54. Salvadori, S.; Deiana, A.M.; Deidda, F.; Lobina, C.; Mulas, A.; Coluccia, E. XX/XY sex chromosome system and chromosome markers in the snake eel *Ophisurus serpens* (Anguilliformes: Ophichtidae). *Mar. Biol. Res.* **2018**, *14*, 158–164. [CrossRef]
55. Schmid, M.; Steinlein, C.; Bogart, J.P.; Feichtinger, W.; León, P.; La Marca, E.; Díaz, L.M.; Sanz, A.; Chen, S.H.; Hedges, S.B. The chromosomes of terraran frogs: Insights into vertebrate cytogenetics. *Cytogenet. Genome Res.* **2016**, *130–131*, 1–568. [CrossRef]
56. Da Costa, M.J.R.; do Amaral, P.J.S.; Pieczarka, J.C.; Sampaio, M.I.; Rossi, R.V.; Mendes-Oliveira, A.C.; Noronha, R.C.R.; Nagamachi, C.Y. Cryptic species in *Proechimys goeldii* (Rodentia, Echimyidae)? A case of molecular and chromosomal differentiation in allopatric populations. *Cytogenet. Genome Res.* **2016**, *148*, 199–210. [CrossRef] [PubMed]
57. Houck, M.L.; Teri, L.; Lear, T.L.; Charter, S.J. Animal cytogenetics. In *The AGT Cytogenetics Laboratory Manual*, 4th ed.; Arsham, M.S., Barch, M.J., Lawce, H.J., Eds.; John Wiley & Sons: Hoboken, NJ, USA, 2017; pp. 1055–1102.
58. Sousa, R.P.C.; Oliveira-Filho, A.B.; Vallinoto, M.; Cioffi, M.B.; Molina, W.F.; de Oliveira, E.H.; Silva-Oliveira, G.C. Cytogenetics description in *Batrachoides surinamensis*, (Batrachoididae: Batrachoidiformes): What does the estuary have to say? *Estuar. Coast. Shelf Sci.* **2018**, *213*, 253–259. [CrossRef]
59. Molina, W.F.; Martinez, P.A.; Bertollo, L.A.C.; Bidau, C.J. Evidence for meiotic drive as an explanation for karyotype changes in fishes. *Mar. Genom.* **2014**, *15*, 29–34. [CrossRef]
60. Krysanov, E.; Demidova, T. Extensive karyotype variability of African fish genus *Nothobranchius* (Cyprinodontiformes). *Comp. Cytogenet.* **2018**, *12*, 387–402. [CrossRef]
61. Völker, M.; Sonnenberg, R.; Ráb, P.; Kullmann, H. Karyotype differentiation in *Chromaphyosemion* killifishes (Cyprinodontiformes, Nothobranchiidae) II: Cytogenetic and mitochondrial DNA analyses demonstrate karyotype differentiation and its evolutionary direction in *C. riggenbachi*. *Cytogenet. Genome Res.* **2006**, *115*, 70–83. [CrossRef]
62. Völker, M.; Ráb, P.; Kullmann, H. Karyotype differentiation in *Chromaphyosemion* killifishes (Cyprinodontiformes,

Nothobranchiidae): Patterns, mechanisms, and evolutionary implications. *Biol. J. Linn. Soc.* **2008**, *94*, 143–153. [CrossRef]

63. Ene, A.-C. Chromosomal polymorphism in the goby *Neogobius eurycephalus* (Perciformes: Gobiidae). *Mar. Biol.* **2002**, *142*, 583–588. [CrossRef]
64. Amores, A.; Wilson, C.A.; Allard, C.A.H.; Detrich, H.W.; Postlethwait, J.H. Cold fusion: Massive karyotype evolution in the Antarctic bullhead notothen *Notothenia coriiceps*. *G3 (Bethesda)* **2017**, *7*, 2195–2207. [CrossRef]
65. Crossman, E.J.; Ráb, P. Chromosome-banding study of the Alaska blackfish, *Dallia pectoralis* (Euteleostei: Esocae) with implications for karyotype evolution and relationships of esocoid fishes. *Can. J. Zool.* **1996**, *74*, 147–156. [CrossRef]
66. Crossman, E.J.; Ráb, P. Chromosomal NOR phenotype and C-banded karyotype of Olympic mudminnow, *Novumbra hubbsi* (Euteleostei: Umbridae). *Copeia* **2001**, *3*, 860–865. [CrossRef]
67. Phillips, R.; Ráb, P. Chromosome evolution in the Salmonidae (Pisces): An update. *Biol. Rev. Camb. Philos. Soc.* **2001**, *76*, 1–25. [CrossRef] [PubMed]
68. Brown, J.D.; O'Neill, R.J. Chromosomes, conflict, and epigenetics: Chromosomal speciation revisited. *Annu. Rev. Genom. Hum. Genet.* **2010**, *11*, 291–316. [CrossRef] [PubMed]
69. Guerrero, R.F.; Kirkpatrick, M. Local adaptation and the evolution of chromosome fusions. *Evolution* **2014**, *68*, 2747–2756. [CrossRef]
70. Lanctôt, C.; Cheutin, T.; Cremer, M.; Cavalli, G.; Cremer, T. Dynamic genome architecture in the nuclear space: Regulation of gene expression in three dimensions. *Nat. Rev. Genet.* **2007**, *8*, 104–115. [CrossRef]
71. Meaburn, K.J.; Misteli, T.; Soutoglou, E. Spatial genome organization in the formation of chromosomal translocations. *Semin. Cancer Biol.* **2007**, *17*, 80–90. [CrossRef]
72. Meaburn, K.J.; Misteli, T. Cell biology: Chromosome territories. *Nature* **2007**, *445*, 379–781. [CrossRef]
73. Cremer, T.; Cremer, M. Chromosome territories. *Cold Spring Harb. Perspect. Biol.* **2010**, *2*, 1–23. [CrossRef]
74. Roukos, V.; Misteli, T. The biogenesis of chromosome translocations. *Nat. Cell Biol.* **2014**, *16*, 293–300. [CrossRef]
75. Fraser, J.; Williamson, I.; Bickmore, W.A.; Dostie, J. An overview of genome organization and how we got there: From FISH to Hi-C. *Microbiol. Mol. Biol. Rev.* **2015**, *79*, 347–372. [CrossRef]
76. Razin, S.V.; Gavrilov, A.A.; Vassetzky, Y.S.; Ulianov, S.V. Topologically-associating domains: Gene warehouses adapted to serve transcriptional regulation. *Transcription* **2016**, *7*, 84–90. [CrossRef] [PubMed]
77. Rosin, L.F.; Crocker, O.; Isenhart, R.L.; Nguyen, S.C.; Xu, Z.; Joyce, E.F. Chromosome territory formation attenuates the translocation potential of cells. *eLife* **2019**, *8*, e49553. [CrossRef] [PubMed]
78. Ghavi-Helm, Y.; Jankowski, A.; Meiers, S.; Viales, R.R.; Korbel, J.O.; Furlong, E.E.M. Highly rearranged chromosomes reveal uncoupling between genome topology and gene expression. *Nat. Genet.* **2019**, *51*, 1272–1282. [CrossRef] [PubMed]
79. Mank, J.E.; Avise, J.C. Phylogenetic conservation of chromosome numbers in Actinopterygiian fishes. *Genetica* **2006**, *127*, 321–327. [CrossRef] [PubMed]
80. Cioffi, M.B.; Bertollo, L.A.C. Chromosomal distribution and evolution of repetitive DNAs in fish. *Genome Dyn.* **2012**, *7*, 197–221. [CrossRef]
81. García-Souto, D.; Qarkaxhija, V.; Pasantes, J.J. Resolving the taxonomic status of *Chamelea gallina* and *C. striatula* (Veneridae, Bivalvia): A combined molecular cytogenetic and phylogenetic approach. *Biomed. Res. Int.* **2017**, *2017*, 7638790. [CrossRef]
82. Weeks, A.R.; Marec, F.; Breeuwer, J.A.J. A mite species that consists entirely of haploid females. *Science* **2001**, *292*, 2479–2482. [CrossRef]
83. Lukhtanov, V.A.; Dincă, V.; Friberg, M.; Šíchová, J.; Olofsson, M.; Vila, R.; Marec, F.; Wiklund, C. Versatility of multivalent orientation, inverted meiosis, and rescued fitness in holocentric chromosomal hybrids. *PNAS* **2018**, *115*, E9610–E9619. [CrossRef]
84. Do Nascimento, V.D.; Coelho, K.A.; Nogaroto, V.; Almeida, R.B.; Ziemniczak, K.; Centofante, L.; Pavanelli, C.S.; Torres, R.A.; Moreira-Filho, O.; Vicari, M.R. Do multiple karyomorphs and population genetics of freshwater darter characines (*Apareiodon affinis*) indicate chromosomal speciation? *Zool. Anz.* **2018**, *272*, 93–103. [CrossRef]
85. Zhu, H.P.; Ma, D.M.; Gui, J.F. Triploid origin of the gibel carp as revealed by 5S rDNA localization and chromosome painting. *Chromosome Res.* **2006**, *14*, 767–776. [CrossRef]
86. Zhang, C.; Ye, L.; Chen, Y.; Xiao, J.; Wu, Y.; Tao, M.; Xiao, Y.; Liu, S. The chromosomal constitution of fish hybrid lineage revealed by 5S rDNA FISH. *BMC Genet.* **2015**, *16*, 140. [CrossRef] [PubMed]

87. Soto, M.Á.; Castro, J.P.; Walker, L.I.; Malabarba, L.R.; Santos, M.H.; Almeida, M.C.; Moreira-Filho, O.; Artoni, R.F. Evolution of trans-Andean endemic fishes of the genus *Cheirodon* (Teleostei: Characidae) are associated with chromosomal rearrangements. *Rev. Chil. Hist. Nat.* **2018**, *91*, 8. [CrossRef]
88. Yano, C.F.; Bertollo, L.A.C.; Ezaz, T.; Trifonov, V.; Sember, A.; Liehr, T.; Cioffi, M.B. Highly conserved Z and molecularly diverged W chromosomes in the fish genus *Triportheus* (Characiformes, Triportheidae). *Heredity* **2017**, *118*, 276–283. [CrossRef] [PubMed]
89. De Oliveira, E.A.; Sember, A.; Bertollo, L.A.C.; Yano, C.F.; Ezaz, T.; Moreira-Filho, O.; Hatanaka, T.; Trifonov, V.; Liehr, T.; Al-Rikabi, A.B.H.; et al. Tracking the evolutionary pathway of sex chromosomes among fishes: Characterizing the unique XX/XY_1Y_2 system in *Hoplias malabaricus* (Teleostei, Characiformes). *Chromosoma* **2018**, *127*, 115–128. [CrossRef] [PubMed]
90. Xu, D.; Sember, A.; Zhu, Q.; de Oliveira, E.A.; Liehr, T.; Al-Rikabi, A.B.H.; Xiao, Z.; Song, H.; Cioffi, M.B. Deciphering the origin and evolution of the X_1X_2Y system in two closely-related *Oplegnathus* species (Oplegnathidae and Centrarchiformes). *J. Mol. Sci.* **2019**, *20*, 3571. [CrossRef] [PubMed]
91. Symonová, R.; Majtánová, Z.; Sember, A.; Staaks, G.B.; Bohlen, J.; Freyhof, J.; Rábová, M.; Ráb, P. Genome differentiation in a species pair of coregonine fishes: An extremely rapid speciation driven by stress-activated retrotransposons mediating extensive ribosomal DNA multiplications. *BMC Evol. Biol.* **2013**, *13*, 42. [CrossRef]
92. Da Silva, A.F.; Feldberg, E.; Carvalho, N.D.M.; Rangel, S.M.H.; Schneider, C.H.; Carvalho-Zilse, G.A.; da Silva, V.F.; Gross, M.C. Effects of environmental pollution on the rDNAomics of Amazonian fish. *Environ. Pollut.* **2019**, *252*, 180–187. [CrossRef]
93. Gornung, E. Twenty years of physical mapping of major ribosomal RNA genes across the teleosts: A review of research. *Cytogenet. Genome Res.* **2013**, *141*, 90–102. [CrossRef]
94. Sochorová, J.; Garcia, S.; Gálvez, F.; Symonová, R.; Kovařík, A. Evolutionary trends in animal ribosomal DNA loci: Introduction to a new online database. *Chromosoma* **2018**, *127*, 141–150. [CrossRef]
95. Milhomem, S.S.R.; Scacchetti, P.C.; Pieczarka, J.C.; Ferguson-Smith, M.A.; Pansonato-Alves, J.C.; O'Brien, P.C.M.; Foresti, F.; Nagamachi, C.Y. Are NORs always located on homeologous chromosomes? A FISH investigation with rDNA and whole chromosome probes in *Gymnotus* fishes (Gymnotiformes). *PLoS ONE* **2013**, *8*, e55608. [CrossRef]
96. Lim, K.Y.; Kovařík, A.; Matyášek, R.; Chase, M.W.; Clarkson, J.J.; Grandbastien, M.A.; Leitch, A.R. Sequence of events leading to near-complete genome turnover in allopolyploid *Nicotiana* within five million years. *New Phytol.* **2007**, *175*, 756–763. [CrossRef] [PubMed]
97. Majka, J.; Majka, M.; Kwiatek, M.; Wiśniewska, H. Similarities and differences in the nuclear genome organization within Pooideae species revealed by comparative genomic in situ hybridization (GISH). *J. Appl. Genet.* **2017**, *58*, 151–161. [CrossRef] [PubMed]
98. Barby, F.F.; Bertollo, L.A.C.; de Oliveira, E.A.; Yano, C.F.; Hatanaka, T.; Ráb, P.; Sember, A.; Ezaz, T.; Artoni, R.F.; Liehr, T.; et al. Emerging patterns of genome organization in Notopteridae species (Teleostei, Osteoglossiformes) as revealed by Zoo-FISH and Comparative Genomic Hybridization (CGH). *Sci. Rep.* **2019**, *9*, 1112. [CrossRef] [PubMed]
99. Kandul, N.P.; Lukhtanov, V.A.; Pierce, N.E. Karyotypic diversity and speciation in *Agrodiaetus* butterflies. *Evolution* **2007**, *61*, 546–559. [CrossRef] [PubMed]
100. Luo, J.; Sun, X.; Cormack, B.P.; Boeke, J.D. Karyotype engineering by chromosome fusion leads to reproductive isolation in yeast. *Nature* **2018**, *560*, 392–396. [CrossRef]
101. Ortiz-Barrientos, D.; Engelstädter, J.; Rieseberg, L.H. Recombination rate evolution and the origin of species. *Trends Ecol. Evol.* **2016**, *31*, 226–236. [CrossRef]

Permissions

All chapters in this book were first published in MDPI; hereby published with permission under the Creative Commons Attribution License or equivalent. Every chapter published in this book has been scrutinized by our experts. Their significance has been extensively debated. The topics covered herein carry significant findings which will fuel the growth of the discipline. They may even be implemented as practical applications or may be referred to as a beginning point for another development.

The contributors of this book come from diverse backgrounds, making this book a truly international effort. This book will bring forth new frontiers with its revolutionizing research information and detailed analysis of the nascent developments around the world.

We would like to thank all the contributing authors for lending their expertise to make the book truly unique. They have played a crucial role in the development of this book. Without their invaluable contributions this book wouldn't have been possible. They have made vital efforts to compile up to date information on the varied aspects of this subject to make this book a valuable addition to the collection of many professionals and students.

This book was conceptualized with the vision of imparting up-to-date information and advanced data in this field. To ensure the same, a matchless editorial board was set up. Every individual on the board went through rigorous rounds of assessment to prove their worth. After which they invested a large part of their time researching and compiling the most relevant data for our readers.

The editorial board has been involved in producing this book since its inception. They have spent rigorous hours researching and exploring the diverse topics which have resulted in the successful publishing of this book. They have passed on their knowledge of decades through this book. To expedite this challenging task, the publisher supported the team at every step. A small team of assistant editors was also appointed to further simplify the editing procedure and attain best results for the readers.

Apart from the editorial board, the designing team has also invested a significant amount of their time in understanding the subject and creating the most relevant covers. They scrutinized every image to scout for the most suitable representation of the subject and create an appropriate cover for the book.

The publishing team has been an ardent support to the editorial, designing and production team. Their endless efforts to recruit the best for this project, has resulted in the accomplishment of this book. They are a veteran in the field of academics and their pool of knowledge is as vast as their experience in printing. Their expertise and guidance has proved useful at every step. Their uncompromising quality standards have made this book an exceptional effort. Their encouragement from time to time has been an inspiration for everyone.

The publisher and the editorial board hope that this book will prove to be a valuable piece of knowledge for researchers, students, practitioners and scholars across the globe.

List of Contributors

Marta Olszewska and Maciej Kurpisz
Institute of Human Genetics, Polish Academy of Sciences, Strzeszynska 32, 60-479 Poznan, Poland

Tomasz Stokowy
Department of Clinical Science, University of Bergen, Postboks 7804, 5020 Bergen, Norway

Nijole Pollock, Andrew Georgiadis, Alexander N. Yatsenko and Svetlana Yatsenko
Department of OBGYN and Reproductive Science, Magee-Womens Research Institute, University of Pittsburgh, Pittsburgh, PA 15213, USA

Nataliya Huleyuk
Institute of Hereditary Pathology, Ukrainian Academy of Medical Sciences, Lysenko Str. 31a, 79000 Lviv, Ukraine

Danuta Zastavna
Institute of Hereditary Pathology, Ukrainian Academy of Medical Sciences, Lysenko Str. 31a, 79000 Lviv, Ukraine
Department of Biotechnology and Bioinformatics, Faculty of Chemistry, Rzeszow University of Technology, Al. Powst. Warszawy 6, 35-959 Rzeszow, Poland

Evgenia Sarrou, Laura Richmond and Karen Keeshan
Paul O'Gorman Leukaemia Research Centre, Institute of Cancer Sciences, University of Glasgow, Glasgow G12 0YN, UK

Ruaidhrí J. Carmody
Centre for Immunobiology, Institute of Infection, Immunity & Inflammation, College of Medicine, Veterinary and Life Sciences, University of Glasgow, Glasgow G12 8TA, UK

Brenda Gibson
Royal Hospital for Children, Glasgow G51 4TF, UK

Basanta Bista and Nicole Valenzuela
Department of Ecology, Evolution, and Organismal Biology, Iowa State University, Ames, IA 50011, USA

Sergey Matveevsky and Oxana Kolomiets
Vavilov Institute of General Genetics, Russian Academy of Sciences, 119991 Moscow, Russia

Aleksey Bogdanov, Elena Alpeeva and Irina Bakloushinskaya
Koltzov Institute of Developmental Biology, Russian Academy of Sciences, 119334 Moscow, Russia

Janine E. Deakin and Tariq Ezaz
Institute for Applied Ecology, University of Canberra, Canberra, ACT 2617, Australia

Sally Potter
Research School of Biology, Australian National University, Acton, ACT 2601, Australia
Australian Museum Research Institute, Australian Museum, 1 William St Sydney, NSW 2010, Australia

Mark D.B. Eldridge
Australian Museum Research Institute, Australian Museum, 1 William St Sydney, NSW 2010, Australia

Rachel O'Neill
Institute for Systems Genomics and Department of Molecular and Cell Biology, University of Connecticut, Storrs, CT 06269, USA

Marcelo B. Cioffi
Laboratório de Citogenética de Peixes, Departamento de Genética e Evolução, Universidade Federal de São Carlos, São Carlos, SP 13565-905, Brazil

Kichi Fukui
Graduate School of Pharmaceutical Sciences, Osaka University, Suita 565-0871, Osaka, Japan

Jennifer A. Marshall Graves
Institute for Applied Ecology, University of Canberra, Canberra, ACT 2617, Australia
School of Life Sciences, LaTrobe University, Melbourne, VIC 3168, Australia

Darren Griffin
School of Biosciences, University of Kent, Canterbury CT2 7NJ, UK

Frank Grutzner and Michail Rovatsos
School of Biological Sciences, The University of Adelaide, Adelaide, SA 5005, Australia

Lukáš Kratochvíl
Department of Ecology, Faculty of Science, Charles University, Viniˇcná 7, 128 44 Prague 2, Czech Republic

Ikuo Miura
Amphibian Research Center, Hiroshima University, Higashi-Hiroshima 739-8526, Japan

Kornsorn Srikulnath
Laboratory of Animal Cytogenetics & Comparative Genomics (ACCG), Department of Genetics, Faculty of Science, Kasetsart University, Bangkok 10900, Thailand

Erik Wapstra
School of Natural Sciences, University of Tasmania, Hobart 7000, Australia

Svetlana A. Romanenko, Dmitry Yu. Prokopov, Olga L. Gladkikh, Natalia A. Serdyukova and Anna S. Molodtseva
Institute of Molecular and Cellular Biology, Siberian Branch of Russian Academy of Sciences, Novosibirsk 630090, Russia

Antonina V. Smorkatcheva and Ivan A. Polikarpov
Department of Vertebrate Zoology, St. Petersburg State University, St. Petersburg 199034, Russia

Yulia M. Kovalskaya
Severtzov Institute of Ecology and Evolution, Russian Academy of Sciences, Moscow 119071, Russia

Vladimir A. Trifonov
Institute of Molecular and Cellular Biology, Siberian Branch of Russian Academy of Sciences, Novosibirsk 630090, Russia
Department of Natural Science, Novosibirsk State University, Novosibirsk 630090, Russia

Malcolm A. Ferguson-Smith and Patricia C. M. O'Brien
Cambridge Resource Centre for Comparative Genomics, Department of Veterinary Medicine, University of Cambridge, Madingley Road, Cambridge CB3 OES, UK

Feodor N. Golenishchev
Zoological Institute, Russian Academy of Sciences, Saint-Petersburg 199034, Russia

Joaquim T. Tapisso, Maria da Graça Ramalhinho and Maria da Luz Mathias
CESAM - Centro de Estudos do Ambiente e do Mar, Faculdade de Ciências, Universidade de Lisboa, 1749-016 Lisboa, Portugal

Sofia I. Gabriel and Ana Mota Cerveira
CESAM - Centro de Estudos do Ambiente e do Mar, Faculdade de Ciências, Universidade de Lisboa, 1749-016 Lisboa, Portugal
Departamento de Biologia, Universidade de Aveiro, Campus Universitário de Santiago, 3810-193 Aveiro, Portugal

Janice Britton-Davidian and Guila Ganem
ISEM - Institut des Sciences de l'Evolution de Montpellier, Université de Montpellier, CNRS, EPHE, IRD, 34095 Montpellier, France

Jeremy B. Searle
Department of Ecology and Evolutionary Biology, Corson Hall, Cornell University, Ithaca, NY 14853, USA

Šárka Pelikánová and Vendula Šlechtová
Laboratory of Fish Genetics, Institute of Animal Physiology and Genetics, Czech Academy of Sciences, Rumburská 89, 277-21 Liběchov, Czech Republic

Terumi Hatanaka and Marcelo de Bello Cioffi
Departamento de Genética e Evolução, Universidade Federal de São Carlos, Rod. Washington Luiz km 235 cep, São Carlos 13565-905, Brazil

Hiep Do Doan
Research Institute of Aquaculture No. 1, Dinh Bang, Tu Son, Bac Ninh 16000, Vietnam

Martin Knytl
Department of Cell Biology, Faculty of Science, Charles University, Viničná 7, 2-128-43 Prague, Czech Republic

Anastasia A. Proskuryakova and Anastasia I. Kulemzina
Institute of Molecular and Cellular Biology, SB RAS, 630090 Novosibirsk, Russia

Polina L. Perelman and Alexander S. Graphodatsky
Institute of Molecular and Cellular Biology, SB RAS, 630090 Novosibirsk, Russia
Novosibirsk State University, 630090 Novosibirsk, Russia

Dmitry V. Yudkin
State Research Center of Virology and Biotechnology "Vector", Federal Service for Surveillance on Consumer Rights Protection and Human Well-being (FBRI SRC VB "Vector", Rospotrebnadzor), 630559 Koltsovo, Novosibirsk Region, Russia

Natalya A. Lemskaya
Institute of Molecular and Cellular Biology, Siberian Branch of Russian Academy of Sciences, Novosibirsk 630090, Russia
State Research Center of Virology and Biotechnology "Vector", Federal Service for Surveillance on Consumer Rights Protection and Human Well-being (FBRI SRC VB "Vector", Rospotrebnadzor), 630559 Koltsovo, Novosibirsk Region, Russia

Innokentii M. Okhlopkov and Egor V. Kirillin
Institute for Biological Problems of Cryolithozone Siberian Branch of RAS, 677980 Yakutsk, Russia

Marta Farré
The Royal Veterinary College, Royal College Street, University of London, London NW1 0TU, UK
School of Biosciences, University of Kent, Canterbury CT2 7NJ, UK

Denis M. Larkin
The Royal Veterinary College, Royal College Street, University of London, London NW1 0TU, UK

Melody E. Roelke-Parker
Frederick National Laboratory of Cancer Research, Leidos Biomedical Research, Inc., Frederick, MD 21701, USA

Stephen J. O'Brien
Theodosius Dobzhansky Center for Genome Bioinformatics, Saint-Petersburg State University, Sredniy Av. 41A, 199034 Saint-Petersburg, Russia
Oceanographic Center, Nova Southeastern University, Fort Lauderdale 3301 College Ave, Fort Lauderdale, FL 33314, USA

Mitchell Bush
The Center for Species Survival, Department of Reproductive Sciences, Smithsonian Conservation Biology Institute, Smithsonian's National Zoological Park, 1500 Remount Road, Front Royal, VA 22630, USA

Atashi Sharma and Vladimir A. Timoshevskiy
Department of Entomology, Virginia Polytechnic and State University, Blacksburg, VA 24061, USA

Nicholas A. Kinney
Genomics Bioinformatics and Computational Biology, Virginia Polytechnic and State University, Blacksburg, VA 24061, USA

Maria V. Sharakhova
Department of Entomology, Virginia Polytechnic and State University, Blacksburg, VA 24061, USA
Laboratory of Evolutionary Genomics of Insects, the Federal Research Center Institute of Cytology and Genetics, Siberian Branch of the Russian Academy of Sciences, 630090 Novosibirsk, Russia
Laboratory of Ecology, Genetics and Environmental Protection, Tomsk State University, 634050 Tomsk, Russia

Igor V. Sharakhov
Department of Entomology, Virginia Polytechnic and State University, Blacksburg, VA 24061, USA
Genomics Bioinformatics and Computational Biology, Virginia Polytechnic and State University, Blacksburg, VA 24061, USA
Laboratory of Evolutionary Genomics of Insects, the Federal Research Center Institute of Cytology and Genetics, Siberian Branch of the Russian Academy of Sciences, 630090 Novosibirsk, Russia
Department of Cytology and Genetics, Tomsk State University, 634050 Tomsk, Russia

Lucas G. Kiazim, Darren K. Griffin and Rebecca E. O'Connor
School of Biosciences, University of Kent, Canterbury CT2 7NJ, UK

Thales Renato Ochotorena de Freitas
Departamento de Genética, Universidade Federal do Rio Grande do Sul, Porto Alegre 91509-900, Brazil

Rafael Kretschmer
School of Biosciences, University of Kent, Canterbury CT2 7NJ, UK
Departamento de Genética, Universidade Federal do Rio Grande do Sul, Porto Alegre 91509-900, Brazil

Ivanete de Oliveira Furo
Instituto de Ciências Biológicas, Universidade Federal do Pará, Belém 66075-110, Brazil
Laboratório de Cultura de Tecidos e Citogenética, SAMAM, Instituto Evandro Chagas, Ananindeua 67030-000, Brazil

Anderson José Baia Gomes
Instituto Federal do Pará, Abaetetuba 68440-000, Brazil

Ricardo José Gunski and Analía del Valle Garnero
Laboratório de Diversidade Genética Animal, Universidade Federal do Pampa, São Gabriel 97300-162, Brazil

Jorge C. Pereira
Animal and Veterinary Research Centre (CECAV), University of Trás-os-Montes and Alto Douro (UTAD), 5000-801 Vila Real, Portugal

Malcolm A. Ferguson-Smith
Cambridge Resource Centre for Comparative Genomics, University of Cambridge Department of Veterinary Medicine, Cambridge CB3 0ES, UK

Edivaldo Herculano Corrêa de Oliveira
Laboratório de Cultura de Tecidos e Citogenética, SAMAM, Instituto Evandro Chagas, Ananindeua 67030-000, Brazil
Instituto de Ciências Exatas e Naturais, Universidade Federal do Pará, Belém 66075-110, Brazil

Sandra Louzada, Mariana Lopes, Daniela Ferreira, Filomena Adega, Ana Escudeiro and Raquel Chaves
Laboratory of Cytogenomics and Animal Genomics (CAG), Department of Genetics and Biotechnology (DGB), University of Trás-os-Montes and Alto Douro (UTAD), 5000-801 Vila Real, Portugal
Biosystems and Integrative Sciences Institute (BioISI), Faculty of Sciences, University of Lisboa, 1749-016 Lisbon, Portugal

Margarida Gama-Carvalho
Biosystems and Integrative Sciences Institute (BioISI), Faculty of Sciences, University of Lisboa, 1749-016 Lisbon, Portugal

Francisca Garcia and Francisco Cortés
Unitat de Cultius Cel.lulars (UCC), Universitat Autònoma de Barcelona, Campus UAB, 08193 Cerdanyola del Vallès, Spain

Luis Biedma and Javier Calzada
Department of Integrated Sciences, Faculty of Experimental Sciences, University of Huelva, Avenida de las Fuerzas Armadas, S/N, 21007 Huelva, Spain

Jacinto Román
Department of Conservation Biology, Doñana Biological Station, CSIC, C. Americo Vespucio 26, 41092 Sevilla, Spain

Alberto Lozano and Aurora Ruiz-Herrera
Departament de Biologia Cellular, Fisiologia i Immunologia, Universitat Autònoma de Barcelona, Campus UAB, 08193 Cerdanyola del Vallès, Spain
Genome Integrity and Instability Group, Institut de Biotecnologia i Biomedicina, Universitat Autònoma de Barcelona, 08193 Cerdanyola del Vallès, Spain

José A. Godoy
Department of Integrative Ecology, Doñana Biological Station, CSIC, C. Americo Vespucio 26, 41092 Sevilla, Spain

Alexandr Sember and Petr Ráb
Laboratory of Fish Genetics, Institute of Animal Physiology and Genetics, Czech Academy of Sciences, Rumburská 89, 277 21 Liběchov, Czech Republic

Ezequiel Aguiar de Oliveira
Departamento de Genética e Evolução, Universidade Federal de São Carlos, São Carlos, São Paulo 13565-905, Brazil
Secretaria de Estado de Educação de Mato Grosso–SEDUC-MT, Cuiabá 78049-909, Brazil

Luiz Antonio Carlos Bertollo, Natália Lourenço de Freitas, Cassia Fernanda Yano and Renata Luiza Rosa de Moraes
Departamento de Genética e Evolução, Universidade Federal de São Carlos, São Carlos, São Paulo 13565-905, Brazil

Patrik Ferreira Viana and Eliana Feldberg
Instituto Nacional de Pesquisas da Amazônia, Coordenação de Biodiversidade, Av. André Araújo 2936, Petrópolis, Manaus 69067-375, Brazil

Manoela Maria Ferreira Marinho
Universidade Federal da Paraíba (UFPB), Departamento de Sistemática e Ecologia (DSE), Laboratório de Sistemática e Morfologia de Peixes, João Pessoa 58051-090, Brazil

Index

Printed in the USA
CPSIA information can be obtained
at www.ICGtesting.com
LVHW080053211023
761596LV00010BA/2035

9 798888 360620